AF412309

Weaponeering for the Warfighter

Weaponeering for the Warfighter

Morris Driels

Paul Park, Editor-in-Chief
University of Texas at Arlington
Arlington, Texas

Published by
American Institute of Aeronautics and Astronautics, Inc.
12700 Sunrise Valley Drive, Reston, VA 20191-5807

American Institute of Aeronautics and Astronautics, Inc., Reston, Virginia

1 2 3 4 5

Library of Congress Cataloging-in-Publication Data
Names: Driels, Morris R., author.
Title: Weaponeering for the warfighter / Morris Driels.
Description: Reston, VA : The American Institute of Aeronautics and
   Astronautics, Inc., [2021] | Series: Library of flight | Includes
   bibliographical references and index.
Identifiers: LCCN 2021028083 | ISBN 9781624106194 (hardcover)
Subjects: LCSH: United States—Armed Forces—Weapons systems—Evaluation.
Classification: LCC UF503 .D75 2021 | DDC 355.820973—dc23
LC record available at https://lccn.loc.gov/2021028083

ISBN: 978-1-62410-619-4

Copyright © 2021 by the American Institute of Aeronautics and Astronautics, Inc. All rights reserved. Printed in the United States of America. No part of this publication may be reproduced, distributed, or transmitted, in any form or by any means, or stored in a database or retrieval system, without the prior written permission of the publisher.

Data and information appearing in this book are for informational purposes only. AIAA is not responsible for any injury or damage resulting from use or reliance, nor does AIAA warrant that use or reliance will be free from privately owned rights.

This book is dedicated to all the men and women who serve in our Armed Forces, past, present, and future. They and their families bear the high cost of the freedom we enjoy to live our lives the way we choose.

Contents

This book been produced with the assistance of many people and organizations and represents an accumulation of their collective knowledge in this subject. In particular, the considerable effort of Bryan Paris and David Chung of the JTCG/ME Program Office in helping bring this work to the widest possible audience is greatly appreciated.

PREFACE

HISTORY OF THE JOINT TECHNICAL COORDINATING GROUP FOR MUNITIONS EFFECTIVENESS (JTCG/ME)

In 1963, an Army/Air Force panel known as the Close Air Support Board issued a report calling attention to large gaps and gross inaccuracies in data then published on air-to-surface (AS) non-nuclear munitions. To remedy these inconsistencies, the board recommended production of a joint service publication containing a comprehensive list of targets with corresponding data on the effectiveness of aerially delivered munitions suitable for defeating those targets. Responding to this challenge, the Joint Chiefs of Staff requested that a joint service working group correct the data deficiencies and prepare a Joint Munitions Effectiveness Manual (JMEM) for air-to-surface weapons.

The Army was tasked to be the lead service. The chairman of the JTCG/ME established an ad hoc group of military and civilian scientists scattered throughout the U.S. Department of Defense (DoD). The group developed the first standardized methodology for evaluating weapons, and this product was a coordination draft of a multiservice manual entitled *Joint Munitions Effectiveness Manual for Air-Delivered Non-nuclear Weapons*. This manual was accepted not only by the scientists and military professionals, but also by the Secretary of Defense. The latter requested that a similar approach be applied to surface-to-surface weapons.

In the fall of 1965, the original ad hoc group was given formal status as the Joint Technical Coordinating Group for Munitions Effectiveness (JTCG/ME) by the Joint Logistics Commanders. By mid-1966, the JTCG/ME was supported by three working groups: Target Vulnerability, Chemical and Biological, and the original Air-to-Surface (JMEM/AS) Group. These working groups, in turn, controlled the activities of subgroups created and tailored for specific aspects of the parent group's mission. The JMEM/AS had subgroups for weapon characteristics, delivery accuracy, flame and incendiary effects, methodology, and publications. A separate Basic Manual Working Group, chaired by the Defense Intelligence Agency, was established to maintain the Weapon Effectiveness, Selection, and Requirements—Basic JMEM/ AS Manual.

"

In January 1967, JTCG/ME brought in a JMEM production contractor to provide support for the development and production of technical handbooks, JMEM, special reports, and related publications. By 1967, the JTCG/ME was concerned with deriving or validating data not only through tests, experiments, and mathematical models, but also from direct inputs of data from the battlefield. The JTCG/ME Wound Data and Munitions Effectiveness Team gathered wound data from Southeast Asia (SEA). Later, the team expanded the scope of its battlefield collection to include materiel and sent specialized teams to SEA under an effort labeled the Battle Damage Assessment and Reporting Program (BDARP). These data are currently stored in the Survivability/ Vulnerability Information Analysis Center (SURVIAC) at Wright-Patterson Air Force Base and are available for study and research by DoD agencies and contractors.

In September 1967, a separate major group—JMEM/SS—was created to examine and produce data on surface-to-surface munitions. Manuals containing data on individual surface-to-surface weapons were published and revised as new targets or munitions were developed. To complete the weapon–target interface, the JTCG/ME established the Anti-Air Working Group in 1976. Additionally, a Red-on-Blue Working Group was formed in 1977 to address the effectiveness of Red munitions on Blue targets. A Special Operations Working Group, initiated in 1983, provided target vulnerability and weapon effectiveness studies for Special Forces.

In 1994, the JTCG/ME was reorganized with four major working groups: Air-to-Surface, Surface-to-Surface, Anti-Air, and Vulnerability (including Special Operations) to cover the spectrum of weapon effects issues. In addition, each working group is supported by a formally chartered Operational Users Working Group (OUWG). The JTCG/ME Program Office is the focal point for all JTCG/ME efforts. They coordinate the efforts of the working groups; the execution of those efforts is the responsibility of the working group chairmen.

In May 1999, the Office of the Secretary of Defense (OSD) revised DoD Directive 5000.2R—Mandatory Procedures for Major Defense Acquisition Programs (MDAPS) and Major Automated Information Systems (MAIS) Acquisition Programs—to require the procuring agency to provide weapon effectiveness data for use in JMEM for weapons in the acquisition process prior to their achieving initial operational capability. OSD also required that these data be prepared using methodology coordinated with the JTCG/ME.

The resources to operate the program are distributed to the JTCG/ME directly from the Director of Operational Test and Evaluation, Office of the Secretary of Defense. Chartered by the Joint Logistics Commanders with two-star or civilian-equivalent Offices of Primary Responsibility (OPR), the JTCG/ ME is guided by a steering committee composed of members from each of the services, the Joint Chiefs of Staff (J-8), the Defense Intelligence Agency, and the Defense Threat Reduction Agency. The Army continues as DoD Executive Agent, as it had been with the original ad hoc group, and the Director of the U.S. Army Materiel Systems Analysis Activity at the Aberdeen Proving Ground, Maryland, presides.

The committee is composed of civilian and military weapon/munitions experts throughout the DoD to ensure objective, scientific guidance in the development and employment of non-nuclear munitions. The organizational chart shown in Figure P.1 reflects the original structure of the JTCG/ME and indicates the domain stovepipes created for individual weaponeering tools were in the three main areas of application:

1. Air-to-surface (AS)
2. Surface-to-surface (SS)
3. Anti-air (AA)

The principal products of the JTCG/ME are known as the Joint Munitions Effectiveness Manuals (JMEM). Given the organizational structure shown in Fig. P.1, it is no surprise to find that each working group developed its own weaponeering tool to serve its customer and service base. The JMEM were developed to provide a set of data and methodologies that would permit a standardized comparison of weapon effectiveness across all service communities. For example, these data and methodologies should enable a Department of Defense planner in the Pentagon, a Navy weapons officer onboard a ship, and an Air Force targeting specialist in Korea to arrive at similar estimates for the effectiveness of a particular weapon employed under specified conditions. This use of standardized data and methodologies facilitates a wide range of activities beyond the planning and execution of operations, including activities such as weapon development and procurement, management of stockpile levels, and prepositioning of war loads.

Of greatest importance to the targeting process, JMEM data and methodologies are crucial for the development of force employment options contained in

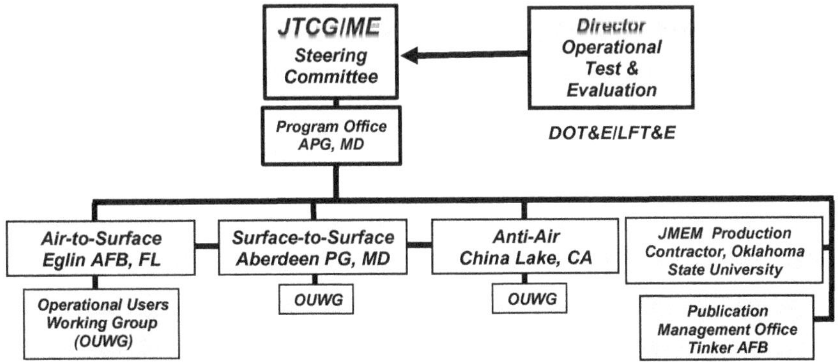

Fig. P.1 Original organization of the JTCG/ME.

the courses of action (COA) nominations developed for theater commanders and the resulting execution tasking orders for tactical units. The JMEM includes detailed data on the physical characteristics and performance of weapons and weapon systems, descriptions of the mathematical methodologies that employ these data to generate effectiveness estimates, software that permits users to calculate effectiveness estimates, and precalculated weapon effectiveness estimates.

Historically, JMEM began with sets of manuals that provided data and methodologies for weaponeering. Over time these manuals were supplemented by computer programs to ease the computational burden on the user, to derive results faster, and to allow more realistic models to be incorporated. In recent years, the JMEM have evolved into CD-ROM products. The first and most widely used was for the air-to-surface community and was known as the Joint Air-to-Surface Weaponeering System (JAWS). Subsequently, the SS Working Group produced a functionally similar product known as the JMEM/ SS Weapons Effectiveness System (JWES). The corresponding product for the Anti-Air group is the Joint Anti-Air Combat Effectiveness Model (J-ACE).

Around 2005, several of the combatant commanders (COCOM) became involved in guiding the development of JMEM tools and moved them towards a target-centric model. From their perspective, it is more important when investigating an attack on a target to estimate the effectiveness of all weapons that could logically be used irrespective of whether they are air launched or ground launched. Up until this point, the scenario-based tools were separate programs, usually running on different computers. As a result, it was decided to integrate the principle tools embodying the JMEM into a single product, the JMEM Weaponeering System (JWS).

This involved a substantial effort in rationalization between the working groups that involved the following subtasks:

- Review target vulnerability data for common targets (e.g., T-72 tank) to ensure consistency.
- Review effectiveness methodologies for common weapon systems (e.g., air and surface machine guns), again to ensure consistency.
- Enhance user transparency regarding the methodologies behind the effectiveness calculations.

This led not only to a different software structure for the single flagship product JWS, but also to a structural reorganization of the JTCG/ME itself in order to ensure a less stovepipe approach to the weapon effectiveness process. This revised organizational structure is shown in Fig. P.2.

As a result of this amalgamation, JWS incorporates both JWES and JAWS together with a common database and allows almost any type of conventional weapon to be played against any type of target supported by the JMEM. Incorporation of J-ACE is also planned for the future.

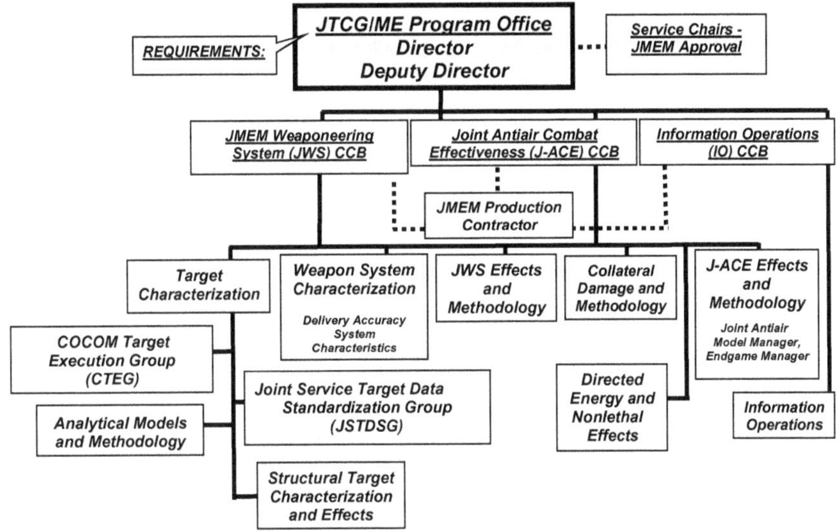

Fig. P.2 Revised organization of the JTCG/ME.

OBJECTIVES OF THE BOOK

The author has been engaged in assisting the JTCG/ME in weaponeering development since the mid-1990s in both a research capacity and as an educator on this subject. As a result, he has published three editions of his *Weaponeering* textbook, appearing in 2001, 2013, and 2019. The objective in writing those books was to archive material resulting from the author's involvement in the subject. As this knowledge grew over time, so did the size of the books, which have sometimes been described as "encyclopedic."

The objective of this book is different. As a result of teaching short courses to a variety of groups in the United States and other parts of the world, two distinct audiences have emerged. The first is mainly civilians working typically in a research and development capacity for Departments of Defense or military services who require a relatively deep understanding of the mathematics and physics driving the underlying methods. For this group, the previous *Weaponeering* textbooks provide much of the information needed to meet this objective.

The second audience may be described as warfighters, because they are usually military personnel in different services and of different ranks who are involved with the execution of military operations, including the planning of offensive strikes against enemy forces. Although tools such as the JWS are available to plan such attacks, and training is provided for using the tools correctly, there is little formal training giving the warfighter an understanding of what lies behind programs such as JWS.

Experience has shown warfighters want to understand more than just the "buttonology" of using weaponeering tools, because that does not provide the background information on the assumptions, limitations, and reliability of the results they produce. There are many examples of trying to improve the probability of damage to the second decimal place when, if the assumptions behind the result were understood, the futility of doing this would become obvious.

This book, therefore, uses much of the knowledge and material utilized in previous weaponeering texts; however, the emphasis is on breadth rather than depth. In addition, the mathematical validation of many of the methods is deemphasized and replaced with more real-world topics and examples to put the underlying methods in the context of how they are applied rather than how they are developed.

Unfortunately, there remains the need to explain some concepts using an approach that still involves some mathematics, physics, and statistics; however, this has been simplified as much as possible to the extent it may be ignored by the casual reader, as long as they are willing to accept the results that are produced.

After presenting chapters on those topics needed to estimate the damage to a target using conventional weapons, an unclassified computer program is described that implements most of the methods described in the book. This is supplemented by a database approved by the JTCG/ME for educational purposes so that several case studies of realistic attacks can be evaluated. Although weaponeers in the United States and selected coalition partners with the appropriate security clearances can obtain copies of the JWS program, and therefore have access to the classified weapon and target vulnerability data, that option will not be available to others. To remedy this to the extent possible, examples are presented indicating how these data may be estimated using open source material.

Chapter 1

OVERVIEW OF WEAPONEERING AND WEAPON EFFECTIVENESS

1.1 INTRODUCTION

In general terms, *weaponeering* can be defined as the process of determining the quantity of a specific type of weapon required to achieve a defined level of target damage, considering target vulnerability, weapon effects, munitions delivery error, damage criteria, probability of kill, weapon reliability, and so forth. In the operational arena, where planners are striving for the most effective use of limited resources, efficiency is a critical factor that must be considered in the weaponeering process.

When faced with planning a strike mission, there is a wide variety of possible weapons to choose from. For example, the F/A-18 Hornet shown in Fig. 1.1 can carry a range of air-to-ground weapons comprising guided and unguided bombs, air-to-surface missiles, guns, rockets, and so on. Other systems, such as the M1A1 main battle tank (shown in Fig. 1.2), may have a smaller selection of munitions that can be fired from the 120-mm main gun (see Fig. 1.3).

Apart from selecting the appropriate weapon to execute a mission, other factors such as whether military personnel are placed in harm's way are also a consideration in the assessment. Therefore, weaponeering is the process of selecting the most efficient weapon to perform an attack against a defined target in order to inflict a specified amount of damage.

1.2 EXAMPLES OF A WEAPONEERING OUTCOME: AIR TASKING ORDER

One product commonly associated with the output of the air-to-surface weaponeering process is the air tasking order (ATO). Table 1.1 shows a sample ATO that would be given to aviators prior to flying a mission.

1

TABLE 1.1   SAMPLE AIR TASKING ORDER (ATO)

MSN: 8088 **UNIT:** 222FS **FORCE:** 2XF/A-18 **C/S:** REGAL55 **TOTZ:** 0835
ORD: SCL-855, FUZE FOR MAX CRATERING
PRITGT: VUL HWY SEG, BV981.335 (GCC REQ FM-668)
DMPI: RTE SEGMENT CROSSING DIKE
RMKS: COODR WITH MSN 7345

How was this ATO generated? Before answering that, let us discuss the ATO in detail, line by line.

MSN	Mission number
UNIT	Operational unit executing the mission
FORCE	Aircraft to be used (two F/A-18s in this case)
C/S	Call sign for the mission
TOTZ	Time over the target (Zulu time)
ORD	Ordnance to use (standard conventional load 855 with fuze settings set to maximum cratering)
PRITGT	Description and coordinates of target (highway)
DMPI	Aimpoint (highway crossing a dike)
RMKS	Remarks (coordinate with mission #7345)

In order to maintain the balance and maneuverability of the attacking aircraft, there is not complete freedom to choose which and how many

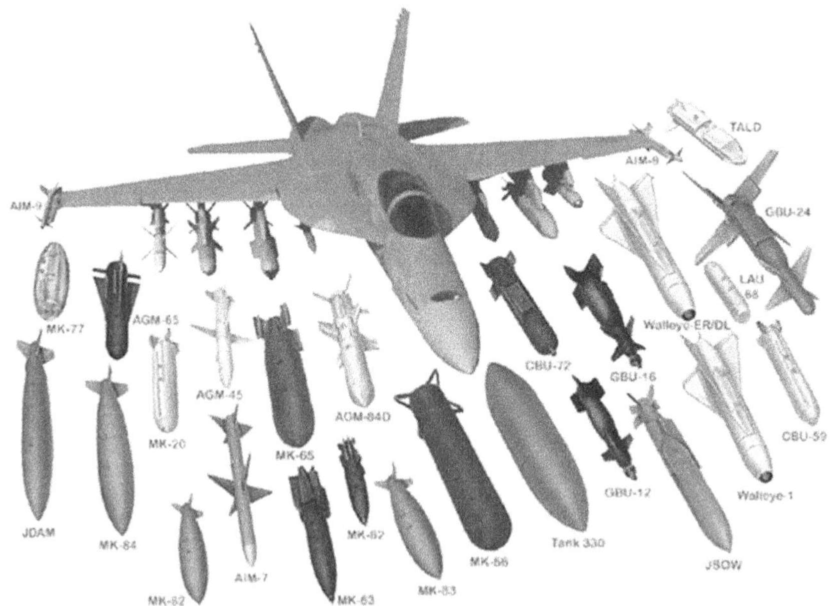

Fig. 1.1   F/A-18 weapon choices.

Fig. 1.2   M1A1 tank.

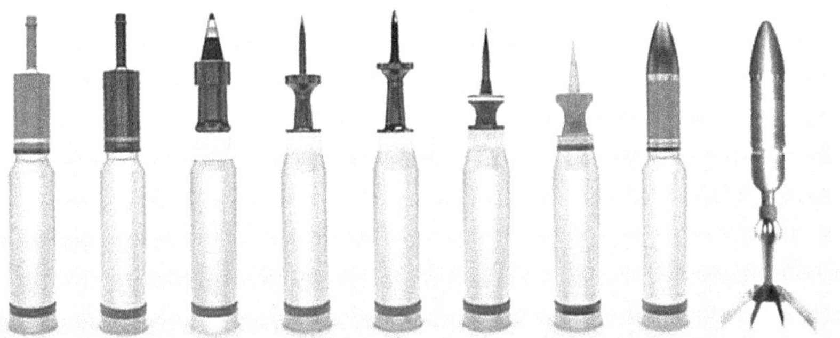

Fig. 1.3   M1A1 main battle tank gun rounds.

weapons are assigned to a mission. Standard conventional loads (SCLs) are defined for different types of attacks; for example, the SCL for attacks against infrastructure targets such as buildings or bridges would be different than the SCL employed against ground troops. Figure 1.4 shows an F/A-18 aircraft with a standard conventional load of eight 500-lb Mk-82 unguided bombs.

Another example of the utility of weaponeering is in the assessment of collateral damage. Consider the image of a planned target area shown in Fig. 1.5. Here the weapon aimpoint is located at the center of the target (building) and a collateral effects radius (CER) circle is centered on it. Weaponeering involves the estimation of lethal effects expected from the warhead and the accuracy with which the weapon is delivered, both of which contribute to the CER. If no strike elements or collateral objects (COs) are identified in the vicinity of the target, the CER helps to support a decision on whether to proceed with the attack. Collateral damage and the calculation of the CER circles is a complex task and is dealt with in detail in Chapter 16.

Fig. 1.4   F/A-18 with a standard conventional load.

The daily air tasking order is the published order that directs all the air missions in a campaign. Weaponeering, together with an affiliated activity known as targeting, will provide the planner with information on the selection of weapons, fuzes, aimpoints, and force levels required. Clearly, both targeting

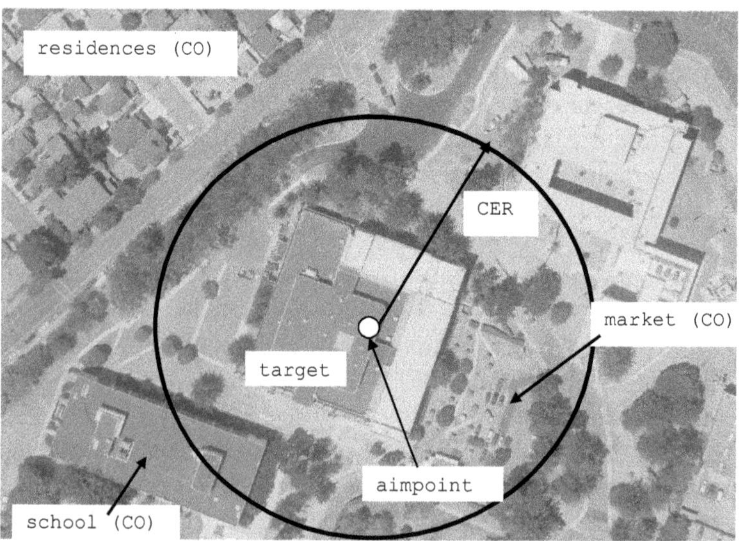

Fig. 1.5   Building target with surrounding collateral objects.

and weaponeering do not take place in isolation from the strategic planning that accompanies warfare, but are an intrinsic part of it. A notional overview of this "bigger picture" is given in the next section.

1.3 WEAPONEERING: PART OF A LARGER PLANNING CYCLE

Although weaponeering is an important piece in the overall attack plan, it is only one part. It is important to recognize that weaponeering lies at the midpoint in the recurring force application cycle, as indicated in Fig. 1.6, requiring inputs from previous steps in the cycle while providing a portion of the inputs to subsequent steps in the cycle.

The individual phases may be characterized by the following activities:

- *Objectives:* Outlined by the immediate commander and senior headquarters in their employment guidance, the mission objectives define which facets of enemy activity are to be affected by the mission. An example of a tactical objective would be to deprive the enemy of the use of electrical power within a given geographical region for a period of 14 days.
- *Target development:* Guided by defined tactical objectives and working with the target list developed by the command structure, targeting specialists select suitable target elements to achieve these objectives. These specialists, known as targeteers, have knowledge of the infrastructure of the region and can identify critical elements relating to the tactical objective. Using the previous example, targeteers would have access to the type of information shown in Fig. 1.7 for the power distribution system in the region of interest [1].

For example, assuming the substations are the focal points for power distribution in the regions shown, to meet the objective listed, substation 2

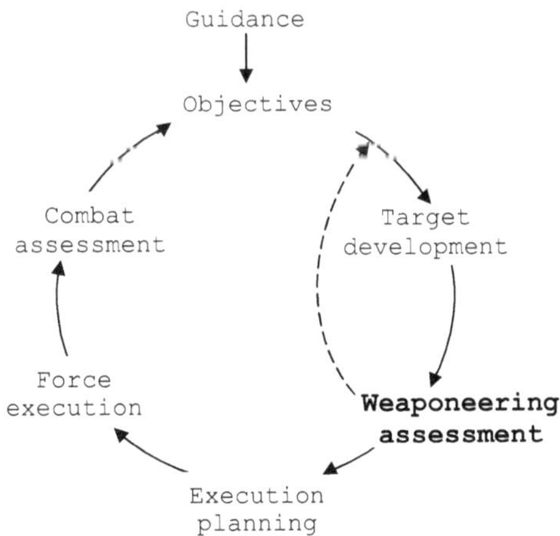

Fig. 1.6   Force application planning cycle.

would be selected for attack to black out the lower left region of Fig. 1.7, and aimpoints in that station (Fig. 1.8) would be defined to produce the desired effect.

In this example, following the attack, repairing the damaged transformers might take about 14 days, but rebuilding the control building would take much

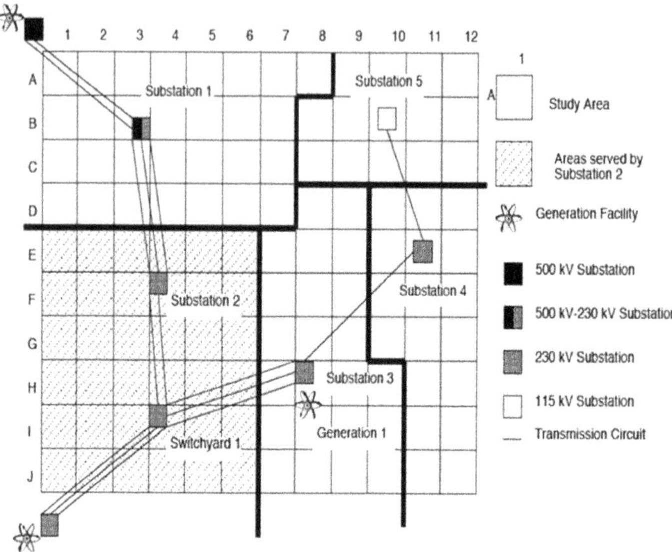

Fig. 1.7   Power distribution system in region of interest.

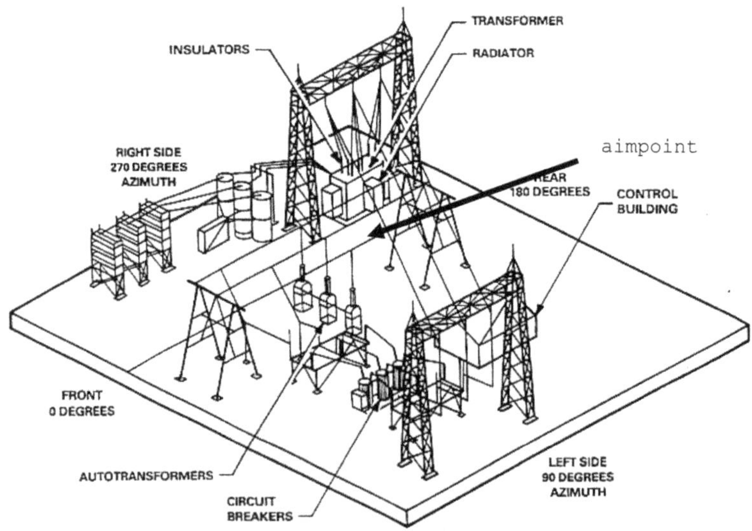

Fig. 1.8   Layout of substation 2.

longer, hence the transformers would be targeted. Note also that selection of substation 2 rather than switchyard 1 achieves the tactical objective while minimizing disruption to other regions, an example of mitigating collateral damage.

- *Weaponeering assessment:* Given the aimpoint and target elements to be attacked, the weaponeer selects the type and quantity of weapons needed to produce the required level of damage. This essentially defines the mission in terms of delivery system (aircraft, surface rockets), warhead, fuzes, sorties, and target, and for air-launched weapons would produce the air tasking order described earlier; for surface-launched rockets it would generate firing orders.
- *Execution planning:* The outputs of the weaponeering process are essential inputs to execution planning, which is conducted on two levels. At the headquarters level, operations staffers assign missions to specific units, perform attack force packaging, task required mission support forces, determine attack timing, and outline communications and coordination requirements; the products of this effort are also embodied in the daily air

TABLE 1.2   KEY FORCE APPLICATION PLANNING QUESTIONS

Planning Step	Questions to Resolve
Objectives	What enemy activity is to be affected?
	Where should effort be focused?
	What effects are desired?
	When and for how long?
Target development	Which targets relate to the objectives?
	Of those, which are most important?
	Of those, which are vulnerable?
	What are the critical target elements?
	Which target defenses must be attacked?
	What is the target prioritization?
	What is the optimum timing of attacks?
Weaponeering assessment	What are the appropriate damage criteria?
	What probability of damage is required?
	How much force and ordnance is needed?
	What are the optimum aimpoints?
	What is the recommended ordnance and fusing?
Execution planning (HQ level)	When should the mission be executed?
	Which units are best suited to the task?
Execution planning (unit level)	What support is required?
	What C^3 is required?
	What are ingress/egress routes and tactics?
	What are the weapon delivery tactics?
Force execution/ combat assessment	What was destroyed or damaged?
	Did the weapons work as expected?
	Did the attack produce the intended effect?

C^3: command, control and communication.

operations or air task orders. At the unit level, the execution-planning phase includes target study and detailed mission planning.

- *Force execution and combat assessment:* Once the mission is planned and executed, the results are assessed to determine whether the objectives have been met or whether a restrike is required.

Note that this cycle may have internal feedback loops depending on the operational level at which the weaponeering is done. For example, after execution of initial sorties, pilot feedback may suggest a more effective weapon is needed for the next sortie or restrike.

Table 1.2 outlines some of the key planning questions that should be resolved at each step in the planning cycle.

1.4   An Example of a Weaponeering Tool

Weaponeering has developed into a highly automated process for most operational users, principally through the availability of products such as the Joint Munitions Effectiveness Manual (JMEM) Weaponeering System (JWS). With such programs it is possible to select a target, choose a weapon, specify release conditions of the weapon from the aircraft, and compute the resulting probability of damage due to a single weapon, denoted as PD_1. JWS may also be used to determine if a surface-launched weapon may be used to attack the same target.

Unfortunately, the JWS product is classified and therefore only available to U.S. personnel with the required clearances. For this book, however, an unclassified weaponeering tool is available to perform attack planning missions in a manner similar to JWS. Figure 1.9 shows a user interface to this program to compute the probability of damage as a function of the number of sorties employed.

In order to use a tool such as this and the JWS program, a weaponeer must understand the nature of the data that have to be input into the program, where to obtain them, and any constraints associated with them. The main areas of interest (and their associated data) needed to compute damage levels are:

- Weapon trajectory characteristics
- Accuracy of weapon delivery on the target
- Warhead damage mechanisms
- Target vulnerability
- A methodology to integrate these into a damage measure

Each of these data requirements will be discussed later in this chapter and in detail in subsequent chapters of this book. The purpose is to provide users with as much background as possible into all the information needed to compute an estimate of damage to the target and also the actual computational process itself. This will allow those who generate damage estimates to be

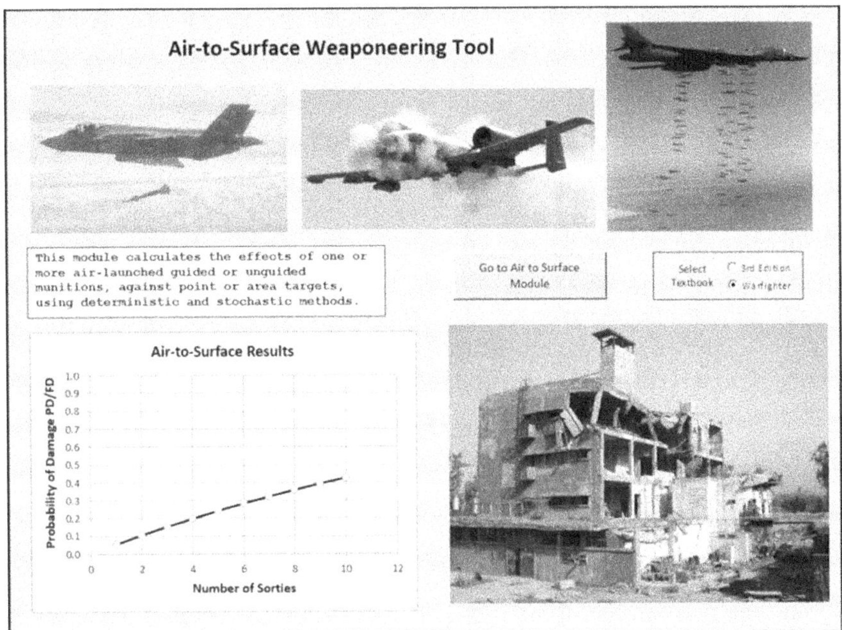

Fig. 1.9   Air-to-surface weaponeering tool.

aware of the assumptions and limitations inherent in the process, and as such to rely on the result to an appropriate degree.

However, before moving into the details of these topics, it is appropriate to look in more detail at the various mechanisms by which conventional weapon warheads can cause damage to any target, whether it be a building, vehicle, or personnel.

1.5   WEAPON DAMAGE MECHANISMS

Weapons may damage targets through a variety of methods; in general, the specification of the target will suggest the appropriate damage mechanism. For example, personnel targets are susceptible to blast and fragments whereas an armored vehicle is susceptible to shaped charge. Most explosive-filled warheads will produce both blast and fragments, although a few are primarily blast weapons. The assessment of the damage-producing ability of a particular warhead is generally known as the process of determining the weapon, or warhead characteristics. The basic damage mechanisms will now be reviewed.

1.5.1   BLAST

Warhead blast effects are particularly damaging against infrastructure targets such as buildings as well as personnel, and inflicts damage with three primary components:

1. Peak pressure
2. Impulse
3. Blast wind

When a high explosive detonates, it is converted almost instantly into a gas at very high pressure and temperature. Under the pressure of the gases generated, the weapon case expands and breaks into fragments. The air surrounding the casing is compressed and a shock (blast) wave is transmitted into it. Typical initial values for a high-explosive weapon are 200 kilobars (kbars) of pressure (1 bar = 1 atmosphere) and 5000°C. As the wave passes a fixed point, the blast wave is perceived as a rapid rise from ambient to some peak pressure followed by an exponential decay to a value below atmospheric, followed by a return to atmospheric, as shown in Fig. 1.10.

The peak pressure is attenuated as the wave propagates from the detonation point, where the rate of attenuation is proportional to the expansion of the volume of gases behind the blast wave (see Fig. 1.11).

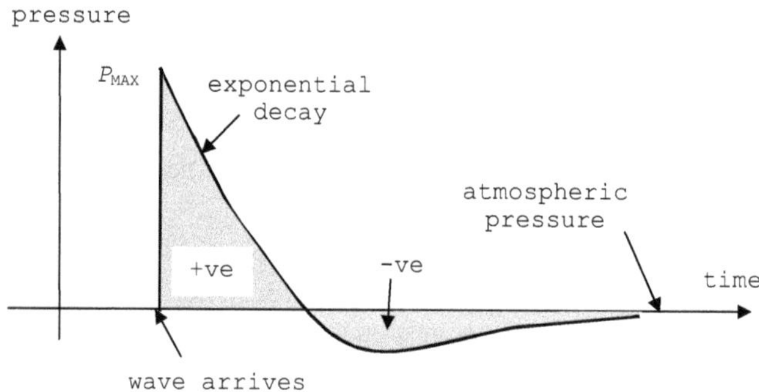

Fig. 1.10   Blast wave pressure–time profile.

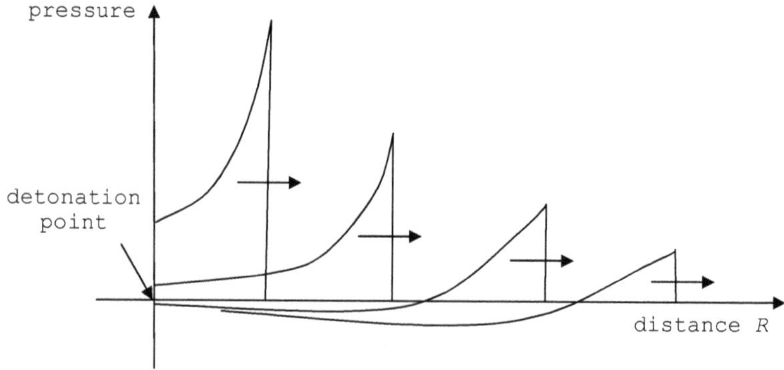

Fig. 1.11   Blast wave attenuation.

Fig. 1.12   Blast wave radiating out from detonation point.

Figure 1.12 shows the spherical propagation of a blast wave radiating outward (above) from a warhead detonation.

Initially the speed of the blast wave is supersonic, but as it moves away from the detonation point it slows down until at a sufficiently large distance it is traveling at the acoustic speed. In addition to the peak pressure, its duration also has an effect on the target. A high pressure exerted over a target for a long time will cause more damage than the same peak pressure exerted over a shorter time. The blast wave impulse is a measure of its duration and is defined mathematically as the integral over time of the pressure wave

$$I = \int_{t=0}^{t=\infty} p(t)\mathrm{d}t \tag{1.1}$$

or more simply, the area under the pressure–time curve shown in Fig. 1.10. From this figure we see a large positive impulse followed by a smaller negative impulse. In the case of structural elements such as floor slabs or supporting columns subjected to an external detonation, failure is a function of both peak pressure and impulse, as indicated in Fig. 1.13.

The same phenomenon is apparent when considering the injury to personnel in the open who are subjected to the passage of a blast wave, as shown by the so-called Bowen curves in Fig. 1.14. Consider a specific injury criterion such as the 1% lethality, 50% incapacitation curve. This graph states that if the combination of impulse and overpressure at a specific location corresponds to a point on the graph that is below the curve, that injury criterion will *not* be met and the target survives, whereas if the combination of impulse and overpressure corresponds to a point above the curve, the target will not survive.

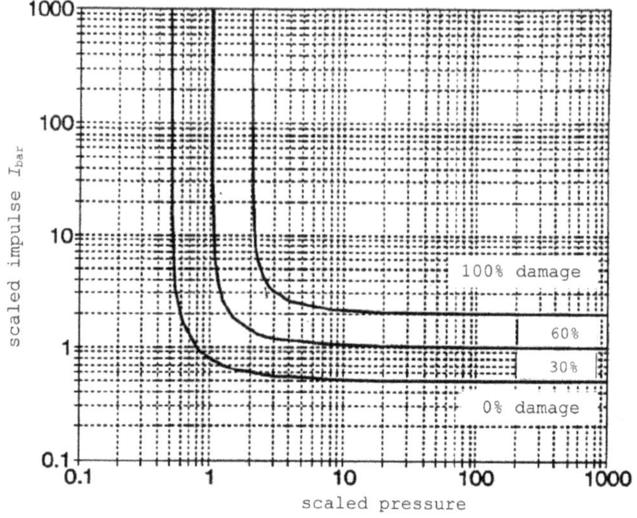

Fig. 1.13 Structural damage due to pressure and impulse.

A third effect of an explosion is called the blast wind. As a blast wave passes a fixed point on the ground, as well as experiencing a rapid rise in pressure and temperature, the air is given a velocity in the same direction as the blast wave. Although this velocity is much less than that of the supersonic blast wave, it can be of the order of several hundred miles per hour (i.e., much

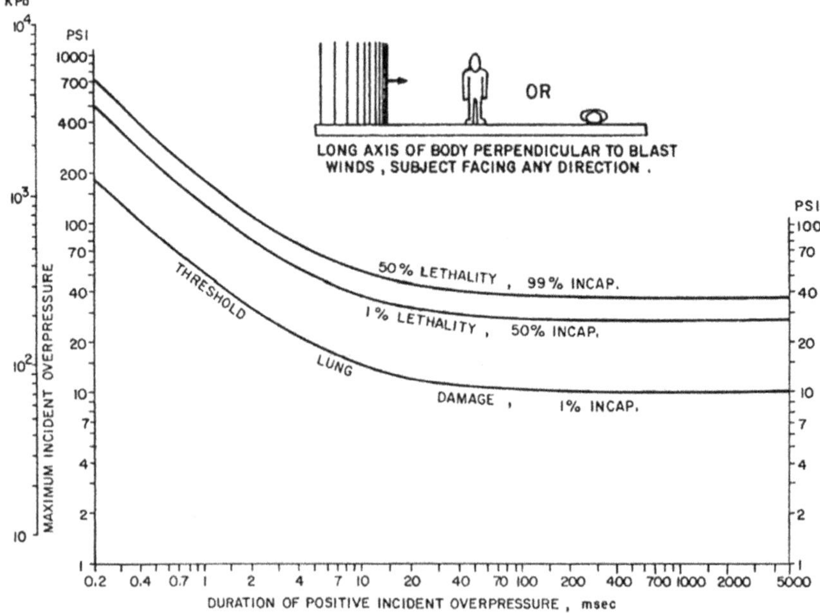

Fig. 1.14 Personnel injury due to pressure and its duration.

greater than the velocity of a severe hurricane). Clearly this can add to the effects of pressure and impulse, particularly for targets such as buildings, as shown in Fig. 1.15.

For buildings, damage to a single floor causing it to collapse may produce structural failure in the whole building through a process called progressive collapse, as shown in the Alfred P. Murrah building in Oklahoma City (Fig. 1.16). In this case, a street-level explosion demolished columns in the lobby supporting the upper floors, which collapsed the front of the structure.

When an explosion occurs within a structure, the peak pressure associated with the initial shock front will be extremely high; this, in turn, may be amplified by reflections from walls, floors, and ceilings within the structure. If the warhead is relatively small, damage may be contained to a particular floor or

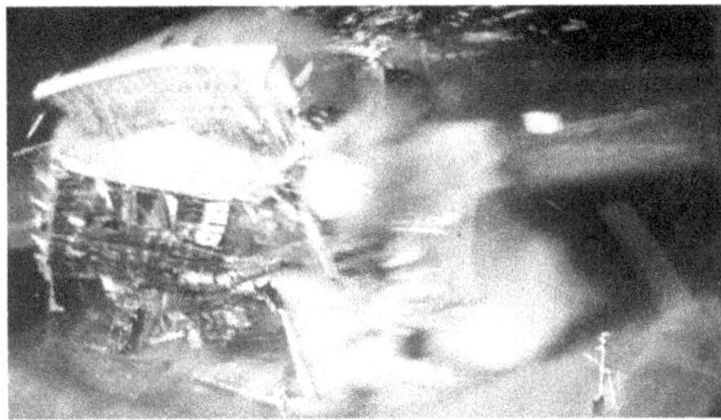

Fig. 1.15   Blast wind adds to building damage.

Fig. 1.16   Building damage due to external blast.

Fig. 1.17   GBU-43 guided bomb with blast warhead.

some rooms on the floor; however, larger warheads may cause sufficient damage to trigger partial or total building collapse.

An example of a weapon designed just to damage targets with blast is the massive ordnance air-blast bomb (MOAB) shown in Fig. 1.17.

Most conventional warheads will produce damage to buildings and personnel due to the blast generated, depending on the amount of explosive fill. Such weapons range from mortar rounds, artillery shells, and multiple-launch rockets to guided and unguided bombs up to the 2000-lb class.

1.5.2 FRAGMENTATION

Approximately 30% of the energy released by the explosive detonation is used to break up the case and impart kinetic energy to the fragments. The balance of available energy is used to create a shock front and blast effects. The fragments are propelled at high velocity, and after a short distance they overtake and pass through the shock wave. The rate at which the velocity of the shock front accompanying the blast decreases is generally much greater than the decrease in velocity of fragments, which occurs due to air friction. Therefore, the advance of the shock front lags behind that of the fragments. The radius of effective fragment damage, although target dependent, thus exceeds considerably the radius of effective blast damage in an air burst.

When a shell or bomb detonates, it expands to 2–3 times its static dimensions before breaking apart, as shown in Fig. 1.18. A general-purpose warhead then disintegrates into irregular or naturally formed fragments having a wide variety of masses and projection velocities (see Fig. 1.19).

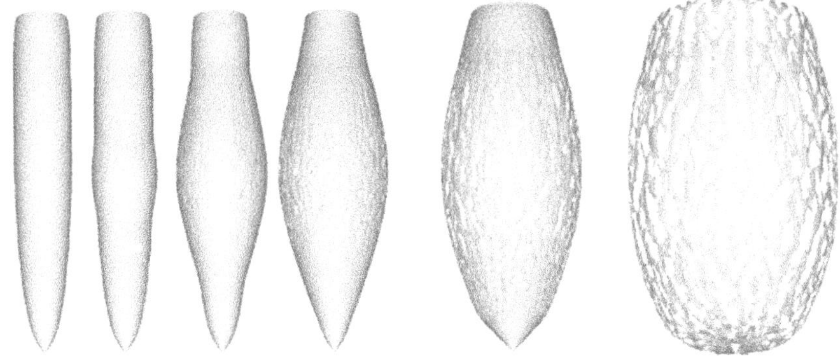

Fig. 1.18   Bomb casing expansion and fragmentation.

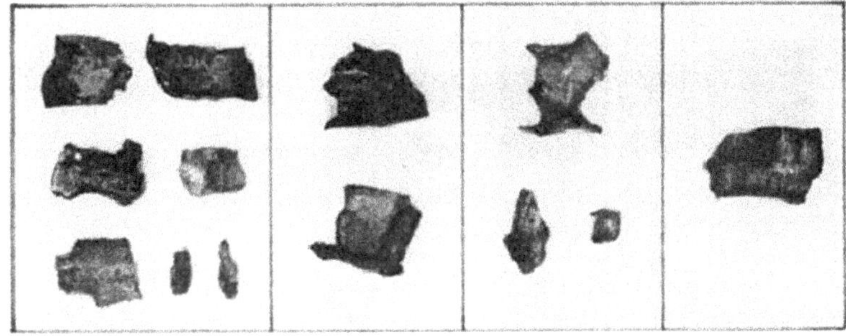

Fig. 1.19   Naturally formed fragments.

In order to tailor the effect of fragments to specific target types, the breakup of the casing may be made more predictable by scoring either the inside or outside prior to filling with explosive, as shown in Fig. 1.20.

The effectiveness of a warhead depends greatly on whether natural or preformed fragments are produced. Table 1.3 shows the differences in the number of fragments and the area over which the fragments cover for a general-purpose Mk-82 500-lb bomb body.

The effects (pressure, impulse, blast wind) of an idealized blast payload are attenuated by a factor roughly equal to $1/R^3$. (R is measured from the detonation point.) The attenuation of idealized fragmentation effects will vary as $1/R^2$ and $1/R$, depending on the specific design of the payload. Herein lies the principle advantage of a fragmentation payload: it can afford a greater miss distance and still remain effective because its attenuation is less.

Fragments are known as primary or secondary fragments depending on their origin. Primary fragments are formed as a result of the shattering of the casing of conventional munitions. These fragments usually are small and travel initially at velocities on the order of thousands of feet per second. Secondary fragments

Fig. 1.20   Internal scoring of warhead casing.

TABLE 1.3   DIFFERENCES DUE TO FRAGMENT TYPE FOR MK-82 BOMB

Fragment Type	Number of Fragments	Effective Area (m)
Natural	3000	80×30
Preformed	17,000	240×80

are formed as a result of high blast pressures on structural components and items in close proximity to the explosion. These fragments are somewhat larger in size than primary fragments and travel initially at velocities on the order of hundreds of feet per second. A commonly used rule of thumb is that a hazardous fragment is one having an impact energy of 58 ft-lb (79 joules) or greater. Figure 1.21 shows fragments projected radially from a detonation.

In the same class of damage mechanisms as fragments are single projectiles such as bullets, armor-piercing (AP) rounds, and long rod penetrators, which damage targets using the same mechanism by displacing the surface layers and penetrating the underlying materials. Figure 1.22 shows the M829 sub-caliber (120-mm) discarding sabot, depleted uranium (DU) round fired by the M1A1 tank.

In this figure, the larger diameter sleeve (sabot) ensures the round fits into the bore, but on exiting it peels back to leave the smaller diameter (about 1 in.) DU dart.

Even very small fragments on the order of a few grains (1 lb = 7000 grains) will cause severe injury to personnel; larger ones may damage unprotected vehicles and aircraft, as shown in Fig. 1.23.

Fig. 1.21   Fragments (outer ring) propagating from a warhead.

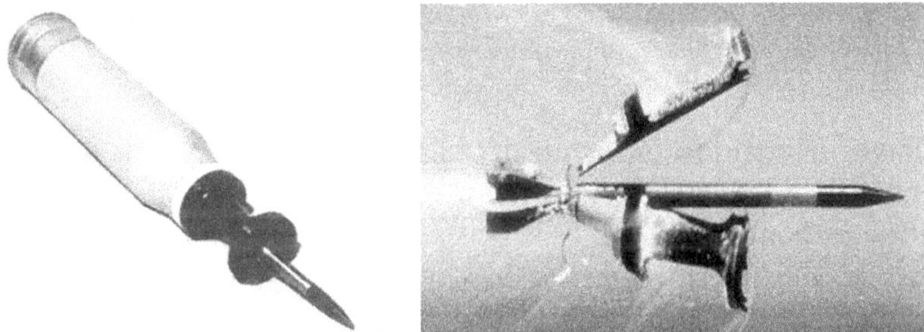

Fig. 1.22   Discarding sabot, kinetic energy penetrator.

Fig. 1.23   Vehicle damage due to fragments.

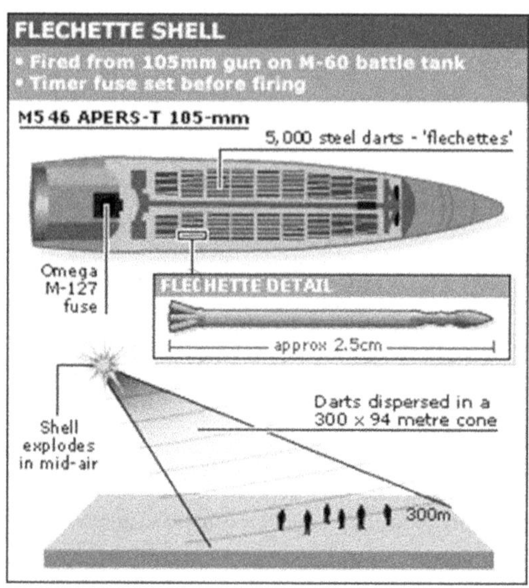

Fig. 1.24   Flechette artillery round.

Because of their random nature, it is difficult to specify exactly the number, size, velocity, and direction of individual, naturally formed fragments generated by a particular warhead. Through extensive testing, however, we are able to produce statistical estimates of these quantities.

Another example of a warhead that produces only penetration effects is a 105-mm round that may be fired by either an artillery cannon or a tank. It produces several thousand 1-in. darts, or flechettes, that are effective against personnel and unprotected materiel targets. An example is shown in Fig. 1.24.

If the flechettes tumble once they penetrate, a greater level of injury will result.

1.5.3 SHAPED CHARGE

A shaped charge warhead consists basically of a hollow liner of metal material, usually copper or aluminum, of conical, hemispherical, or other shape, backed on the convex side by explosive. A container, fuze, and detonating device complete the round. Figure 1.25 shows the M77 81-mm submunition, which has a shaped charge in the lower portion and also a steel case for producing fragments.

When this warhead is in close proximity to the target, the fuze detonates the charge from the rear. A detonation wave sweeps forward and begins to collapse the metal cone liner at its apex. The collapse of the cone results in the

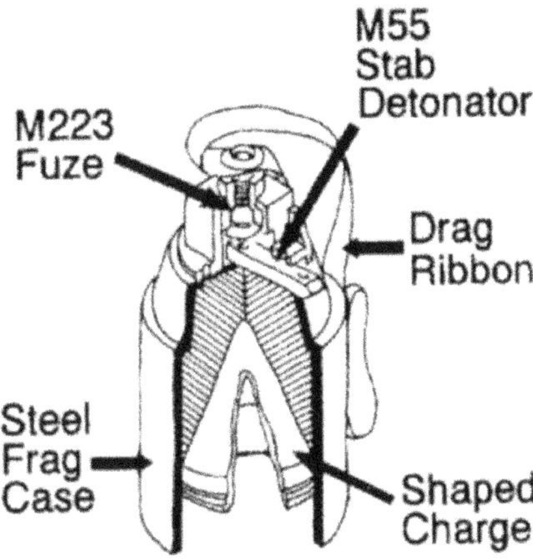

Fig. 1.25   Dual-purpose fragment/shaped charge warhead.

formation and ejection of a continuous high-velocity solid jet of liner material. At the pressures produced by the explosion, the liner material flows like a fluid, and is therefore referred to as a *jet*.

The use of copper as a shaped charge jet will be discussed in Chapter 7, but one important issue is the phase transformation from solid to liquid has to take place very rapidly; therefore, highly conductive materials are required.

The velocity of the tip of the jet is around 8500 m/s, whereas the trailing end of the jet has a velocity of around 1500 m/s. This produces a velocity gradient that tends to stretch out or lengthen the jet. The jet is then followed by a slug that consists of about 80% of the liner mass. The slug has a velocity on the order of 600 m/s.

When the jet strikes a target of armor plate or mild steel, pressures in the range of hundreds of kilobars are produced at the point of contact. This pressure produces stresses far above the yield strength of steel, and the target material also flows like a fluid out of the path of the jet, as shown in Fig. 1.26. This phenomenon is called *hydrodynamic penetration*, and the effect on a thick steel plate is shown in Fig. 1.27. In displacing the steel to produce the hole, the kinetic energy of the jet is expended.

The goal for damage to a target such as an armored vehicle is for the jet to produce a hole such as that shown in Fig. 1.27 while still having sufficient jet material remaining to enter the interior of the target and cause fire, injure the crew, and detonate munitions.

The depth of penetration is maximized for a particular standoff (i.e., the distance between the warhead and target at detonation), and shaped charged warheads are fuzed at this optimum by a trigger (see Fig. 1.28). In this design, the critical standoff is achieved by mounting a trigger on a hollow cone ahead of the charge with an electrical connection to the fuze located behind.

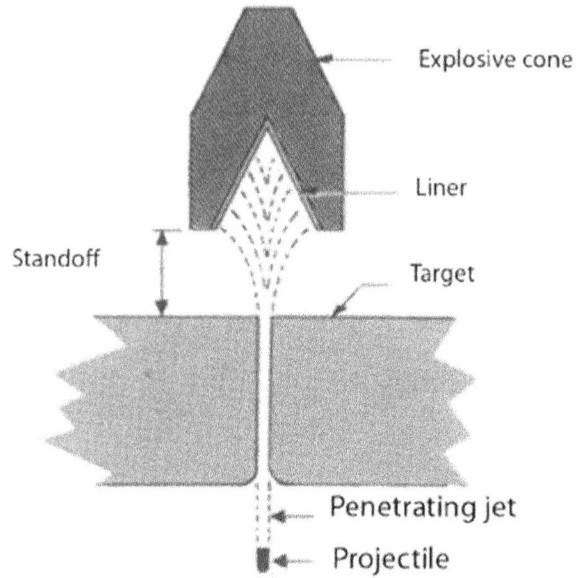

Fig. 1.26   Shaped charge damage mechanism.

Unlike blast or fragmentation, a shaped charge weapon has to hit the target in order to damage it; as a result, a simple protective metal chain curtain suspended a few inches from the side of an armored vehicle will often cause the weapon to detonate harmlessly away from the hull.

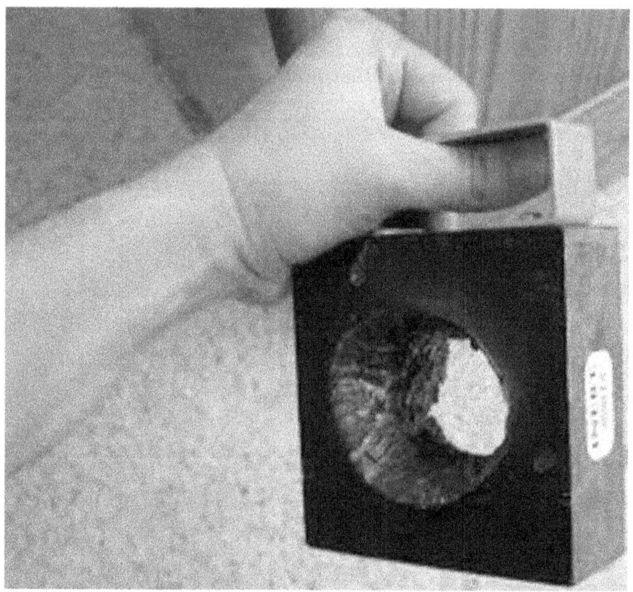

Fig. 1.27   Effect of a shaped charge on a steel plate.

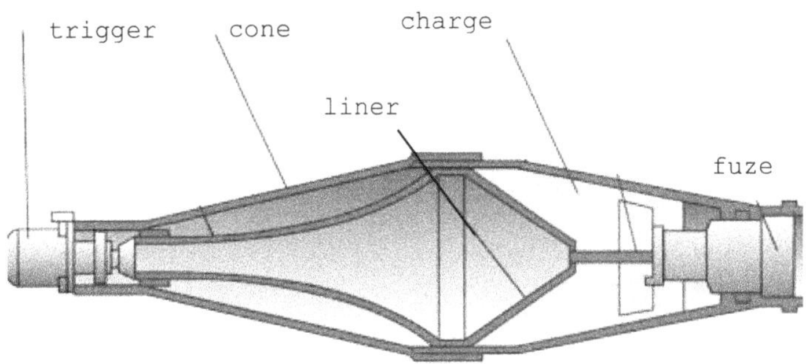

Fig. 1.28 Shaped charge (RPG-7) warhead.

1.5.4 EXPLOSIVELY FORMED PROJECTILE (EFP)

An EFP is similar to a shaped charge but is designed to travel over a larger standoff distance. A conventional shaped charge generally has a conical metal liner that projects a hypervelocity jet of metal able to penetrate to great depths into steel armor; however, in traveling over a long distance, the jet breaks up along its length into particles that drift out of alignment, greatly diminishing its effectiveness.

An EFP, on the other hand, has a liner in the shape of a shallow dish. The force of the blast molds the copper plate into any of a number of configurations, depending on how the plate is formed and how the explosive is detonated. Sophisticated EFP warheads have multiple detonators that can be fired in different arrangements causing various types of waveforms in the explosive, resulting in a long rod penetrator, an aerodynamic slug projectile, or multiple high-velocity fragments. These generally remain intact and are therefore able to penetrate armor at long range, delivering a wide spray of fragments of liner material and vehicle armor back-spall into the vehicle's interior, injuring its crew and damaging other systems.

As a rule of thumb, an EFP will perforate a thickness of armor equal to only about the diameter of its charge, whereas a typical shaped charge will go through six or more diameters.

Figure 1.29 shows the formation of the EFP following detonation of the warhead, and Fig. 1.30 shows the final form of the projectile.

Standoff distances for EFPs are in feet rather than the few inches for a shaped charge.

1.6 FUZING SYSTEMS

The fuze is not directly a damage mechanism like a blast or fragments, but it has a considerable effect on such mechanisms. For example, injury to a group of personnel standing on the ground will be different if the same

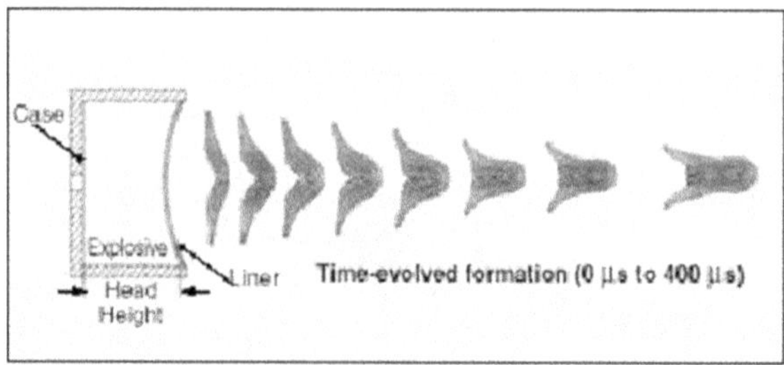

Fig. 1.29   Formation of an EFP.

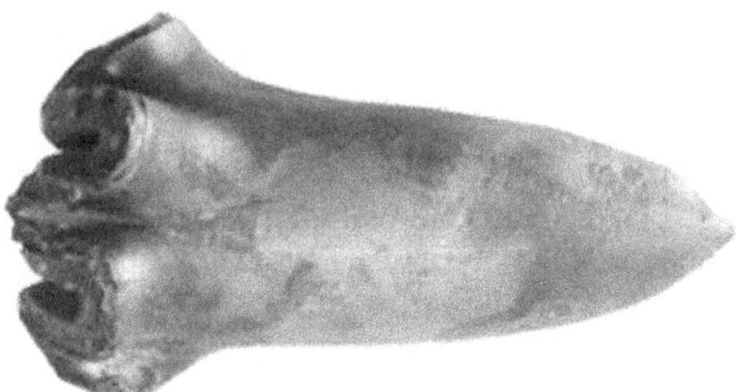

Fig. 1.30   Projectile shape.

weapon is detonated in the air close to the group compared to detonating the weapon after it has buried itself 20 ft into the ground. Therefore, the effect of fuzing must be considered in any damage calculation.

A fuze initiates warhead detonation when a particular event occurs such as elapsed time from firing, a certain height above the ground is achieved, or a large negative acceleration (impact) is measured. Fuzes may be located in the nose or tail of the weapon, or both. The most common type of fuze is a delay, and there are two types of delays associated with general purpose fuzes:

1. The arming delay is the time after release/firing when the weapon becomes armed.
2. The functioning delay is the time after impact when the warhead detonates.

Weapons are armed by one or more of the following methods:

- For bombs, the *arming vane*, a small propeller, is rotated by airflow after weapon release. A specified number of rotations arm the fuze.
- The *arming pin* is ejected or withdrawn by a spring action or lanyard, releasing the arming mechanism and allowing the fuze to arm.
- The *inertia fuze* is armed by abrupt changes in the velocity of the weapon.
- The *electric fuze* is armed by a time-delay circuit powered by a thermal battery activated by extraction of the arming lanyard upon bomb release.

Fuzes may be mechanical, electrical, or electronic in design.

- A *mechanical fuze* operates in a manner similar to the hammer firing a cartridge in a gun. When the weapon impacts a target, it drives a striker into the detonator, which sets off a train of explosives. More sophisticated mechanical components in the fuze produce functioning delays.
- An *electrical fuze* is similar to a mechanical fuze, but a series of resistor/capacitor networks discharge over set times to produce both arming and functioning delays.
- An *electronic fuze* is a solid state electrical fuze with some selectable delays.

Different options are obtained by mating different warheads to one of several fuzes. Common fuze functions may be described in one of the following ways:

- An impact, instantaneous, or point detonating (PD) fuze is designed to function on impact or a preset time after impact (impact delay). Detonation upon impact is selected for targets such as supply dumps when the main destructive energy desired is blast. For a building, a delayed detonation might be selected so a bomb can penetrate several floors before exploding. Figure 1.31 shows an artillery fuze that may be set for variable time, delay, or point detonating modes before firing.
- A proximity (PX) fuze is designed to detonate the warhead at some time after firing. A type of PX fuze is the variable time (VT) variety, in which a fixed time after firing is set and the fuze will initiate at that time irrespective of its location (see Fig. 1.32). This mechanism is a way to achieve an airburst over a target if accurate trajectory information is known. The fixed time after firing may be achieved using mechanical or electrical timers.

```
NOMEN: M782
FUZE SETTER: EPIAFS
TIME: 0.5 to 199.9 sec
VT: 1.0 to 199.0 sec
DLY: Inductively Set
PD: Inductively Set
DODIC: NA09
RESTRICTIONS: Not compatible w/
base-ejecting projectiles.

Improvements
• Last inductive fuze setting
will remain on fuze unless
reset.
• Has a battery life of 30 yrs.
• PD back-up if fuze fails to
function properly.'
• Compatible with all 105mm and
155mm bursting projectiles.
```

Fig. 1.31   Fuze setting options.

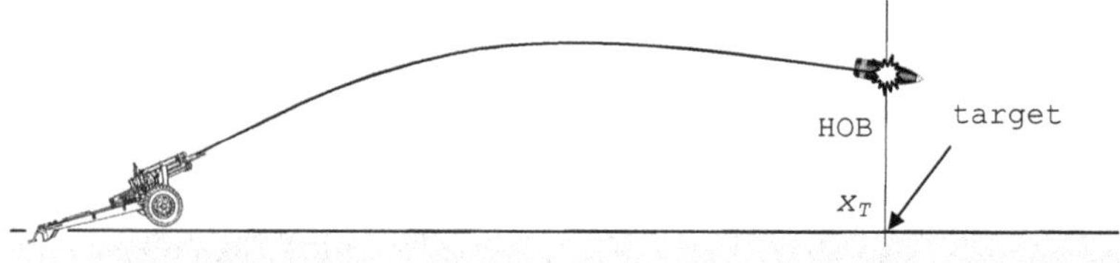

Fig. 1.32   Airburst with VT fuze.

Other types of PX fuze are sensor based, where closeness to the target will initiate detonation. A height of burst (HOB) fuze contains an altimeter (barometric or radar) that is designed to detonate the warhead a specific height above the ground. Another sensor-triggered fuze contains a miniature Doppler radar set that senses distance from an object. When the explosion occurs above the ground it is called an airburst, and most of the destructive effect is caused by the bomb casing fragments. Proximity-fuzed bombs are used against targets such as troops in trenches, radars, vehicles, and in-flight aircraft. Most air-to-air missiles have proximity and impact fuzes.

- A hydrostatic fuze is employed in depth bombs used for underwater demolition work. The MK-36/40 Destructor is a special fuze with a sensor that can be mated to a bomb. It senses the presence of metallic objects such as trucks or ships, making it in effect a mine. These weapons can be used against either land or water targets.
- The Hard Target Smart Fuze (HTSF) is a microcontroller-based fuze designed to be physically and electrically compatible with several guided and unguided weapons. The HTSF was designed for current and future penetrator weapons to define the fuze function point as either a desired distance within a desired void or a depth of burial beneath a hard layer. It operates in one of three modes: hard-layer detection, void detection, and path-length integration. It also has an adjustable backup time delay that is set in 1-ms increments with a maximum delay of 250 ms. The HTSF uses a void-sensing technique to count layers within a structure to initiate fuze function, a depth of burial mode that causes the fuze to function a preset distance after it senses a hard layer, and an integral time delay backup. The layer counting method is shown in Fig. 1.33 for a weapon that, for example, has to detonate in a particular floor of a building.

Fuzes are commonly nose mounted; however, warheads penetrating hard materials will usually be tail mounted to avoid damage during the penetration process.

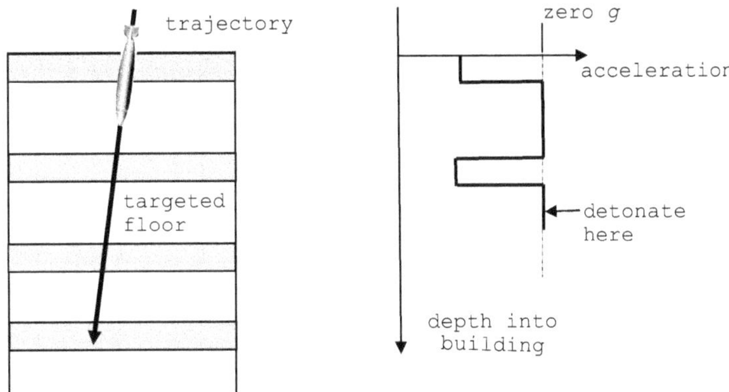

Fig. 1.33 **Void counting fuze.**

1.7 COMPUTING TARGET DAMAGE

Reconsider the weaponeering tool shown in Fig. 1.9. Clicking the Go to Air to Surface Module button brings up the more detailed screen shown in Fig. 1.34.

In order to compute the probability of damage due to a single weapon PD1 shown at the top of the program, the user has to supply certain data that are grouped into different sections, the most important of which are as follows:

- *Accuracy:* A CEP of 20 ft is assumed.
- *Effectiveness Index:* An MAE_F of 1000 ft^2 is used.
- *Release and Terminal conditions:* An impact angle of 65 deg has been calculated from release altitude and speed.
- *Target:* A single unitary target is selected.
- *Number of weapons:* One weapon is used in the attack.
- *Cluster munition:* The weapon contains several smaller submunitions rather than a single high explosive (HE) warhead.

For this case, clicking the Calculate button produces the result PD1 $= 0.3483$. This means for the weapon selected against a defined target, a single attack will produce a probability of damage to the target of 34.83%.

Readers may or may not be familiar with the data terminology used; however, that doesn't matter because the purpose of this book is to explain what these terms mean, how the data are obtained, and the role they play in calculating the result. Figure 1.34 is therefore a useful road map of the various components of the weaponeering process and will be used for the remainder of this book.

Although each of these items will be discussed in detail in the following chapters of the book, the reader should be aware that we also need a method to mathematically combine them to produce the result we need.

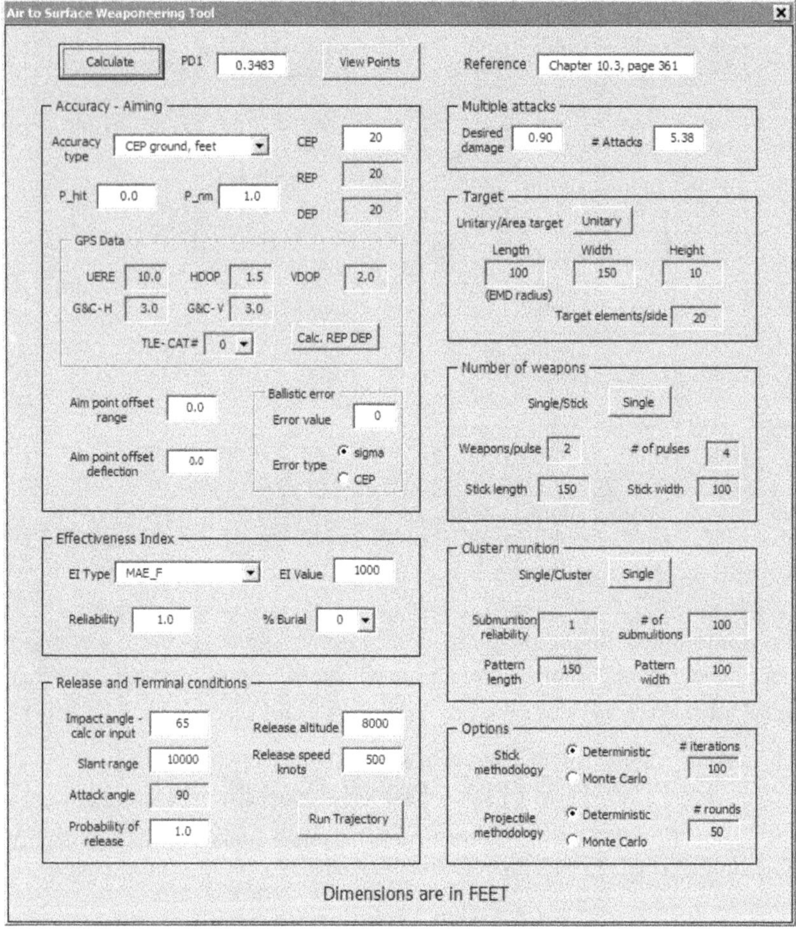

Fig. 1.34   Air-to-surface weaponeering module.

1.8   Chapter Summary

- We have seen that weaponeering is an activity that forms part of a larger force application cycle in which targeteers provide specific aimpoints on particular targets in order to meet strategic objectives. Weaponeers then select the most effective weapon and how many to use in order to achieve a required level of damage.

- Conventional weapons may damage targets with one or more mechanisms, including:

  - Blast
  - Fragments
  - Projectile penetration
  - Shaped charge

- Computational tools exist to estimate the amount of damage sustained by the target in such an attack but need the user to supply data in order to compute the result. The data needed fall into the following categories:
  - Weapon trajectory characteristics
  - Accuracy of weapon delivery on the target
  - Warhead damage mechanisms
  - Target vulnerability
  - The number and type of weapons used

In the next chapters we will discuss in more detail the types of data needed to estimate target damage.

REFERENCE

[1] Eidinger, J. M., and Ostrom, D., "Earthquake Loss Estimation Methods, Technical Manual, Electric Power Utilities," Rept. 23.04, National Institute for Building Sciences, June 1994.

Chapter 2

Introduction to Statistical Methods

2.1 Why Are Statistics Needed?

Consider the situation shown in Fig. 2.1 where a self-propelled 155-mm artillery cannon is repeatedly firing single, unguided rounds at a target.

Experience tells us that each round fired does not land in exactly the same place on the ground but follows a slightly different trajectory contained within the curved conical envelope shown in the figure. This results in an apparently random pattern of impact points, also shown in the figure. The random nature of each impact point relative to the aimpoint is a function of the variability of many parameters, such as

- The changing atmospheric properties (temperature, density, wind) the projectile flies through
- The precise shape and surface roughness of the projectile
- The amount of propellant in the charge
- The manner in which the propellant burns when ignited

Considering, for example, the last two of these items, the amount of propellant may vary from round to round by 1–2 g, and the manner in which it burns may be very similar but not identical; however, both or these effects will vary the pressure-time history at the back of the projectile as it is pushed down the gun tube.

Because of the variability of these parameters, successive rounds will land in different locations on the ground. Even if we could model how these features would affect a projectile trajectory, it is not practically feasible to measure any of them prior to firing a particular round.

We have to accept that it is not possible to model all of the complex mechanisms that influence the trajectory of typical projectiles, which in turn leads to unique impact points for what are assumed to be identical projectiles. Instead, we have to realize that the compound effect of these and other unknown

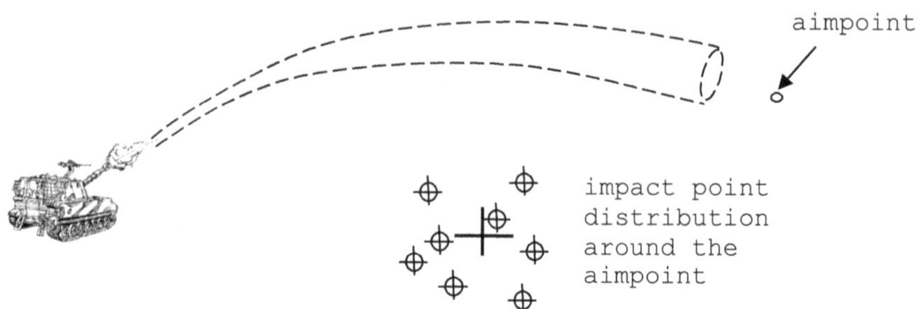

Fig. 2.1   Artillery firing at a target.

mechanisms is a distribution of impact points on the ground, and the characterization of this distribution allows us to estimate the effect of a particular round on the damage sustained by the target.

The branch of mathematics known as statistics deals with the characterization of random variables, such as the impact points described previously, and allows us to estimate the probability that a shell will land sufficiently close to the target in order to inflict the required amount of damage.

We will take a very basic approach to statistics by limiting the material presented to the bare minimum needed to describe those random processes required to understand various weapon characteristics. Readers interested in a more detailed discussion of this subject are referred to the *Introductory Weaponeering* textbook.

2.2   RANDOM SAMPLES AND STATISTICAL DISTRIBUTIONS

Consider the situation shown in Fig. 2.2, where an aircraft is dropping unguided bombs on a road. In this scenario, the aircraft flies in the direction

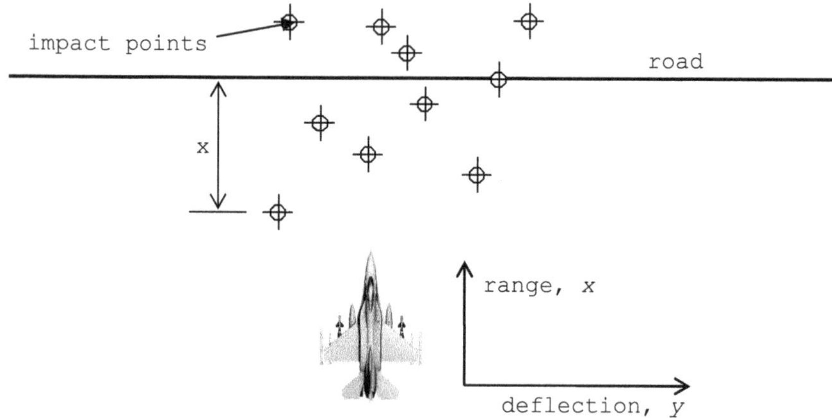

Fig. 2.2   Aircraft bombing a road.

shown and drops a single bomb. It then circles around and drops another bomb, and so on.

A commonly used convention is to call the direction the aircraft is heading the range (or along-track) direction and the direction perpendicular to this the deflection (cross-track) direction. The following assumptions are made regarding the scenario shown in Fig. 2.2:

- The attack direction is perpendicular to the road.
- The road has negligible width.
- The only variable of interest is the miss distance of individual weapons in the range direction, x.

Suppose the aircraft, or others like it, make n bombing runs, and the miss distance x (x_1, x_2, ..., x_n) is recorded for each impact. Two variables may be calculated that describe the collective nature of the miss distances.

- The *mean* or average miss distance of a set of data is obtained by adding them up and dividing by the number of samples defined mathematically by

$$\mu = \frac{1}{n}(x_1 + x_2 + \dots + x_n) \tag{2.1}$$

- The *variance* is obtained from the calculated mean and the data set by subtracting the mean from each sample, squaring it, adding up all the terms, and dividing by the number of samples, less one. Mathematically this is defined by

$$\text{VAR} = \frac{1}{(n-1)}[(x_1 - \mu)^2 + (x_2 - \mu)^2 + \dots + (x_n - \mu)^2] \tag{2.2}$$

A more familiar quantity is the square root of the variance, the *standard deviation* σ.

$$\sigma = \sqrt{\text{VAR}} \tag{2.3}$$

If the units of x are feet, then so are the units of the mean and standard deviation. Consider the following observations:

- The mean is what we can expect, on average, for the miss distance. If the aircraft fire control system is functioning properly, we might expect μ to be zero or close to it.
- The standard deviation represents the spread of the distribution, where a large value indicates larger miss distances whereas a smaller value indicates a more accurate system. We might expect the standard deviation of a guided weapon to be smaller than that of an unguided weapon.

A third statistic is often encountered, the *median*. This is defined as the midpoint of the data set such that half of the data are larger than the median and half of the data are smaller.

In order to avoid evaluating these formulae by hand, we will perform our statistical analysis using Microsoft Excel. So, assuming we have 100 samples of the miss distance and they are entered into a spreadsheet in cells A1 to A100, we can use:

- Mean = AVERAGE(A1:A100)
- Sigma = STDEV(A1:A100)
- Median = MEDIAN(A1:A100)

Random variables may be represented by a statistical distribution that, in turn, may be described by a mathematical function called a *probability density function*. At this point we make the assumption that the miss distances follow what is called a *normal or Gaussian distribution*. Having calculated the mean and standard deviation, the probability density function (PDF) of a normally distributed random variable is given by the following equation:

$$f(x) = \frac{1}{\sigma\sqrt{2\pi}} \exp\left[-(x - \mu)^2 / 2\sigma^2\right] \tag{2.4}$$

where the mean and standard deviation are described by Eqs. (2.1) to (2.3). If we plot the PDF as a function of x, we get the bell-shaped curve shown in Fig. 2.3.

This curve is symmetrical about the mean value, which is shown to be a negative value in Fig. 2.3, but for the road bombing problem may be assumed to be zero. A small sigma would produce a sharp, higher peak curve (good accuracy), whereas a large sigma would flatten and spread it out (poor accuracy).

The complexity of this equation is not important because we can use its properties without manipulating the formula itself. For our purposes, the most important property of the PDF is that the probability a random sample of x lies

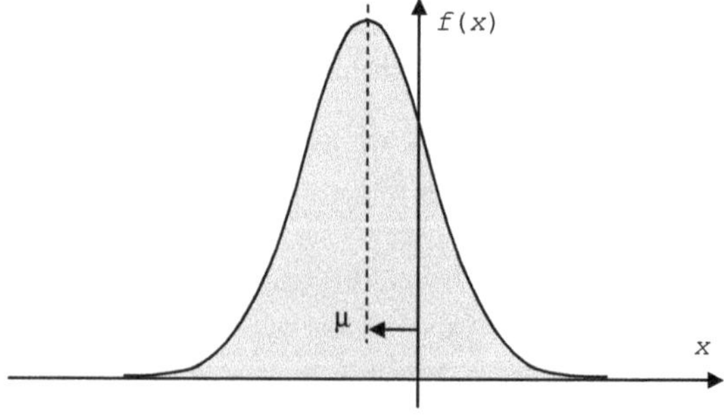

Fig. 2.3   Normal probability density function.

between two values a and b is given by the integral of the PDF between these values. Mathematically this can be written as follows:

$$\text{Probability}\{x \text{ lies between } a \text{ and } b\} = P(a < x < b) = \int_{x=a}^{x=b} f(x)\mathrm{d}x \quad (2.5)$$

Graphically this probability is the area under the PDF between a and b, as indicated in Fig. 2.4.

As well as the PDF, there is a *cumulative density function (CDF)* associated with a normally distributed random variable. The CDF is defined as the integral of the PDF from minus infinity up to some value of x denoted by X.

$$F(x) = \int_{x=-\infty}^{x=X} f(x)\mathrm{d}x \quad (2.6)$$

It may be seen that the area under the curve in Fig. 2.4 is the difference between two CDF values.

$$P(a < x < b) = \int_{x=a}^{x=b} f(x)\mathrm{d}x = F(b) - F(a) \quad (2.7)$$

Excel has the NORMDIST function for the CDF; therefore, Eq. (2.7) is simply evaluated with the following Excel formula:

$$P(a < x < b) = \text{NORMDIST}(b,\mu,\sigma,1) - \text{NORMDIST}(a,\mu,\sigma,1) \quad (2.8)$$

The conclusion, therefore, is that the probability a sample lies between two values is the difference between two CDFs evaluated using the NORMDIST function at those values.

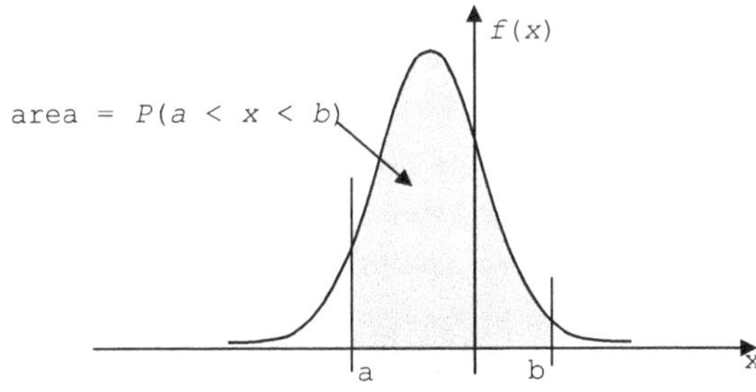

Fig. 2.4   Calculating probabilities from the PDF.

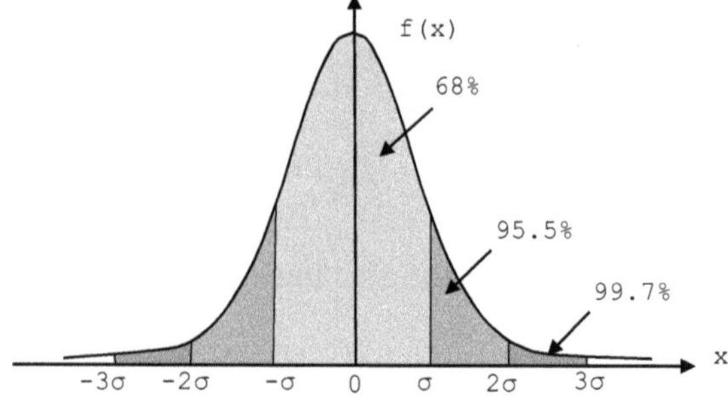

Fig. 2.5   Proportion of area under the normal distribution.

EXAMPLE 2.1

We are now in a position to use what we have learned so far. Suppose test data indicate that the mean value for the bombing problem is 0 and the standard deviation is 20 ft. We can calculate the probability a bomb lands within 50 ft of the road as follows:

$$P\,(-25<x<25) = \text{NORMDIST}\,(25,0,20,1) - \text{NORMDIST}\,(-25,0,20,1)$$
$$= 0.78 \tag{2.9}$$

One other feature of the normal distribution needs to be discussed for future reference. See Fig. 2.5, which shows a PDF with the areas beneath the curve for $\pm\sigma$, $\pm 2\sigma$, and $\pm 3\sigma$ intervals.Considering the $\pm\sigma$ boundary, for example, we can state the following:

- Sixty-eight percent of a data set distribution (e.g., miss distances) is contained within the boundary.
- There is a 68% chance that a single sample from the distribution will lie within the boundary.

What we have described so far is called a univariate, normal distribution, because it is function of one random variable, the miss distance x. We now extend this to two random variables, resulting in a bivariate distribution.

2.3   BIVARIATE NORMAL DISTRIBUTION

Suppose we revisit the bombing problem in Fig. 2.2, but now we are interested in hitting a particular point on the road, perhaps a short overbridge. In

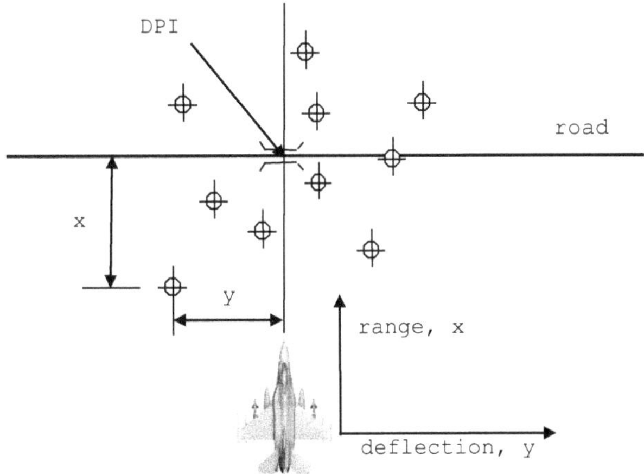

Fig. 2.6　Range and deflection miss distances.

this case, not only the miss distance in the range direction is important, but also the miss distance perpendicular to it, y, in the deflection direction, as shown in Fig. 2.6.

Now that the aircraft is aiming at a specific point, the term *desired point of impact (DPI)* is introduced to define this location.

Just as the univariate distribution in the range direction could be approximated by a normal distribution, so it could be assumed that the miss distance in the deflection direction is also normal, although there is no reason to expect that the statistics (μ and σ) are the same in both directions. In fact, it is usually assumed that the probability of a particular miss in the range direction is different and independent of the miss in the deflection direction.

Although we will not use it extensively, for completeness we will define the PDF for this bivariate, normal distribution by the following equation:

$$f(x,y) = \frac{1}{2\pi\sigma_x\sigma_y}\exp-\left[\frac{(x-\mu_x)^2}{2\sigma_x{}^2} + \frac{(y-\mu_y)^2}{2\sigma_y{}^2}\right] \qquad (2.10)$$

This may be represented by Fig. 2.7 which is analogous to the bell-shaped curve in Fig. 2.3.

In practice, the bivariate distribution is usually split into two independent univariate distributions for the purposes of calculating probabilities, as illustrated by the following example.

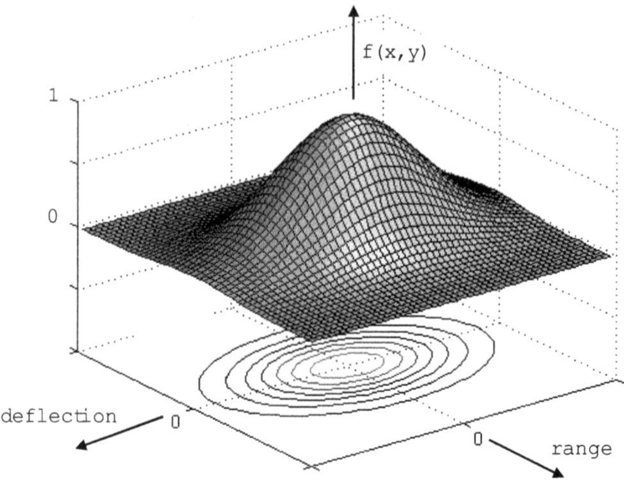

Fig. 2.7   Bivariate normal distribution.

EXAMPLE 2.2

Suppose x and y are normally distributed with $\mu_x = +3$ ft, $\sigma_x = 25$ ft, $\mu_y = -10$ ft, and $\sigma_y = 100$ ft. Find the probability a bomb will fall inside the rectangle shown in Fig. 2.8.

Considering the miss distances as independent of each other, we can calculate the probability the impact point is inside the rectangle by calculating separately the probability of an impact in each direction. In the range direction, we can write the following:

$$P(-15 < x < +15) = \text{NORMDIST}(15,3,25,1) - \text{NORMDIST}(-15,3,25,1)$$
$$= 0.4486$$

Similarly, in the deflection direction,

$$P(-25 < y < +25) = \text{NORMDIST}(25,10,100,1) - \text{NORMDIST}(-25,10,100,1)$$
$$= 0.1964$$

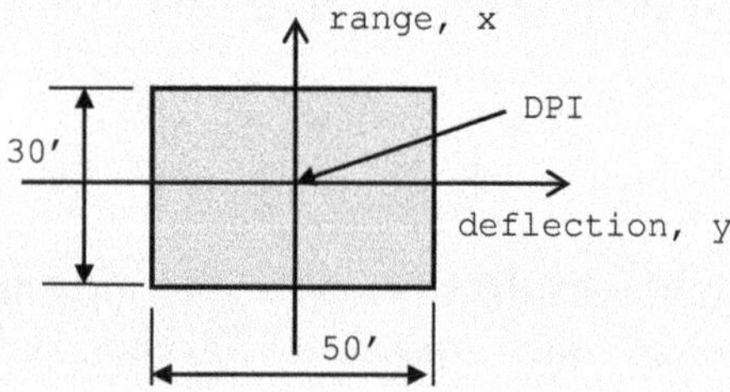

Fig. 2.8   Desired area for impact.

Clearly, to be inside the rectangle we have to have $(-15 < x < +15)$ AND $(-25 < y < +25)$, where the "AND" is interpreted as a multiplication.

$$P(-15 < x < 15 \text{ AND } -25 < y < 25) = P(-15 < x < 15) \times$$
$$P(-25 < y < 25) \tag{2.11}$$

This gives

$$P(-15 < x < 15 \text{ AND } -25 < y < 25) = 0.4486 \times 0.1964 = 0.0881$$

2.4 CIRCULAR NORMAL AND RAYLEIGH DISTRIBUTIONS

Although the previous example showed how a bivariate normal PDF can be used to predict the probability a round lands within a rectangle centered on the target, a more common requirement is to find the probability a round lands within a certain distance from the target. This implies a shift from Cartesian coordinates (x, y) to radial or polar coordinate r, as shown in Fig. 2.9.

So, if we have 100 impact points, we have 100 samples of x and 100 samples of y. If the radial miss distance is given by

$$r = \sqrt{x^2 + y^2} \tag{2.12}$$

then we can calculate 100 values of r. The question arises: If x and y are normally distributed, what is the distribution of r? To answer this, we first consider a *circular normal* distribution of x and y defined as one where the two mean values are zero and the standard deviations are equal.

$$\mu_x = \mu_y = 0 \tag{2.13}$$

$$\sigma_x = \sigma_y = \sigma \tag{2.14}$$

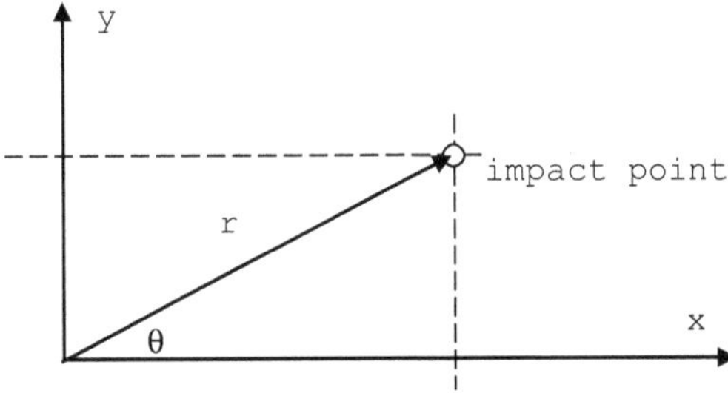

Fig. 2.9   Transformation of Cartesian to polar coordinates.

If these conditions are true, it may be shown that the distribution of radial miss distances is not normal, but follows what is known as a *Rayleigh distribution*, where the PDF is given by

$$f(r) = \frac{r}{\sigma^2} \exp\left[\frac{-r^2}{2\sigma^2}\right] \tag{2.15}$$

This distribution, unlike the normal distribution, is asymmetrical, and sample distributions are shown in Fig. 2.10.

This is sometimes referred to as a skewed (not symmetrical), one-sided ($+ve$ r only) distribution. We know that the CDF is the integral from minus infinity up to some value of the random variable [Eq. (2.6)]. This also applies to the radial distribution and may be obtained by direct integration of the PDF.

$$F(R) = \int_{x=0}^{x=R} f(r)\,dx = 1 - \exp\left[\frac{-R^2}{2\sigma^2}\right] \tag{2.16}$$

Because the integral can be evaluated in the form of Eq. (2.16), there is no need for a special function like NORMDIST that was used for the normal distribution. The probability that r lies between two values is, as before, obtained by the difference of two CDF functions.

$$P(r_1 < r < r_2) = \exp\left[\frac{-r_2^2}{2\sigma^2}\right] - \exp\left[\frac{-r_1^2}{2\sigma^2}\right] \tag{2.17}$$

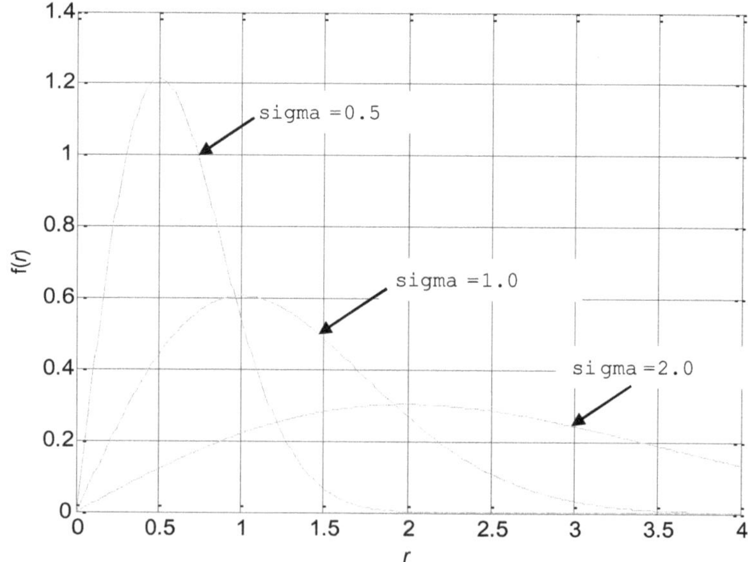

Fig. 2.10   Rayleigh PDF for some sample σ values.

It is important to realize that in Eqs. 2.15–2.17, σ is the common standard deviation of the x and y data set. Of course, for real test data, the two means will never be exactly zero, and σ_x will never be exactly equal to σ_y, so the question arises: How close is close enough? As long as the two means are small compared to the standard deviations, and the largest σ is no greater than twice the smallest, the Rayleigh distribution is a good approximation. If either of these conditions is not met, the Rayleigh distribution should not be used. In practice, the sigma used is the mean of σ_x and σ_y.

$$\sigma = \frac{\sigma_x + \sigma_y}{2} \tag{2.18}$$

EXAMPLE 2.3

Suppose x and y are normally distributed with $\mu_x = 1$ ft, $\sigma_x = 24$ ft, $\mu_y = -0.5$ ft, and $\sigma_y = 18$ ft. Find the probability a bomb will fall within 10 ft of the aimpoint.

We assume the two mean values are sufficiently small as to be considered zero and the 2:1 test for the ratio of standard deviation passes. However, because they are different, we will use an average of $\sigma = 21$ ft. From Eq. (2.17) we get

$$P(0 < r < 10) = \exp\left[\frac{-0^2}{2 \times 21^2}\right] - \exp\left[\frac{-10^2}{2 \times 21^2}\right] = 1 - \exp[-0.113] = 0.107$$

$$\tag{2.19}$$

2.5   UNIFORM DISTRIBUTION

Many random processes exhibit a uniform probability density function or may be approximated by one. In this distribution, there is an equal probability a sample may lie between two values a fixed distance apart within the bounds of the distribution.

For example, a cluster munition that opens up and releases mines attempts to distribute them uniformly along a road so that the probability of encountering one is equal anywhere between the beginning and end of the minefield. This may be represented by a *continuous univariate uniform distribution*, as shown graphically in Fig. 2.11.

Mathematically, the continuous univariate uniform PDF is written as follows:

$$f(x) = \frac{1}{(b - a)} \quad \text{for } a < x < b \tag{2.20}$$

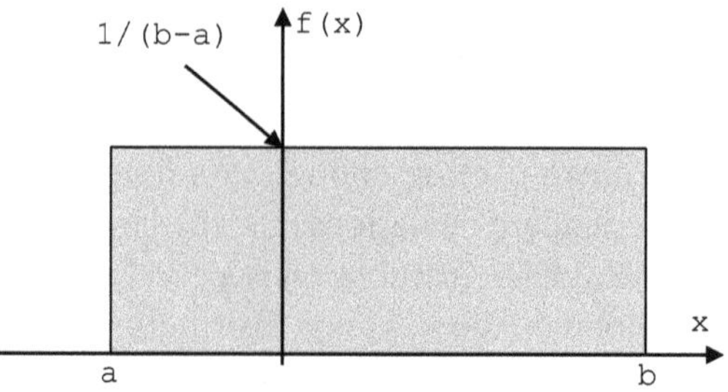

Fig. 2.11   PDF for uniform distribution.

The corresponding cumulative density function is obtained by direct integration of the PDF, resulting in the following:

$$F(X) = \int_{x=-\infty}^{x=X} f(x)\, dx = \frac{X-a}{b-a} \tag{2.21}$$

EXAMPLE 2.4

An aircraft makes a strafing run along a 200-ft-long road. What is the probability a bullet hits a truck 15 ft long?

We assume that the bullet impacts x are uniformly distributed along the road where $0 < x < 200$; therefore, $a = 0$ and $b = 200$. Assuming the truck occupies the space $100 < x < 115$, the probability a bullet hits the truck is again the difference between two CDF functions.

$$P_{\text{HIT}} = F(115) - F(100) = \frac{115-0}{200} - \frac{100-0}{200} = \frac{15}{200} = 0.075 \tag{2.22}$$

Note that this result of 7.5% is independent of where the truck is located along the road.

EXAMPLE 2.5

A ship passes along the center of a 1000-ft-wide channel (see Fig. 2.12). Suppose an aircraft has flown across the channel and dropped a mine such that the mine location may be anywhere between the banks with an equal (uniform) probability. If the ship is 100 ft wide, and the mine will detonate and damage the ship if it comes within 50 ft of it, what is the probability the ship will be damaged?

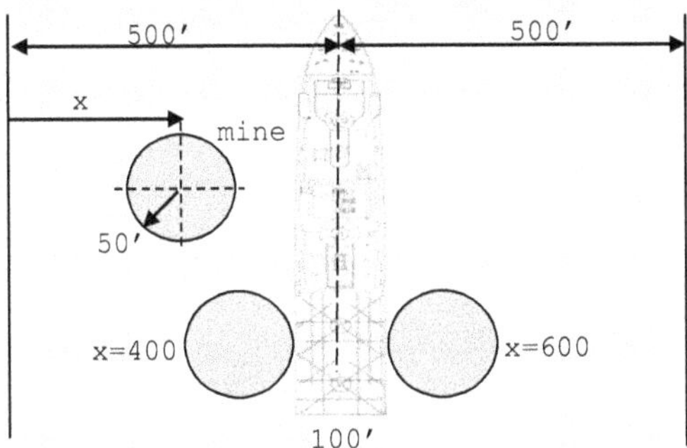

Fig. 2.12   Ship and mine geometry.

Suppose the location of the mine from the left bank is x. We observe that the ship will be damaged if the mine location is between 400 ft and 600 ft from the left bank. We will use Eq. (2.21) to answer the question: What is the probability a random sample lies between these values? Again, it is seen to be the difference between two cumulative density functions (CDFs) evaluated at these values.

$$P(400 < x < 600) = F(600) - F(400) = \left[\frac{600 - 0}{1000 - 0}\right] - \left[\frac{400 - 0}{1000 - 0}\right]$$

$$= \frac{200}{1000} = 0.2 \qquad (2.23)$$

It is left to the reader to consider the result if the pilot aims for the center line of the channel.

2.6   BINOMIAL DISTRIBUTION SURVIVOR RULE

There is another distribution of significance for weaponeers called the *binomial distribution*. Although we will not describe this in detail, there is an important conclusion that we will use later.

If the result of a single attack produces a damage probability of PD_1, the damage from n identical but independent attacks PD is given by

$$PD = 1 - (1 - PD_1)^n \qquad (2.24)$$

Equation (2.24) is sometimes referred to as either the *powering up* or *survivor* rule. As an example, suppose the probability of damaging a target with a single attack is 0.3. We may ask: What is the damage sustained by the target

TABLE 2.1   EXAMPLE OF POWERING UP

n	PD
1	0.300
2	0.510
3	0.657
4	0.760
5	0.832
7	0.918
10	0.972

due to one, two, three, or more attacks? Using Eq. (2.24), Table 2.1 may be generated for n attacks on the target.

An important condition of using this equation is that each attack has to be independent of all the other attacks; otherwise, the result will not be correct. We will return to the topic of independent events later.

Another use of Eq. (2.24) is to determine n in order to achieve a required amount of damage. For example, suppose the probability of damage due to a single bomb is $PD_1 = 0.2$, and we require a damage level of at least 60%. How many weapons are needed? Transforming Eq. (2.24),

$$1 - PD = (1 - PD_1)^n \qquad (2.25)$$

$$\log_e (1 - PD) = n \log_e (1 - PD_1) \qquad (2.26)$$

Resulting in

$$n = \frac{\log_e (1 - PD)}{\log_e (1 - PD_1)} \qquad (2.27)$$

Substituting the data, we get

$$n = \frac{\log_e (1 - 0.6)}{\log_e (1 - 0.2)} = 4.1 \qquad (2.28)$$

Therefore, five weapons would give at least 60% damage.

2.7   TESTING DATA FOR A PARTICULAR DISTRIBUTION

It is usual in analyzing experimental data to assume that the variable is random with a particular distribution (e.g., normal, Rayleigh, uniform, etc.). How can we measure the confidence of this assumption? In the field of statistics, this subject is known as *hypothesis testing*. The first test to apply is to simply

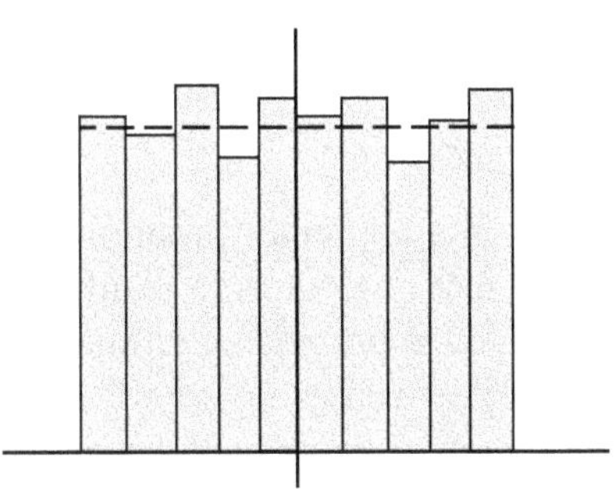 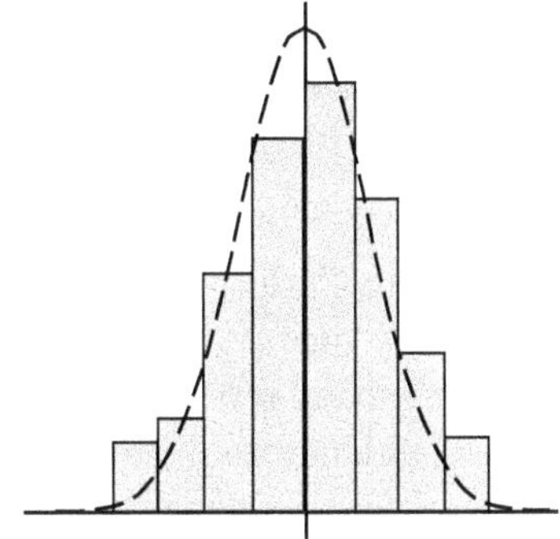

Fig. 2.13   Sample data sets grouped into histograms.

observe the distribution of data in a histogram to give an idea of what analytical distribution may be present. For example, the histograms shown in Fig. 2.13 suggest perhaps uniform and normal distributions, respectively.

It is useful to obtain quantitative measures of how close the data are to the assumed distribution when results are presented. One example of such a process is the *chi-squared test*, which we will described briefly here. By way of an example, 50 data points were generated from a normal random number generator and the test applied. The results were compiled into Table 2.2.

Detailed explanation of the chi-squared process may be found in the *Introductory Weaponeering* textbook; however, the final result in the bottom-

TABLE 2.2   CHI-SQUARED TEST ON EXAMPLE DATA SET

Range Error			
Bin Range	**Observed**	**Predicted**	**Chi Square Error**
Less than -9.71	5	2.91	1.496
-9.71 to -5.73	5	6.12	0.204
-5.73 to -3.16	5	6.59	0.383
-3.16 to -1.30	5	5.76	0.101
-1.30 to -0.04	5	4.13	0.185
-0.04 to 0.92	5	3.14	1.105
0.92 to 2.74	5	5.64	0.074
2.74 to 4.40	5	4.50	0.057
4.40 to 7.20	5	5.65	0.075
More than 7.20	5	5.56	0.057
Sum	**50**	**50**	**3.74**
		Chi-square test	**0.9279**

right indicates a 92.79% probability the distribution agrees with the hypothesis (i.e., the data appear to be normally distributed).

2.8 MONTE CARLO SIMULATIONS

A Monte Carlo process is usually embodied in a computer program that repetitively models the physical process of interest to produce a result. By executing many repetitions, the results are averaged to provide an estimate of what would happen if the process were physically executed.

Reconsider the mine problem discussed earlier where it was recognized that the ship would be damaged if the mine position were between 400 ft and 600 ft from the left bank.

$$(PD_1 = 1) \text{ IF } (400 < x < 600) \tag{2.29}$$

An alternate way to solve this problem would be through simulation of the actual events that take place when the aircraft drops the mine and the ship traverses the channel. This would involve the following repetitive process:

1. Get a sample of x from a uniformly distributed random number generator, in the range $0 < x < 1000$.
2. Check whether x lies between 400 and 600.
3. If it does, increase a counter by 1; otherwise, don't.
4. Repeat from step 1 a large number of times.

Suppose we perform 1000 repetitions, and the value of the counter at the end is 476. This means that 476 samples of x resulted in damage and 524 did not; therefore, the probability a single sample resulting in damage is given by the following equation:

$$PD_1 = \frac{476}{1000} = 0.476 \tag{2.30}$$

The following code segment is a Visual Basic program to do this simulation in which a random number is drawn from a uniform distribution (Rnd) for the mine location, which is then checked against Eq. (2.29) to determine whether that location damages the ship. Repeating this process for a large number of trials enables us to determine the proportion of mine locations that damage the ship (i.e., PD_1).

```
' -------------------------------------------------------------------
Sub Mine_Example()
    n=100000                    ' number of iterations
    Pk=0                        ' initialize counter
    i=1                         ' first iteration
```

```
While i<n                  'start conditional loop
    x=1000 * Rnd            ' get mine position
    If (x>400) And (x<600) Then ' inside
      boundaries?
        Pk=Pk+1             ' ship damaged
    End If
    i=i+1                   ' increment trial number
Wend                        ' done with iterations

    PD1=Pk / n             ' calculate average
    MsgBox PD1             ' display result
End Sub
'------------------------------------------------------------
```

In this programming language, the function Rnd gets a uniformly distributed random number in the range 0 to 1. Running the program for the 100,000 trials indicated produces the result $PD = 0.2009$. This shows that the expected value theorem produces approximately the same result as the simulation.

A question always arises with this type of problem: How many Monte Carlo samples should be taken to get the correct result? This cannot be answered ahead of time, but Table 2.3 shows the results from the same program, which is run three consecutive times for a fixed number of iterations.

We can see that for 100 iterations, the simulation produces results that vary quite a bit but are in the region of the known answer. As the number of iterations increases, there is less spread in successive runs of the simulation until for 100,000 iterations the result is "stable" to two decimal places. Two observations may be drawn:

1. The number of iterations needed increases with the need for higher precision in the answer.
2. The Monte Carlo method may take considerable time to execute if high precision is required.

2.9 COMBINING RANDOM VARIABLES

Sometimes random variables have to be combined with other random variables, and it is of interest to find the statistics that describe the combined

TABLE 2.3 RESULTS OF MONTE CARLO SIMULATION

Number of Iterations	Run 1	Run 2	Run 3
100	0.230	0.210	0.170
1000	0.197	0.171	0.213
100,000	0.19863	0.19987	0.20078

variable. One of the more common combinations is where two random variables are added together. Suppose x and y are random variables with means μ_x and μ_y, and standard deviations σ_x and σ_y. We now form a third variable z such that

$$z = x + y \tag{2.31}$$

The variable z will also be a random variable such that if x and y are uncorrelated (i.e., independent), then the mean and standard deviation of z are given by

$$\mu_z = \mu_x + \mu_y \tag{2.32}$$

$$\sigma_z = \sqrt{\sigma_x^2 + \sigma_y^2} \tag{2.33}$$

This latter equation is sometimes referred to as a root sum squared (RSS) process for independent random variables. Apart from this particular case, it is often difficult to predict the statistics of derived variables.

2.10 APPLICATION OF STATISTICS TO WEAPONEERING

To summarize, weaponeering uses statistical principles to describe physically observed phenomena that cannot be modelled, hence predicted. Examples include the following:

- The impact point of a particular weapon, leading to statistical descriptions of miss distance. This results in delivery accuracy data being obtained from tests and the results expressed statistically.
- Distribution of submunitions from cluster bombs or artillery shells within a specified ground pattern.
- Errors associated with weapon fire control systems and instruments contained therein [e.g., artillery shell muzzle velocity, laser rangefinder measurements, Forward Looking InfraRed (FLIR) pointing angle measurements, aircraft state variables (position, velocity orientation, etc.)].
- For artillery fire, the location of a target being attacked and the location of the gun(s) firing at them.
- The effect of terrain on the outcome of an engagement.

If we use statistics to calculate an effect, that result represents what we might reasonably *expect* to happen, not what *will* happen on a particular occasion.

For example, suppose we fire a missile at a target, and we predict the probability of hitting the target is 0.9. In an actual attack, the outcome is either hit or miss—we cannot predict which, just that a hit is more likely.

For the same example, if the weapon is so expensive that we test fire only one at the target and it misses, the probability of the hit number might be questioned. The response is: If you fire 10 weapons at the target, we would expect, on average, 9 would hit and 1 would miss. Unfortunately, the only test conducted resulted in the one miss that was observed.

2.11 CHAPTER SUMMARY

- The chapter began by explaining why it is necessary to study statistics and then went on to develop probability density functions and cumulative density functions for several random distributions commonly found in weaponeering. It was shown that the probability of a random sample falling between two values for a particular distribution is the difference between two CDFs evaluated at these values.
- An example of hypothesis testing was shown, followed by a description and use of Monte Carlo simulations.

Chapter 3

WEAPON TRAJECTORY

3.1 WHY TRAJECTORY IS IMPORTANT

The trajectory of a weapon can have considerable influence on its effectiveness, and the important data needed to determine this are known as the *terminal conditions* (i.e., the speed and orientation of the weapon when it arrives at the target). Some weapons require specific terminal conditions in order to penetrate into a target (see Fig. 3.1).

In order to penetrate the concrete aircraft shelter shown in the figure, the weapon needs to impact as close to perpendicular as possible; otherwise, it may ricochet. It also needs a minimum impact velocity in order to penetrate the concrete roof; however, if the velocity is too high, the weapon casing may break up. Notice that the impact angle is defined relative to the local surface, so if the weapon comes in vertically on the roof of the shelter, the impact angle is 90 deg; however, if it strikes the sloping walls, the impact angle is 45 deg.

In addition, the terminal impact conditions affect the area on the ground subjected to the fragment spray pattern. This is because most of the fragments that are projected come off the side of the weapon, compared to the number from the front and rear sections. Therefore, the area on the ground subjected to the fragments will be short but wide if the impact angle is shallow, whereas the area will be more circular or uniform for a near-vertical detonation, as shown in Fig. 3.2.

Other quantities are available from a weapon trajectory program that assist the weaponeer to make the correct selection for a particular attack, including the following:

- The down-range distance traveled determines whether the target can be reached.
- When multiple munitions are released from an aircraft or fired from a cannon, the trajectory and intervals between individual warheads determine

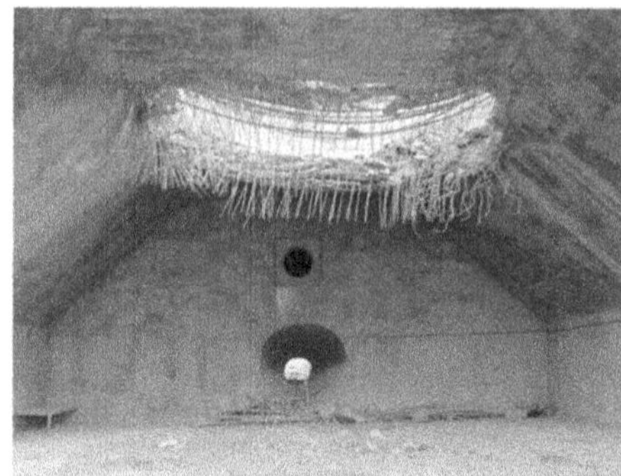

Fig. 3.1   Weapon penetration.

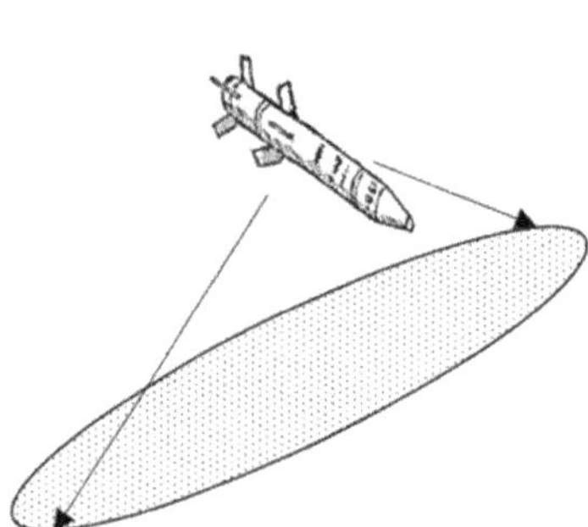
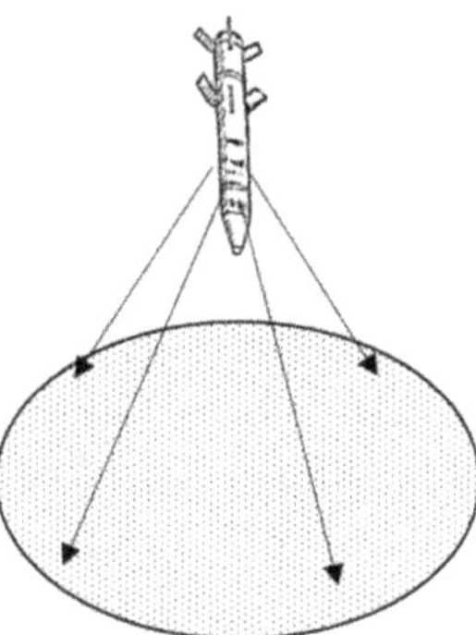

Fig. 3.2   Fragment spray pattern for shallow and steep impacts.

the area on the ground covered by all of the munitions. This area may or may not have overlapping regions producing different effects on areas of target elements.
• The time of flight of the weapon may determine whether the enemy can detect it and employ countermeasures to mitigate its effect.

These variables will be available from a weapon trajectory; however, computing trajectory data may be a difficult task, particularly for guided weapons, depending upon the level of detail contained within the model. As usual with physical modeling problems, there is a tradeoff among the level of complexity of the model, the ease with which results may be computed, and the correspondence between those results and experimental or test data.

Reasonable answers may be obtained from fairly simple models, at least in order to illustrate some basic principles, and therefore simple models will be developed and used for this purpose. For more accurate work, highly detailed and validated models exist, and these are also described. Much of this chapter details the trajectory of unguided weapons, a subject usually termed *ballistics*, although a discussion of guided weapon trajectory is also presented.

3.2   A BASIC EXAMPLE

Recall a simple example from high school physics: A ball is thrown with a speed of 10 m/s at an angle of 45 deg from the top of a 30 m-high building (see Fig. 3.3). Find:

- The time it takes to hit the ground
- The distance from the building at impact
- The velocity of the ball when it hits the ground
- The impact angle relative to the ground
- The maximum height reached by the ball

We assume the motion in the horizontal and vertical directions are independent of each other, and the ball has zero drag. We define the x, y coordinate axes as shown in Fig. 3.3, and the following equations for the acceleration, velocity, and position apply to these directions[1]:

x-direction:

$$a_x = 0 \tag{3.1}$$

$$v_x = v_{x0} \tag{3.2}$$

$$x = v_x t \tag{3.3}$$

y-direction:

$$a_y = -g \tag{3.4}$$

$$v_y = v_{y0} + gt \tag{3.5}$$

$$y = y_0 + v_{y0}t + \frac{1}{2}gt^2 \tag{3.6}$$

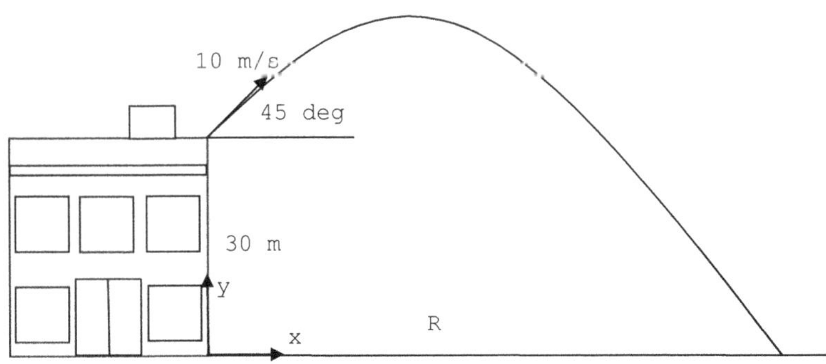

Fig. 3.3   Simple trajectory.

---

[1] See the *Introductory Weaponeering* textbook for these derivations.

where

v_{x0}, v_{y0} = initial velocities in x and y
g = acceleration due to gravity (-9.81 m/s^2)
y_0 = initial value of y (30 m)

The initial velocities in x and y are given by

$$v_{y0} = v_0 \sin \theta \tag{3.7}$$

$$v_{x0} = v_0 \cos \theta \tag{3.8}$$

where

v_0 = initial speed of the ball (10 m/s)
θ = launch angle (45 deg)

The calculations begin by recognizing that the ball hits the ground when $y = 0$, so substituting into Eq. (3.6), we get

$$\frac{1}{2} g t^2 + v_{y0} t + y_0 = 0 \tag{3.9}$$

This is a quadratic equation in time t and is of the form

$$ax^2 + bx + c = 0 \tag{3.10}$$

which we know has the solutions

$$x = \frac{-b \pm \sqrt{b^2 - 4ac}}{2a} \tag{3.11}$$

or, in terms of our problem

$$\text{TOF} = \frac{-v_{y0} \pm \sqrt{v_{y0}^2 - \dfrac{4 g y_0}{2}}}{g} = \frac{-v_{y0} \pm \sqrt{v_{y0}^2 - 2 g y_0}}{g} \tag{3.12}$$

Here we define the time to impact as the time of flight (TOF). Once this has been determined, most of the required results can be calculated. From Eq. (3.3),

$$R = v_{x0} \times \text{TOF} \tag{3.13}$$

From Eq. (3.5), the vertical velocity at impact v_{yI} is

$$v_{yI} = v_{y0} + g \times \text{TOF} \tag{3.14}$$

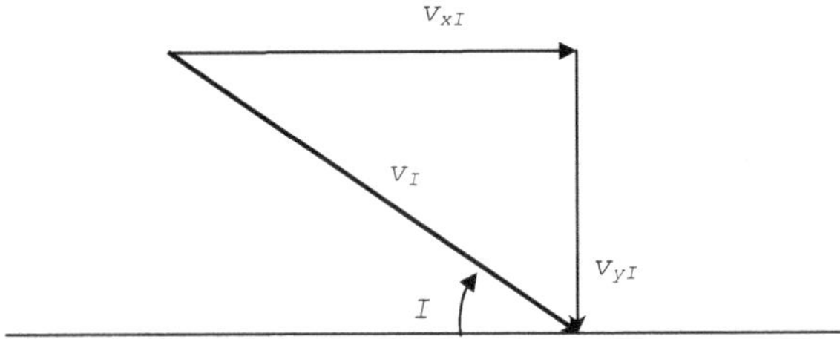

Fig. 3.4 Impact velocities.

The horizontal velocity at impact is the same as the constant horizontal velocity from Eq. (3.2)

$$v_{xI} = v_{x0} \tag{3.15}$$

At impact, the horizontal and vertical velocities may be represented by vectors, as shown in Fig. 3.4.

The total velocity of the ball at impact is given by

$$v_I = \sqrt{v_{xI}^{\,2} + v_{yI}^{\,2}} \tag{3.16}$$

The impact angle I measured relative to the horizontal is obtained from

$$I = \tan^{-1}\left[\frac{v_{yI}}{v_{xI}}\right] \tag{3.17}$$

To determine the maximum height reached (apogee), we recognize that the vertical velocity is zero; therefore, from Eq. (3.5), the time to apogee is

$$t_A = \frac{-v_{y0}}{g} \tag{3.18}$$

and the maximum height reached is obtained from Eq. (3.6)

$$y_A = y_0 + v_{y0}t_A + \frac{1}{2}gt_A^{\,2} \tag{3.19}$$

The actual trajectory is parabolic in shape, or in this example part of a parabola, as shown in Fig. 3.3.

Instead of calculating the results by hand, we develop an Excel spreadsheet to do the computations for us, as shown in Table 3.1.

A few points arise from the results.

- Because gravity acts down and y is positive up, $g = -9.81$.
- The vertical velocity at impact is negative (i.e., downward).

TABLE 3.1   SIMPLE TRAJECTORY CALCULATOR

Zero drag trajectory calculator		
Inputs		
Initial velocity	10	
Angle of launch	45	
Initial height	30	
Acceleration due to gravity	-9.81	
Intermediate calculations		
Initial horizontal velocity	7.07	
Initial vertical velocity	7.07	
Results		Eq #
Time of flight	3.30	3.12
Range	23.31	3.13
Vertical velocity at impact	-25.27	3.14
Horizontal velocity at impact	7.07	3.15
Total impact velocity	26.24	3.16
Impact angle	74.92	3.17
Time to apogee	0.72	3.18
Maximum height	32.55	3.19

- The solution of Eq. (3.12) actually produces two values due to the +/-; however, the second value is −1.86, so we always take the positive value of time.
- Any system of consistent units may be used; however, the value of the acceleration due to gravity g should be changed accordingly.

Now that the basic spreadsheet has been developed, it is easy to apply it to common weapon systems.

3.3 APPLICATION TO AIR-LAUNCHED UNGUIDED BOMBS

Consider the case shown in Fig. 3.5, in which an aircraft flying at 450 kt (759.6 ft/s) at an altitude of 12,000 ft in a 25-deg dive releases an unguided bomb. We will use the spreadsheet to examine the bomb trajectory.

The corresponding spreadsheet is shown in Table 3.2. Note that all that is needed is to insert the correct initial or release conditions, height, speed, and angle.

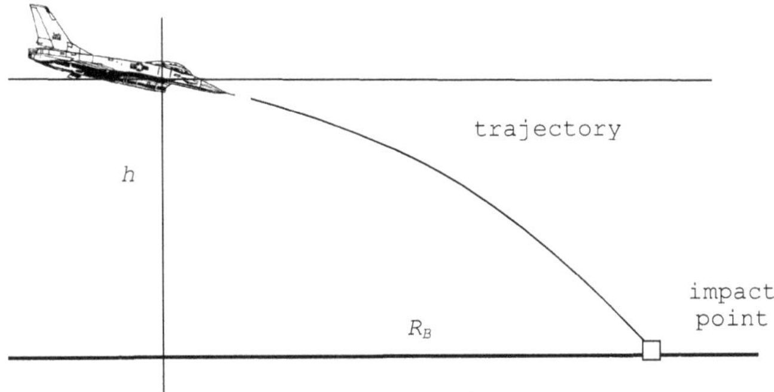

Fig. 3.5   Air-launched unguided bomb.

TABLE 3.2   TRAJECTORY OF AIR-LAUNCHED UNGUIDED BOMB

Zero drag trajectory calculator		
Inputs		
Initial velocity	759.6	
Angle of launch	-25	
Initial height	12000	
Acceleration due to gravity	-32.2	
Intermediate calculations		
Initial horizontal velocity	688.43	
Initial vertical velocity	-321.02	
Results		Eq #
Time of flight	19.09	3.12
Range	13145.41	3.13
Vertical velocity at impact	-935.87	3.14
Horizontal velocity at impact	688.43	3.15
Total impact velocity	1161.81	3.16
Impact angle	59.35	3.17
Time to apogee	-9.97	3.18
Maximum height	13600.22	3.19

Again, note the following entries in the spreadsheet:

- The speed of the aircraft is converted from kt to ft/s.
- Acceleration due to gravity is -32.2 ft/s^2.
- The dive angle is entered as a negative number, because y is measured positive up.
- There is no apogee for this trajectory, so the last two results should be ignored.[2]

---

[2] The astute reader may see what these numbers actually represent.

3.4 APPLICATION TO SURFACE-LAUNCHED UNGUIDED PROJECTILES

Now consider the case of an artillery shell fired from a cannon, as shown in Fig. 3.6. In this case, the muzzle velocity is 800 m/s fired at an angle of 25 deg.

The corresponding spreadsheet is illustrated in Table 3.3. Note for the zero-drag case, the impact conditions (angle, velocity) are the same as the launch conditions.

In this example, we specify the elevation angle (quadrant elevation, QE) and the range is calculated; however, a more common requirement is that the range is specified, and we have to calculate the QE to achieve it. It should also be noted that for this case, two elevation angles may be calculated to give that range, one below 45 deg and one above. This is shown in Fig. 3.7.

These angles are referred to as the high and low angle firing solutions. If $y_0 = 0$, then Eq. (3.12) reduces to

$$\text{TOF} = \frac{2v_{y0}}{g} = \frac{2v_0 \sin \theta}{g} \tag{3.20}$$

From Eq. (3.13),

$$d = v_{x0} \times \text{TOF} = v_0 \cos \theta \frac{2v_0 \sin \theta}{g} = \frac{v_0{}^2}{g} 2 \sin \theta \cos \theta = \frac{v_0{}^2}{2} \sin 2\theta \tag{3.21}$$

This gives the result

$$\theta = \frac{1}{2} \sin^{-1}\left[\frac{gd}{v_0{}^2}\right] \tag{3.22}$$

This leads to two solutions in the ranges 0–90 deg and 90–180 deg, corresponding to angles θ_1 and θ_2, respectively, in Fig. 3.7. From a tactical

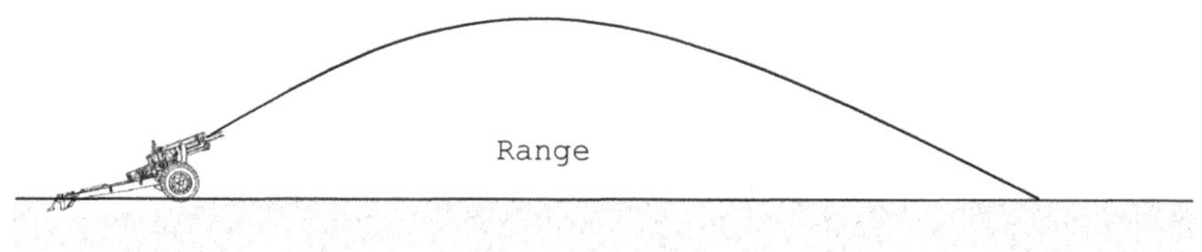

Fig. 3.6 Surface-launched unguided projectile.

TABLE 3.3   TRAJECTORY OF SURFACE-LAUNCHED UNGUIDED SHELL

Zero drag trajectory calculator		
Inputs		
Initial velocity	800	
Angle of launch	25	
Initial height	0	
Acceleration due to gravity	-9.81	
Intermediate calculations		
Initial horizontal velocity	725.05	
Initial vertical velocity	338.09	
Results		Eq #
Time of flight	68.93	3.12
Range	49976.40	3.13
Vertical velocity at impact	-338.09	3.14
Horizontal velocity at impact	725.05	3.15
Total impact velocity	800.00	3.16
Impact angle	25.00	3.17
Time to apogee	34.46	3.18
Maximum height	5826.09	3.19

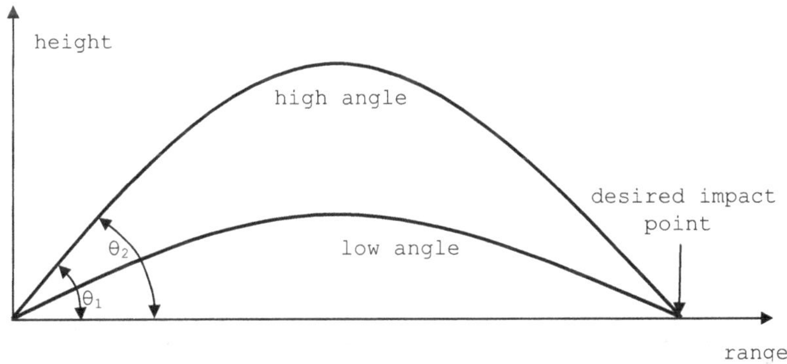

Fig. 3.7   Two elevation angles for the same range.

perspective, low angle is preferred unless the projectile has to clear an obstruction.

3.5   THE EFFECT OF AIR RESISTANCE: DRAG

If air resistance is now considered, the equations for the trajectory become more difficult to solve. Consider the case of an unguided bomb, shown in Fig. 3.8, showing the weight and drag force acting on the bomb.

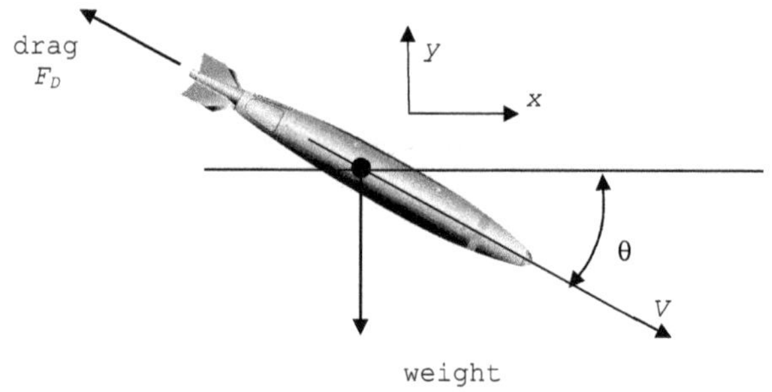

Fig. 3.8   Free body diagram of projectile.

Clearly the weight acts down, and the drag force F_D acts opposite to the velocity vector, assumed aligned with the weapon axis.[3] By putting F_D equal to 0, the resulting equations of motion may be solved, resulting in the equations shown in Section 3.2; however, now the drag force is computed as

$$F_d = \frac{1}{2}sC_d\rho V^2 \tag{3.23}$$

where

s = maximum cross-sectional area of the munition = $\pi D^2/4$
C_d = coefficient of drag
ρ = density of the air
V = munition velocity

The drag coefficient C_d is not constant, but a function of Mach number while the air density varies with altitude. This leads to a nonlinear differential equation with varying coefficients that may only be solved using numerical methods.[4]

This high fidelity model may be implemented in a spreadsheet, which can accommodate a free fall bomb, an artillery projectile or a surface launched rocket. This is illustrated in Fig. 3.9 for the unguided bomb case shown in Fig. 3.8, while the same program for the surface launched artillery projectile is shown in Fig. 3.10. Note the trajectories are no longer parabolic.

Considering the unguided bomb example again, it is important to realize that the trajectory has to be computer very fast in order to provide the pilot with visual indications of where the bomb will impact. This may be illustrated in Fig. 3.11 showing the pilot's view through the head-up display (HUD) while delivering an unguided bomb.

---

[3] This implies the angle of attack is zero.
[4] See the *Introductory Weaponeering* textbook.

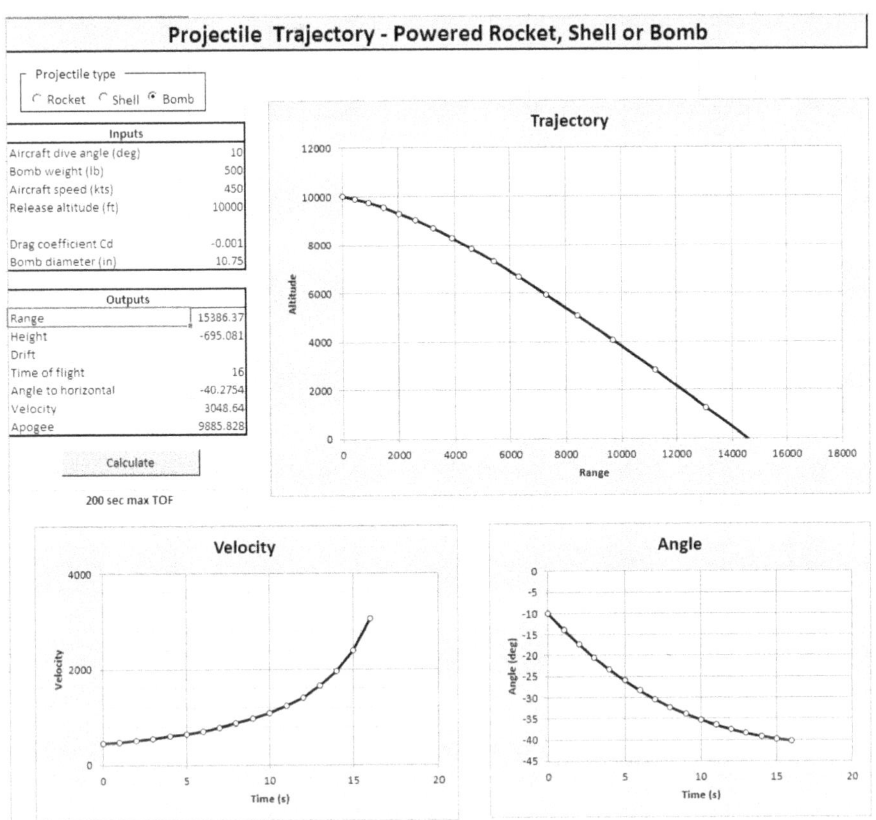

Fig. 3.9   Free fall bomb trajectory.

The fire control system gets input from the aircraft's avionic system regarding altitude, speed, roll and pitch and assuming the bomb is released at that instant, calculates where on the ground it would impact. This point is displayed on the HUD together with the plane containing the trajectory. A short time later, the process is repeated and the impact symbol updated. The pilot's job is to locate the target on the ground and maneuver the aircraft so that the impact symbol and target coincide, at which point the pilot releases the bomb.

This bombing mode is called continuously computer impact point (CCIP) and in order to display smoothly changing symbology on the HUD, the trajectory calculations have to be performed many times per second.

Regarding the artillery shell, the trajectory will be different from the unguided bomb case in that the shell is spinning in order to stabilize its motion, as explained in the next section.

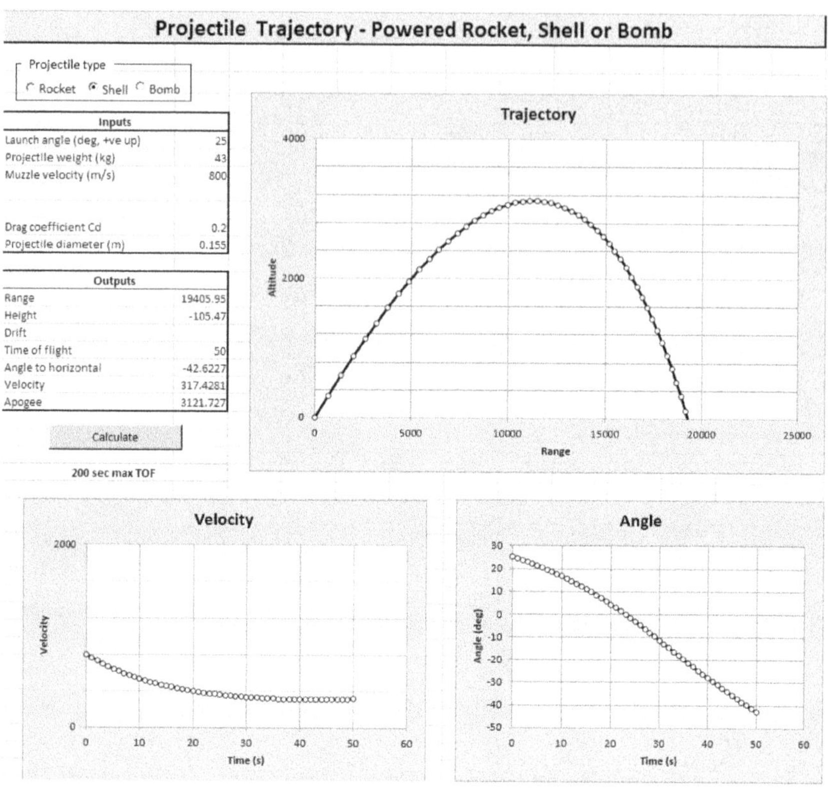

Fig. 3.10   Surface launched rocket trajectory.

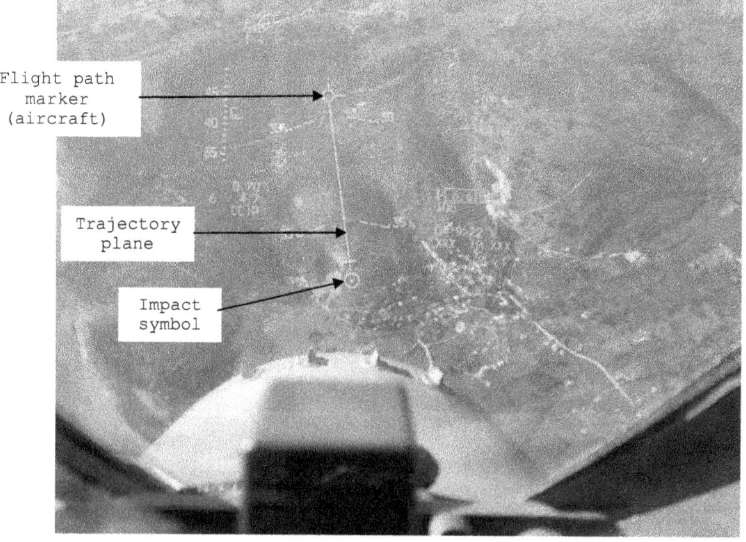

Fig. 3.11   Pilot's view through the HUD.

3.6 GENERAL STABILITY OF PROJECTILES

As a projectile moves through the air mass, all of the aerodynamic forces may be resolved into a single force and moment acting at any prescribed point along the axis of the projectile. If this point is selected such that there is no resultant moment, then this point is known as the *center of pressure*. The stability of a projectile is determined by the position of the center of pressure (CP) relative to the center of gravity (CG), as shown in Fig. 3.12.

If the CP is behind the CG, then the projectile is stable; otherwise, it is unstable and will tumble end-over-end along the trajectory.

This has been recognized ever since the development of crude weapons: metal-tipped spears are stable because the mass is concentrated at the front, and arrows are stabilized by placing feathers (fins) at the rear. In the first case, the CG is moved ahead of the CP, whereas in the second, the CP is moved behind the CG.

Projectiles having the tapered shape shown in Fig. 3.12 will always be unstable. Because it is desirable to keep the nose of the weapon pointed along the trajectory and have the axis tangential to the trajectory at all times, a stable motion is required. A modern projectile is usually stabilized in one of the following two ways:

- *Fin stabilized:* In this method, similar to the arrow principle just described, fixed or moveable fins are located at the rear of the projectile to provide an additional aerodynamic force aft of the center of gravity so that perturbations to the flight cause steering corrections to be made automatically. Such systems are usually used for bombs, as illustrated in Fig. 3.13a.
- *Spin stabilized:* This method is usually used for artillery projectiles and smaller caliber bullets. This is because the projectile has to sustain the violent explosion of the propellant, and therefore the rear of the shell is thick and heavy. This design feature moves the CG towards the rear and aft of the CP. Because the shell has to fit inside the bore, any fin design must be deployable once the shell exits the tube, and moveable fins are difficult to design for the harsh environment of the breach of a cannon when detonation takes place. For these reasons, projectiles are usually spin stabilized (see Fig. 3.13b).

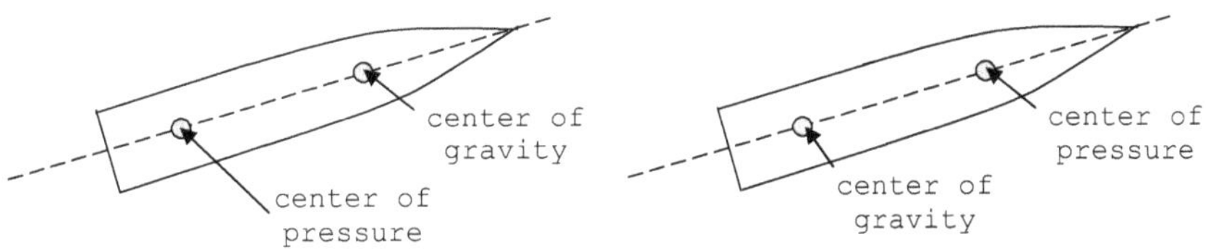

Fig. 3.12   Stable (left) and unstable (right) projectiles.

Fig. 3.13   a) Fin- and b) spin-stabilized munitions

A projectile that is rotating about its axis of symmetry resists disturbances due to conservation of angular momentum about this axis, thereby stabilizing the projectile. The cyclist shown in Fig. 3.14 remains upright as long as the rotating wheels generate sufficient angular momentum.

Spin is imparted on a projectile as it moves along the firing tube by rifling grooves machined inside the barrel, and a matching rotator ring on the shell. Figure 3.15 shows the internal rifling on the gun tube and the copper rotating band near the base of the artillery shell, with which it engages. Under the back pressure of the detonation, the oversize shell is forced along the tube, making the soft rotator band conform to the grooves. The grooves are machined as a

Fig. 3.14   Example of spin stabilization.

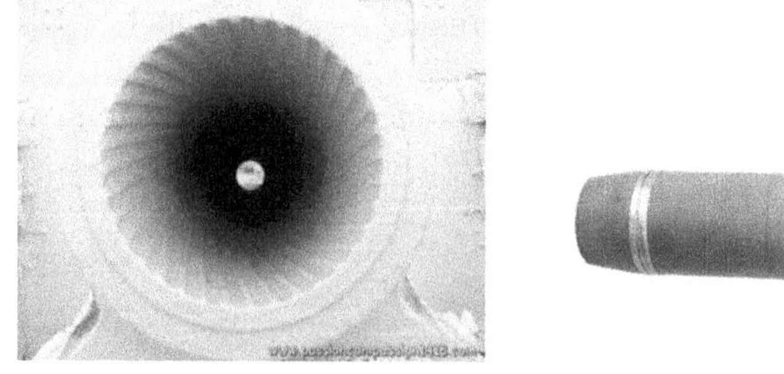

Fig. 3.15   Rifling and projectile rotator band.

spiral, which causes the shell to rotate as it moves down the barrel. Typical spin velocities are on the order of 250 rpm.

Most artillery pieces are rifled, but some mortars are smooth bore, and therefore their munitions have fins.

In the absence of a cross wind, the trajectory of a fin-stabilized munition is contained in a vertical plane that includes the release platform (cannon or aircraft) and the target, as shown in Fig. 3.16. A spinning projectile, however, generates several forces other than gravity and drag that will move the round out of this vertical plane, causing the impact point to be to the right of the target. This lateral displacement is called *drift*.

For the latter, it is important to have an accurate trajectory model so that this drift can be calculated, and therefore the gun may be traversed to the left to compensate. Three effects cause the spinning projectile to move out of plane; these are defined as follows:

1. *Gyroscopic forces:* Because the spinning projectile is rotating about two axes, it precesses about the third axis, causing an angle of attack (yaw of repose) to appear. This generates a horizontal lift force to the right, which causes a displacement in the same direction.
2. *Magnus force:* The same yaw of repose inclines the spinning projectile to generate another force to the right—the Magnus force—causing further displacements.
3. *Force due to Coriolis acceleration:* This is a fictitious acceleration introduced to account for the fact that the Earth is spinning while the projectile is in the air.

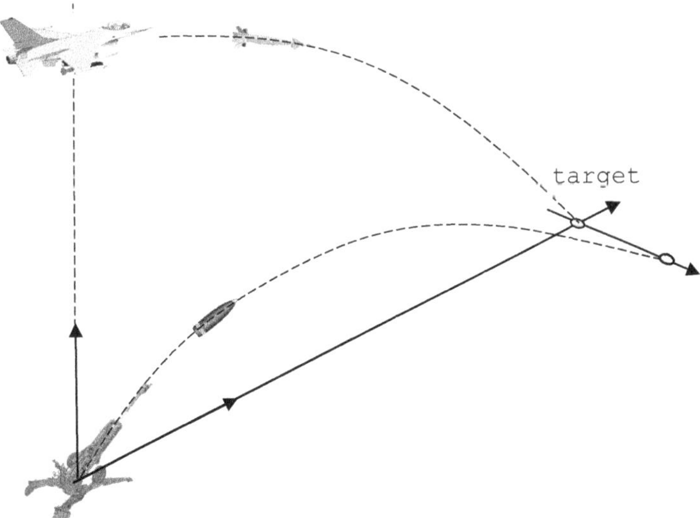

Fig. 3.16 Fin- and spin-stabilized trajectories.

These effects are somewhat complex to analyze, so the resulting trajectory model, known as the Modified Point Mass Trajectory Model (MPMTM), can only be solved using numerical methods. The interested reader should refer to the *Advanced Weaponeering* textbook for further details.

3.7 ARTILLERY TABULAR FIRING TABLES

Because of the complexity of solving the MPMTM trajectory equations, solutions for artillery projectiles were historically computed on mainframe computers and the results stored in tabular form in books called tabular firing tables (TFTs) by the Army and range tables by the Navy. The format for artillery firing tables is based on standardized agreements, and with small exceptions, are very similar between services and countries.

The main purpose of the TFTs is to provide the data to bring effective fire on a target under any set of conditions. Data for firing tables are obtained from running the MPMTM for all projectiles, charges, and ranges, supplemented by validation test firings.

The tables contain considerable data regarding adjustments that have to be made to various inputs to the fire control system that is used to compute a firing solution. Of particular interest here are the data shown in Table 3.4.

Column 1 of this table shows the required range and the corresponding QE to obtain that range, in milliradians (mrad). Column 7 lists the computed time of flight (for variable time fuze settings), and the two rightmost columns list deflection errors due to drift (rotational effects) and cross wind. This allows the gun crew to correct the gun pointing angle to the left for these effects and place fire more accurately on target.

3.8 TRAJECTORY FOR GUIDED WEAPONS AND MUNITIONS

The preceding sections addressed unguided munitions only, because these follow a ballistic trajectory. For guided munitions, a different approach is taken depending on what kind of weapon is being used. Theoretically we may assume that an air-launched weapon such as a guided missile follows a straight-line path from the launch aircraft to the target, as shown in Fig. 3.17.

It is assumed the target provides some optical or electromagnetic signal that the weapon seeker recognizes and guides it towards the signal source. The signal may be a visual image of the target or a reflected radar or laser transmission. A simple guidance algorithm might be to make the velocity vector point at the target.

This figure shows how the weapon attempts to follow the straight line corresponding to the line of sight (LOS) from launch to impact. The quality of the control system determines the fluctuation of the path from the LOS, which, if

TABLE 3.4   PART OF THE TABULAR FIRING TABLES (TFTs)

1	2	3	4	5	6	7	8	9
Range	ELEV	FS for graze burst Fuze M582	DFS per 10 m Dec	DR per 1 mil D ELEV	Fork	Time of flight	Azimuth corrections	
							Drift (corr to L)	CW of 1 knot
M	MIL			M	MIL	SEC	MIL	MIL
10,500	332.2	29.8	0.07	18	4	29.8	9.8	0.52
10,600	337.6	30.2	0.07	18	4	30.2	10.0	0.52
10,700	343.2	30.6	0.07	18	4	30.6	10.2	0.53
10,800	348.8	31.0	0.07	18	4	31.0	10.4	0.53
10,900	354.4	31.5	0.07	18	5	31.5	10.6	0.53
11,000	360.1	31.9	0.07	17	5	31.9	10.7	0.54
11,100	365.9	32.3	0.07	17	5	32.3	10.9	0.54
11,200	371.8	32.7	0.07	17	5	32.7	11.1	0.55
11,300	377.7	33.2	0.06	17	5	33.2	11.3	0.55
11,400	383.7	33.6	0.06	17	5	33.6	11.5	0.55
11,500	389.7	34.0	0.06	16	5	34.0	11.7	0.56
11,600	395.9	34.5	0.06	16	5	34.5	11.9	0.56
11,700	402.1	34.9	0.06	16	5	34.9	12.2	0.57
11,800	408.4	35.4	0.06	16	5	35.4	12.4	0.57
11,900	414.8	35.8	0.06	16	5	35.8	12.6	0.58
12,000	421.2	36.3	0.06	15	6	36.3	12.8	0.58

considerable, results in an unpredictable impact angle. In low-cost munitions such as legacy laser-guided bombs produced in large quantities, the control system is typically of the inexpensive "bang-bang" type where the control surfaces are hard over in one direction or the other as it attempts to align the weapon with the LOS. Bang-bang control systems produce considerable excursions from the assumed straight-line path. Newer munitions, such as cruise missiles and second generation laser-guided bombs (LGBs), tend to have more accurate, proportional control systems that reduce considerably the deviations from the LOS.

In practice, aircrew may release unpowered guided munitions, and LGBs in particular, as close to the ballistic point as possible (i.e., treating the weapon as unguided). At the same time, the designating laser is pointed away from the target so as not to be detected, moving to the desired impact point a few seconds before impact, as shown in Fig. 3.18. In this case, the trajectory may be assumed to be purely ballistic and calculated as in Section 3.2.

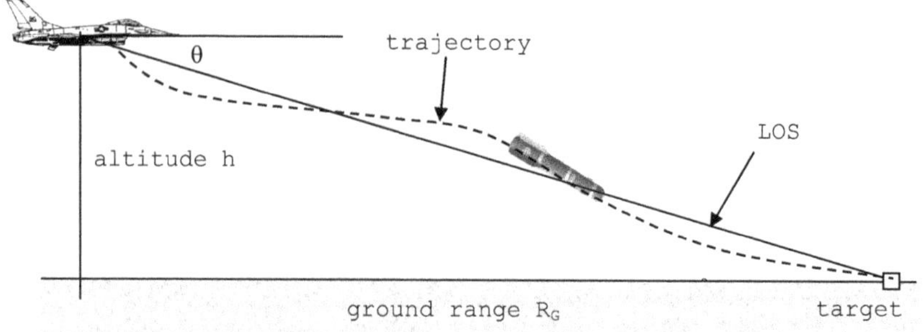

Fig. 3.17 Assumed trajectory of a missile to target.

Some munitions have a trajectory shaping capability that enables the desired impact angle to be input to the control system before launch. An example of this type of trajectory, which is common for weapons that need high impact angles to penetrate a hard target, is shown in Fig. 3.19.

Such pop-up maneuvers are usually initiated by an on-board altimeter so the deviation from the straight line-of-sight trajectory occurs at a fixed altitude. This capability enables the munition to impact the target at a high incidence angle, thereby maximizing its penetration capability.

For guided munitions in general, the impact angle is computed as either the LOS angle θ shown in Fig. 3.17, the desired impact angle I as indicated in Fig. 3.19, or the impact angle resulting from a quasi-ballistic trajectory if that delivery technique is utilized.

If a guided weapon does not have its impact conditions specified, doesn't follow an essentially ballistic trajectory, or doesn't move in a straight line from the release point to the target, the impact conditions are usually difficult to predict. Munitions that fall into this category might be inertial navigation system/global positioning system (INS/GPS) guided munitions or missiles attacking a moving target. Under these circumstances, the weaponeer must

Fig. 3.18 Example of offset lasing for guided weapon. DPI: desired point of impact.

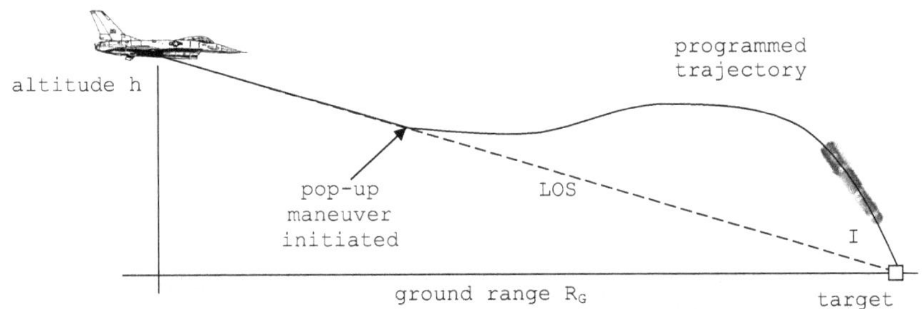

Fig. 3.19   Trajectory shaping for required impact angle.

estimate the impact angle (and any other parameters that might be needed), making one of the three listed assumptions. If precise impact conditions are required, a trajectory program for the specific weapon employed (fly-out model) may be needed for the planned attack.

It may be appreciated that for an unguided delivery at a fixed altitude, such as that shown in Fig. 3.5, there is a single ballistic point at which the weapon must be released in order to fall onto the target. For guided weapons, however, there is a region in space called the launch acceptable region (LAR) such that if the weapon is released inside the LAR, it will reach the target. An example is shown in Fig. 3.20.

In fact, there are two types of LAR—the in-range and in-zone contours. The in-range LAR defines a region on the ground that the weapon can reach, whereas the in-zone LAR defines the region of the ground the weapon can reach and achieve given terminal conditions (e.g., with an impact angle of at least 80 deg. LARs are displayed in the cockpit along with an indication of aircraft position, so aircrew know when the target is reachable.

The LAR shown in Fig. 3.20 is a horizontal slice for a fixed altitude; however, a vertical LAR also exists, as shown in Fig. 3.21.

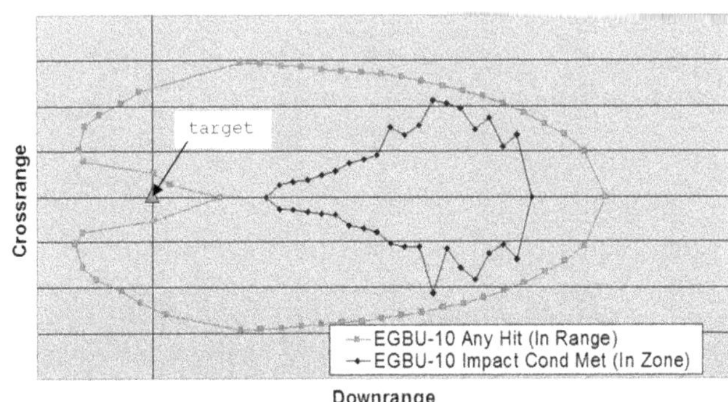

Fig. 3.20   Guided weapon LAR.

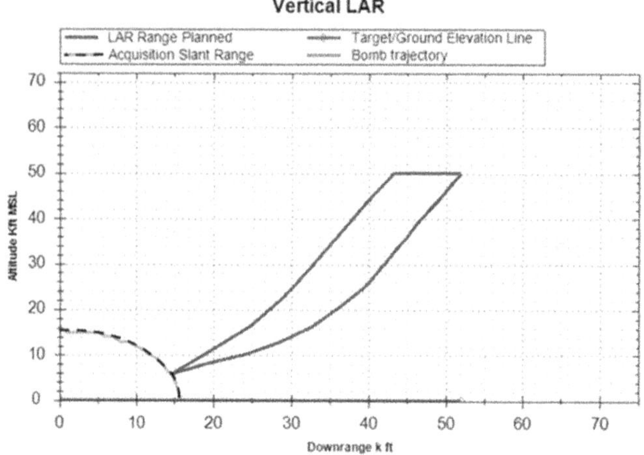

Fig. 3.21   Guided weapon vertical LAR.

LARs such as the ones shown in Figs. 3.20 and 3.21 are generated for a fixed altitude by considering a grid of weapon release points in the horizontal plane, running a trajectory model, and determining for each point whether the weapon can reach the target. The LAR is then simply the region containing all the launch points where the weapon can reach the target. Weaponeering tools such as the Precision Munitions Planning Tool (PMPT), shown in Fig. 3.22, are used to generate LARs.

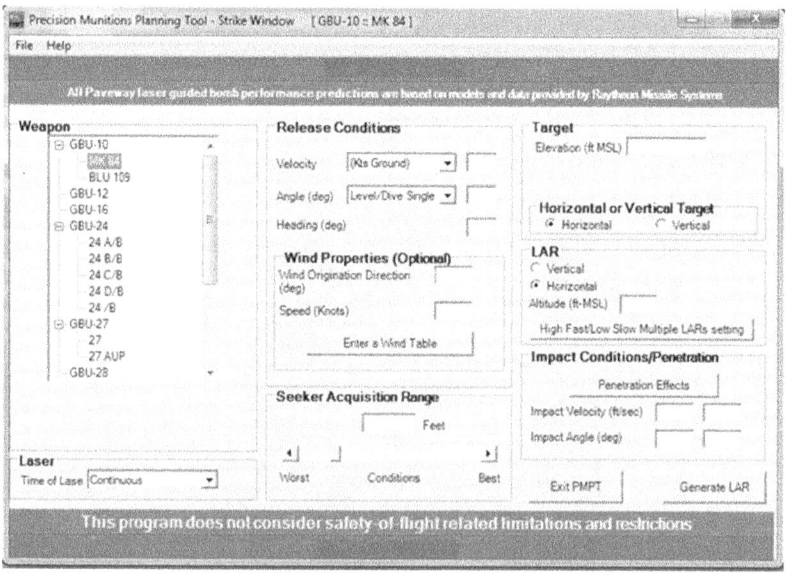

Fig. 3.22   PMPT LAR generation tool.

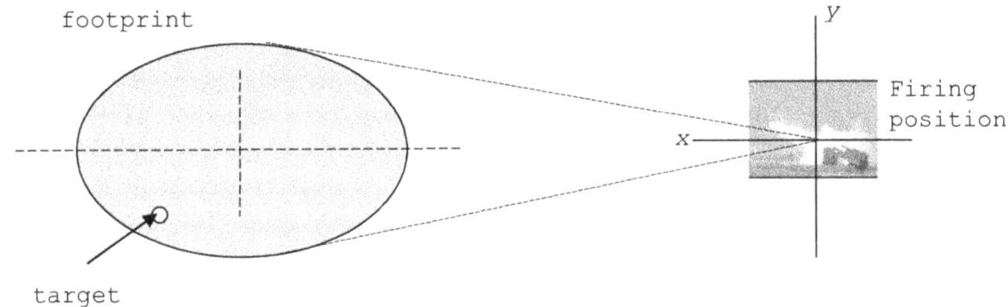

Fig. 3.23 Footprint for surface-launched guided weapon.

There is an equivalence to the LAR for surface-launched guided weapons called a footprint, as shown in Fig. 3.23, which shows a region on the ground relative to the firing position that the weapon can reach.

Note the LAR is centered on the target and shows when the aircraft is inside it, whereas the footprint is centered on the gun and shows whether the target is inside it.

3.9 CHAPTER SUMMARY

- The solution of the equations of motion for an air- or surface-launched, unguided, nonspinning projectile is known as the point mass model; if air resistance is ignored, their solution is simple.
- The solution of the equations of motion provides velocity and position as functions of time in the horizontal and vertical directions.
- When aerodynamic drag is included, the resulting equations of motion can only be solved numerically.
- In order to provide stability in flight, some munitions are spin stabilized, which leads to complications in their trajectory. Other weapons are fin stabilized, for which relatively simple equations of motion apply.
- Guided weapon trajectories are difficult to predict without a complex manufacturer's fly-out model; however, some guided weapons may be programmed to impact at a predetermined angle.
- An important consequence of knowing the trajectory of guided weapons is the ability to produce a launch-acceptable region or footprint in order to know if the weapon can reach the target.
- For guided weapons, the user may have to estimate and enter the impact angle directly.

REFERENCES

[1] Farrar, C. L., and Fleming, D. W., "Military Ballistics—A Basic Manual," *Brassey's Battlefield Weapons Systems and Technology,* Vol. X, Brassey, Dulles VA, 1983.

[2]   Hill, P. and Peterson, C., *Mechanics and Thermodynamics of Propulsion*, 2nd ed., Addison-Wesley, Reading, MA, 1992.

[3]   U.S. Marine Corp., "Tactics, Techniques and Procedures for Field Artillery Manual Cannon Gunnery," FM 6-40, Headquarters, Department of the Army/U.S. Marine Corps, Washington DC, 23 April 1996, Chapter 3.

[4]   Bray, D., "External Ballistics—Trajectory Modeling," Cranfield University, 2006, https://www.cranfield.ac.uk/courses/short/defence-and-security/fundamentals-ofballistics

[5]   Rowles, S., *Joint Gun Effectiveness Model with Air-Bursting Methodology—Users Manual*, Systems Concepts and Analysis Branch, Naval Surface Warfare Center, Dahlgren, VA, 1995.

Chapter 4

DELIVERY ACCURACY

4.1 INTRODUCTION

Delivery accuracy may be defined as a quantitative measure of the capability of a weapon system to place ordnance on its intended target. It is measured using a variety of statistics that will be described in the following sections, and it is assumed that readers have a working knowledge of the statistical methods described in Chapter 2. The purpose of this chapter is to clarify the meaning of various measures of accuracy so the reader can provide relevant data needed to calculate target damage.

Basically, we are interested in the random dispersion of rounds directed at a target. If the results of a test or simulation produce impact points on the ground such as those shown in Fig. 4.1, we could express the accuracy of such a weapon system in terms of the dispersion in x and y (i.e., as σ_x and σ_y). The smaller the value of sigma, the more accurate the weapon system is.

In this chapter, we will discuss the accuracy of various weapon systems by dividing the discussion into the following sections:

- Single unguided and guided weapons
- Multiple weapon deliveries, where each weapon may be delivered in a dependent or independent manner

4.2 DISPERSION FOR SINGLE WEAPON DELIVERIES

Suppose as a result of testing we are presented with a set of impact points on the ground plane as shown in Fig. 4.2, where each point represents the impact of a single weapon. We will restrict the discussion to where we know each impact point results from a weapon that is independently aimed at the target, that is, as would be obtained by one aircraft dropping one bomb followed by another aircraft with another bomb and so on. It is also assumed that all impacts occur in a relatively short time (i.e., minutes rather than hours).

71

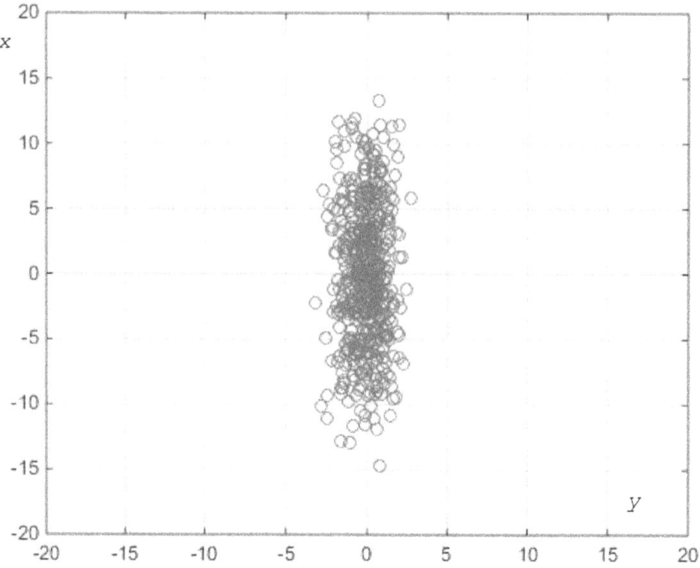

Fig. 4.1 Impact point distributions on the ground plane.

Clearly, it is impossible to predict what happens to a particular bomb; all that may be done is to describe, in a statistical sense, the behavior of the sample of impact points. The desired point of impact (DPI) is the location of the single weapon aimpoint on the ground and corresponds to the origin of x, y axes shown in Fig. 4.2.

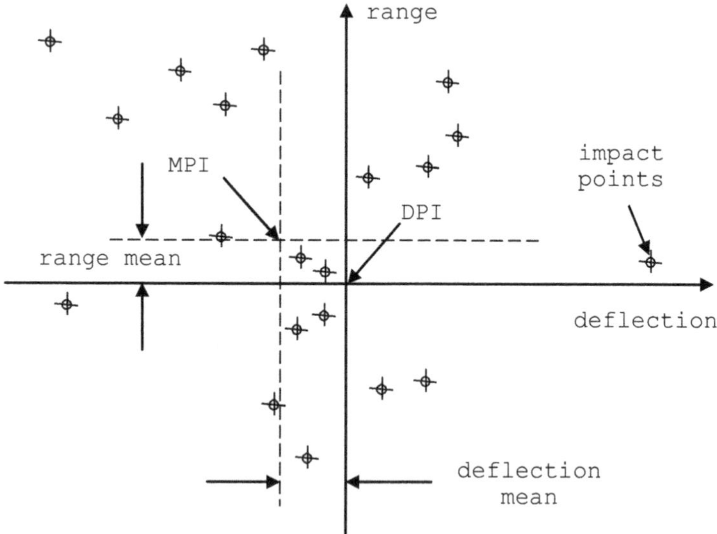

Fig. 4.2 Distribution of impact points around the desired point of impact (DPI).

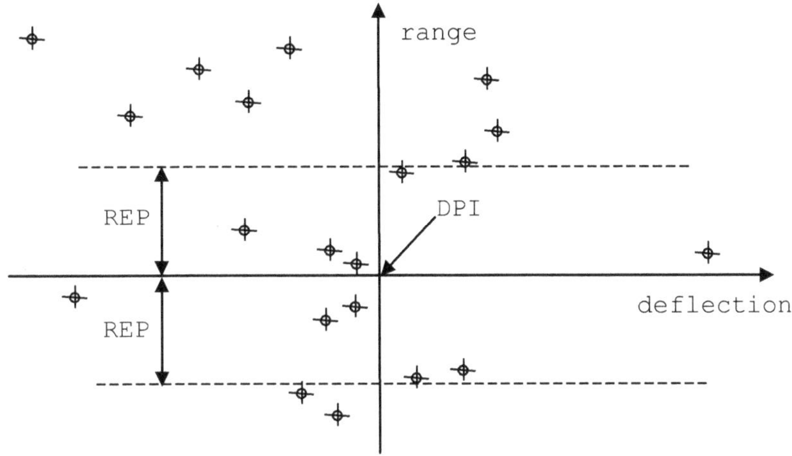

Fig. 4.3   Definition of REP.

As described in Chapter 2, these axes in the ground plane are the range and deflection directions. The range direction is defined as the projection of the velocity vector of the weapon at impact/detonation onto the ground plane. The deflection direction is perpendicular to it, to the right. The resulting z-axis points upward.

One of the simplest statistical tasks is to find the mean of the range and deflection coordinates for all impact points, defining the mean point of impact (MPI). The MPI may be offset from the DPI by the range and deflection mean values, as shown in Fig. 4.2. Although not part of the test results shown here, if multiple weapons are released against the target in a single attack, we define the desired mean point of impact (DMPI) as that location where the mean of all weapon impacts is ideally located.

To accompany the mean values outlined previously, it would be useful to have some measure of the dispersion or distribution of the impact points. Although the variance or standard deviation (discussed in Chapter 2) might appear a suitable choice, weaponeers prefer to use the range error probable (REP) and the deflection error probable (DEP).

REP is the distance from the DPI to one of a pair of lines perpendicular to the range direction, equidistant from the DPI such that 50% of the impact points lie between them. This is shown in Fig. 4.3.

Similarly, the DEP is the distance from the DPI to one of a pair of lines parallel to the range direction, equidistant from the DPI such that 50% of the impact points lie between them, as shown in Fig. 4.4.

Although REP and DEP give an idea of dispersion in the range and deflection directions, another commonly used measure is the circular error probable (CEP). The CEP is the radius of a circle, centered on the DPI such that 50% of the impact points lie within it, as shown in Fig. 4.5.

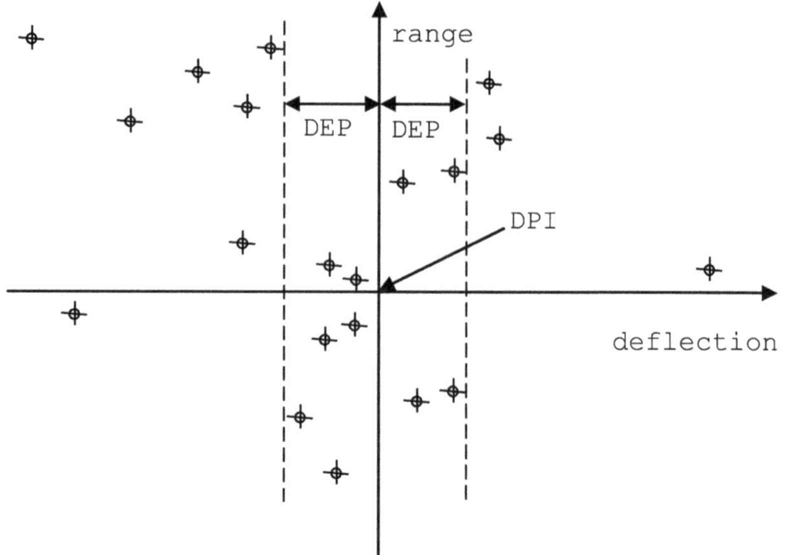

Fig. 4.4   Definition of DEP.

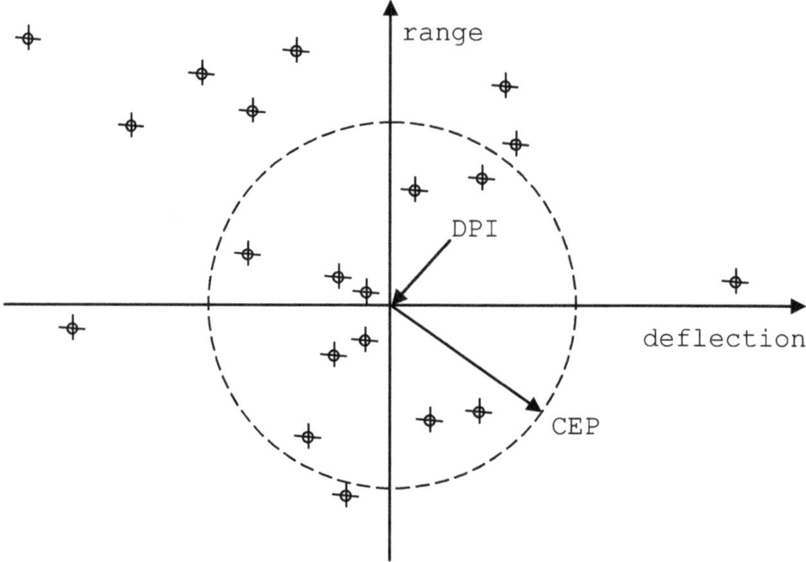

Fig. 4.5   Definition of circular error probable (CEP).

The error probable may be interpreted in two ways:

1. The distance inside of which 50% of the weapons may be expected to fall
2. The distance inside of which a single weapon has a 50% chance of falling

These metrics do not assume any specific distribution associated with the miss distances and may be determined graphically, as indicated in the preceding figures, from impact points plotted on graph paper. Note also that the error probable corresponds to the median of the data set, as defined in Chapter 2.

4.3 VERTICAL ERRORS

So far, only miss distances in the ground plane have been considered, but some targets may be represented by a vertical target plane, in which case the distribution of impacts vertically has to be considered. This is shown in Fig. 4.6.

In this case, the height error probable (HEP) is defined as the distance between two horizontal lines equidistant from the DPI enclosing 50% of the impacts.

As stated earlier, all the measures of accuracy discussed so far may be estimated from experimental data without any assumptions being made about the statistical nature of the data. If some assumptions are made, however, the computational analysis becomes simpler. In the next section we will discuss the types of assumptions that may be made and how the statistics of accuracy may be calculated.

We need to say a few words about the possible difference between the MPI and DPI shown in Fig. 4.2. In a real test, it is highly unlikely that the MPI is zero. This raises the question of whether this offset should be considered as part of the weapon accuracy. The way to answer this is to conduct several tests on different days and conditions and examine the MPI. If it is different for each test and apparently random in location, we can argue it is a product of the test, and because in a properly designed fire control system we would expect

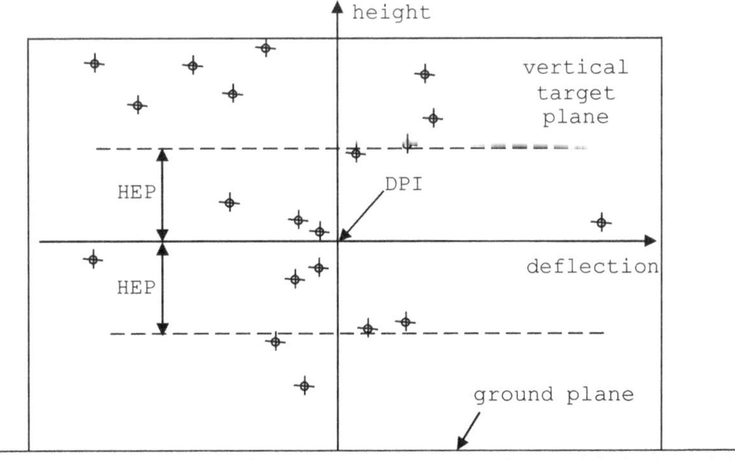

Fig. 4.6   Example test results on vertical target.

it to be zero, it can be ignored. However, if the MPI is in a consistent position relative to the target, it has to be accepted as a characteristic of the weapon and must be considered when aiming at the target.

4.4 COMMONLY USED EQUATIONS BASED ON NORMALLY DISTRIBUTED DATA

The usual assumption made in the analysis of accuracy data, supported by analysis of test data, is that the distribution of impact points in range, deflection, and height is Gaussian, or normal; therefore, the statistical analyses described in Chapter 2 may be applied. For example, we know that $\pm\sigma$ contains 68% of the samples, and because $\pm$REP contains 50% of the samples, then REP $< \sigma$, as shown in Fig. 4.7.

In fact, it may be shown that REP $= 0.6745\sigma$, so we can write for all normally distributed random variables

$$REP = 0.6745\sigma_x \tag{4.1}$$

$$DEP = 0.6745\sigma_y \tag{4.2}$$

$$HEP = 0.6745\sigma_z \tag{4.3}$$

In the ground plane, if $\sigma_x = \sigma_y = \sigma$ (circular normal), and the mean values of miss distance are both zero, the radial miss distance may be described by a Rayleigh distribution. The CEP obtained from Eq. (2.17) in Chapter 2, by setting $P = 0.5$, $r_1 = 0$, and $r_2 = $ CEP, results in

$$CEP = 1.1774\sigma \tag{4.4}$$

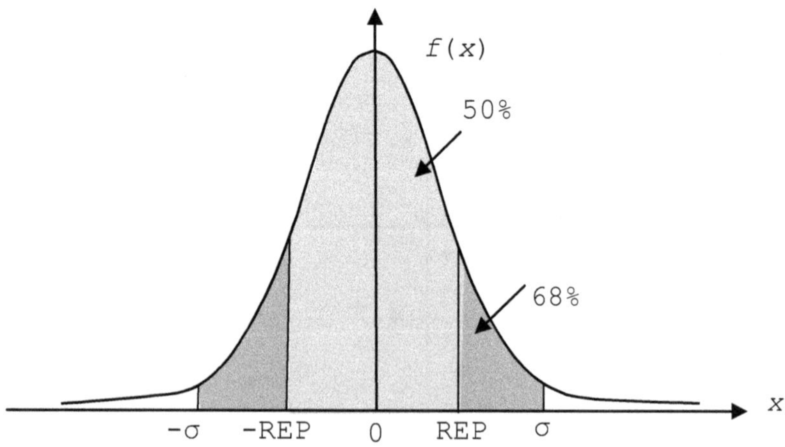

Fig. 4.7   Relationship between error probable and standard deviation.

We can find the relationship among REP, DEP, and CEP by writing

$$\text{CEP} = 1.1774\sigma = 1.1774\frac{\text{REP}}{0.6745} = 1.7456\text{REP} \tag{4.5}$$

Because REP and DEP are equal in a circular normal distribution, we also obtain

$$\text{CEP} = 1.7456\text{DEP} \tag{4.6}$$

Alternatively, we may write

$$\text{REP} = \text{DEP} = 0.573 \times \text{CEP} \tag{4.7}$$

For some applications, it is necessary to find the radius of a circle that contains 90% of the impacts, sometimes referred to as the CEP_{90} or CE_{90}. Again, from Eq. (2.17) in Chapter 2, by setting $P=0.9$, $r_1=0$, and $r_2=\text{CEP}_{90}$, we get

$$\text{CEP}_{90} = 2.146\sigma \tag{4.8}$$

From Eqs. (4.4) and (4.8) we get the useful scaling for the circular errors.

$$\text{CEP}_{90} = 1.8227 \times \text{CE}_{50} = 1.8227 \times \text{CEP} \tag{4.9}$$

In a linear direction, REP, DEP, and HEP are given, for example, by

$$\text{REP}_{90} = 2.483 \times \text{REP}_{50} \tag{4.10}$$

A question regarding semantics: What is the difference between CE_{50} and the CEP? Usually, the P notation means the number is derived from data by assuming a specific probability density function, such as Gaussian or Rayleigh. We have seen, however, that the CEP may be calculated directly from the data set without using any assumed distribution, as shown in Fig. 4.5. Strictly speaking, therefore, the CEP is derived from a data set using an assumed, specific distribution whereas a CE_{50} is derived from raw data. The same applies to the 90% metric as well.

4.5 NONCIRCULAR DISTRIBUTIONS

We saw in the discussion in Chapter 2, Section 2.4 on the Rayleigh distribution that it is only valid if the ratio of the maximum to minimum standard deviations is less than about 2. In many instances, the distribution of miss distances may be Gaussian in range and deflection, but is not circular, so the assumption that the radial miss distance is Rayleigh distributed is not valid.

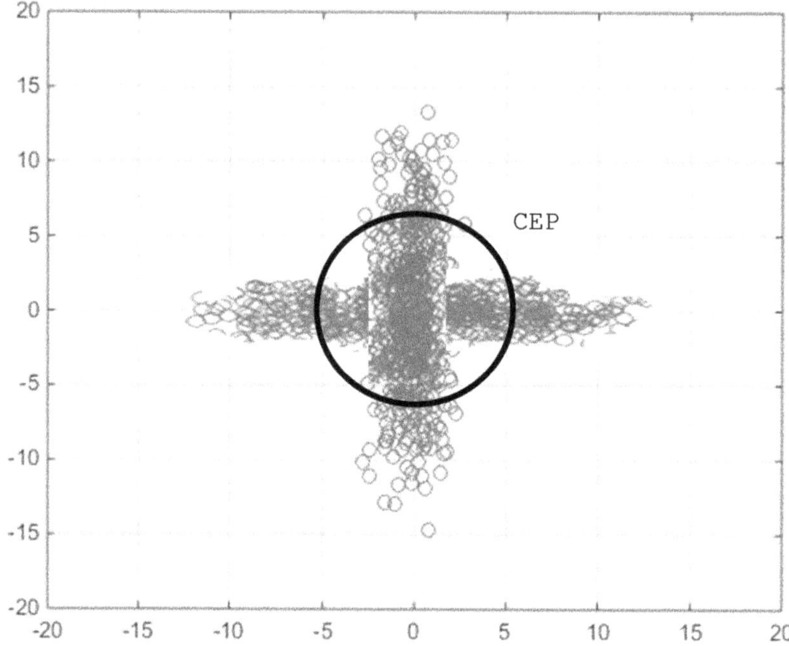

Fig. 4.8   Different distributions of impact points.

An example is shown in Fig. 4.1, where the dispersion in range is many times that in deflection.

In such a case, a useful empirical approximation for calculating CEP is given by Pitman's equation, as follows:

$$\text{CEP} = 0.562 \times \text{MAX}(\sigma_x, \sigma_y) + 0.617 \times \text{MIN}(\sigma_x, \sigma_y) \qquad (4.11)$$

It is important to realize, however, that the CEP is only a useful measure of accuracy if the distribution of impact points is close to circular normal. Consider the situation shown in Fig. 4.8, which shows two distributions, one with large REP and small DEP, the other with small REP and large DEP. However, the CEP might be the same for both distributions and is therefore a poor descriptor of each.

It is therefore preferable to specify a bivariate distribution with REP and DEP rather than CEP to avoid this ambiguity. A CEP may be found from REP and DEP, but the reverse is not true.

4.6   ACCURACY IN THE NORMAL PLANE

So far, we have assumed that accuracies are measured in the ground plane, and this is usual for weapons that fall to the ground (i.e., bombs, artillery

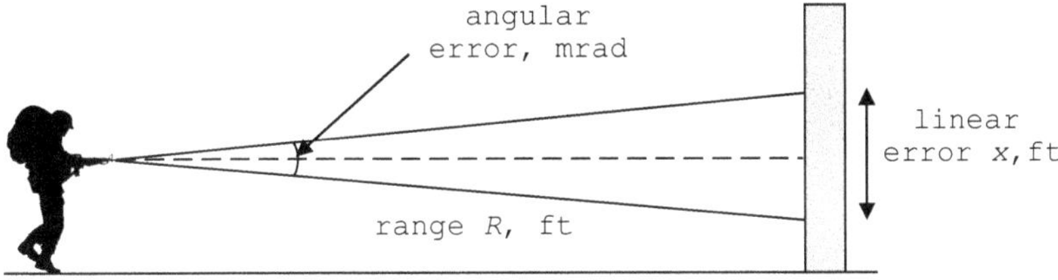

Fig. 4.9   Normal plane angular and linear errors.

shells, rockets, etc.); however, for what are called *direct fire weapons*, such as guns, the attack usually takes place in the normal plane.

Accuracy in the ground plane is measured in feet or meters, but the metrics used when impacts occur in the normal plane are angular units such as the milliradian (mil), one thousandth of a radian. Consider the soldier shooting at a target on the vertical wall shown in Fig. 4.9.

The normal plane is defined as a plane perpendicular to the velocity vector of the projectile at impact, so if we assume a straight-line trajectory of the bullet, the vertical plane is the wall.

If θ is the angular error in mil, the linear error x in the normal plane at a target range R is given by the following equation:

$$x = R \times \frac{\theta}{1000} \tag{4.12}$$

One advantage of angular measures for the error is that they scale linearly with distance, as shown in Fig. 4.10.

Recall the basic assumption that the ranges are sufficiently small such that the projectile trajectory is a straight line. Because this is not true for air-launched weapons, in calculating the actual distance traveled by the bomb we should use the path length along the trajectory; instead, we use the length of the line connecting the release point with the target, as shown in Fig. 4.11.

This line is called the *slant range* and may be calculated using Pythagoras' theorem from the release altitude and the bomb fall range R_B.

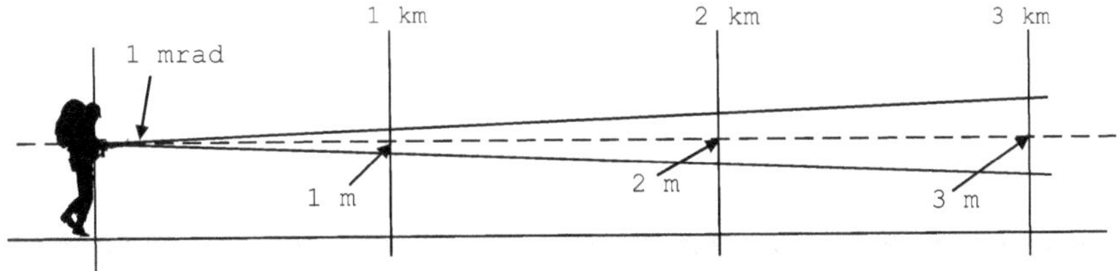

Fig. 4.10   Linear error scales with range.

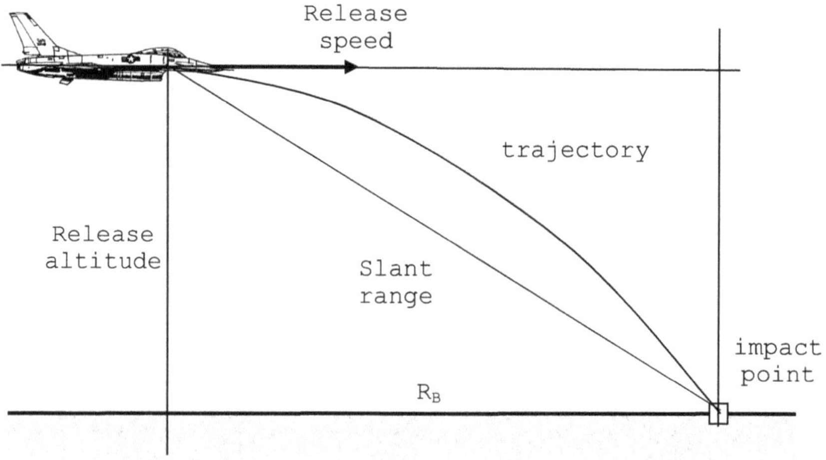

Fig. 4.11   Slant range for air-launched weapon.

4.7   NON-NORMAL DISTRIBUTIONS

In the analysis so far, we have assumed that miss distances can be represented by normal, or Gaussian, distributions. When testing guided weapons, however, it became apparent that the data from the test range did not always fit a normal distribution and that further characterization was required. It should be apparent that any attempt to predict the effectiveness of a weapon will depend on its accuracy distribution, so mathematical ways to characterize non-normal distributions are necessary.

Guided weapons test data have resulted in range and deflection miss distance distributions shown by the histogram in Fig. 4.12 and, as confirmed by hypothesis testing, is clearly non-normal. This distribution has the following features:

- A number of gross errors
- A portion of impacts that appear to have a normal distribution
- A number of direct hits on the target in excess of the second item in this list

Guided weapons may exhibit gross errors characterized by impacts a considerable distance from the target. Such errors are usually due to some catastrophic failure of the guidance system, such as loss of seeker lock shortly after firing a missile, disruption of the laser guidance beam for laser-guided bombs (LGBs), or mechanical failure of one of the control surfaces. Gross errors are sometimes referred to as an in-flight reliability issue.

The question often arises: How is a gross error defined? By considering Fig. 2.5 in Chapter 2 we may conclude that an impact outside $\pm 4\sigma$ of the mean of a normal distribution is highly unlikely; therefore, this is used as our criterion.

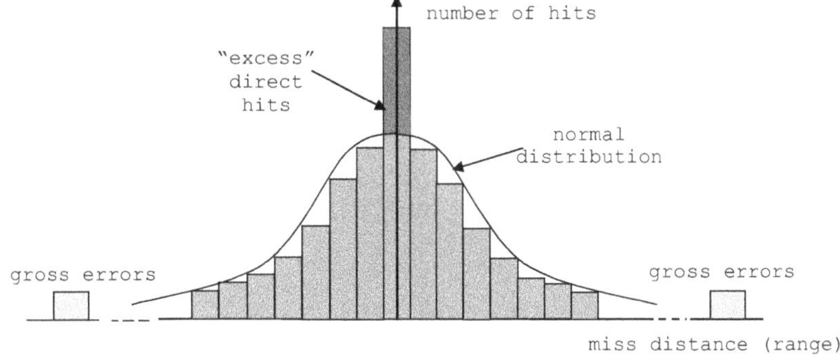

Fig. 4.12   Non-Gaussian distribution of guided weapons.

Any impact outside these limits cannot be part of the Gaussian distribution; therefore, that impact is a gross error.

In addition, there is a disproportionate or unexpected number of direct hits (i.e., weapons that impact within a histogram bin that is relatively small, centered on the aimpoint). This is a good thing. After all, this is why guided weapons were developed—they hit the target more often than unguided weapons.

Considering the data represented in Fig. 4.12, our approach will be to split the distribution into three parts.

1. The gross errors
2. The excess direct hits (i.e., those in excess of what would be expected of a normal distribution)
3. The remaining, normally distributed data

The resulting analytical process is called the $P_{\mathrm{HIT}}/P_{\mathrm{NM}}$ methodology and is best explained by means of an example.

Suppose we have 100 impact points [i.e., 100 sets of (x, y) coordinates].

1. Calculate the standard deviation in range and deflection σ_x and σ_y.
2. Any impact point where either x or y is greater than 4 times its corresponding sigma value is defined as a gross error and is removed from the data set. Suppose there are 5 (N_{GE}) such points. The remaining data set is then reduced to 95 samples.
3. Suppose the center bin in the histogram shown in Fig. 4.12 is 10 ft wide and σ_x is the standard deviation of the remaining 95 data points. The probability a single sample will be in this bin will be the area under the normal distribution between -5 and $+5$, which from Eq. (2.8) in Chapter 2 is

$$P(-5 < x < 5) = \mathrm{NORMDIST}(5,0,\sigma_x,1) - \mathrm{NORMDIST}(-5,0,\sigma_x,1)$$

$$(4.13)$$

4. Therefore, the number of samples out of the remaining 95 expected in this bin will be called direct hits and are given by the following equation:

$$N_{\text{CENTER}} = 95 \times [\text{NORMDIST}(5,0,\sigma_x,1) - \text{NORMDIST}(-5,0,\sigma_x,1)]$$

$$(4.14)$$

5. If the actual number of test data in this center bin is N_{ACTUAL} then the number of "excess" hits in the center bin (N_{HIT}) is given by

$$N_{\text{HIT}} = N_{\text{ACTUAL}} - N_{\text{CENTER}} \tag{4.15}$$

We now define N_{NM} to be the number of data points once the gross errors and excess direct hits have been removed. Suppose $N_{\text{HIT}} = 25$ and P refers to "the proportion of" rather than "the probability of," we can then calculate the following:

$$P_{\text{GE}} = \frac{N_{\text{GE}}}{N} = \frac{5}{100} = 0.05$$

$$P_{\text{HIT}} = \frac{N_{\text{HIT}}}{N} = \frac{25}{100} = 0.25$$

$$P_{\text{NM}} = \frac{N_{\text{NM}}}{N} = \frac{100 - 5 - 25}{100} = 0.70 \tag{4.16}$$

Note the sum $P_{\text{GE}} + P_{\text{HIT}} + P_{\text{NM}} = 1$, so we only need to specify two of these numbers to obtain the third. We now calculate REP, DEP, or perhaps the CEP of the 70 data points making up the normally distributed part of the data set. The resulting data would be reported in terms of the four parameters REP, DEP, P_{HIT}, and P_{NM}, or in terms of the three parameters CEP, P_{HIT}, and P_{NM}. A typical specification might be

$$P_{\text{HIT}} = 0.25 \quad P_{\text{NM}} = 0.70 \quad \text{CEP} = 25\,\text{ft} \tag{4.17}$$

How these data are used to calculate effectiveness will be dealt with later, but first consider the following specifications for a particular weapon: CEP $= 15$ ft, $P_{\text{HIT}} = 0.05$, $P_{\text{NM}} = 0.95$ would indicate that the weapon is about 95% Gaussian, and the CEP value may be used as it would be for an unguided weapon.

If, however, we have CEP $= 15$ ft, $P_{\text{HIT}} = 0.95$, $P_{\text{NM}} = 0.05$, it would indicate that the weapon has about a 95% probability of hitting the target, and to a first approximation will not miss. The value of CEP is therefore of little consequence because it will only miss the target 5% of the time.

In this formulation it is a common mistake to regard the quoted CEP as the weapon CEP; it is not. It is the CEP of the normal portion of the accuracy distribution only. If, for the previous example, we ask how many impacts lie within the stated CEP of 25 ft, the answer is given by the following:

$$\text{Impacts inside CEP} = (50\% \text{ of } 70) + 25 \text{ direct hits} = 60 \tag{4.18}$$

Here we see the apparent contradiction that 60% of impacts lie within the CEP; however, recall that the expected 50% value was obtained from circular normal data, which this is not.

4.8 MULTIPLE WEAPON DELIVERIES

When a target is attacked with multiple weapons, those weapons may be delivered independently or dependently. Consider the attack shown in Fig. 4.13, which shows four aircraft successively attacking a ground target with unguided bombs.

This is defined as a series of independent attacks, because each pilot will aim their weapon independently of the other three. On the other hand, the same number of the same bombs may be released against the same target by a stick dropped from a single aircraft, as shown in Fig. 4.14, an example of dependent weapon delivery.

Fig. 4.13　Several A-10 aircraft attacking a ground target.

Fig. 4.14　Stick of bombs dropped on a ground target.

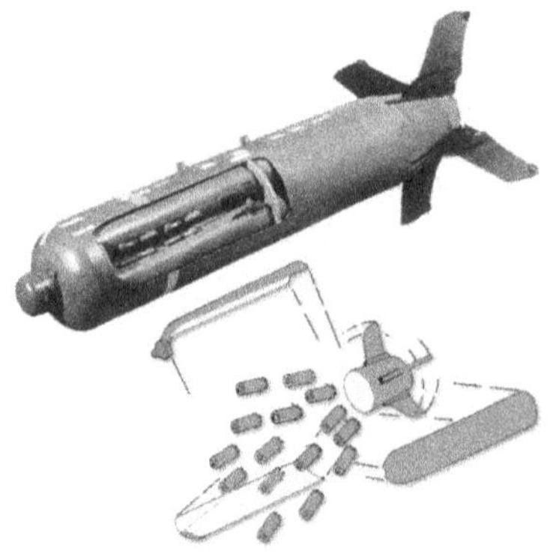

Fig. 4.15 Multiple weapon deliveries.

These attacks may be distinguished from each other by considering what happens if a pilot, for whatever reason, hesitates in releasing the weapon(s). For the independent sequence, only the one bomb will miss the target due to the single pilot's hesitation, whereas for the dependent attack, all four bombs will miss.

The basic test is that for independently delivered weapons, each is independently aimed at the target, whereas for dependent delivery, multiple weapons share the same aiming error.

This issue of dependent/independent attacks applies to other weapon systems, too, as shown in Fig. 4.15, including

- An aircraft dropping a cluster bomb
- A machine gun firing a burst of rounds
- An artillery battery typically comprising six guns firing one or more volleys

These weapon systems are all examples of dependent, multiple weapon delivery.

For all these examples, we will see that there are now two error statistics that have to be measured. Consider the situation shown in Fig. 4.16 of a soldier firing an automatic weapon at a target in the normal plane.

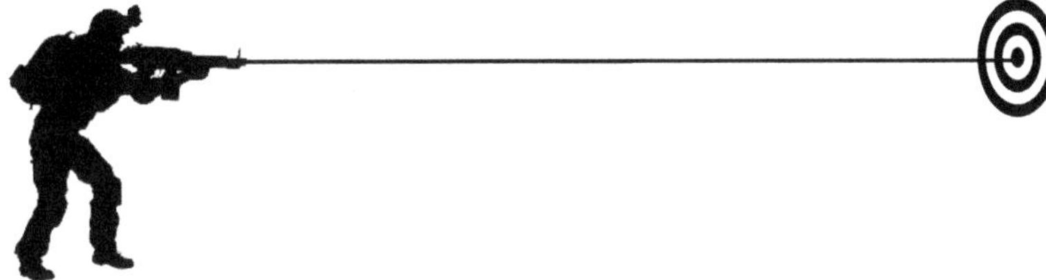

Fig. 4.16   Firing bursts at a target.

Here, the shooter aims at the target, fires a burst, inspects the impacts, fires another burst, and repeats. The impacts in the normal plane will look like those shown in Fig. 4.17. This is an example of dependent fire.

For each burst, the mean point of impact (MPI) may be determined by averaging the (x, y) coordinates for all the rounds in that burst. Its distance from the origin, the MPI error, represents the ability of the shooter to correctly aim the gun at the center of the target. As expected, not all shots in the burst disappear into the same hole, but six individual impact points on the wall will be observed. The dispersion of these six points about the MPI is called the *precision error* and represents the relative accuracy of each round with respect to the MPI.

In terms of accuracy measures, we now have a dispersion in the x and y directions as before, but corresponding to both the MPI and precision errors. The accuracy statistics characterizing the scenario are therefore $\sigma_{x\text{-MPI}}$, $\sigma_{y\text{-MPI}}$ and $\sigma_{x\text{-prec}}$, $\sigma_{y\text{-prec}}$.

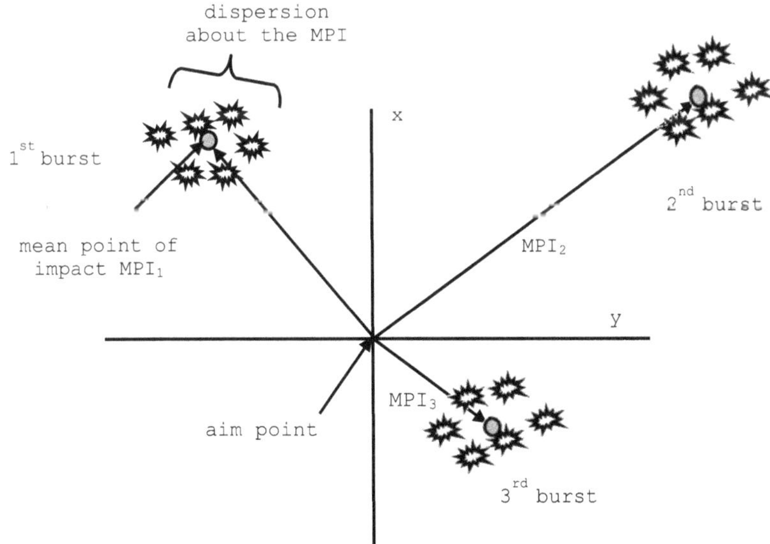

Fig. 4.17   Pattern of shots on the vertical plane.

The concept of two error types (MPI and precision) is also found in other weapon systems. Consider the ground plane impact points due to firing three rounds from an artillery cannon in the space of about 1 min, as shown in Fig. 4.18. The next day, the cannon is towed to the same firing position, aimed at the same target, and another three rounds fired, and so on. The change in the MPI is due to varying weather (MET) conditions such as air density, temperature, and wind, and the accuracy of aiming the cannon at the same spot on consecutive days.

Clearly, both error types are present for this case, as they will be for air-launched weapons. For example, Fig. 4.19 shows an aircraft dropping a stick of four unguided bombs on a target where each bomb is designed to be at the

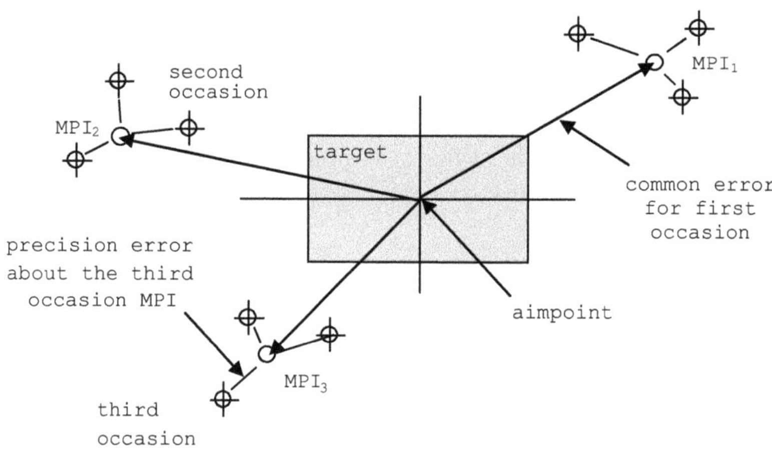

Fig. 4.18   MPI and precision errors for artillery.

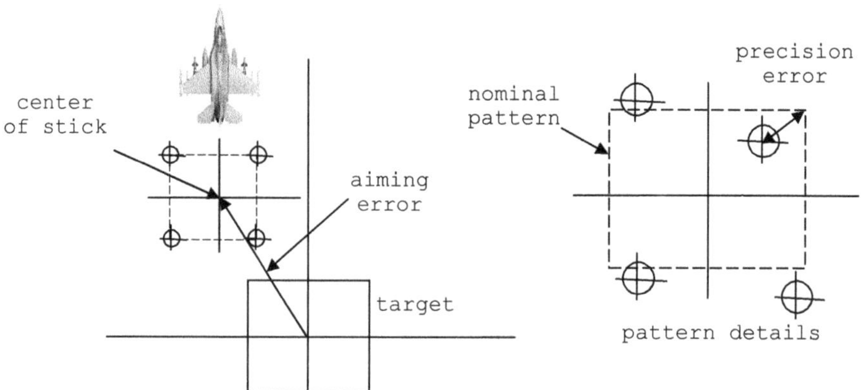

Fig. 4.19   Aiming and precision errors.

corner of a square. The center of the stick, which should ideally be located on the target, will be offset due to inaccurate aiming. Notice how the whole stick pattern is displaced by the MPI error, called here the aiming error.

Also shown is the effect of the precision errors—called *ballistic dispersion*—perturbing the impact locations of the individual bombs from their ideal locations at the corners of the square to some individual, randomly located impact points.

A further example would be when multiple global positioning system–guided (GPS-guided) weapons are directed to a set of coordinates that are thought to point to the target of interest. Depending on how such coordinates are obtained, there is usually an error associated with their specification, known as the *target location error (TLE)*. The resultant impacts might look as shown in Fig. 4.20.

In this case, the MPI error is called the TLE whereas the round-to-round (precision) error is attributable to differences in the guidance and control (G&C) system performance of each weapon, resulting in different flight paths to the specified common coordinates.

The reader may have noticed some differences in terminology in describing the same type of error (e.g., MPI, aiming error, and TLE all describe the same thing). For historical and service-specific reasons, different communities use different expressions to describe the same types of error. Efforts for different operational communities to agree on a common terminology have not succeeded. Among analysts, however, MPI error types are usually referred to as *dependent errors* whereas precision type errors are called *independent errors*. Table 4.1 lists terminology among the different operational communities.

Some care has to be exercised in using the word *dependent* because this is not accurate in the strict statistical sense. Within a burst, or occasion, one

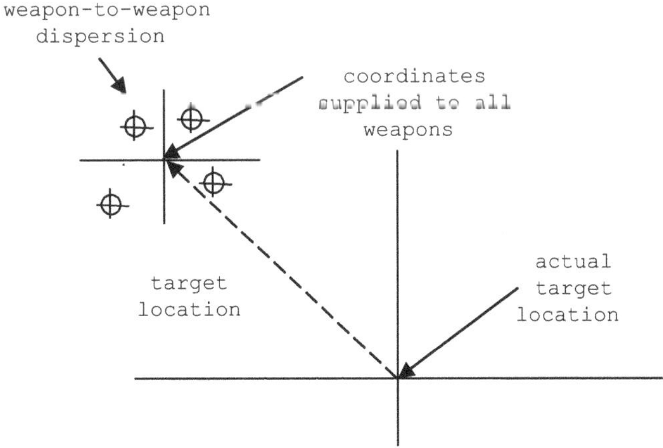

Fig. 4.20 GPS-guided weapons with TLE.

TABLE 4.1   ALTERNATIVE DESCRIPTIONS OF ERROR TYPES

Community	Dependent Error	Independent Error
Artillery	Mean point of impact (MPI)	Precision
Air-to-surface	Aiming	Ballistic dispersion
Direct fire	Burst-to-burst	Within burst
GPS	Target location error (TLE)	Guidance and control
Surface-to-surface	Occasion-to-occasion	Round-to-round

round does not depend on where a previous round landed, but they do share a common MPI error, such as the same target coordinates or manually sighted aimpoint. A better description is to say the rounds are correlated; however, due to common usage, the term *dependent* will be used.

It is important to note that both dependent and independent errors will not be constant for a given weapon system but have to be further qualified before meaningful accuracy measures can be used. Consider, for example, an M240 machine gun firing the same round but with different mounts, as shown in Fig. 4.21.

Of the two cases shown, the vehicle-mounted weapon will be more stable, and therefore the dispersion of bullets on the target should be smaller. Note the aiming system that governs the MPI error is also different for the two cases shown. Therefore, both MPI and precision errors will be different for these systems, even though it is the same gun firing the same round.

In each case, the different precision error will contain the same fundamental ballistic dispersion of the round, which is dependent only on the type of projectile being fired. It is worth asking: What is the actual dispersion of a 7.62-mm caliber bullet, and how is it measured? It is determined in a test setting by using a Mann barrel, an example of which is shown Fig. 4.22.

This is a piece of test equipment composed of a heavy, thick-walled barrel of the correct caliber for the munition under study, and a breach mechanism to load and fire the round. The complete firing mechanism is fixed to a heavy

Fig. 4.21   Different mountings of an M240 machine gun.

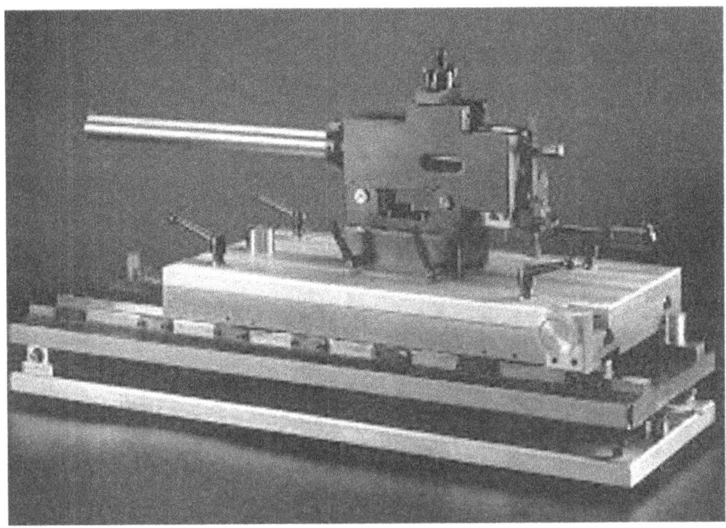

Fig. 4.22 Mann barrel used to determine ballistic dispersion.

plate, which in turn is attached to a large, rigid mount, usually made of concrete. Single rounds are fired at targets located at fixed ranges, and the ballistic dispersion is determined from impact data. This would be the smallest ballistic dispersion of the round, independent of the gun used to fire it or how the gun was mounted or aimed.

As we have seen, independent errors may be measured using the same parameters as dependent errors and combined in a way that depends on the scenario being considered. Both errors may be considered bivariate random variables in the ground or normal plane, an example of which for a *single* shot is shown in Fig. 4.23.

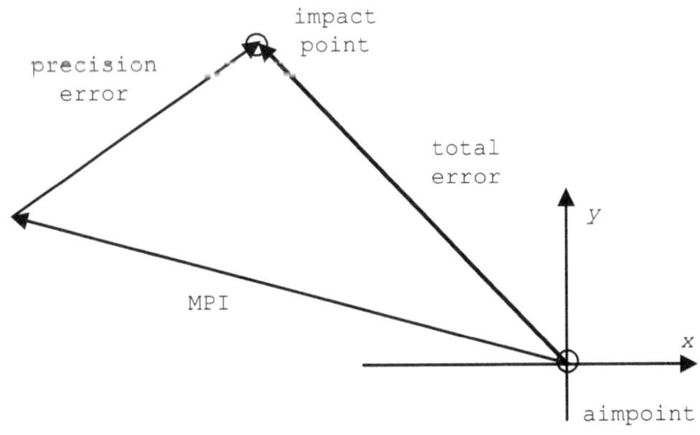

Fig. 4.23 Single-shot aiming, precision, and total errors.

The aimpoint is usually the target when it is small, or the center of a larger target. When single munitions are used against a target, the miss distances actually measured are due to both MPI and precision errors, as indicated in the figure, and would be added as follows:

$$x_{\text{DEP}} + x_{\text{IND}} = x_{\text{MISS}} \tag{4.19}$$

$$y_{\text{DEP}} + y_{\text{IND}} = y_{\text{MISS}} \tag{4.20}$$

Because the MPI and precision errors are assumed to be independent random variables, the standard deviation of the measured miss distances would be obtained from the root sum squared (RSS) process, as described in Chapter 2, Section 2.9.

$$\sigma_{\text{DEP}}^2 + \sigma_{\text{IND}}^2 = \sigma_{\text{MISS}}^2 \tag{4.21}$$

In calculating effectiveness, Eq. (4.21) may be used only for the delivery of a single munition, not for multiple munitions such as air-delivered sticks, volleys fired from artillery cannons, or multiple rounds from a machine gun. In those cases, the two errors have to be treated separately.

4.9 *EFFECT OF VERTICAL ERRORS ON GROUND PLANE ACCURACY*

Suppose we have a GPS-guided weapon that is used to attack a target on the ground plane. We might know, for example, that the ground plane CEP of such a weapon is 10 ft; this would indicate an accuracy that would result in a high probability of damaging the target. However, a GPS weapon guides to a point in space in terms of (x, y, z) coordinates, and although the accuracy in x and y is well represented by the CEP, inaccuracies in z may degrade the expected ground plane accuracies. Consider the terminal phase of this weapon's trajectory, as shown in Fig. 4.24.

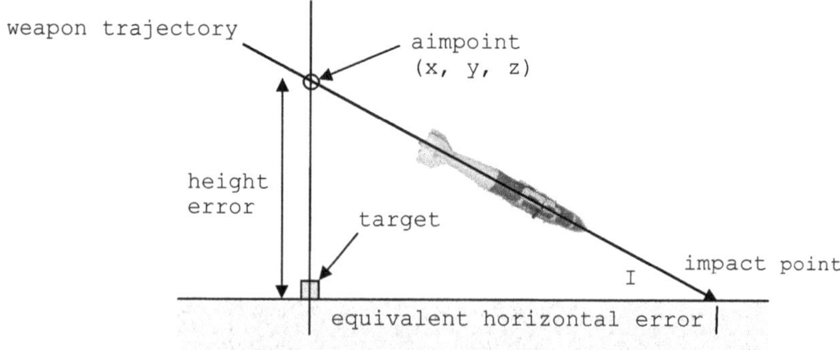

Fig. 4.24   Resolving the elevation error into the ground plane.

In this figure, the weapon is impacting at a relatively shallow angle. Although it navigates precisely to the on-board target coordinates (x, y, z), an error in the z-coordinates (HEP) of the target results in the weapon passing the aimpoint to impact on the ground plane downrange of the actual target. The specified ground plane range error (REP) therefore has to be aggregated with this additional, projected range error and is represented by REP′. Although a trajectory program should be used to translate the vertical error into a horizontal error, it is usually assumed that the distances involved are small, so the terminal trajectory is approximately a straight line. This allows the quantities to be related through the impact angle as follows:

$$\mathrm{REP}' = \frac{\mathrm{HEP}}{\tan I} \tag{4.22}$$

The HEP produces a projected error in the ground plane in the range direction only REP′, which may then be RSS with the existing REP value.

$$(\mathrm{REP}_{\mathrm{TOTAL}})^2 = (\mathrm{REP}')^2 + (\mathrm{REP})^2 \tag{4.23}$$

Figure 4.25 shows the effect of 1000 impacts of a weapon, first with zero elevation error and then a 50-ft elevation error (sigma) on the impact points on the ground. Clearly these height errors are magnified for small impact angles and are mitigated for high impact angles. It is usual to plan for GPS-guided weapons to impact at high angles to reduce this effect as much as possible.

Note that the errors in the deflection direction are unaffected by impact angle; therefore, we may write

$$\mathrm{DEP}_{\mathrm{TOTAL}} = \mathrm{DEP} \tag{4.24}$$

Note that a weapon that may have been circular normal in the ground plane before considering the height error may no longer be circular normal, as indicated in Fig. 4.25.

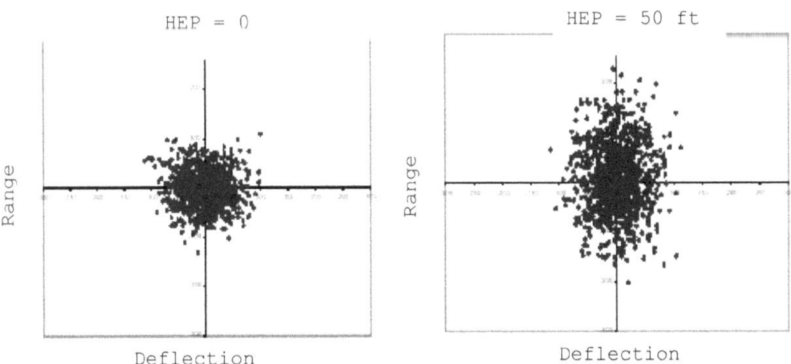

Fig. 4.25 Effect of height error on ground plane impacts.

4.10 GENERATING WEAPON SYSTEM ACCURACIES

Weapon systems currently in use may be categorized at the most general level as being either air- or surface-launched, guided or unguided. In this context, surface-launched applies to ground-fired, surface ship-launched, and submarine-launched whereas air-launched encompasses fixed and rotary wing platforms, man in the loop or autonomous. For all these categories, there are several sources from which accuracy data may be obtained.

- Contractor-specified data (spec data)
- Field tests or trials
- Predictions from analytical models

In weapon development work, we assume the spec value is always available and forms part of the milestone requirements typically used to allow the weapon procurement process to proceed. In developmental and operational testing, the spec value is checked against limited test data to ensure compliance with contract requirements. This value may be specified for one or a few cases of weapon use, but probably not all possible ways to use it.

For example, reconsider Fig. 4.21, where the weapon is the M240 machine gun. The spec value may be obtained from the Mann barrel test; however, in operational use this gun may have a dozen different mounts, each resulting in a different accuracy. The best solution would be an analytical model that represents each mount by a characteristic "stiffness" to obtain the accuracy for that particular gun mount.

4.10.1 UNGUIDED MUNITIONS

The need for a predictive model to produce delivery accuracy (DA) data, at least for unguided weapons, is further reinforced by considering the issue of determining the accuracy of an aircraft delivering unguided bombs on a target, as illustrated in Fig. 4.11. Suppose we consider a test in which a single Mk-82 bomb is dropped from an F/A-18 traveling at 400 kt, 5000 ft altitude, straight and level flight. In order to generate REP and DEP, we would have to perform a number of drops to produce impact points relative to a DPI such as those shown in Fig. 4.2. This raises the following issues regarding the test procedure:

- If we need 30–50 impact points to form a set of data sufficiently large to get values from which REP and DEP could be calculated, it is very unlikely that each drop would have identical release conditions.
- Even if that issue can be overcome, it would give us DA data for just that release condition. What about releases at any altitude between 500 ft and 20,000 ft, release speeds between 100 kt and 500 kt, and any loft or dive angle?

- The only way to address these problems is to have a predictive model (computer program) that can take any point values of release conditions from a spectrum of possible values and calculate a unique accuracy.

Although it is beyond the scope of this book to describe predictive models to overcome these issues, some are described in the *Introductory Weaponeering* and *Advanced Weaponeering* textbooks; the interested reader is referred there for more information. The JMEM Weaponeering System (JWS) includes many of these models dealing with air- and surface-launched unguided projectiles, GPS-guided weapons, laser-guided weapons, and A/C-130 gunship weapon systems and allows estimation of accuracy data over the whole spectrum of operational conditions.

Even with the availability of these models, the operational community often requires some field testing to ensure the models give reasonable results, at least for selected delivery conditions. It is easily realized that these tests are expensive to perform, at least for large weapon systems like bombs, missiles, and rockets, although small caliber gun–type systems are less expensive.

These models deal mostly with unguided weapons such as bombs, artillery projectiles, and gun systems, where in order to get good accuracy, several parameters need to be known accurately:

- Where the releasing/firing platform is located
- The direction and velocity of the weapon when released/fired
- Where the target is located when the weapon arrives
- A trajectory model to connect these two locations

Predictive models usually are based on what is called an *error budget methodology* in which the variability in each of the four items listed are estimated and then statistically combined with an RSS process to predict total system accuracy.

4.10.2 GUIDED MUNITIONS

Guided weapons are different from unguided in several important ways. The existence of a launch-acceptable region (LAR) rather than a ballistic point means that guided weapon accuracy is independent of the three requirements listed previously and depends only on the ability of the guidance system to navigate the weapon to the target. This implies, for example, a GBU-38 500-lb Joint Direct Attack Munition (JDAM) GPS-guided bomb has the same accuracy at the target whether it is released from a B-2 Stealth bomber or a Cessna 150, assuming it can be released efficiently from the latter.

Additionally, unknown meteorological (MET) conditions, which are usually the largest contributor to unguided weapon accuracy predictions, do not influence guided weapon accuracy because the guidance system can compensate in real time for this effect. The trajectory of a guided weapon is less important than for an unguided weapon because as long as it is released in the LAR, it should get to the target. This statement is, of course, conditional on any required impact conditions needed to penetrate a target.

All of these conditions imply that guided weapons are very robust regarding their employment, and therefore field testing becomes more practical in terms of measuring their accuracy. Guided weapons may be released under a wide range of delivery conditions. As long as the release is inside the LAR and no specific terminal conditions are required, each impact point is valid for calculating accuracy for any release condition.

One final point regarding weapon accuracies needs to be considered. There may be one component of weapon system accuracy that has nothing to do with the weapon itself; for example, a GPS-guided weapon may have excellent accuracy (<1 m), but if imprecise target coordinates are provided (>10 m), the overall system accuracy will be the larger value. It may seem unfair to characterize the weapon accuracy by including effects beyond the influence of the weapon, but in most cases the operational user is only concerned with overall accuracy when planning a mission; therefore, the larger value must be used.

4.10.3 *IDENTIFYING THE CORRECT ERROR TYPE*

Whether predicted with an analytical model or measured with field testing, it is important to consider what accuracy type is actually being generated. For example, towing an artillery cannon out onto the test range, firing 50 rounds, measuring the impact points, and calculating REP and DEP is measuring which parameter? The answer is the precision error. If the cannon were being aimed at a specific location, the difference between the DPI and MPI would be a measure of the MPI error, albeit a single sample.

Such considerations may help to specify the testing procedure. Continuing the example of field artillery, we might ask: If weaponeers ruled the world with unlimited resources, what kind of test would we perform? Such a test is shown in Fig. 4.26.

This test would be conducted over perhaps several days and consist of

1. Take the cannon out onto the range, aim it at the target, fire 12 rounds, and then tow the cannon back to the depot.
2. Repeat the process on successive days to produce the fall of shot shown in the figure.

Each day produces a sample of the MPI error and 12 samples of the precision error. If the test is repeated over a sufficient number of days, we will have sufficient data to calculate REP and DEP for both MPI and precision error types, and perhaps calculate an overall CEP as shown.

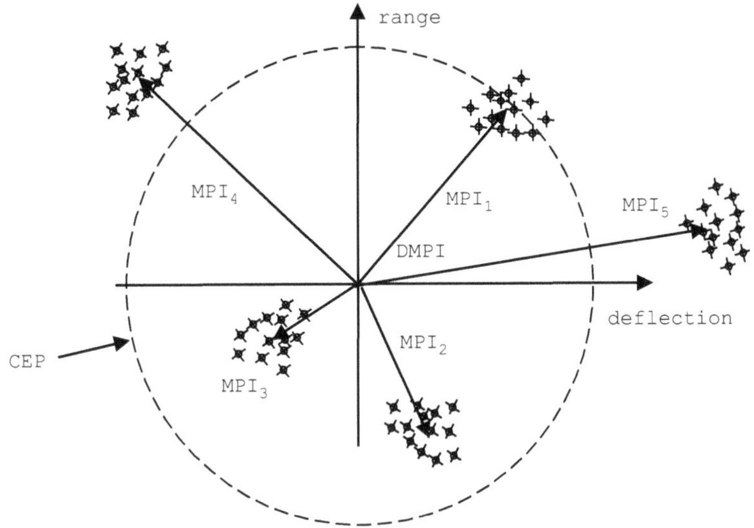

Fig. 4.26 Test to measure MPI and precision errors.

This illustrates that when a test is conducted, it is important to consider which accuracy metric is being measured, and if only one is being measured (MPI or precision), where will the other come from?

4.10.4 MODE-DEPENDENT ACCURACY

Weapon systems may be used in several different ways, or modes, and a different accuracy is associated with each. For example, artillery fire may be implemented using nonadjusted or predicted fire where

- The location of the cannon is known.
- The location of the target is known.
- Given a good trajectory model of the projectile, gun orders are generated telling the gun crew how to slew and elevate the gun to hit the target.

Several rounds may be fired, and unless later observation of the target is available, the testers will never know how accurate their rounds were. In observer adjust mode, a different procedure is employed using a forward observer (FO) who can observe the target. This mode takes the following form:

1. One artillery cannon, using predicted fire mode, fires a single round at the target. This is known as *ranging*.
2. The FO reports back how much the round missed the target in range and deflection.
3. The gun crew adjusts the slew and elevation of the gun to correct for the reported error and fires another ranging round.

TABLE 4.2   SAMPLE INDIRECT FIRE DELIVERY ACCURACY DATA

Impact Conditions				Probable Error (50%)					
				Precision Error (m)		MPI Error (m)			
						Observer Adjusted		Nonadjusted	
Range	Charge	Angle	Velocity	Range	Defl.	Range	Defl.	Range	Defl.
3000	4(M67)	13	246	13.5	1.3	22.3	7.4	29.7	10.1
5000	5(M67)	19	263	12.1	2.7	21.6	7.4	44.5	12.8
7000	6(M67)	26	269	11.5	1.3	21.6	6.7	70.1	20.2
9000	7(M67)	31	278	15.5	2.0	22.9	7.4	80.3	33.7
11,000	7(M67)	44	283	18.2	3.4	25.0	8.1	108.6	51.9
13,500	8(M200)	49	300	35.0	8.0	40.0	12.0	131.0	73.0

4. The process is repeated until the FO decides the last round is sufficiently accurate to be effective.
5. The ranging gun reports corrections to the predictive fire solution to the other cannons in the battery.
6. Several volleys are fired at the target in fire for effect mode.

Because observer adjust mode is correcting for estimated MET conditions, for example, its accuracy is better than for predicted fire mode, as shown in Table 4.2.

Although this example is for unguided artillery fire, different modes also exist for air-launched unguided bombs and some GPS weapon deliveries and, as explained earlier in this chapter, different accuracies will apply to different modes.

4.11   FUZES

Although a brief overview of fuze types was given in Chapter 1, more details regarding the influence trajectory has on fuzing will now be discussed, based on [1]. Although the fuze is a separate device from the weapon, it is important to include any errors in when it detonates, because this affects the position of the weapon relative to the target and therefore represents a component of delivery error.

So far, only the path the projectile takes as it moves through the atmosphere toward the target has been considered, but the precision with which the fuze operates will affect this relative position and hence the effectiveness of the munition. In treating how fuze function affects the location of the warhead at detonation, we will frame the discussion in terms of a Monte Carlo approach; that is, we will observe where the warhead is located when the randomness of the fuze function is allowed to vary over the principle features of the fuze mechanism. In general, the three types of fuzes fitted to a warhead that we will consider here are

- Impact, or point detonating (PD)
- Proximity (PX)
- Time delay

4.11.1 POINT-DETONATING (PD) FUZE

A PD contact or direct action fuze is one that triggers the warhead when it comes into contact with something reasonably hard, like the ground, a building, or a vehicle. It may be located in the nose of the weapon, as in the case of a general-purpose Mk-82 unguided bomb, or in the tail for a bomb designed to penetrate hard materials such as concrete where a nose fuze would be damaged. Different fuze locations produce different detonation waves in the explosive fill and can result in varying damage effects. It may be assumed that there will be a delay between when the weapon first makes contact with the ground and when it detonates due to possible electromechanical components delays, accelerometer response, and signal transmission times. Testing would provide a mean value (say 5ms) for this inherent delay, however there will be a distribution of delay times associated with this mean. Figure 4.27 shows the straight line penetration of a shell into the ground with the distribution of fuze delays along the trajectory.

Depending on the intrinsic delay, impact velocity and trajectory, the weapon may be partially or totally buried when it bursts, with a corresponding mitigation of blast and fragment damage to the target.

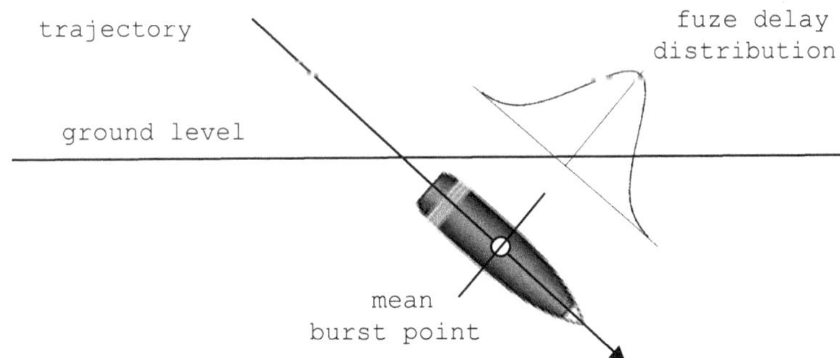

Fig. 4.27 PD fuze with intrinsic delay.

4.11.2 PROXIMITY (PX) FUZE

This fuze type triggers when a particular threshold is exceeded and is further divided into the following categories:

- Variable time (VT)
- Sensor trigger

For the VT fuze, the threshold is fixed after firing and the timer may be mechanical (MT) or electronic. For the MT fuze, the time is set manually prior to firing the round, whereas an electronic timer may be set with software. Early VT fuzes were designed for air-bursting artillery shells in World War I, when it was observed that lethality was increased considerably compared to PD fuzed shells. In order to correctly set the fuze, some knowledge of the shell trajectory and desired height of burst (HOB) would need to be known, as illustrated in Fig. 4.28.

In sensor fuzes, the threshold needed to detonate the warhead comes from a more sophisticated device compared to a simple timer. In many cases, the sensor is a small radar system that estimates the distance to the target (or ground) and initiates detonation when this distance is smaller than the set threshold value. These fuzes are typically used in antiaircraft weapons or the air-bursting scenario shown in Fig. 4.28, where there is less dependency on having accurate trajectory information.

Given the trajectory of the projectile as in Fig. 4.28, the time of flight to the point above the target may be calculated; this would be the VT fuze setting. The variability is therefore in the accuracy of the timer in the fuze about its nominal, or desired, value. Assuming this variability is Gaussian in nature, it may be represented by Fig. 4.29.

Given the randomness of delivery shown in Fig. 4.1 and adding in the variability in fuze timing, the expected detonation points for a Monte Carlo simulation would be as shown in Fig. 4.30. In these graphs, the effect of fuze variability is shown in the horizontal and vertical planes with the nominal burst at coordinates $(0, 0, 2)$, that is, 2 m above the ground.

Note that for long fuze time errors, an airburst is not achieved, and the round impacts the ground plane.

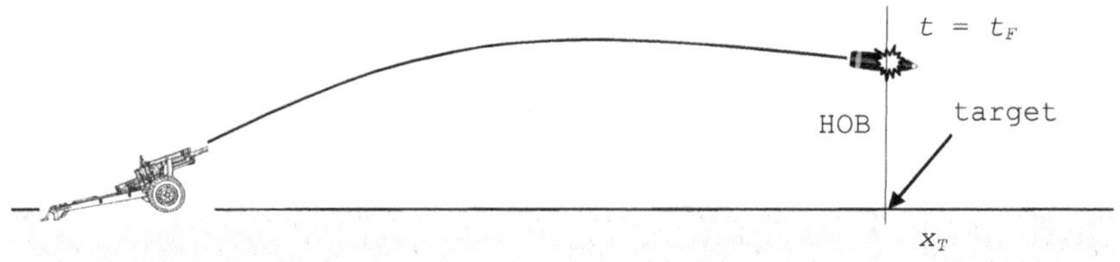

Fig. 4.28 Airburst with PX fuze.

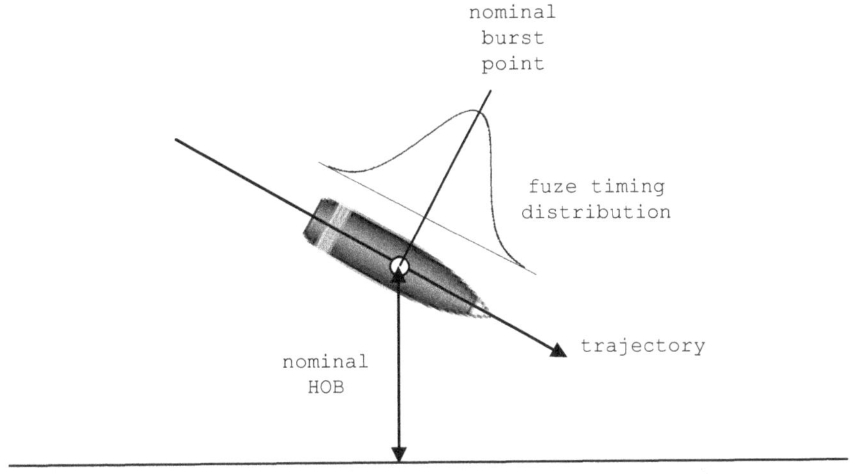

Fig. 4.29 VT fuze model.

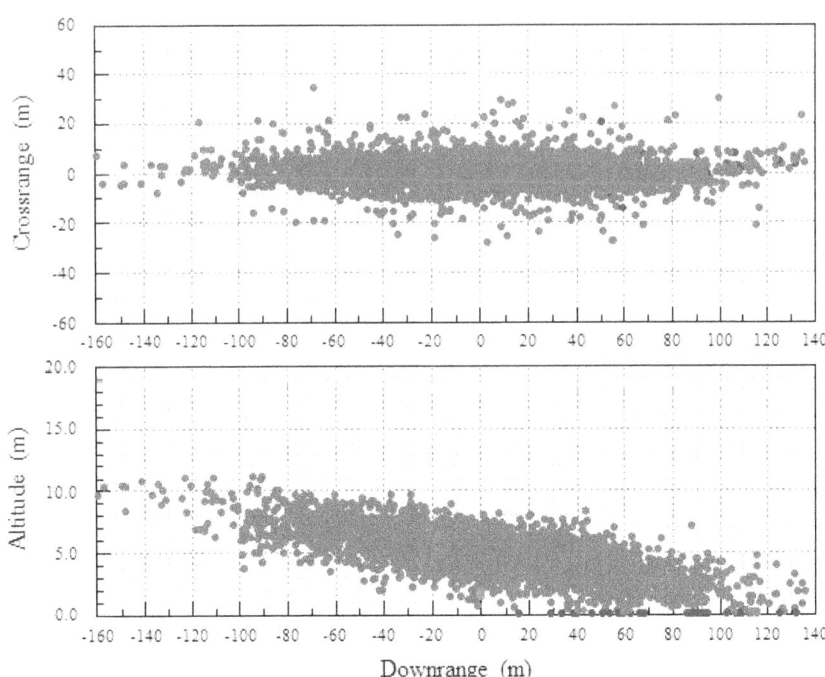

Fig. 4.30 Simulated detonation points for VT fuze model.

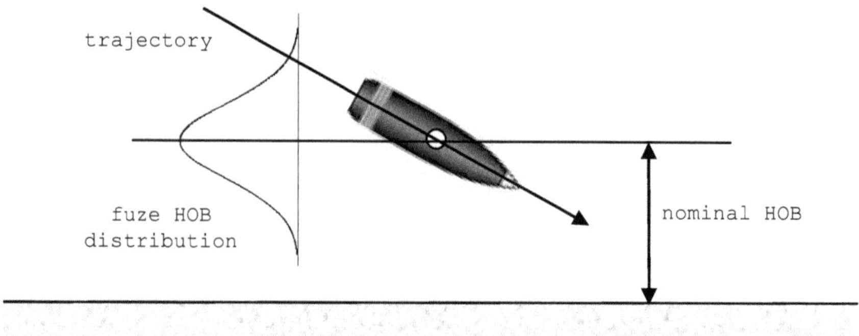

Fig. 4.31   HOB fuze model.

For a sensor-triggered air-bursting round, again the objective is to achieve an airburst of the type shown in Fig. 4.29, but the employment of an altitude sensor may give better accuracy at detonation, as shown in Fig. 4.31.

Again, the variability of detonation point in the horizontal and vertical directions may be simulated, as shown in Fig. 4.32 for a nominal 10-m HOB.

For the case shown, where the desired burst point is at (0, 0, 10), there is greater spread for the HOB fuze compared to the VT fuze, as would be expected for shallow approach angles of the warhead.

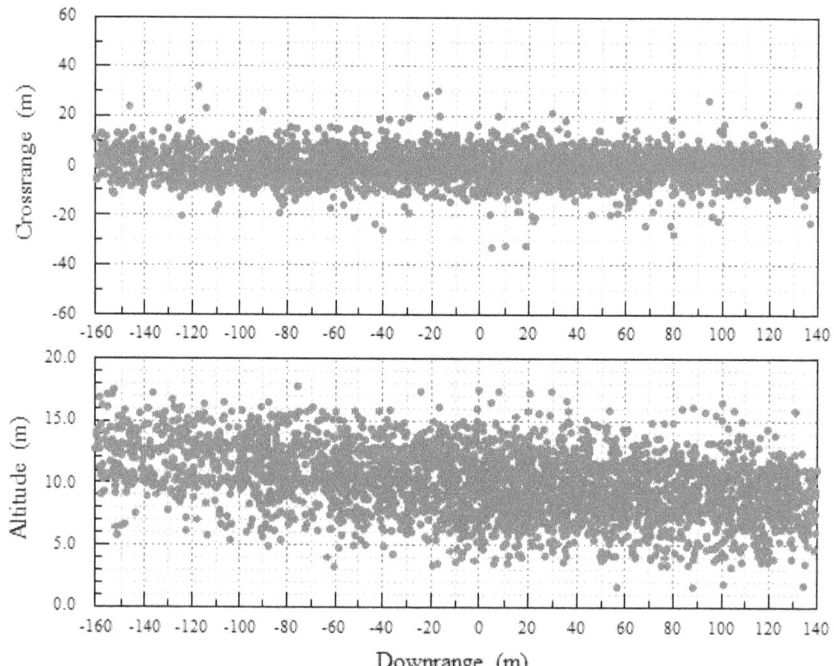

Fig. 4.32   Simulated detonation points for HOB fuze model.

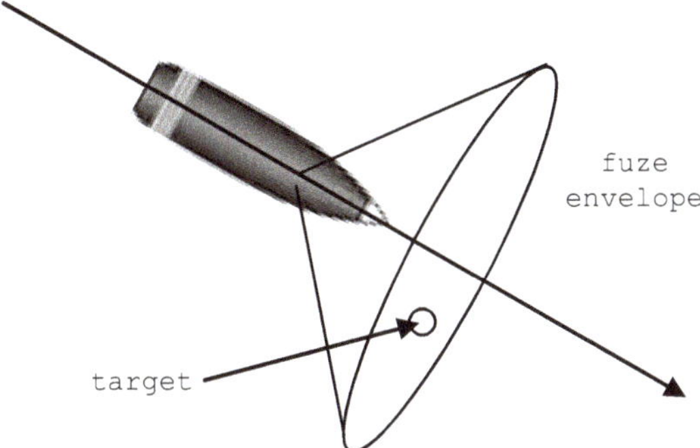

Fig. 4.33   Proximity fuze detection envelope.

For an antiaircraft round, the model is more complex than the previous ones because we need to represent the probability the target is detected within the fuze sensing envelope. The assumption here is that the detection envelope may be represented by a cone by specifying a cone angle and maximum radius, and the target is detected when it is inside the cone, as illustrated in Fig. 4.33.

In this figure, the target may consist of a physical object, in which case its geometry is required to determine the first part of the target to trigger the fuze, or simply the target centroid.

Clearly, the process of determining when the fuze will function is more complex than for other fuze types. Typically, the velocity vector of the weapon in its terminal state is determined from trajectory calculations. Then the weapon is considered fixed and its velocity imposed on the target. The fuze functions when the target intersects the conical volume of the envelope.

Figure 4.34 shows simulated detections where the solid area represents the detection volume, and the individual dots represent target locations that do not fuze the warhead.

4.11.3   TIME DELAY FUZE

A delay fuze is basically one with a fixed time delay following impact and the subsequent initiation of the PD fuze. Again, the delay mechanism may be set mechanically or electrically. These fuzes are used against targets where a small delay following impact can increase the effectiveness of the warhead against the target. Examples include the following:

- For a building target, delaying detonation until the warhead is inside the structure rather than detonating at impact on the roof. This may result in preventing the functions inside the building without its total demolition.
- For a bridge target, detonation just below the span initially lifts it up, but then the downward motion imparts a greater loading than an identical

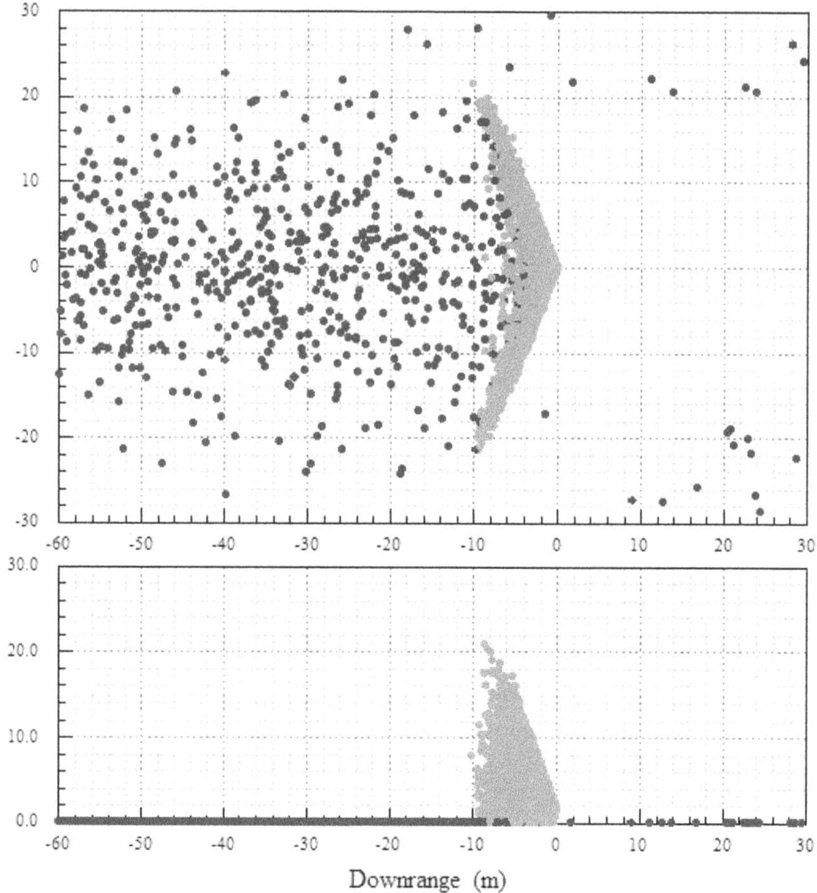

Fig. 4.34   Simulated detonation points for PX fuze model.

detonation on top of the span. This may result in the whole span collapsing rather than simply making a hole in the deck.

- For a ship target, greater damage will result from an internal detonation compared to a surface contact detonation and may initiate additional damage mechanisms such as fire or secondary explosions from a magazine or fuel tanks.

The required delay is usually achieved electrically by charging a capacitor circuit initiated by impact, as shown in Fig. 4.35.

It may be seen from this discussion that the type of fuze used to detonate a warhead has to be specified along with the variability in fuze mechanism it uses to operate. As an example, a warhead traveling at 1000 m/s with a variable time fuze having a 50-ms delay error in a shallow approach angle will overshoot the target by 50 m. This will have a considerable effect on the damage sustained by the target.

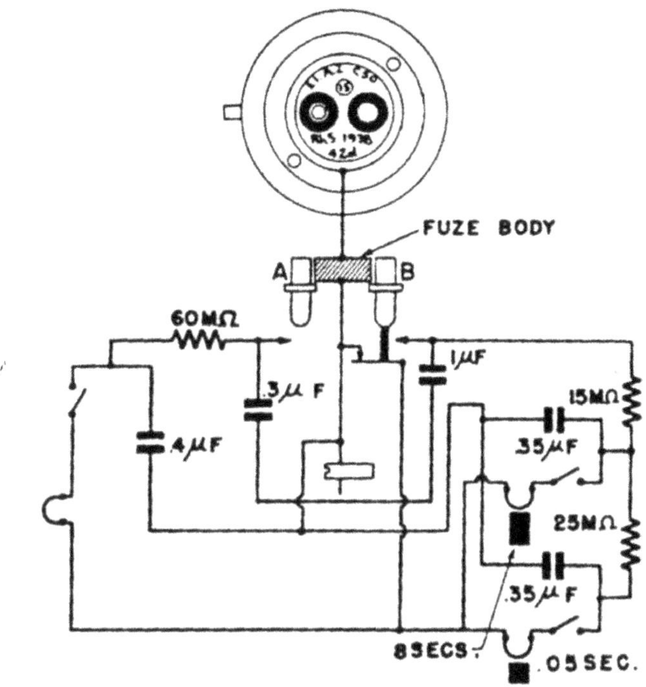

Fig. 4.35   Electrical circuit for time delay fuzes.

4.12   CHAPTER SUMMARY

- In its simplest form, a weapon accuracy in the ground plane may be described in terms of the standard deviations in range and deflection; however, the use of error probable measures is more common.

- It is usually assumed that range and deflection accuracy data are normally distributed; however, the bivariate distribution is not necessarily circular.

- For projectiles that do not normally land in the ground plane (guns), a normal plane accuracy in the angular measure mil is more appropriate.

- A means to describe non-normal delivery accuracy data was described and termed the P_{HIT}/P_{NM} methodology.

- The concepts of launch acceptable region (LAR) and footprint were described.

- It was explained why most weapon deliveries are described by two distributions, MPI and precision, and that in order to calculate effects due to multiple weapon deliveries, these accuracies have to be known and used separately.

- The accuracies of several air-to-surface and surface-to-surface weapons were described through examples.

- The weapon fuze affects its position relative to the target at detonation, so fuze functioning accuracy is considered part of the overall delivery error, and several common fuzes were described.

REFERENCE

[1] Rowles, S., *Joint Gun Effectiveness Model with Air-Bursting Methodology—User Manual*, Systems Concepts and Analysis Branch, Naval Surface Warfare Center, Dahlgren, VA, 1995.

Chapter 5

ACCURACY OF GPS/INS GUIDED MUNITIONS

5.1 OVERVIEW OF THE GLOBAL POSITIONING SYSTEM (GPS)

The GPS is a system of 24 satellites, operated by the U.S. Department of Defense (DOD), that orbit the Earth at an altitude of approximately 12,000 miles, repeating the same path over the ground twice a day. At each location on the Earth's surface there is a minimum of 5 and a maximum of 11 satellites in view at any one time. Each satellite transmits a microwave signal that contains information including the satellite location and precise time the signal was sent. This signal is received by a GPS receiver, and the time of arrival is determined by the clock in the receiver. The microwave transmission travels at the speed of light c, so the distance, or range from the satellite to the receiver, is given by

$$L = c \times \Delta t \tag{5.1}$$

Here Δt is the transmission time as determined by the difference in the two clock times including Einstein's relativistic effects due to the motion of the satellite and receiver. The satellite orbits are precisely controlled so that at any instant the position of each satellite is known relative to a set of coordinate axes that are Earth centered (geocentric), Earth fixed (ECEF). In this frame, which rotates with the Earth, the x-axis points from the center of the Earth through the point of intersection of the equator and the prime meridian (0 deg longitude). The y-axis is at right angles to the x-axis in the equatorial plane, and the z-axis makes up the right-handed orthogonal frame (see Fig. 5.1).

The satellite signal also contains other information, such as its identification number, its "health," and where it is in x-y-z coordinates relative to the ECEF frame. The GPS receiver takes these signals from the satellites and

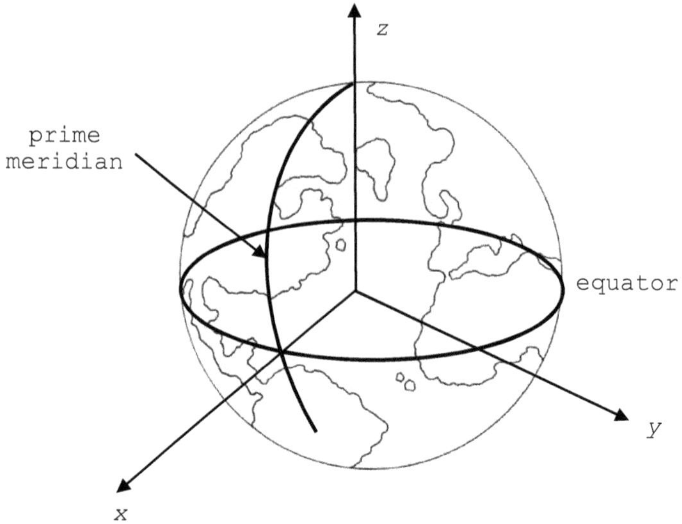

Fig. 5.1   The ECEF frame.

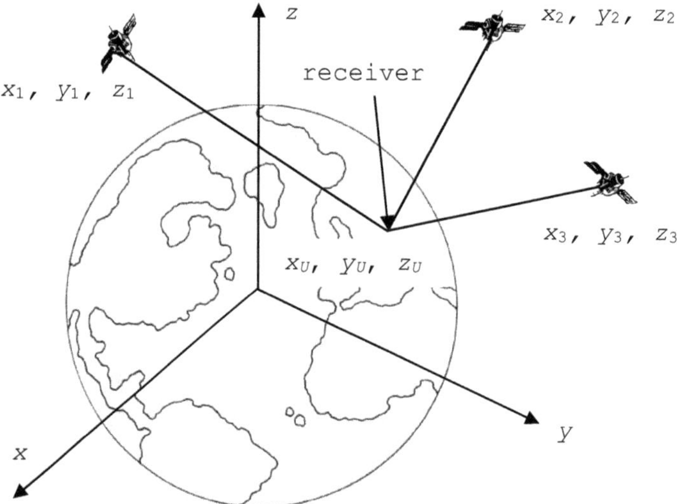

Fig. 5.2   Locating the receiver.

calculates the range to each one using Eq. (5.1). This can then be used to trilaterate[1] the position of the user receiver (U) with respect to the satellites, and hence with respect to the ECEF coordinate axes, as indicated in Fig. 5.2.

If the ranges are calculated accurately, then only three satellites are needed for the trilateration calculation.

---

[1] Trilateration means using three distances to determine a position, as opposed to triangulation, which uses three angles.

5.2 DETERMINING POSITION WITH NO ERRORS PRESENT

Figure 5.3 shows the ECEF frame, three satellites, a user receiver, and ranges calculated by the receiver to the three satellites. The analysis that is presented follows that of Tsui [1].

The distance between each satellite and the receiver is called the *pseudorange* ρ and is calculated using the Pythagoras theorem in three dimensions. If (x, y, z) are the coordinates of a satellite, then

$$(x - x_U)^2 + (y - y_U)^2 + (z - z_U)^2 = \rho^2 \tag{5.2}$$

Using Eq. (5.2), and in the absence of any ranging errors, we may write the following relationships for the location of the receiver (required), the locations of the three satellites (known), and the pseudoranges (calculated):

$$\rho_1 = \sqrt{(x_1 - x_U)^2 + (y_1 - y_U)^2 + (z_1 - z_U)^2} \tag{5.3}$$

$$\rho_2 = \sqrt{(x_2 - x_U)^2 + (y_2 - y_U)^2 + (z_2 - z_U)^2} \tag{5.4}$$

$$\rho_3 = \sqrt{(x_3 - x_U)^2 + (y_3 - y_U)^2 + (z_3 - z_U)^2} \tag{5.5}$$

If the position of the satellites and ranges are known, Eqs. (5.3) through (5.5) may be solved for the three unknowns x_U, y_U, and z_U. The actual solution of these equations is somewhat complex and will not be dealt with here.

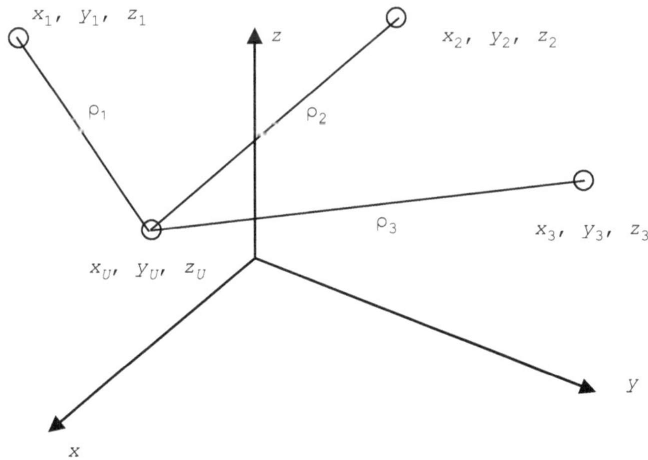

Fig. 5.3   Ranging from three satellites to a user receiver.

5.3   PSEUDORANGE AND CLOCK ERRORS

Considerable effort is expended to ensure that the atomic clocks in each of the satellites are synchronized, but it is impossible to do the same for the clocks in the user's receiver. From Eq. (5.1), it may be seen that very accurate measurement of Δt is needed; in fact, a 1-ns error in this quantity will result in about 1 ft of error in the range calculation. It is inevitable that although the receiver clock will have adequate precision, it will not be synchronized with the satellite clocks. Therefore, the constant error between the receiver and satellite clocks has to be considered an unknown because it cannot be determined from any data the satellite transmits.

We assign this error the symbol b_{UT} as the bias in the user receiver clock (in seconds) that will result in an error in the range calculation. Equation (5.3) has to be rewritten as

$$\rho_1 = \sqrt{(x_1 - x_U)^2 + (y_1 - y_U)^2 + (z_1 - z_U)^2} + b_U \qquad (5.6)$$

Here, b_U is the error in pseudorange due to the receiver clock bias, measured in units of distance, hence

$$b_U = c \times b_{UT} \qquad (5.7)$$

Because b_{UT} is unknown, so is b_U; therefore, we are going to need four, not three, equations to solve for the position of the receiver and the clock bias. This requires pseudoranges from at least four satellites. We can still use Eq. (5.6) applied to the original three satellites, but we must write another equation for the fourth satellite to range to.

$$\rho_4 = \sqrt{(x_4 - x_U)^2 + (y_4 - y_U)^2 + (z_4 - z_U)^2} + b_U \qquad (5.8)$$

This set of four equations is now solved for the four unknowns, the (x_U, y_U, z_U) position of the receiver, and the common, unknown receiver clock bias b_U.

5.4   USER POSITION IN GEODETIC AND LOCAL COORDINATES

By itself, x-y-z data as derived in the previous section are of little use because we need to know the location of the receiver with respect to Earth's surface, not with respect to the ECEF frame, so the receiver converts them to more convenient latitude, longitude, and height. This geodetic frame may be considered a 3-D polar coordinate frame, as opposed to the Cartesian ECEF frame (see Fig. 5.4). This geodetic frame is also referred to as a global reference frame.

In order to do this transformation, we must make some assumptions regarding the shape of the Earth. The Earth is not a sphere; it is slightly "flattened"

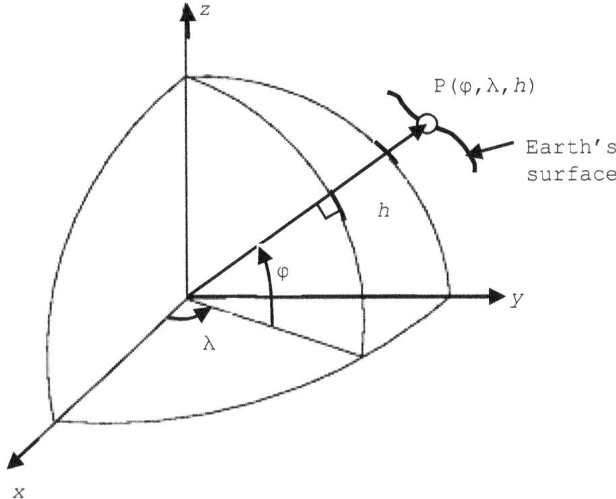

Fig. 5.4 Latitude φ, longitude λ, and height h coordinate frames.

at the poles, so a convenient mathematical representation of the Earth is an ellipsoid. There are many geodetic ellipsoidal representations of the Earth going back to the early 1800s, but the one used by GPS is the World Geodetic System–1984 (WGS-84). This ellipsoid may be considered to be a smooth version of the Earth's surface and is defined by the equatorial and polar radii a and b respectively, which for WGS-84 are $a = 6{,}378{,}137.0$ m and $b = 6{,}356{,}752.3$ m. From these data, two other properties of the ellipsoid can be calculated, the flattening factor f and eccentricity e.

$$f = \frac{a - b}{a} \tag{5.9}$$

$$e^2 = 2f - f^2 \tag{5.10}$$

This allows the calculation of latitude φ, longitude λ, and height above ellipsoid (HAE) h to be calculated from the ECEF (x, y, z) coordinates as follows. First, we calculate

$$p = \sqrt{x^2 + y^2}, \quad \theta = \tan^{-1}\left(\frac{z \times a}{p \times b}\right), \quad N(\varphi) = \frac{a}{\sqrt{1 - e^2 \sin^2 \varphi}} \tag{5.11}$$

Then the latitude φ, longitude λ, and height h are derived as follows:

$$\varphi = \tan^{-1}\left[\frac{z + e^2 b \sin^3 \theta}{p - e^2 a \cos^3 \theta}\right] \tag{5.12}$$

$$\lambda = \tan^{-1}\left(\frac{y}{x}\right) \qquad (5.13)$$

$$h = \frac{p}{\cos\varphi} - N(\varphi) \qquad (5.14)$$

Note that height is height above ellipsoid (HAE), so in order to specify the coordinates of a specific point on the Earth's surface, we would need to know the height of that point relative to the WGS-84 ellipsoid. In fact, some points on the Earth's surface are below the ellipsoid whereas others are above, which is why the ellipsoid may also be thought of as a smoothed average representation of the surface of the Earth.

It is worth considering the vertical reference further when attempting to specify a location on or above the Earth's surface. The WGS-84 ellipsoid is the datum used by GPS, and all height data that the GPS calculates are HAE, which may be positive or negative. HAE is measured along a normal to the ellipsoid passing through the point (latitude, longitude) of interest. Many systems, such as aircraft, reference mean sea level (MSL) as the vertical datum. Actual sea level at a particular location will vary with time due to tides and also due to variations in gravity over the surface of the Earth.[2] The WGS-84 ellipsoid does not coincide with MSL. One of the functions of the National Geospatial-Intelligence Agency (NGA, formerly National Imagery and Mapping Agency) is to provide data on HAE for all points over the surface of the Earth. These reference heights are illustrated in Fig. 5.5.

Another useful frame of reference needed to analyze GPS/inertial navigation system (INS)-guided weapons is termed the *local frame*. This reference frame origin is located at some reference point of interest, such as a target on the surface of the Earth with axes pointing east, north, and up (ENU). This frame is needed when discussing weapon accuracy, because a statement such

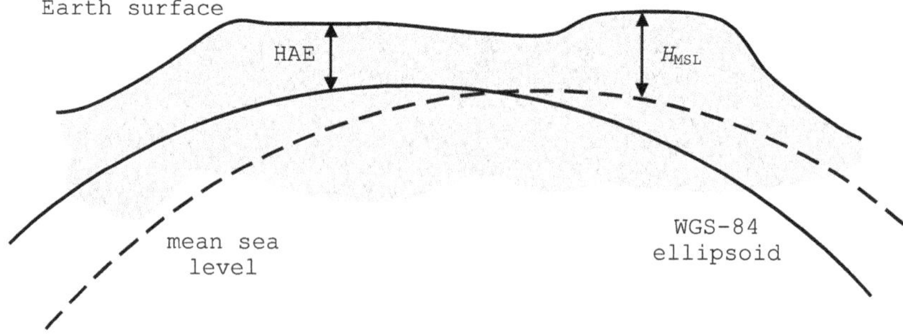

Fig. 5.5   Height relative to different datum.

---

[2]Even in the absence of waves, the surface of the ocean is not locally flat.

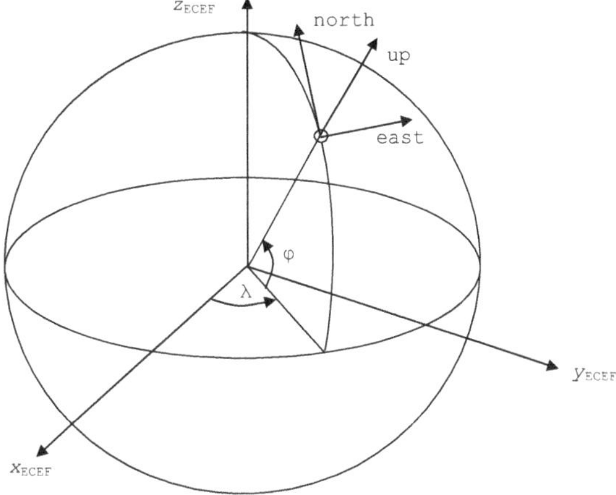

Fig. 5.6 ECEF, global, and local coordinate frames.

as "the weapon missed the target by 50 ft" is meaningless in anything other than the local frame. Sometimes an airborne local frame may be specified as north, east, and down (NED). The three most common frames used are ECEF (x, y, z), geodetic (latitude, longitude, HAE), and local (east, north, up). These are shown in Fig. 5.6.

The transformation from ECEF to geodetic coordinates is defined in Eqs. (5.9) through (5.14). To transform from ECEF coordinates to the local coordinates, we need a local reference point, such as the location of a target. If the origin of the ENU frame is located at the target, then the ENU coordinates are obtained from the latitude, longitude, HAE coordinates as follows:

$$\begin{bmatrix} x \\ y \\ z \end{bmatrix} = \begin{bmatrix} -\sin\lambda & \cos\lambda & 0 \\ -\sin\varphi\cos\lambda & -\sin\varphi\sin\lambda & \cos\varphi \\ \cos\varphi\cos\lambda & \cos\varphi\sin\lambda & \sin\varphi \end{bmatrix} \begin{bmatrix} x_p \\ y_p \\ z_p \end{bmatrix} \qquad (5.15)$$

Considerable care must be taken in specifying target and weapon coordinates because their reference frames must also be defined, and if necessary converted to the common WGS-84 axes used by the GPS system.

5.5 ACCURACY OF GPS-MEASURED LOCATIONS

The accuracy with which a GPS receiver supplies coordinates will depend on the satellite geometry from which it receives ranging. Better accuracy is obtained if the ranging satellites are widely spaced apart rather than close together, as shown in Fig. 5.7.

This effect is called *dilution of precision (DOP)*. Because the satellites all move around the Earth, for a given location the satellite constellation relative

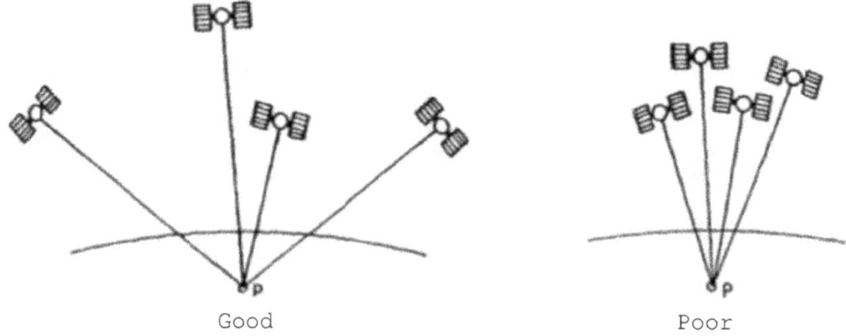

Fig. 5.7   Effect of satellite geometry on GPS accuracy.

to that point will vary; therefore, so will GPS accuracy within, for example, a 24-h period. Because the orbits are accurately known, it is possible to predict DOP values for a particular point on the Earth's surface for a given time based on the constellation at that location.

The accuracy with which the pseudorange is calculated is a random variable because it will vary with range and the amount of atmosphere the signal travels through. We can therefore calculate a standard deviation σ_ρ of the pseudorange variability. The DOP value relates this standard deviation to that of the (x, y, z) local coordinates as follows:

$$\sigma_x = x_{\text{DOP}}\sigma_\rho \tag{5.16}$$

$$\sigma_y = y_{\text{DOP}}\sigma_\rho \tag{5.17}$$

$$\sigma_z = z_{\text{DOP}}\sigma_\rho \tag{5.18}$$

Some more commonly used dilution of precision parameters are h_{DOP} and v_{DOP} for horizontal and vertical DOP values, respectively. These are derived as follows:

$$h_{\text{DOP}} = \frac{\sigma_h}{\sigma_\rho} = \frac{\sqrt{\sigma_x^2 + \sigma_y^2}}{\sigma_\rho} \tag{5.19}$$

$$v_{\text{DOP}} = z_{\text{DOP}} = \frac{\sigma_z}{\sigma_\rho} \tag{5.20}$$

The positional dilution of precision is defined as the following:

$$p_{\text{DOP}} = \sqrt{h_{\text{DOP}}^2 + v_{\text{DOP}}^2} \tag{5.21}$$

TABLE 5.1 QUALITY OF DOP VALUES

DOP Value	Rating	Description
<1	Ideal	Highest possible confidence level to be used for applications demanding the highest possible precision at all times.
1–2	Excellent	At this confidence level, positional measurements are considered accurate enough to meet all but the most sensitive applications.
2–5	Good	Represents a level that marks the minimum appropriate for making accurate decisions. Positional measurements could be used to make reliable in-route navigation suggestions to the user.
5–10	Moderate	Positional measurements could be used for calculations, but the fix quality could still be improved. A more open view of the sky is recommended.
10–20	Fair	Represents a low confidence level. Positional measurements should be discarded or used only to indicate a very rough estimate of the current location.
>20	Poor	At this level, measurements are inaccurate by as much as 300 m with a 6-m-accurate device (50 DOP×6 m) and should be discarded.

Note also that the velocity of the receiver may be determined from the Doppler shift of the transmitted signal carrier frequency, which is particularly useful for aircraft avionic systems and for weapons attacking a moving target.

Clearly, we want the DOP values to be as small as possible; sample recommended DOP values are shown in Table 5.1.

The DOP is constellation specific, so approximate location and date/time is required for its calculation. For some attack situations, such as time-sensitive targets (TSTs) or targets of opportunity, this information may not be available in a timely manner. A single "average" value of DOP may be made by averaging DOP values over the surface of the Earth at varying times [3]. The results are shown in Table 5.2, which indicates that DOP values less than unity are achievable. This implies that the accuracies in the ENU frame are better than the ranging accuracy to a single satellite.

At a fixed location on the Earth's surface, the DOP values will vary with time due to the changing constellation around that point. A typical variation in h_{DOP} over a fixed point on the Earth's surface is shown in Fig. 5.8.

For particularly critical strike missions that use GPS/INS guided weapons, and where the variation of DOP is in excess of 2:1, the time of attack may be influenced by this variation [i.e., attacking when the weapon accuracy is maximized (minimum DOP)]. For the example shown in Fig. 5.8, an attack at 0300 will have GPS accuracies three time better than an attack at 1300.

**TABLE 5.2   AVERAGE OVER
SPACE AND TIME DOP VALUES**

DOP Type	Average Value
x_{DOP} (east)	0.6
y_{DOP} (north)	0.7
v_{DOP} (up)	1.8
t_{DOP}	1.1
h_{DOP}	1.0
p_{DOP}	2.0
g_{DOP}	2.3

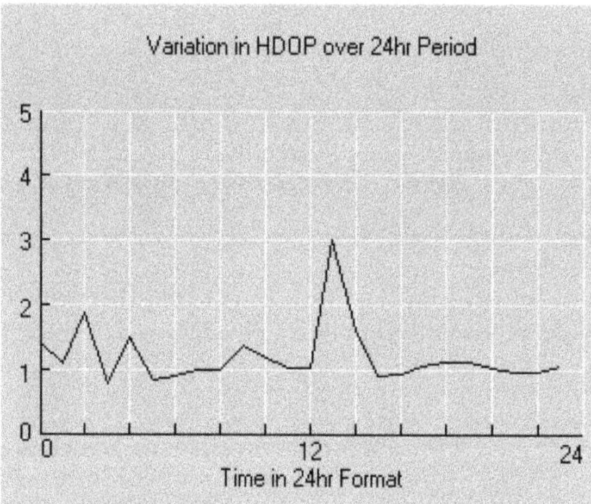

Fig. 5.8   Variation of h_{DOP} with time.

5.6   DIFFERENTIAL GPS

Differential GPS (DGPS) is used to increase the accuracy of a GPS receiver by correcting some of the error in calculating the pseudorange. To implement DGPS, a receiver station, sometimes referred to as a beacon, is located at an accurately mensurated point on the ground, and the error between where GPS says the location is and the known, correct position is determined. Once the calculated error is known, it can be transmitted to all receivers capable of accepting DGPS signals. Alternatively, the beacon may be a high quality GPS receiver that ranges to a large number of satellites over a period of several hours and therefore determines a mean position through self-survey.

The accuracy of GPS receivers depends on their cost—clearly a system for one-time use, as in a bomb, is cheaper than a high precision aircraft navigation system that has to operate for years. Table 5.3 shows how the total accuracy is made up of several contributing components and indicates typical commercial and military system accuracies.

TABLE 5.3   MILITARY GPS ACCURACY

Receiver Error Source	Commercial GPS Ranging Error (m)	Military GPS Ranging Error (m)	Military DGPS Ranging Error (m)
Satellite clock	1.0	0.4	0.0
Orbit—ephemeris	1.0	0.4	0.0
Troposphere	1.0	0.5	0.0
Ionosphere	10.0	0.5	0.0
Receiver noise	1.0	0.4	0.4
Multipath	0.5	0.5	0.5
Total RSS error	10.2	1.11	0.64

The accuracy of GPS systems such as those shown in Table 5.3 is very good; however, this navigation (NAV) error is but one component that contributes to the overall accuracy of a GPS-guided weapon, as discussed in the next section.

5.7   GPS/INS-GUIDED MUNITIONS

Although some weapons are described as GPS-guided, they are actually GPS/INS-guided, which means that the weapon is steered to the target by a conventional inertial navigation/guidance system. The shortcoming of INS-guided systems is their drift over time, so the GPS receiver constantly updates the INS with accurate position and velocity information, thereby eliminating the drift problem. The term *GPS-aided guidance* seems more appropriate for this arrangement. Functionally, the guidance system is as shown in Fig. 5.9.

Typically, the update rate for what is known as a tightly coupled system is several hundred times per second, but obviously in the presence of GPS jamming the accuracy will degrade to that of the INS system alone, drifting over the time from when jamming began.

GPS/INS weapons will maneuver to a specified target with a total error that is determined largely by three independent components:

1. The inherent accuracy of the on-board weapon GPS system, termed the
   navigation error (NAV)

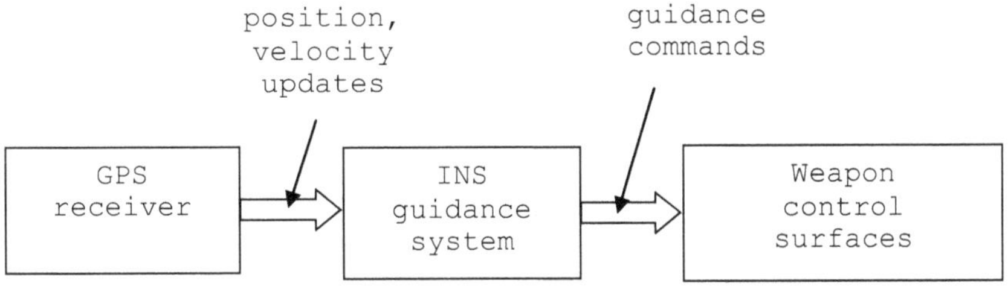

Fig. 5.9   GPS/INS guidance system.

2. The ability of the *guidance and control (G&C)* system of the weapon to maneuver the weapon to the target coordinates
3. The error with which the coordinates of the target are "known," termed the *target location error (TLE)*

Estimates of all three error sources in both horizontal and vertical directions are needed in order to predict the total accuracy of GPS-guided munitions. It may seem unfair to include the effect of TLE because this is weapon independent, implying that however accurate the weapon is, a bad estimation of target coordinates may render it ineffective. However, operational users need an overall accuracy for the weapon system; therefore, TLE is always included in this estimate.

Because the three error sources are independent, we may combine them in an RSS process and write the equation for the total error as follows:

$$\text{Error}_{\text{TOTAL}}^2 = (\text{NAV})^2 + (\text{G\&C})^2 + (\text{TLE})^2 \tag{5.22}$$

In this equation, the error components must be in consistent units such as standard deviations (sigma), circular error probable (CEP), or CE90.

The first of these error sources, NAV, is dependent on the location and number of satellites from which the position of the weapon GPS derives its location, as well as the quality of the receiver, as discussed and characterized in the previous sections of this chapter.

The second error source is the G&C error. Typically, the weapon guidance system determines the difference between its position using GPS and that of the target and sends this information to the control system, which modifies the aerodynamic surfaces in order to adjust the trajectory. Some munitions have small control surfaces at the front of the projectile known as canards, such as the 155-mm Excalibur artillery round shown in Fig. 5.10. Other

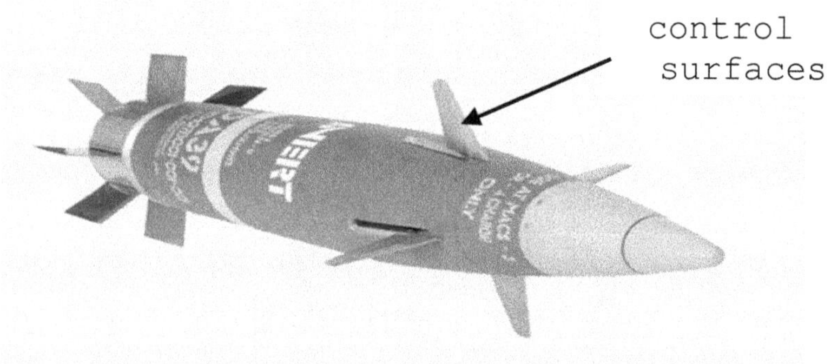

Fig. 5.10   Canard control surfaces on XM982 Excalibur round.

Fig. 5.11   Control surfaces on the GBU-43/B (MOAB).

weapons, such as the GBU-43 shown in Fig. 5.11, have tail-mounted control surfaces.

The ability of the control surfaces to change the trajectory of the weapon will depend on their surface area and the inertial properties of the weapon. The inability of the weapon to maneuver in response to such commands, particularly in the terminal phase when relative position changes quickly, may result in missing the aimpoint. These errors are difficult to quantify or measure, so the weapon manufacturer usually supplies them, based on aggregating many high-fidelity engagement simulations.

The third error source, TLE, is dependent on the source of target coordinates. This may vary considerably from several hundred feet if targets are identified from a map to a few feet if the target has been previously mensurated. Obviously, in combat situations where targets are located in enemy territory, accurate target GPS coordinates may be difficult to obtain. Several interesting schemes have been developed to mitigate the TLE.

Operationally, TLE accuracy is partitioned into several ranges or categories as a way of indicating the quality of the TLE source. These categories (CAT) are defined in Table 5.4.

In this table, the numerical values correspond to both CE90 (circular) and vertical error, 90th percentile (VE90—linear) measures. This terminology often relates to the rules of engagement where, for example, a target may not be engaged with a coordinate-seeking weapon unless the TLE is no larger than a specified CAT level. This becomes important if the attack is in an area that has one or more collateral damage concerns.

TABLE 5.4   TLE CATEGORY (CAT) DEFINITIONS

Category (CAT)	Range (m)
1	0–6
2	7–15
3	16–30
4	31–91
5	92–305
6	>305

5.8   EFFECT OF VERTICAL ERRORS ON TOTAL GROUND PLANE ACCURACY

This topic was discussed in general terms in Chapter 4, Section 4.9; the only modification necessary to expand this approach to include GPS-guided weapons is to recognize that each error contribution (NAV, G&C, TLE) will have its own horizontal and vertical error component. Recall Fig. 4.24 as applied to GPS TLE.

The vertical error is resolved into the ground plane through the impact angle, as shown in Fig. 5.12, and combined (RSS) with the horizontal error to obtain the total range error.

$$\sigma_{R-\text{TOTAL}}{}^2 = \sigma_H{}^2 + \left(\frac{\sigma_\nu}{\tan I}\right)^2 \tag{5.23}$$

Note that this only increases the range error; the deflection error is unchanged.

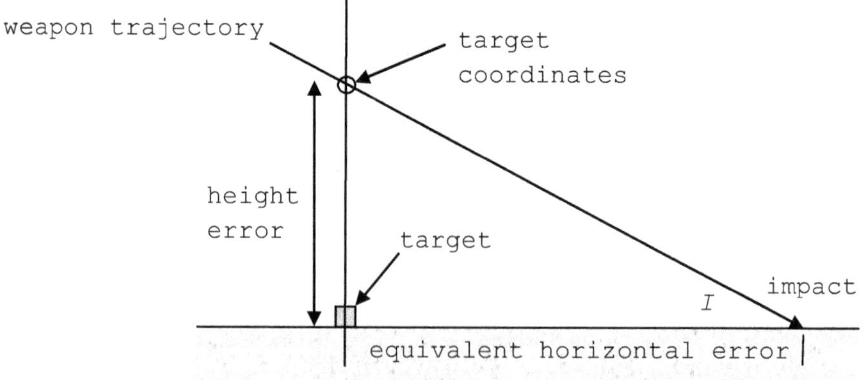

Fig. 5.12   Resolving vertical TLE into horizontal error.

5.9 OBTAINING TARGET COORDINATES

Target coordinates may be obtained in different ways, but the simplest way is shown in Fig. 5.13; it is termed *bombing on coordinates (BOC)* if the attack is from the air or *predicted fire* if it is from the ground. These two modes are analogous to their unguided weapon counterparts.

Figure 5.13 shows a target being attacked by either a surface-launched or air-launched GPS/INS-guided weapon, where it is assumed that target coordinates have been previously loaded into the weapon. This approach works well for fixed targets such as buildings or bridges.

Obtaining target coordinates in this manner commonly uses imagery from either an aircraft or a satellite. An example of a synthetic aperture radar (SAR) image is shown in Fig. 5.14.

Although it is relatively simple to identify a target in this or a satellite image, it is less obvious what the WGS-84 coordinates are in order to load them into the weapon. From recent maps or other sources, the latitude and longitude of the selected target may be known to some level of accuracy; however, what is the height above ellipsoid (HAE)? To solve this issue, the image is usually registered with a digital terrain elevation database (DTED), which provides latitude, longitude, and HAE coordinates for any selected point in the image.

A different approach is to designate the target in real time from either an aircraft or an observer on the ground. Figure 5.15 shows airborne and ground-based targeting systems where the airborne system uses pointing angles and a laser range finder whereas the ground-based system has a digital compass and laser range finder. In both cases, the operator points the device at the target in much the same way as a laser designator, but in this case the laser range finder together with the pointing angles are used to determine the position of the target relative to the designator.

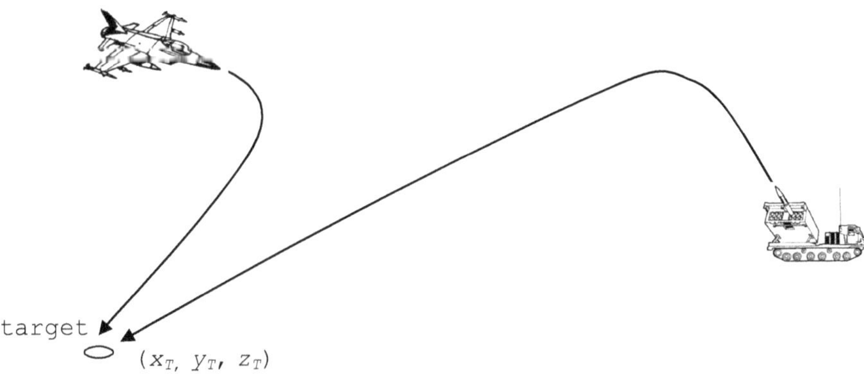

Fig. 5.13   Bombing on coordinates/predicted fire mode.

Fig. 5.14   High altitude SAR image.

Fig. 5.15   Airborne and ground-based targeting pods.

Because the designator knows its WGS-84 coordinates from its onboard GPS system, the coordinates of the target may be computed as an offset, as shown in Fig. 5.16. The calculated GPS coordinates are then loaded into the munition, and it flies to the coordinates supplied. This approach is used for time-sensitive targets or targets of opportunity.

For the airborne system, designation may be performed by another aircraft or the one launching the weapon. If it is the launching aircraft, the designator may communicate directly with the weapon without passing through the air-craft avionics or pilot.

There are literally dozens of other ways to obtain target coordinates includ-ing locating the source of electronic signals emanating from it (SIGINT),

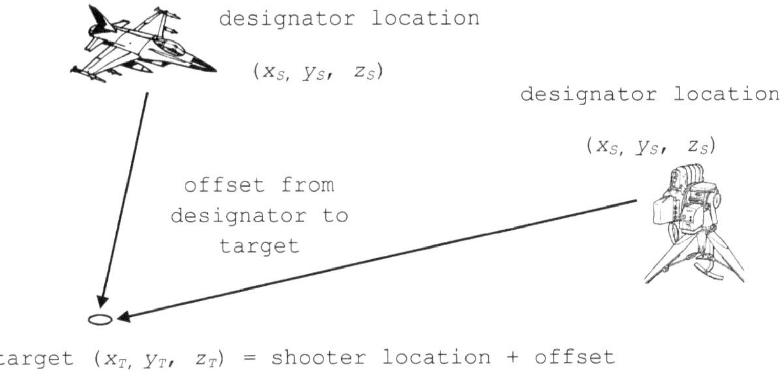

Fig. 5.16 Designating the target.

counterfire radar, or SOF personnel standing next to the target prior to the attack with a personal GPS system acquiring coordinates.

5.10 GPS JAMMING

It is relatively easy to jam signals received by a GPS device because they reach the Earth's surface at very low power levels. These signals propagate from satellites with limited power storage and recharging capabilities, resulting in transmitted signals approximately in the 35-W power level. Because the satellite orbit is around 20,000 km and some loss occurs due to transmission through the ionosphere and troposphere, signal strength on the Earth surface is very small, about the same as that received from a 100-W lightbulb on the surface of the Moon.

Although this signal is below the level of naturally occurring background radiation, digital signal processing techniques allow the GPS signal to be extracted by the receiver from this noisy background. Even so, it is not difficult for an external white noise transmitter to emit a signal that even at relatively low power overcomes the GPS signal the receiver is trying to detect. This is the basis of the simplest form of GPS jamming.

In discussing the capability of jammers to disable a GPS receiver, there are several parameters to consider.

- The jammer-to-signal strength *J/S* expressing the signal power generated by the jammer relative to the power of the satellite signal at the receiver. This is the inverse of the more familiar signal-to-noise ratio.
- The maximum power output of the jammer—the more powerful the jammer, the more effective it is over a larger range.
- The distance from the receiver over which the jammer has an effect. This is a maximum at the jammer and decays exponentially with separation distance.

- The code the receiver is utilizing—civilian C/A code or military P(Y). The latter is more resistant to jamming.
- What function of the receiver is disabled—initial acquisition of the satellites or the ability to block signals once acquisition has occurred.

5.10.1 RECEIVER–JAMMER INTERACTION

Figure 5.17 shows the relationship between the parameters listed in the previous section [4]. The horizontal dotted lines correspond to representative J/S values to inhibit receiver functions that utilize different codes. For example, the upper horizontal dotted line shows about a 55-db jammer-to-signal ratio is needed to cause the loss of a P(Y) receiver to track GPS satellites once acquisition has occurred. This must now be read in conjunction with the diagonal lines to determine what range a jammer of known power output will be effective for that jamming capability. From the graph it is seen that a 1-W jammer will be effective over a distance of about 4.5 km in order for a P(Y) receiver to lose GPS once acquired. Given this conclusion, it is instructive to look at commercially available jammers.

5.10.2 GPS JAMMER PERFORMANCE

GPS jamming devices basically emit white noise signals that attempt to mask the GPS signal. In most countries the public use of jamming devices is

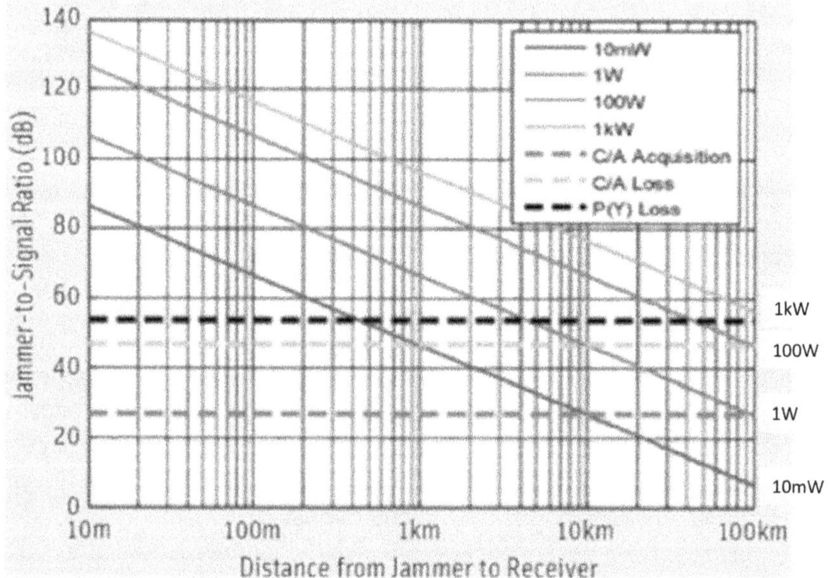

Fig. 5.17 Receiver/jammer performance.

illegal, but this has not stopped such devices from being available from the usual retail outlets at a low price (see Fig. 5.18).

The devices shown can be either handheld or fit into a vehicle and are about the size of a cell phone. The in-car unit on the left jams the L1 band (C/A) receiver and has a power of 125 mW, which from Fig. 5.17 gives it an effective radius in excess of 1 km. The other jammer shown in the center of Fig. 5.18 has a power output of 500 mW. Although P(Y) receivers are more resistant to jamming, it should be obvious that GPS jamming is easily achieved by relatively high-power, simple, compact, and inexpensive devices.

The effect increased noise level has on the positional accuracy of a receiver may be interpreted using Table 5.3 by observing the contribution to the error budget by noise in the receiver. Increasing the power output of the jammer would degrade the positional accuracy of the receiver, as indicated in Fig. 5.19. This map shows the area of effects of jammers for two different power levels.

Spoofing is different from jamming and is more challenging. Here, an attempt is made to decode the signals being transmitted by the satellites to the receiver and then replace them with correctly coded false signals. This allows the weapon to be diverted from its original target to a different, harmless location. Spoofing requires sophisticated electronics to track the weapon trajectory, decode the satellite signals, and generate false data all while the weapon is inbound to the target, and is greatly impaired if the encoded P(Y) code is in use.

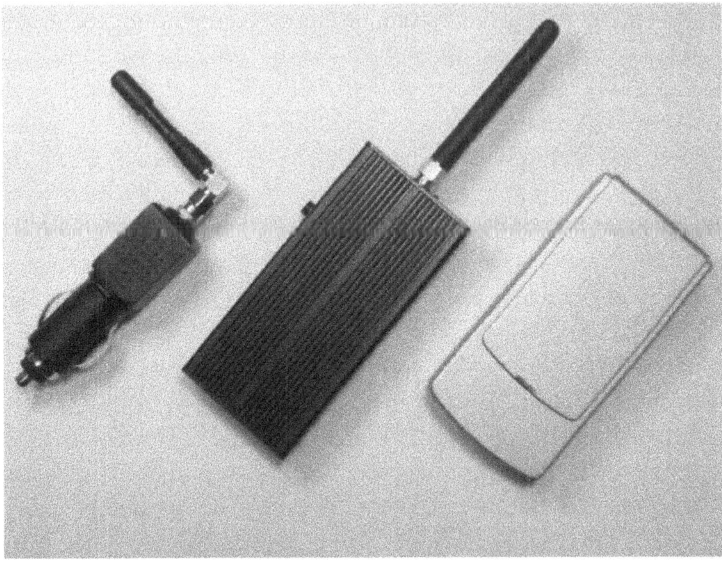

Fig. 5.18 Commercial GPS jamming devices.

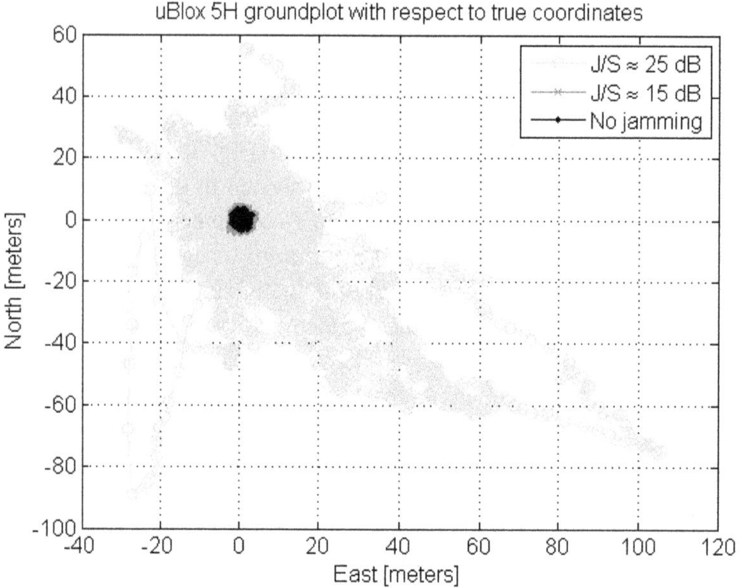

Fig. 5.19   Degradation of positional accuracy with jamming.

5.10.3 *JAMMING MITIGATION*

There are several improvements to the initial GPS system that attempt to make receivers more resistant to jamming signals, particularly for military systems.

- *M-code:* The U.S. military is developing a new encrypted signal called the M-code. It provides better antijamming capability by using higher power levels than the P(Y) code. The challenge is to ensure this higher power signal does not interfere with existing signals on the same carrier frequency.

- *Directional antenna:* As discussed earlier, the minimum DOP (accuracy) is achieved by reading as many spatially diverse satellites as possible; therefore, GPS antennas observe a full hemisphere of the sky. Unfortunately, this allows jamming signals from any direction to influence the receiver. A directional antenna receives signals from the known directions of the broadcasting satellites while generating null-reception areas everywhere else, thereby blocking any jamming signals. Electrically steerable directional antennas, also known as controlled-reception pattern antennas (CRPAs) or "smart" antennas, are also effective.

- *INS navigation.* As we have learned, GPS-guided weapons are really INS-guided with the GPS updating the INS to periodically eliminate gyroscopic drift. It will be shown in the next section how such drift increases the CEP over time, but the amount of increase depends on the quality (hence cost) of the INS. This is shown in Fig. 5.20.[3]

---

[3]Based on work by Gilmore and Delaney of Lincoln Laboratory, Lexington, Massachusetts.

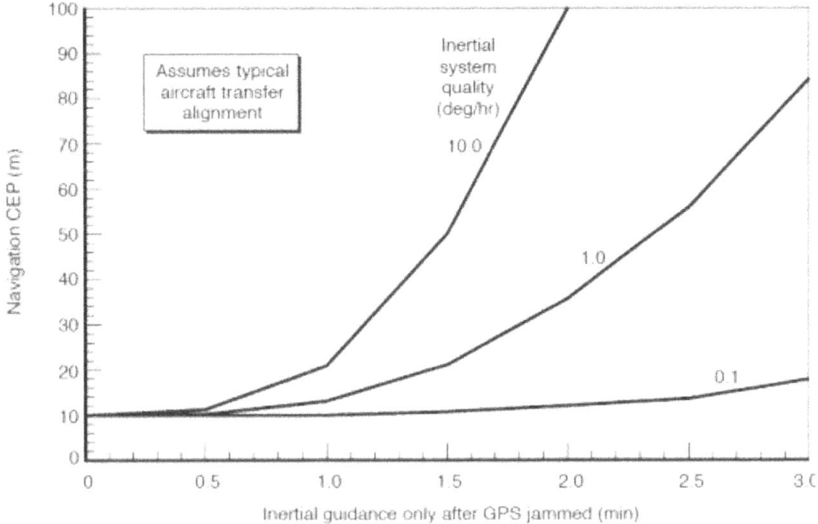

Fig. 5.20   Effect of INS drift on weapon CEP.

As a point of reference, the cost of these INS systems is shown in Table 5.5 for different drift rates. Because the high-cost INS systems are those found in military aircraft, of which there are relatively few, the INS systems used in high-production weapons with limited flight time will have to consider a tradeoff between accuracy and cost.

- *Mission planning tools:* Tools currently exist that allow the user to lay out a map of the engagement scenario so that mobile GPS receivers can be moved through an environment of user-defined GPS jammers. As the receiver moves along a specified path, typically an aircraft ingress path to the target, degradation of receiver performance can be determined. The GPS Interference and Navigation Tool (GIANT) is one such a system (see Fig. 5.21), simulating many receivers vs many jammers and producing maps and charts of DOP and CEP as well as J/S values along the ingress route. GIANT requires knowledge of adversarial resources along the path in order to provide useful data.

TABLE 5.5   COST OF INS WITH PERFORMANCE

Drift rate (deg/hr)	Cost ($)
10	1000
1	2000–5000
0.1	20,000–50,000
0.01	100,000–200,000

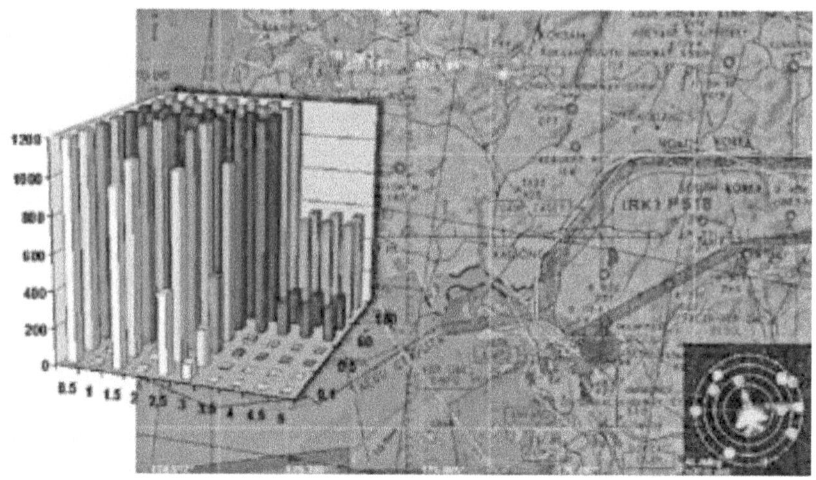

Fig. 5.21   GIANT mission planning tool.

5.11   INS-ONLY ACCURACY

A weapon's GPS systems can monitor the signal-to-noise ratio or the J/S ratio if jamming is present, and if it exceeds a prescribed level it will ignore subsequent GPS location data until it drops to an acceptable level. If GPS is switched out or denied to the weapon guidance system, then it must navigate using its internal INS. As mentioned previously, the INS system generally is less accurate than when GPS is present due to inherent drift; therefore, additional errors are introduced. In order to estimate the magnitude of such errors, a model of the weapon-specific INS is required. As a first approximation, the dominant errors are initial velocity and accelerometer bias, leading to the following simplified quadratic form:

$$s(t) = s_0 + \nu_0 t + \frac{1}{2} a_b t^2 \tag{5.24}$$

This may be written in the following equation, where a_0 and a_1 characterize the INS drift:

$$\text{CEP} = \text{CEP}_{\text{GPS}} + a_0 t + a_1 t^2 \tag{5.25}$$

Note different values of a_0 and a_1 for different quality receives in Fig. 5.20. It should also be recognized that GPS/INS-guided weapons may begin their movement toward the target guided by INS only. This is because the receiving antenna on the weapon may not be exposed to the satellites until the munition is fired from a cannon or released from an aircraft, where it is shielded by the wing or fuselage. Figure 5.22 shows how the equivalent ground plane CEP of a GPS/INS weapon varies with time from release.

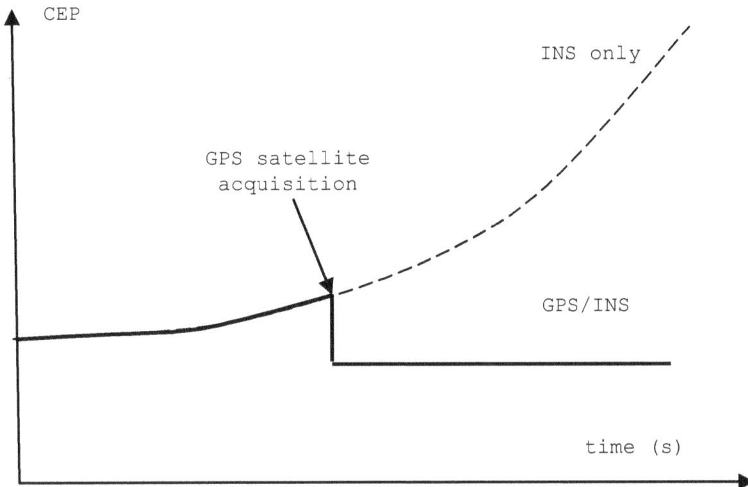

Fig. 5.22   Acquiring GPS after launch.

In this figure, the improvement in accuracy may be seen once GPS satellites have been acquired, and positional information is used to update the INS system. If satellite acquisition does not occur, the weapon is guided by INS control only where the quadratic degradation of accuracy is governed by Eq. (5.25).

This obscuration problem may be avoided if, prior to releasing the weapon, the launch system can upload its current GPS position by a process called transfer alignment (TXA).

The same quadratic function may also be used to predict the effect of GPS jamming on accuracy by determining the errors incurred over that portion of the weapon's time of flight subject to jamming, as indicated in Fig. 5.23.

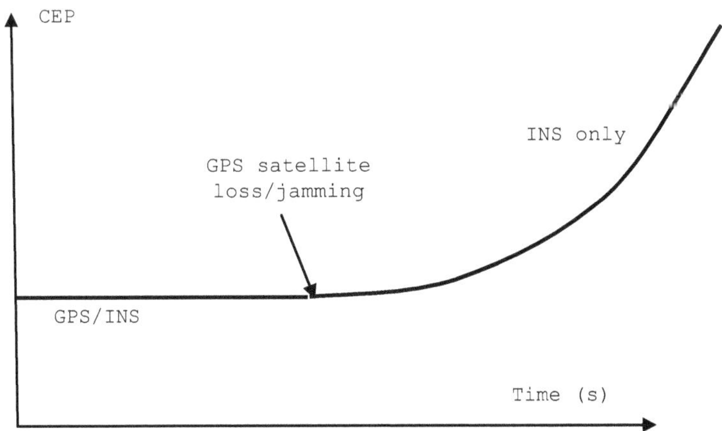

Fig. 5.23   INS drift after GPS jamming.

Clearly, the longer the flight time before impact that GPS is not available, the worse the weapon accuracy will be. For error estimation purposes, the INS drift may be used in the following form:

$$\sigma_{\text{DRIFT}-X} = \sigma_{\text{DRIFT}-Y} = s_0 + v_0 t + a_b t^2 \tag{5.26}$$

The coefficients s_0, v_0 and, a_b have to be obtained from contractor or program offices for the weapon of interest. As usual, some care needs to be exercised in interpreting weapon data because some weapon systems show a growth of CEP with time whereas others predict a growth of a standard deviation, as shown in Eq. (5.25) and Eq. (5.26).

5.12 COMPUTATIONAL TOOLS TO PREDICT GPS ACCURACIES

A simple spreadsheet that illustrates the combination of the NAV, G&C, and TLE errors is shown in Table 5.6. In this spreadsheet, the inputs are shown in shaded cells and include the weapon impact angle together with the NAV, G&C, and TLE errors. These errors are specified as circular ground plane and linear vertical measures, where the TLE is provided at the more common 90% level. The calculations proceed as follows:

TABLE 5.6   SIMPLE GPS WEAPON ACCURACY CALCULATOR

GPS Weapon accuracy calculator 1				
Impact angle (deg)	65	tangent (impact)	2.14	
Errors	Ground plane (ft)		Vertical (ft)	
NAV (with DOP)	15	Circular 50%	20	Linear 50%
G&C	5	Circular 50%	5	Linear 50%
TLE	25	Circular 90%	35	Linear 90%
Convert TLE to 50%	13.72	Circular 50%	14.10	Linear 50%
Convert CEP to REP, DEP	REP	DEP		
NAV	8.59	8.59		
G&C	2.86	2.86		
TLE	14.32	14.32		
Projection of height error				
NAV vertical	9.33			
G&C vertical	2.33			
TLE vertical	6.57			
RSS total	20.54	16.95		
CEP	32.62			

1. The 90% TLE values are converted to 50% values using Eq. (4.9) for the circular measure and Eq. (4.10) for the vertical, linear measure.
2. The 50% circular measures are converted to REP and DEP using Eq. (4.7).
3. The three vertical error components are projected into the ground plane using the tangent of the impact angle and Eq. (4.23) to produce equivalent errors in the range direction only.
4. The component REP and DEP errors are RSS together and a CEP calculated using Pittman's equation.

Although REP and DEP are preferred in any effectiveness calculations, determining in which directions these point is not obvious. Because the weapon approach to the target is not easily estimated without a high fidelity trajectory program, and given the range direction is defined by the approach velocity vector, the range and deflection directions are equally difficult to determine. Under these circumstances, using the CEP is an acceptable alternative, since the miss distance distribution is the same in any direction.

The next spreadsheet shown in Table 5.7 is a more detailed version of the previous one in that it provides the error budget for the NAV component and

TABLE 5.7   DETAILED GPS WEAPON ACCURACY CALCULATOR

GPS Weapon Accuracy Calculator 2 - (inputs are green cells) Distances in meters

	Bias	Random			
Satellite clock	0.400				
Ephemeris	0.400		HDOP		1
Troposphere	0.500		VDOP		1.8
Ionosphere	0.500		GDOP		2.06
Noise		0.400	Impact θ		70
Multipath		0.500			
Receiver	0.954				
Space and clock (S&C)	0.566				
RSS	0.906	0.640			
UERE (pseudo range sigma)	1.109				

	Horizontal			Vertical		
	Bias	Random	Total H	Bias	Random	Total V
Nav error H (sigma)	0.640	0.453	0.784	1.630	1.153	1.996
G&C H (sigma)		1.000	1.000		1.000	1.000
TLE H (sigma)	3.000		3.000	5.000		5.000
RSS (sigma)	3.068	1.098	3.258	5.259	1.526	5.476

	Range			Deflection		
Summary-horizontal plane	Bias	Random	Total R	Bias	Random	Total D
REP/DEP H plane	3.616	1.230	3.819	3.068	1.098	3.258
Single weapon CEP H plane	4.157					

	Vertical			Deflection		
Summary-vertical plane	Bias	Random	Total V	Bias	Random	Total D
HEP/DEP V plane	9.934	1.526	10.051	3.068	1.098	3.258
Single weapon CEP V plane	7.659					

user-supplied values for certain DOP quantities. It implements the methodologies discussed in this chapter to determine GPS weapon accuracy, and allocates each error into a bias or random (MPI and precision) component, which may be combined for single weapon attacks but kept separate for multiple weapon deliveries.

The user again has to supply data in the shaded cells, although representative values discussed earlier in the chapter have been used. The results are valid for bombing on coordinates/predicted fire mode with no GPS jamming.

Finally, many of the features discussed in this chapter may be combined into a single computational tool such as the one shown in Fig. 5.24. It consists of four separate modules:

1. A DOP tool to determine the NAV accuracy
2. A simple ballistic trajectory program for air-launched weapons to give an estimate of impact angle and time of flight
3. An integrated GPS accuracy tool that combines NAV, G&C, and TLE data
4. A GPS jamming tool

The flow of data between these sections is indicated by the arrows connecting them.

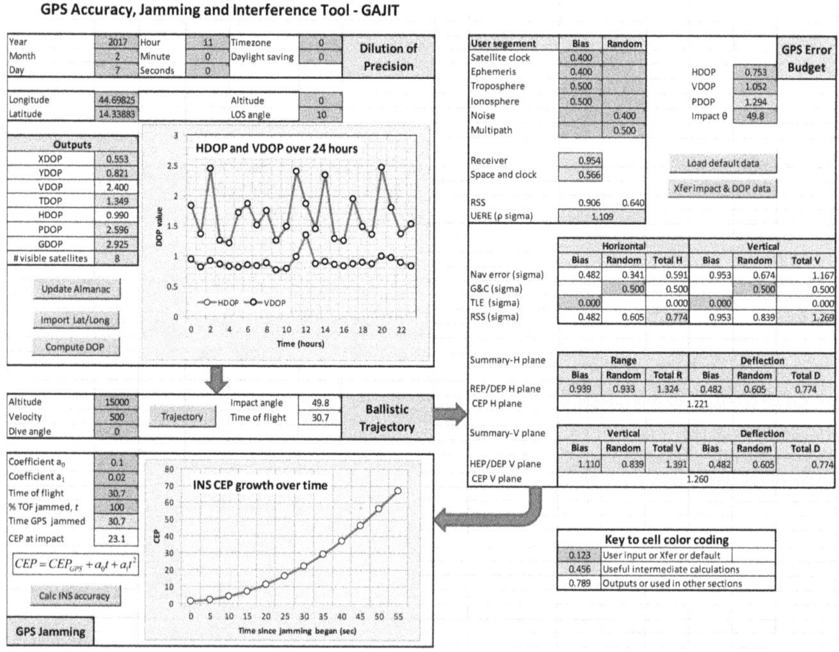

Fig. 5.24 Integrated GPS weapon accuracy calculator.

This program has the following features:

- Starting in the top left, the user can specify the time, date, and location of the planned attack, and the point value as well as the variation in DOP is generated. This includes being able to obtain the latest almanac data and the ability to import target locations from resources such as Google Earth or Google Maps.
- Moving down, the user inputs trajectory information. This may be a simple ballistics calculation using supplied inputs of altitude, velocity, and dive angle, or specific weapon terminal conditions obtained offline.
- The next step on the right is the GPS accuracy calculator shown in Table 5.7 where the user supplies the GPS receiver parameters or uses the default. In addition, the DOP and impact data may be transferred from the previous sections, producing the various accuracy metrics REP, DEP, and CEP in both the horizontal and vertical directions.
- The final section allows accuracy in a GPS-denied environment to be calculated and plotted as a function of time. The user has to supply the INS drift characteristics in Eq. (5.26) and the duration of the INS-only navigation. This is expressed as a percentage of the weapon flight time calculated in Chapter 3.2.

5.13 CHAPTER SUMMARY

1. GPS/INS weapons depend on the constellation of satellites supporting the system, and their accuracy varies with the position of the satellites relative to the weapon as it navigates to the target. This is the NAV component of accuracy, and it is subjected to dilution of precision (DOP) degradation.
2. As for any other guided weapon, part of the accuracy depends on the ability of the guidance surfaces to change the trajectory of the weapon. This is known as the G&C component.
3. The third component of accuracy is the accuracy of the target coordinates, the target location error (TLE), to which the weapon is directed, a quantity over which the weapon has no control.
4. GPS signals are easily jammed, in which case the weapon reverts to INS-only guidance. This is inferior to GPS accuracy and increases over time due to INS drift.
5. Although GPS/INS-guided weapons have complexities not seen in other weapons (DOP, jamming, etc.), their accuracy may be represented by a REP/DEP or CEP that takes into account these effects. From this point on, at least in terms of calculating weapon effects, the weapon is indistinguishable from any other.

REFERENCES

[1] Tsui, J. B., *Fundamentals of Global Positioning System Receivers*, Wiley Interscience, Hoboken, NJ, 2000.

[2] Kaplan, E. D., (ed.), *Understanding GPS Principles and Applications*, Artech House, Boston, MA, 1996.

[3] Yuen, M. F., "Dilution of Precision (DOP) Calculations for Mission Planning Purposes," MS Thesis, Naval Postgraduate School, Monterey, CA, March 2009.

[4] Jones, M., "The Civilian Battlefield: Protecting GNSS Receivers from Interference and Jamming," *Inside GNSS*, March/April 2011, pp. 40–49.

Chapter 6

Introduction to Vulnerability and Effectiveness

6.1 Introduction

This chapter investigates the destructive effects that weapons have on targets and involves both the lethality of the weapon and the vulnerability of the target.

There is a considerable amount of material to be explained, so this topic is divided among three chapters.

- This chapter provides an overview of the subject by making some liberal assumptions in order to get broad concepts understood.
- Chapter 7 looks in more detail at target vulnerability independent of specific weapons.
- Chapter 8 deals with the effect of specific warhead damage mechanisms on targets with defined vulnerabilities.

In this introductory chapter, weapon effectiveness is discussed in terms of an *effectiveness index (EI)*. This chapter will explain two types of EIs in detail; the others are deferred until later chapters

1. A *mean area of effectiveness (MAE)* is used for weapons that can damage a target without hitting it (e.g., bomb, artillery shell, rocket, mortar round, and grenade). Damage is caused by blast, fragments, or both. These appear as MAE_F and MAE_B in the EI section of the weaponeering program.
2. A *vulnerable area (A_V)* is used for weapons that have to hit the target in order to cause damage [e.g., a penetrator, shaped charge, or explosively formed projectile (EFP)]. This is labeled VA_N in the EI section of the weaponeering program.

In the case of an MAE, the EI is an area (lethal area) measured in square feet or square meters on the ground or normal plane. Figure 6.1 shows a typical MAE as a circle on the ground within which the truck target will be damaged by an air-bursting artillery shell. If the target is outside the lethal area, it survives.

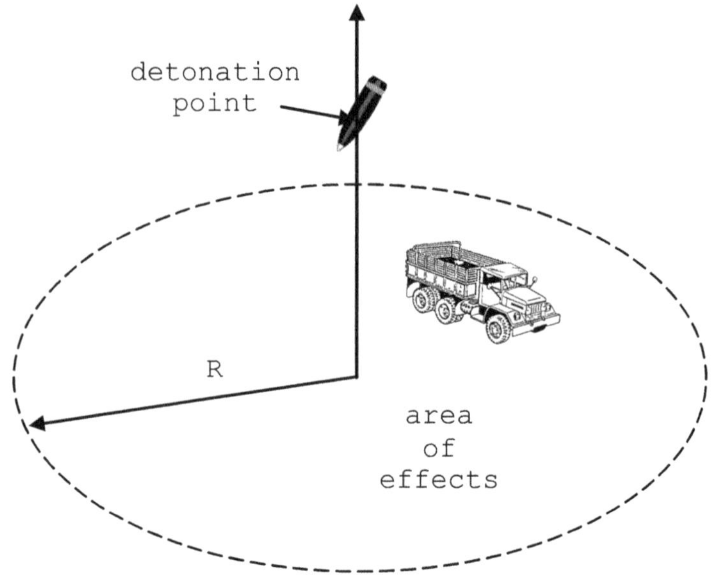

Fig. 6.1   Effects area for general purpose warhead.

For this example, we would get the following for the area of effects, which may be due to either fragments (MAE_F) or blast (MAE_B):

$$\mathrm{MAE} = \pi R^2 \tag{6.1}$$

6.2   EFFECTIVENESS INDICES

The effectiveness index (EI) may also be referred to as a lethal area or a damage function. An EI is defined only for a given degree of damage by a specific weapon to a particular target.

The specific nature of damage inflicted on the target is usually referred to as a *damage definition*, and it varies for different types of targets. Sometimes the word *kill* is used instead of *damage*, and the words *level* or *criterion* are used instead of *definition*. Therefore, phrases such as *damage definition*, *kill level*, and *damage criterion* are synonymous and are used interchangeably.

The kill definition varies with target; for example, those relating to an armored vehicle are not the same as those for personnel targets. Examples of kill definitions for some target types are shown in Table 6.1.

The Appendix at the end of this book contains a table listing most of the Joint Technical Coordinating Group for Munitions Effectiveness (JTCG/ME) target types vulnerable to conventional weapons and the kill definitions associated with them. The table also contains recommendations for the types of weapon considered to be effective against a specific target type.

TABLE 6.1   EXAMPLES OF KILL DEFINITIONS

Target Type	Kill Type	Kill Definitions
Land vehicles	K-kill M0-kill M40-kill F-kill	Catastrophic kill (not repairable) Mobility kill (cannot move, immediately) Mobility kill (cannot move within 40 minutes) Firepower kill (cannot fire)
Parked aircraft	PTO PTO_4 PTO_{24}	Repairs requiring at least 5 minutes Repairs requiring at least 4 hours Repairs requiring at least 24 hours
Personnel (standing)	30-s defense 30-s assault 5-min assault 12-h supply	Prevents defense after 30 s Prevents assault after 30 s Prevents assault after 5 min Prevents supply after 12 h

TABLE 6.2   EXAMPLE EI VALUES

Target ▶	Tank	Tank	Troops
Kill definition ▶	M-kill	K-kill	5-min assault
Weapon ▼			
Mk-82	600	450	3000
Mk-84	1200	550	12,000

To illustrate the variability of EI with target, weapon, and kill criteria, consider the effect of both an Mk-82 500-lb bomb and an Mk-84 2000-lb bomb. Representative MAE_F values (in ft^2) are shown in Table 6.2 for various target types and kill definitions.

Readers should clearly understand the trends associated with the EI value as the weapon, target, and kill definition vary.

6.3   REQUIREMENTS FOR THE COMPUTATION OF EFFECTIVENESS INDICES

Calculating an EI requires considerable time, knowledge, expertise, and data; in general, the following elements will be needed:

- Definition of the kill levels required
- Collection of the physical, geometrical, and functional descriptions of the target and the components that are contained within it
- Knowledge of the critical components of the target [i.e., which components need to be killed in order to achieve the kill level(s) required]
- The warhead characteristics (blast, fragments, penetrator, shaped charge jet) of the weapon to be used against the target

- An understanding of what damage level is required to kill each of the critical components
- A methodology to compute the appropriate EI for the given combination of weapon, target, and kill level

To determine the effect of a particular weapon on a target of interest, two individual studies are usually done for each kill definition of interest:

- *Vulnerability assessment:* The target vulnerability to a wide range of *user-selected* warhead characteristics is determined.
- *Effectiveness assessment:* This follows the vulnerability assessment. The computation of an EI for a *specific weapon* and detonation scenario is computed.

To understand why this two-step process is adopted, consider damage to targets by a variety of warheads due to the effect of just fragments. Each warhead (bomb, shell, grenade) will project different numbers of fragments having different masses traveling with different velocities. Each weapon type will have its own fragment characteristics.

The vulnerability assessment is target specific; however, it uses generic fragment masses and velocities not related to a particular warhead. If the vulnerability assessment is performed for a wide range of fragment masses and velocities, and a table is generated showing the probability of target damage due to all possible combinations of fragment masses and velocities, it would look like that shown in Fig. 6.2.

From this table, all current and future warheads can be "mapped" to a specific region of the matrix, possibly with overlapping areas. This implies that the vulnerability assessment only ever has to be done once, and then all current and future warheads can be located somewhere in the table.

In order to study the vulnerability of specific targets to conventional weapons, we will consider the effects of the primary damage mechanisms a weapon may produce.

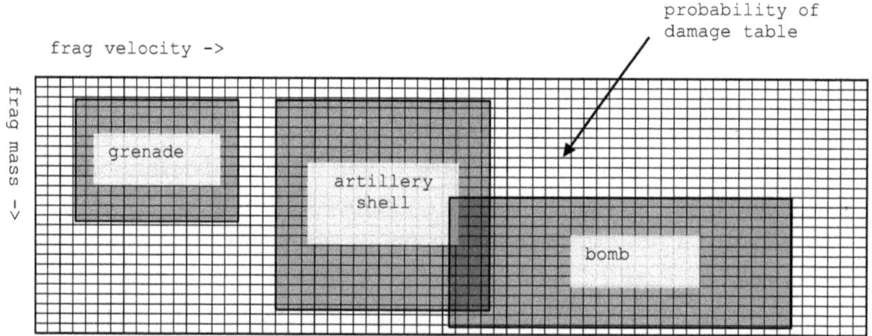

Fig. 6.2   Relationship between vulnerability data and specific weapon characteristics.

- Fragments
- Blast
- Penetrator, shaped charge, and EFP

The general nature of these damage mechanisms was described in Chapter 1; however, a particular warhead may produce one or more of these effects, so the combined effect of each damage mechanism on the target will have to be determined. In terms of modeling complexity, fragments present the most challenges, so they will be dealt with first. Before embarking on the vulnerability study, we review how real warhead fragment characterization is performed.

6.4 COLLECTING AND CHARACTERIZING WARHEAD FRAGMENT DATA

In order to determine the effectiveness of fragments on materiel and personnel targets, the characteristics of such fragments have to be measured for each warhead type. This characterization is often achieved through testing, and the process has matured into a standardized approach. The process is referred to as *arena testing*, because the static warhead is surrounded by various devices that determine the number, weight, velocity, and region of origin of fragments produced when it detonates.

Although testing is used to characterize fragments, there are some theoretical methods that predict these characteristics and are often employed to plan the test. These are known as the *Gurney equations* and may be summarized as follows.

6.4.1 THEORETICAL PREDICTIONS OF FRAGMENT CHARACTERISTICS

The theoretical velocity of fragments immediately following detonation is given by

$$V_0^2 = \frac{2(C/M)A^2}{2 + (C/M)} \tag{6.2}$$

where
  V_0 = initial fragment velocity (ft/s)
  A = a constant depending on the explosive ($\sim$8400)
  C/M = ratio of the weight of explosive to the weight of the metal casing calculated from the following:

$$\frac{C}{M} = \frac{P_C}{P_M \left[(D_0^2/D_i^2) - 1 \right]} \tag{6.3}$$

  D_0 = outside diameter of shell (in.)
  D_i = inside diameter of shell (in.)
  P_C = density of explosive (lb/in.^{3})
  P_M = density of metal parts (lb/in.^{3})

Fragment masses are usually measured in grains where 7000 grains=1 lb. Although there will be a distribution of fragment masses, the average mass is given by

$$m_0 = \frac{CD_0^2}{V_0^2} \tag{6.4}$$

where
  m_0 = average mass of fragments (grains)
  C = a constant = 60.10^6

For a standard shell, this does not allow for the large fragments that come from the nose or base. The number of fragments having a specific mass comes from what is usually known as the two-dimensional Mott distribution

$$N = \frac{M \exp[-(2m/m_0)^{0.5}]}{m} \tag{6.5}$$

where
  N = number of fragments
  M = weight of metal case (grains)
  m = weight of smallest fragment considered (grains)

6.4.2 MEASUREMENT OF FRAGMENT CHARACTERISTICS

For real warheads, fragments are not evenly distributed around the warhead because the main spray from the side has the greatest number of fragments, whereas the nose and base have a smaller number of large or very large fragments.

The fragment characteristics are defined within zones that are conical regions defined by boundaries radiating out from the warhead centroid, as shown in two dimensions in Fig. 6.3. If each zone covers 10 degrees of solid angle, there will be 18 zones total.

In three dimensions for an air-bursting munition, the conical zones will spray fragments onto the ground plane in the manner shown in Fig. 6.4.

It is assumed that the fragment spray pattern is radially symmetrical about the weapon axis; therefore, the test arena is configured in the manner shown in Fig. 6.5, where one side is designed to collect the fragments themselves and the other side determines their velocity.

The fragment recovery area consists of fiberboard panels in which the fragments become embedded. The fragments are then removed, counted, and weighed. The velocity panels are coated with a piezoelectric material that produces an

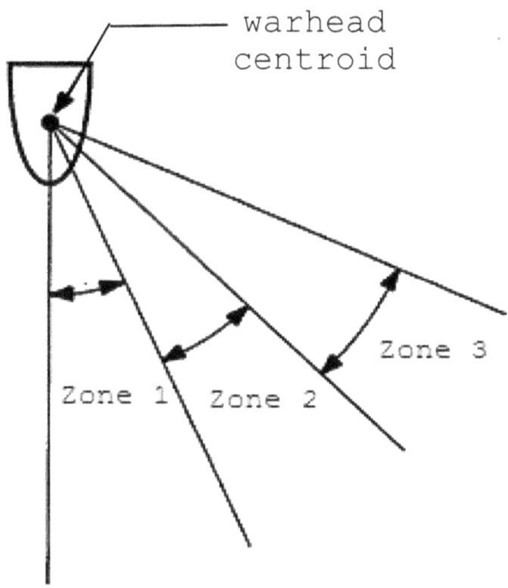

Fig. 6.3   Weapon static fragmentation zones.

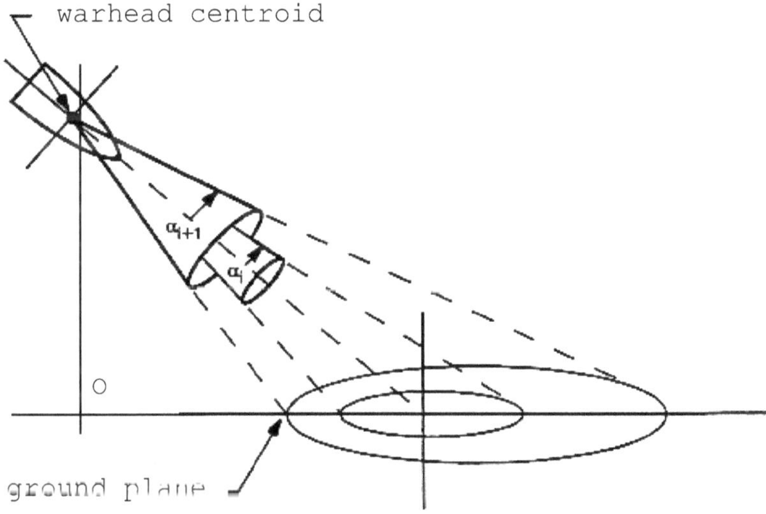

Fig. 6.4   Fragment spray pattern on the ground.

electrical signal when impacted by a fragment. The time taken for a fragment to travel from the warhead to the panel may therefore be calculated.

This process is also recorded by high speed cameras behind the flash panels, which allow the fragment velocity to be determined for a given frame rate. Of course, the panels have to be sufficiently far from the detonation so they are not damaged by the blast wave.

The collection panels have the fragmentation zones projected onto them to determine which fragments belong in which zones, as indicated in Fig. 6.6.

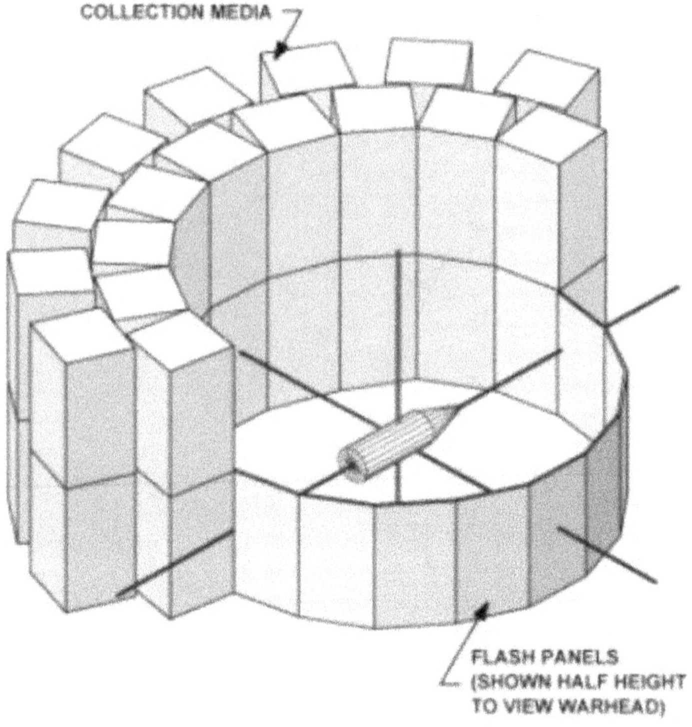

Fig. 6.5   Typical test arena configuration.

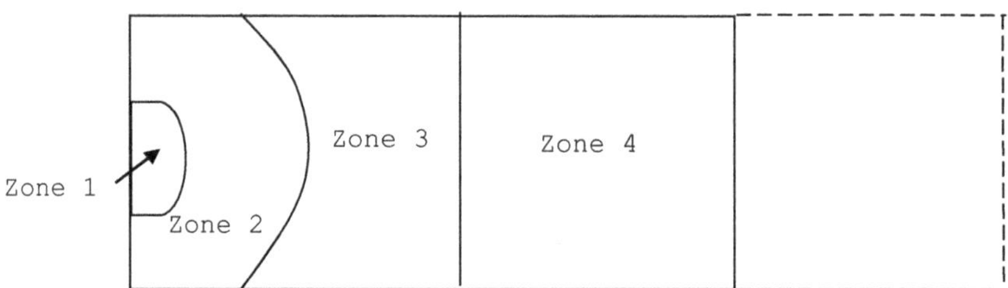

Fig. 6.6   Fragment zones projected onto the recovery panels.

Although the measured velocity is defined as the distance traveled by the fragment divided by the time it travels, the actual fragment velocity varies over this short distance. What is needed is the velocity of fragments as they leave the warhead immediately after detonation, so idealized trajectory models of individual fragments allow this to be calculated from the measured velocities.

The mass and number of fragments are also grouped—aggregated for more than one test—into each zone. Then, within each zone, the weights are grouped into bins and the average weight in each bin is determined. The average number of fragments again is averaged over several tests, and this average is then used to calculate and record weight.

The data for each zone have to be scaled to account for the fact that only a proportion of fragments are captured. This is calculated using the three-dimensional geometry of the fragment zone and fragment capture panel; a two-dimensional illustration of the concept is illustrated in Fig. 6.7. This shows a view along the horizontally mounted munition axis at detonation. A vertical capture panel of height h is located at a distance r from the munition axis. For example, if the angle the panel makes at the warhead is 60 deg, and assuming the fragments are uniformly distributed around the axis, the number of fragments collected by the panel has to be multiplied by 6 to determine the total number of fragments for that zone.

The specification of fragment mass, velocity, and number of fragments is used to define the *zonal or Z-data file* for each zone. The Z-data file is the end result of the test and is a required input for the effectiveness calculation. The Z-data file has the format shown in Table 6.3.

The first three numbers (α) give the angular boundaries of the zone; for this example this starts at 0 deg (weapon axis-nose) and extends to 20 deg with a midpoint of 10 deg. It is assumed that the fragment velocities depend only on the angular offset into the zone, and the fragment velocities at these angles are 5500, 4770, and 5010 ft/s, respectively. The midpoint velocity of 5010 ft/s corresponds to that measured from the test as described previously, and the outer velocities are interpolated from the next zone. Their use will be explained later in this chapter.

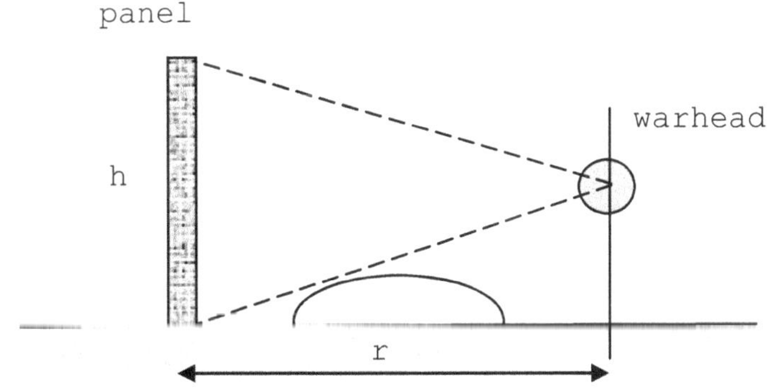

Fig. 6.7   Partial fragment capture.

TABLE 6.3   FRAGMENT CHARACTERISTICS FOR ONE ZONE

α's			Velocities			# wts
0.000	20.000	10.000	5500	4770	5010	5

Frag weight →	0.100	4.336	16.303	33.862	104.337
# of frags →	2.927	52.692	40.128	34.273	16.709

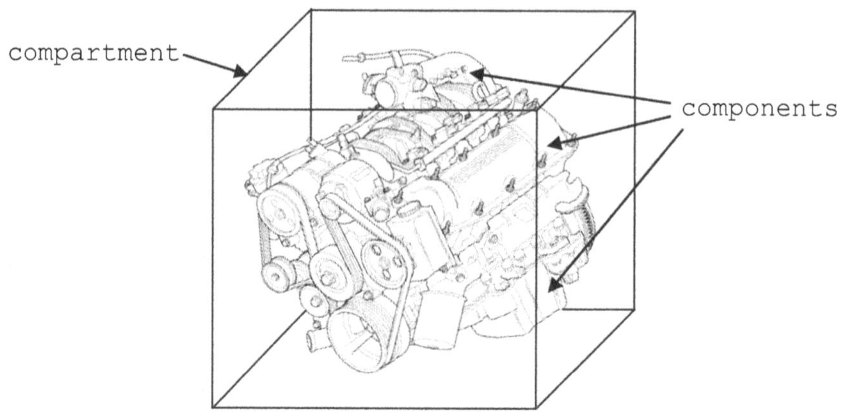

Fig. 6.8   Engine compartment with internal components.

The number 5 indicates there are five fragment weight groups in this zone. The first group has an average weight of 0.1 grains and there are, on average, 2.927 fragments at this weight; the next group comprises an average of 52.692 fragments weighing an average of 4.336 grains, and so on. Clearly, fractional values of the average number of fragments indicate data from more than one test; in fact, the JTCG/ME produces a testing document recommending the number of weapons to be tested, their orientation, analysis procedure, and so on.

6.5   COMPARTMENT VS. COMPONENT VULNERABILITY

Early vulnerability models divided whole targets into relatively large compartments so that, for example, a vehicle would have a crew compartment, an ammunition compartment (magazine), an engine compartment, and so forth, as shown in Fig. 6.8.

In such a representation, the probability of killing the engine is essentially an SME generated empirical value. As computer capability increased, it became possible to model the vulnerability to a higher fidelity by representing the actual components within the compartment. The engine, for example, could be represented by its constituent parts such as the oil pump, main block, fuel injectors, and so on. This allowed more accurate assessment of kill levels associated with the compartment, such as mobility and firing ability.

6.6   VULNERABILITY ASSESSMENT FOR FRAGMENTATION WARHEADS

The general approach to vulnerability assessment for a whole target is to assess whether the kill level has been achieved by determining how many and which components that make up the target have been individually killed. The definition of exactly what comprises a component is somewhat vague but depends on the level of accuracy needed in the analysis. For example, if we consider the target to be a vehicle, one component of the target is the propulsion system; however, this may be divided into smaller components such as the engine, gearbox, oil cooling system, electrical controls, and so on. As we have seen in the previous section, the engine may be further divided into its subcomponents.

One important parameter produced by a vulnerability analysis is $P_{K/H}$. This is the probability that the object is killed (to a specified level) if it is hit. The word *object* is used because $P_{K/H}$ could refer to an individual component or to the whole target. Another term in common usage is *survivability*. For example, the probability that the target survives a hit, $P_{S/H}$, is the complement of the probability the target is killed if it is hit, $P_{K/H}$. We write this as

$$P_{S/H} = 1 - P_{K/H} \qquad (6.6)$$

At this point we will introduce some notational conventions used in this chapter; these are the same ones used by Ball [1]. They allow us to differentiate between complete targets and the components contained within them. In this convention, uppercase symbols refer to a whole target, whereas lowercase ones refer to a particular component. Some sample terms used later are listed in Table 6.4.

The most basic data used in a vulnerability assessment are $P_{ki/hi}$, and this value may be determined experimentally by firing single fragments of different weights at different velocities at the component from different directions. An alternative approach to testing is to run analytical models that predict penetration of a component and its subsequent loss of function. However it is determined, typical data that might result from this process are shown in Fig. 6.9; this is usually known as a fragility curve. For a particular critical component, there will be one fragility curve for each kill level.

TABLE 6.4 DEFINITION OF VARIABLES USED IN VULNERABILITY ANALYSIS

Description	*i*th Component	Target
Probability of killing the *i*th component given a hit on the *i*th component, or probability of killing the target given a hit on the target	$P_{ki/hi}$	$P_{K/H}$
Probability of survival of the *i*th component/target given a hit on the target	$P_{si/H}$	$P_{S/H}$
Probability of hitting the *i*th component given a hit on the target	$P_{hi/H}$	
Probability of killing the *i*th component given a hit on the target	$P_{ki/H}$	
Vulnerable area of the *i*th component/target	A_{vi}	A_V
Presented area of the *i*th component/target	A_{pi}	A_P

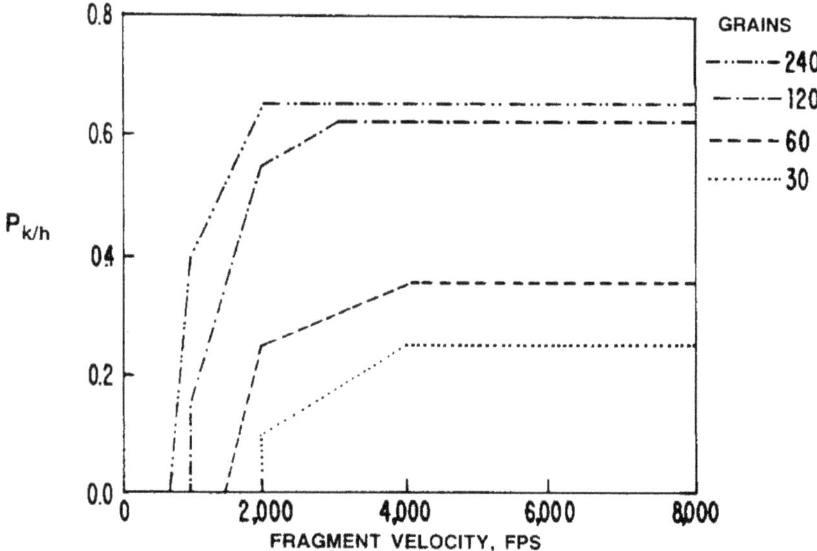

Fig. 6.9   **Fragility graph for a single component.**

Another common form for the fragility curve is shown in Fig. 6.10 for a given fragment mass.

This form is generally used when minimum energy needed to cause any damage and maximum energy needed to totally fail the component are estimated, and a linear relationship in between is assumed.

It cannot be overemphasized that the effectiveness of a weapon against a specific target *requires* the fragility curves for all components inside the target.

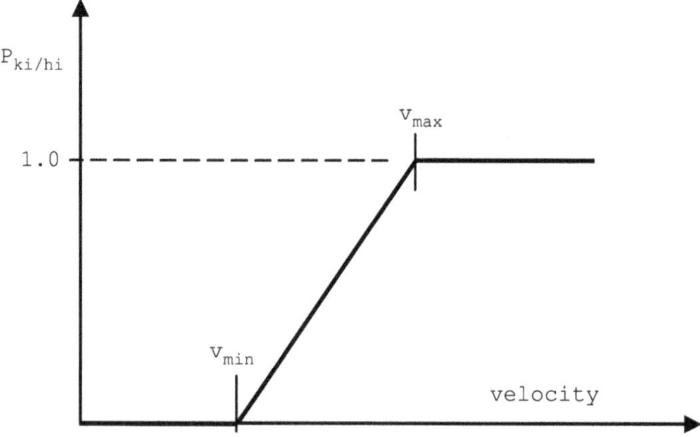

Fig. 6.10   **Simplified fragility graph.**

6.7 PRESENTED AND VULNERABLE AREAS

For weapons that have to hit the target in order to damage it, we use a concept called the vulnerable area, which may be written as A_V or VA_N. Figure 6.11 shows a tank target viewed from a particular direction showing the vulnerable area as a generalized "blob"; however, if we chose to represent it as a circle of radius R, then the A_V would be given by Eq. (6.1).

Just because we hit the tank doesn't necessarily mean we kill it, leading to the important observation that the probability of hit is not the same as the probability of kill.

$$P_H \neq P_K \tag{6.7}$$

However, if a weapon impacts within the vulnerable area, the target is killed. The vulnerable area is therefore always smaller than the presented area A_P of the target, and they are related by the following equation:

$$A_V = P_{K/H} \times A_P \tag{6.8}$$

where A_V is the vulnerable area, $P_{K/H}$ is the probability of kill given a hit, and A_P is the presented area.

The presented area is simply the area enclosing the target silhouette when viewed from any given aspect. Also known as the projected area, it may be visualized by placing a sheet of paper behind the component and projecting onto it the boundary of the component with parallel rays. The presented area is the area enclosed by the projected boundary.

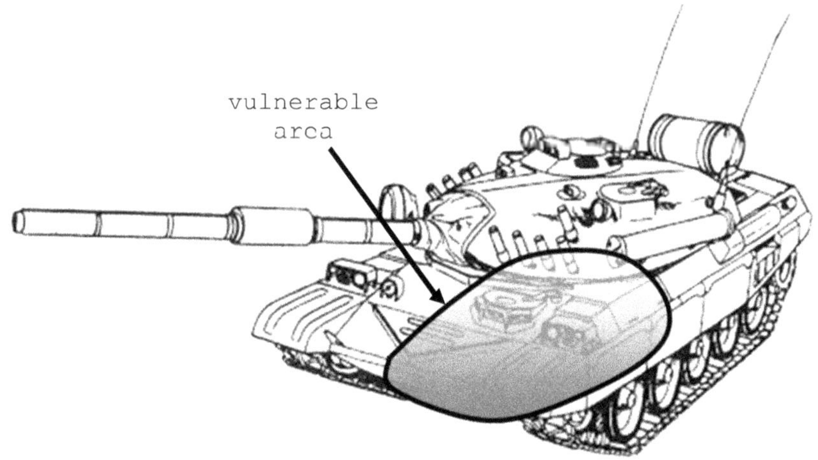

Fig. 6.11 Vulnerable area for penetrator round.

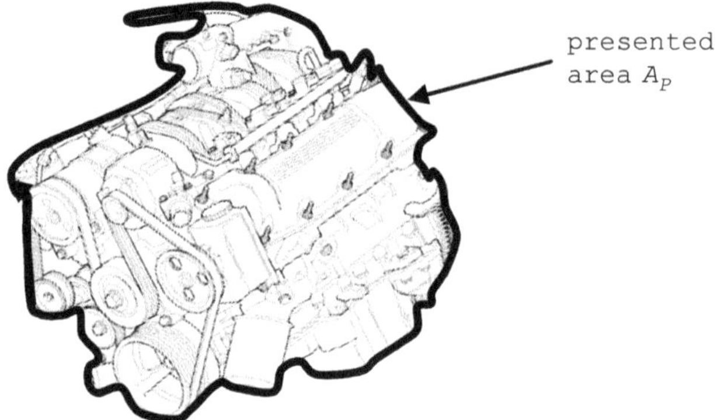

Fig. 6.12   Presented area of a component.

The vulnerable area is a quantitative measure of vulnerability that may be applied to a single component or a complete target. When applied to a whole target, it is an EI, specifically A_V or sometimes VA_N. An example of the presented area would be the area inside the boundary of the engine, shown in Fig. 6.12.

When applied to a component rather than the whole target, Eq. (6.8) becomes

$$A_{vi} = A_{pi} \times P_{ki/hi} \tag{6.9}$$

The relationship between vulnerable areas and presented area for a target and a component is shown graphically in Fig. 6.13.

Note that in this figure, the target and component presented areas correspond to a physical object (i.e., the target and the component); however, the vulnerable area has no assumed location or shape at this point.

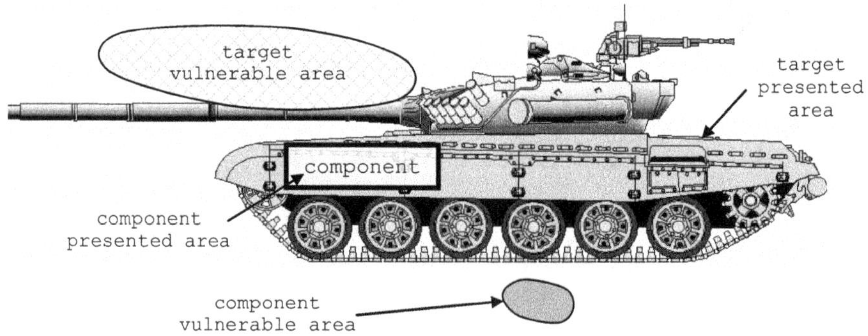

Fig. 6.13   Presented and vulnerable areas.

6.8 CRITICAL AND NONCRITICAL COMPONENTS

Components of a target may be defined as either critical or noncritical. A critical component is one that, if killed, causes a target kill, whereas a noncritical component does not lead directly to a target kill. It is important to include noncritical components in a vulnerability analysis, however, because they may shield critical components to such an extent that the critical component does not absorb sufficient damage to kill it, thereby preventing a target kill. A good example of a noncritical component is armor plate.

6.9 REDUNDANT AND NONREDUNDANT CRITICAL COMPONENTS

A component is said to be nonredundant if killing that component alone results in a target kill. Redundant components require that several be killed before a target kill results. Redundant components are usually referenced as a group, or system, such that a group kill results in a target kill; however, it may not be necessary for all redundant components to be killed in order to produce a group kill. For example, an aircraft may have four engines and is able to fly on two or more. The engines comprise the propulsion group and are individually redundant, because killing one or two does not result in a group kill. Killing three engines, however, does result in a group kill, and therefore a target kill. These components are linked by a kill tree, sometimes referred to as a disablement diagram or failure analysis and logic tree (FALT), which will be discussed in the next section.

6.10 CHARACTERIZING THE TARGET

Now we focus on characterizing the target in terms of the geometry and function of its components. This involves two separate activities.

1. The *target geometric model (TGM)* that represents the physical size, constituent materials, and location of each critical and noncritical component in the target, has to be generated. For real targets this might involve thousands of individual parts, but in our current study we will make simplifying assumptions, limiting this number to just three. In a practical sense, the TGM resides in a computer-aided design (CAD) representation of the target, which provides the necessary data relating to each component.

2. In order to determine whether a particular target kill has been achieved, we will be investigating which critical components associated with that kill definition have been killed. It is therefore necessary to have a formal association of kill definitions and those critical components needed to achieve it. This

association may be described by several terms such as a failure analysis logic tree (FALT), failure modes and effects analysis (FMEA) diagram, damage modes and effects analysis (DMEA), or a disablement diagram.

Although it may be appreciated that a given target will contain many critical components, it is important to understand that different critical components are associated with different kill definitions. For example, Fig. 6.14 shows some critical components associated with a mobility kill, such as the driver's compartment, engine, and track drive system. On the other hand, Fig. 6.15 shows a different set of components corresponding to a firepower kill. In this case, these components might be the main gun, magazine, engine (for power), and the gunner and loader in the main crew compartment.

The FALT tree is implemented by a logic flow diagram in which the critical components are connected by paths typically defined by the AND and OR operations. For example, if we consider the mobility kill shown in Fig. 6.14 and assume the critical components are

- One driver
- Two fuel tanks
- One engine
- Two track systems (track, drive sprocket, idler wheels, etc.)

Fig. 6.14   Critical components for a mobility kill.

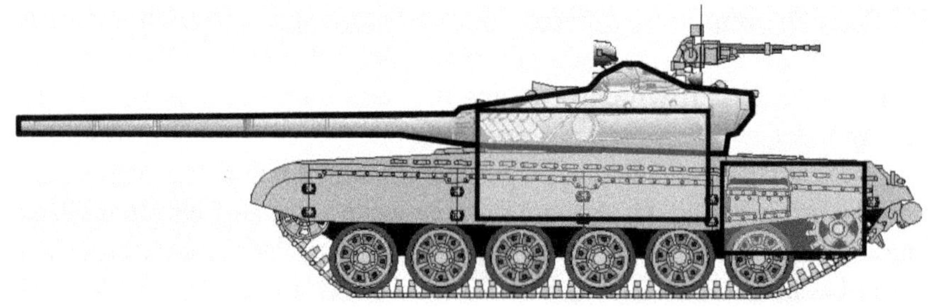

Fig. 6.15   Critical components for a firepower kill.

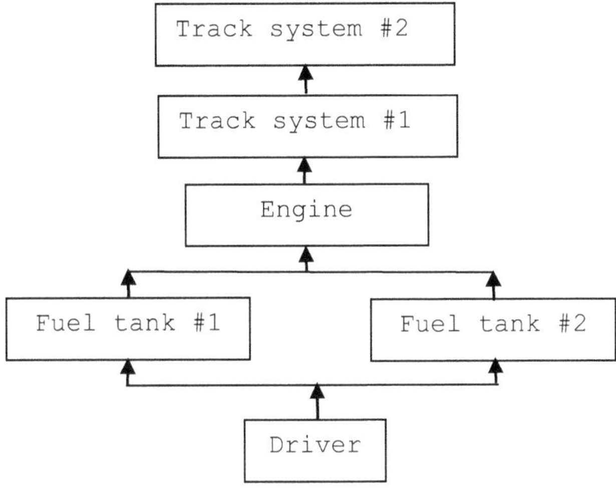

Fig. 6.16   FALT for a mobility kill.

then a mobility kill is achieved if

- The driver is killed, OR
- Both fuel tanks are killed OR
- The engine is killed, OR
- Either track system is killed.

The FALT tree would take the logical flow form shown in Fig. 6.16.

Traditionally the flow starts from the bottom and moves upwards; if the flow is unbroken, the target survives. One can see that the OR operation translates to elements in series whereas the AND operation produces parallel paths.

6.11   TARGET VULNERABILITY TO SINGLE FRAGMENTS

We will now analyze the vulnerability of a target to the penetration of a single fragment, and hence calculate $P_{K/H}$. In doing so, it is understood that the following conditions apply:

- The appropriate kill level has been defined.
- The direction the fragment approaches the target is given.
- The mass and velocity of the fragment are known.

Consider the target shown in Fig. 6.17 comprising an armored vehicle viewed from the left side with three nonredundant critical components: the magazine, the engine, and a fuel tank.

Suppose we are interested in a K-kill and have identified the three critical components that will independently produce this level of damage. It is possible to define a kill tree for this target in terms of those critical components shown in Fig. 6.18.

Fig. 6.17 Target with three critical components.

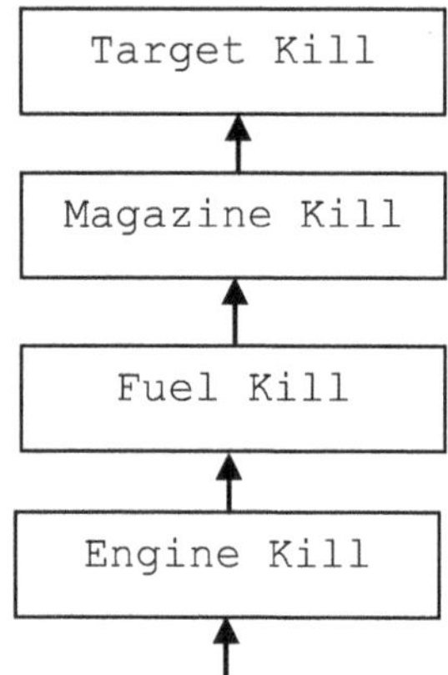

Fig. 6.18 Kill tree for target.

This indicates that the target is killed if the magazine is killed OR the fuel system is killed OR the engine is killed. We are now in a position to compile a table to define the target vulnerability, as in Table 6.5. Here we list for each critical component the presented area, probability of kill given a hit, vulnerable area, and that component's contribution to the probability of killing the complete target. Table 6.5 shows both input data and computed values for a specific fragment and assumes that we have a single fragment of known mass and velocity impacting the target from the left-hand side.

TABLE 6.5 COMPUTATIONS FOR TARGET VULNERABILITY

Component	A_{pi} (ft^2)	$\times P_{ki/hi}$	$= A_{vi}$ (ft^2)	$P_{ki/H}$
1. Magazine	20.0	1.0	20.0	0.0667
2. Engine	50.0	0.6	30.0	0.1000
3. Fuel	60.0	0.3	18.0	0.0600
$A_P = 300.0$ ft^2			$\Sigma A_{vi} = 68.0$	$P_{K/H} = \Sigma P_{ki/H} = 0.2267$

In this table, the presented areas are obtained from the TGM and $P_{ki/hi}$ entries from the fragility curves representing known data; the rest is computed. The column for A_{vi} is computed from Eq. (6.9). The total A_V is obtained from the individual component vulnerable areas.

$$A_V = \sum_{i=1}^{i=3} A_{vi} \tag{6.10}$$

The last column is the probability of killing the ith component given a hit on the target. It is computed by assuming the single fragment impacting the target has an equal (uniform) probability of striking anywhere within the presented area. The $P_{ki/H}$ is therefore just the ratio of the vulnerable area of each component to the presented area of the target.

There are only three critical components, so the probability of killing the target is the probability the magazine is killed OR the engine is killed OR the fuel system is killed.

$$P_{K/H} = \sum_{i=1}^{i=3} P_{ki/H} \tag{6.11}$$

Note that this result can be calculated by dividing the total target vulnerable area (68) by the presented area (300). Also, it is implicitly assumed that the fragment velocity and mass are constant and unchanged as it travels from the point of detonation to the target and then penetrates the target to reach the critical components.

The end result of this process is the computation of vulnerable area for the specified fragment mass, velocity, and direction (i.e., 68 ft^2), although the value of $P_{K/H}$ is also sometimes used. The value of 68 ft^2 would be one entry in the probability of damage table shown in Fig. 6.2 corresponding to whatever fragment mass and striking velocity was used to derive it. Clearly, at some point we will have to vary both of these variables in order to populate the whole table.

6.12   MULTIPLE-HIT VULNERABILITY

Now consider what happens when multiple, identical fragments hit the target, rather than just one. Assuming each impact is independent of all others, we can use the survivor rule discussed in Chapter 2, Section 2.6.

$$P_{K/H}^{(n)} = 1 - (1 - P_{K/H})^n \tag{6.12}$$

This is shown for increasing n in Table 6.6.

TABLE 6.6 $P_{K/H}$ FOR MULTIPLE HITS

Multiple-Hit $P_{K/H}$						
$n \rightarrow$	2	3	4	5	10	20	30
$P_{K/H} \rightarrow$	0.402	0.538	0.642	0.732	0.924	0.994	0.999

This completes the vulnerability assessment phase, which calculates target vulnerability (68 ft^2) for a single generic fragment.

6.13   EFFECTIVENESS ASSESSMENT FOR A SPECIFIC WEAPON

So far, the analysis has focused on a vulnerability assessment of the target; that is, calculating the vulnerability to a hypothetical single fragment, or multiple fragments of the same mass and velocity, as specified by the analyst. Now attention is turned to the second stage of the analysis defined in Section 6.3, the effectiveness study.

Knowing the $P_{K/H}$ as a function of the number of hits, it is possible to calculate the probability of kill for a particular fragmentation warhead. To do this, it is necessary to calculate the expected number of fragments impacting the target, and clearly the characteristics of the warhead and its proximity to the target at detonation have to be taken into account. In order to emphasize the nature of the process, again a simplified treatment of weapon/target interaction will be given here; a more detailed analysis will be provided in the next two chapters.

The fragment pattern projected from a particular warhead is specified in terms of the Z-data file discussed previously and reproduced for convenience in Table 6.7.

Now consider the relationship between the weapon at detonation and the target, shown in a plan view in Fig. 6.19.

We make the following simplifying assumptions regarding the scenario shown in Fig. 6.19:

- The target is totally enclosed in fragmentation zone 1 (i.e., the Z-data shown in Table 6.7), except we further assume that fragments smaller than 5 grains do not penetrate the outer armor of the target and may be ignored.
- From the Z-data file, we assume a total of $K = 90$ fragments $(40 + 34 + 16)$ are contained in the fragment zone that includes the target.
- The target corresponds to that discussed earlier having $A_V = 68$ ft^2.
- Although Table 6.7 shows fragments of different weights within the zone, we assume that all fragments have the same mass and velocity corresponding to the values used to derive Table 6.5. This assumption is questionable, but we make it in order to simplify and explain the process.

TABLE 6.7   FRAGMENT CHARACTERISTICS

α's			Velocities			# wts
0.000	20.000	10.000	5500	4770	5010	5

frag weight →	0.100	4.336	16.303	33.862	104.337
# of frags →	2.927	52.692	40.128	34.273	16.709

- The weapon detonates at a distance of $r = 80$ ft from the target and is pointed toward the target at detonation.
- The velocity of a fragment is not reduced as it travels from the detonation point to the target, or as it penetrates the target to reach any critical components.

It will be assumed that all fragments emanating from the warhead are contained on the surface of a sphere (fragmentation front), which propagates

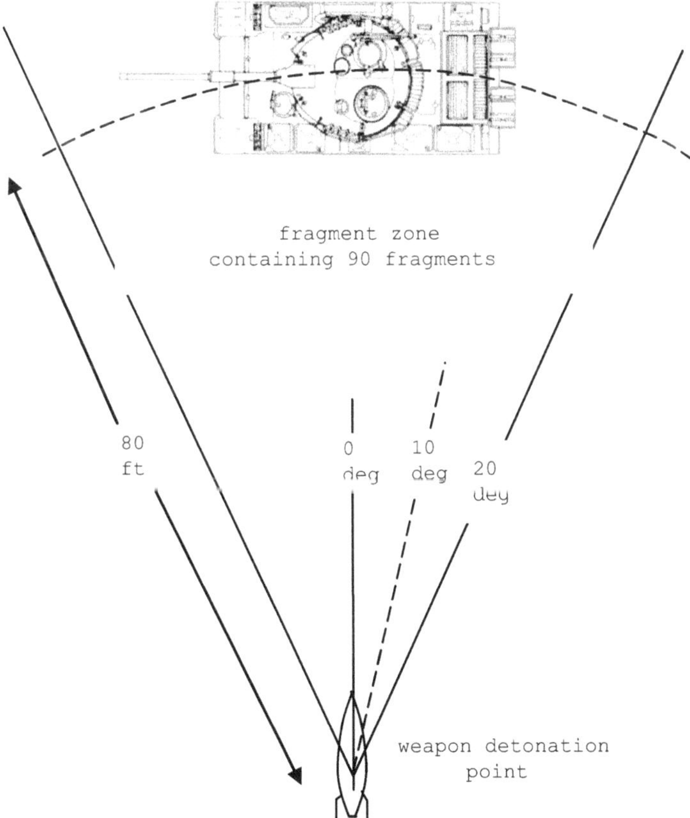

Fig. 6.19   Engagement geometry.

outward from the detonation point in much the same way a blast wave does. It is further assumed that the fragments are uniformly distributed over the surface of this sphere; because the number of fragments is fixed, the space between them increases as the sphere gets bigger.

In calculating the number of hits on the target, the three-dimensionality of the fragment spray pattern has to be considered by using some simple elements of solid geometry. Consider a cone formed by a solid angle Ω and a spherical cap on it, as shown in Fig. 6.20.

The solid angle Ω measured in steradians in terms of the conical half angle φ in radians is defined by the following:

$$\Omega = 2\pi(1 - \cos\varphi) \tag{6.13}$$

The area of the spherical cap subtending the solid angle Ω at the apex is

$$A_S = \Omega r^2 \tag{6.14}$$

If we have a hollow cone, such as fragment zone 2, where the angular boundaries are defined by φ_1 and φ_2, Eqs. (6.13) and (6.14) and lead to

$$A_S = 2\pi r^2 (\cos\varphi_1 - \cos\varphi_2) \tag{6.15}$$

where φ_1 and φ_2 are the angles defining the zone measured from the warhead axis. It is assumed that all fragments in the zone lie on the surface of the spherical cap.

For the case considered in Fig. 6.19, the 90 fragments are contained within a 40-deg cone. This cone corresponds to a solid angle [Eq. (6.13)] of $\Omega = 0.379$ steradians. At a range of 80 ft from the detonation point, this zone will project a spherical area A_S where, from Eq. (6.14), $A_S = 2425$ ft^2. Recognize, however, that not all of the 90 fragments contained on the spherical cap will

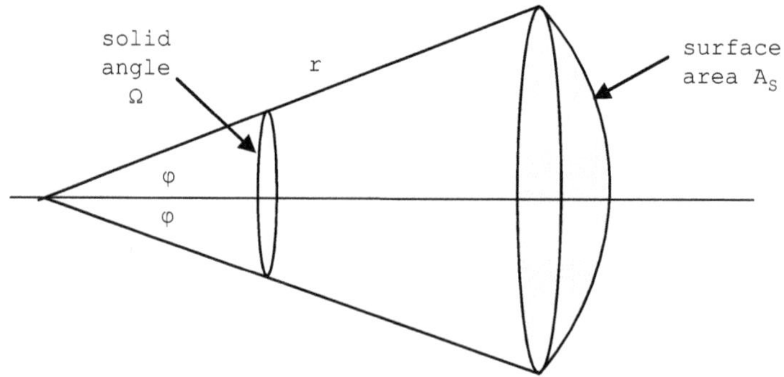

Fig. 6.20   Solid angle of a cone.

impact the target. Recalling the fragments are uniformly distributed over the spherical cap area, the number hitting the target is seen to be

$$n = \frac{A_P}{A_S} \times K = \frac{300}{2425} \times 90 \approx 11 \tag{6.16}$$

Now we know the number of impacting fragments, the $P_{K/H}$ can be computed from the survivor rule in Table 6.6. Instead of using this method, a different approach will be given here by considering the fragment density ρ having units of fragments/ft^2, which is calculated as follows:

$$\rho = \frac{\text{\# of fragments}}{\text{area containing fragments}} = \frac{K}{A_S} = \frac{90}{2425} = 0.0371 \tag{6.17}$$

We can rewrite the survivor rule to produce the following equation:

$$P_{K/D} = 1 - (1 - P_{K/H})^n = 1 - \left(1 - \frac{A_V}{A_P}\right)^n = 1 - \exp\left(-n\frac{A_V}{A_P}\right) \tag{6.18}$$

$$= 1 - \exp(-\rho A_V) = 0.920$$

Note we change $P_{K/H}$ to the probability of kill given a detonation $P_{K/D}$ because not all fragments from the detonation hit the target. This completes the effectiveness study, at least for the specific geometry shown in Fig. 6.19. Recall the two-stage process described earlier:

1. The vulnerability assessment uses the target description to produce $A_V = 68$ ft^2 for the whole target but a generic user-defined fragment.
2. The effectiveness combines this with specific warhead data in the form of a Z-data file and a specific positional relationship between the weapon and target to produce a $P_{K/D}$.

We now look at issues regarding some of the assumptions used in the analysis and how they may be relaxed.

6.14 CENTROID OF VULNERABILITY AND DIFFERENT FRAGMENT WEIGHTS

In the preceding analysis, there should be a couple of issues troubling the reader concerning the relationship between the target and warhead shown in Fig. 6.19.

- From the Z-data file it is seen that even if we assume a single fragment mass for the 90 fragments, the fragment velocity depends on the angular offset into the zone α. Because the target spans the zone, the fragment velocity will vary; however, the vulnerable area of 68 ft² assumes a single velocity. What velocity should be used?

- If the target is moved closer to the detonation point, it will eventually spill over into the next fragment zone, and we have to consider how we will account for part of the target being in zone 1 while part is in zone 2. It becomes mathematically intractable to divide the target into the zones defined for the warhead.

To resolve both of these issues, we now change the physical representation of the target from a three-dimensional object of fixed dimensions to a single point, known as the centroid or center of vulnerability (COV), together with the associated vulnerable area given in Table 6.5. This transformation is shown in Fig. 6.21.

In doing this transformation, the target loses all geometric and functional form developed in the vulnerability study; however, because it retains its associated vulnerable area, the $P_{K/D}$ may still be calculated using Eq. (6.18).

$$P_{K/D} = 1 - \exp(-\rho A_V) \tag{6.19}$$

This transformation resolves the two issues previously listed: as a point, all fragments will have a common velocity defined by a single offset into the zone, and as a point, it cannot span two zones.

We are now in a position to revisit the rather drastic assumption that all fragments hitting the target have the same mass, contradicting the variable masses shown in the Z-data file. We can, in fact, use all five weight groups leading to five different fragment densities.

$$\rho(j) = \frac{K(j)}{r^2 \Omega} \tag{6.20}$$

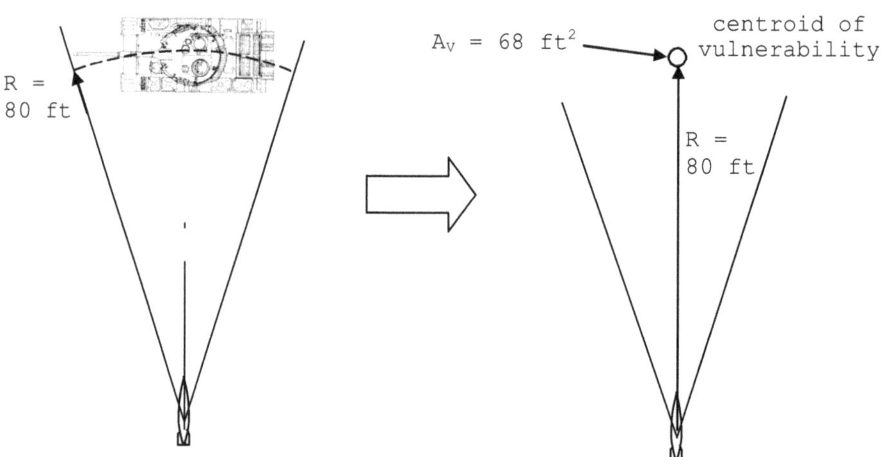

Fig. 6.21 Transforming target to the centroid of vulnerability.

The densities are different because there are different numbers of fragments belonging to each group, and this leads to a different $P_{K/D}$ for the weight group given by Eq. (6.19). Note, however, they all share the common spherical cap area A_S. Generalizing Eq. (6.19) by applying it to each weight group in the Z-data file, rather than a single aggregated fragment weight for the whole warhead, we can write

$$P_{K/D}(j) = 1 - \exp(-\rho(j)A_V(j)) \qquad (6.21)$$

Observe that not only the fragment density is dependent on the weight group due to varying numbers of fragments, but also the vulnerable area of the target due to the varying weights. For example, the 68-ft^2 assumed vulnerable area for all common fragment weights would now have five vulnerable areas corresponding to the five average fragment weights shown in the Z-data file. These would have to be evaluated by repeating the process given in Section 6.11 for the fragment weights in the Z-data file. The total $P_{K/D}$ would be given by using the survivor rule for all fragment weight groups (5), as shown here:

$$P_{K-\mathrm{FRAG}} = 1 - \prod_{j=1}^{N}(1 - P_{K/D}(j)) \qquad (6.22)$$

where there are a total of N fragment weight groups. Note that this process still uses just the fragment density for each weight group ρ and the vulnerable area for that group A_V; the presented area of the target is not used. The result is still only valid for the geometry shown in Fig. 6.19, however, so now we look at what happens when this geometry changes.

6.15 DAMAGE MATRIX AND LETHAL AREA

The value of $P_{K/D}$ computed previously is only valid at the detonation stand-off of 80 ft with the target side-on to the warhead. At a larger range, the $P_{K/D}$ will decrease because fewer fragments impact the target (i.e., the fragment density will decrease). For example, at 120-ft range we get $P_{K/D} = 0.455$. It is possible to plot $P_{K/D}$ against detonation range for this case and obtain the graph shown in Fig. 6.22.

Suppose we consider the warhead fixed and the location of the target to vary along the weapon axis. This figure shows how the probability of killing the target varies with distance from the point of detonation to the centroid of vulnerability in the direction of the weapon axis.

The question now arises: What if the target, again represented by the center of vulnerability, is located relative to the weapon at some point off the weapon axis? This situation is shown in Fig. 6.23, which is a plan view of the ground plane and the weapon/target engagement scenario. The weapon is fixed at the

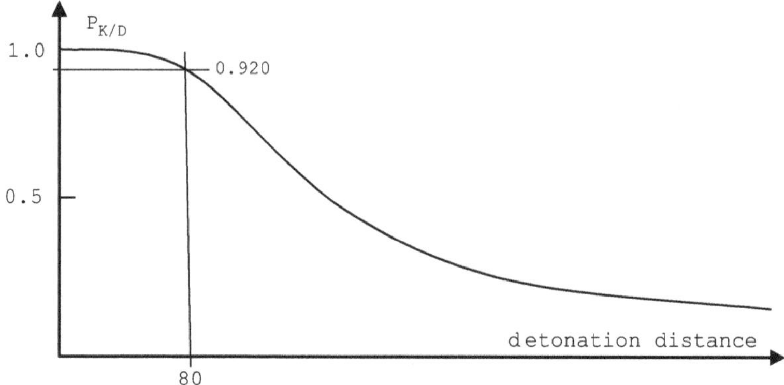

Fig. 6.22 $P_{K/D}$ **plotted against detonation distance.**

origin and the target placed relative to it. Consider covering the ground plane with a grid of cells of selected dimensions. We place the target at coordinates (x, y) at the center of an arbitrary cell having a fixed size dx by dy feet.

For this location, the value of $P_{K/D}$ is calculated in the same way as before except that now the target may be located in a different fragment zone. If the cell size is small, it may be assumed that the calculated value of $P_{K/D}$ is constant within it; however, in an adjacent cell, the value of $P_{K/D}$ would probably be different. Representing the target by the centroid of vulnerability ensures that it will be located within a single fragmentation zone and a single cell. By placing the target at the center of all cells in the ground plane and calculating the $P_{K/D}$ in each cell, we generate the so-called damage matrix, or lethal area matrix, shown in Fig. 6.24.

Note how, in practice, the value of $P_{K/D}$ becomes zero at considerable distance from the detonation point. This is because when air resistance on the fragments is accounted for, at large distances from the detonation point they slow to such a small velocity they fail to penetrate the target. The lethal area matrix will therefore be bounded.

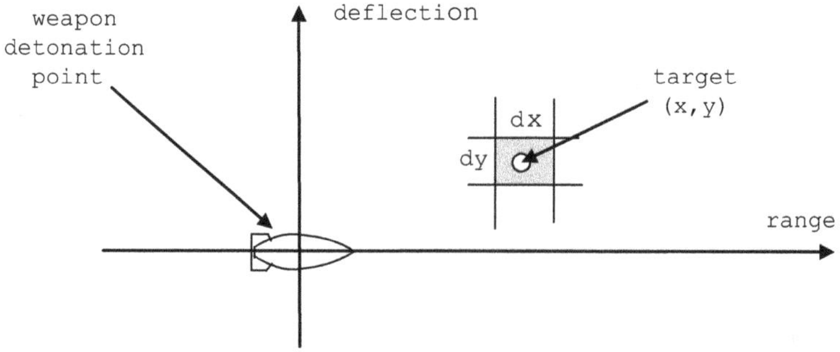

Fig. 6.23 **Calculation** $P_{K/D}$ **at a point** (x, y)**.**

0	0	0	0	0	0	0	0	0	0	0	0	0	0
0	0	0	0	0	.001	0	0	.001	0	0	0	0	0
0	0	0	0	.001	.002	.003	.003	.002	.001	0	0	0	0
0	.001	.002	.003	.004	.007	.014	.014	.007	.004	.003	.002	.001	0
0	0	0	.002	.008	.04	.227	.227	.04	.008	.002	0	0	0
0	0	0	0	.001	.002	.049	.049	.002	.001	0	0	0	0
0	0	0	0	0	0	.002	.002	0	0	0	0	0	0
0	0	0	0	0	0	0	0	0	0	0	0	0	0

Fig. 6.24   Damage matrix.

The lethal area is calculated by multiplying each cell area by the value of $P_{K/D}$ inside it and summing over all cells with a nonzero $P_{K/D}$. This area may be written mathematically as

$$A_L = \text{MAE}_F = \sum_{x=x_{\min}}^{x=x_{\max}} \sum_{y=y_{\min}}^{y=y_{\max}} P_{K/D}\Delta y \Delta x \qquad (6.23)$$

This represents the final result from our vulnerability study and produces a single lethal area number from Eq. (6.23) with units of square feet. These are how the lethal area numbers shown in Table 6.2 are derived and how the lethal area is equated to the mean area of effectiveness MAE_F, one of the EI types discussed earlier.

Observe that the effectiveness has been represented in two ways.

1. The lethal area matrix shows the distribution of effectiveness of the weapon against the target for specific placement of the latter.
2. The EI is a single number representation of the lethal area lacking any information about the shape of the effectiveness distribution.

It is also important to remember the following points:

- The vulnerable area, $P_{K/H}$, and lethal area are dependent upon the fragment weight and velocity. Recall Fig. 6.10 to realize that changing the weight and velocity will change $P_{ki/hi}$, changing all the computations in later sections.
- All computations are dependent on the aspect of the target. The cases considered assume fragments approach the target from the left. If this direction changes to, say, from above the target, the $P_{ki/hi}$ will also change because different noncritical components may shield the critical components to a greater or lesser degree, depending on the physical internal layout of the target.
- All the measures discussed are dependent on the kill level of interest. Recall that the fragility curves shown in Fig. 6.10 are defined for a specific kill level. If the kill level becomes more severe, for example, the $P_{k/h}$ values will change and probably decrease.

Clearly, some way has to be found to take these factors into account when determining the lethal area values required by the weaponeering process.

Unfortunately, this requires employing more detailed techniques than those described in this chapter, and numerical answers may only be obtained by using computer programs to implement these more complex methodologies. That being said, the next chapter will show that these methodologies have their basis in the somewhat simple methods described here.

6.16 VULNERABILITY ASSESSMENT FOR BLAST WARHEADS

So far, we have focused on damage due to fragments, so in this section a target that is vulnerable only to blast is considered. Superficially, blast appears simpler to model than fragments because the blast wave propagates spherically outward from the point of detonation with decreasing amplitude. It is reasonable to argue, therefore, that the amount of damage sustained by a target is a function only of its distance from the detonation point, as shown in Fig. 6.25. Note again that the target will be represented by a point, so we don't need to consider if parts of the target are closer to the warhead than others.

For a given combination of warhead, target, and kill definition, we define a distance called the blast radius R_B within which a kill is achieved and outside of which the target survives (i.e., a sphere centered on the detonation point). For an air burst weapon, the equivalent ground plane area would be a circle of radius R_{BG} where

$$R_{BG}^2 = R_B^2 - \text{HOB}^2 \tag{6.24}$$

where HOB is weapon height of burst. Clearly for HOB>R_B, there will be no ground plane effect due to blast; if the weapon detonates on the ground, then $R_{BG} = R_B$. If there is a ground plane effect, then we can define another effectiveness index—the mean area of effectiveness due to blast (MAE_B), where

$$\text{MAE}_B = \pi(R_{BG})^2 \tag{6.25}$$

The next chapter will present details of how R_B is calculated.

6.17 CHAPTER SUMMARY

* The concept of an effectiveness index (EI) was introduced and it was shown how its valued depends on the weapon, target, and kill definition.
* Calculating the EI involves two processes, a vulnerability study and an effectiveness study.
* Arena tests were described as an accepted way to collect weapon fragmentation data, which are then represented in a Z-data file.

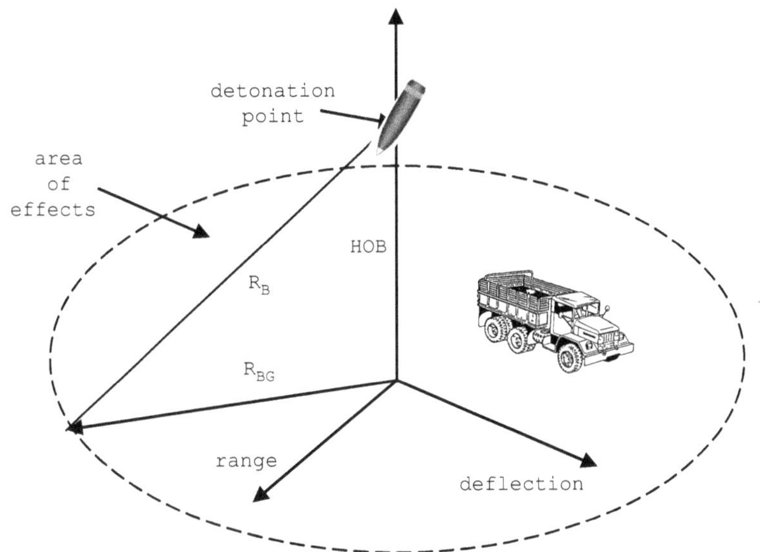

Fig. 6.25   Effects area for blast warhead.

- Targets were represented by a target geometric model (TGM) specifying all internal components, their construction, and their location within the target. A fragility curve is required for all critical components. The target was also represented by a FALT diagram showing which components lead to a specific kill definition.
- An example of a vulnerability study was given leading to the target vulnerable area for user-specified fragments. The effect of multiple fragments was also explained.
- The effectiveness study calculated the damage when the target is placed a fixed distance from a specific weapon, characterized by the Z-data file.
- The concept of the centroid of vulnerability was defined to make the calculation of P_K more tractable and to relax some of the assumptions made previously.
- Keeping the weapon fixed and moving the target to the center of a grid of ground plane cells produces the lethal area matrix.
- The lethal area matrix was used to calculate a single value measurement of effectiveness. This is the EI value and is defined as the MAE_F.
- Vulnerability due to blast effects was discussed, leading to the MAE_B effectiveness index.

REFERENCE

[1] Ball, R., *The Fundamentals of Aircraft Combat Survivability Analysis and Design*, 2nd ed., AIAA Education Series, AIAA, Reston, VA, 2003.

Chapter 7

Vulnerability Implementation

7.1 Introduction

The previous chapter provided an overview of the concepts and methodology to determine the effects of fragments and blast warheads on single targets comprising critical and noncritical components. This was divided into a vulnerability study that produced vulnerable area data for user-specified fragments followed by an effectiveness assessment where the effect of a specific warhead was introduced. Because of the complexity in implementing these two phases, this chapter will focus on only the first of these; the following chapter will deal with effectiveness calculations.

In the examples used previously, the target was considered to comprise three critical components only, and simplifying assumptions were made regarding the penetration of the target by fragments. Some of these assumptions were:

- The effects of noncritical components were not included.
- No velocity or mass reduction as the fragment penetrated the target was allowed for.
- The fragments penetrated the target from one direction only.

Clearly, these restrictions will have to be addressed when real targets are considered, as will the target/weapon geometry.

The material presented in this chapter will transition the simple, generalized methods introduced previously into those dealing with more realistic target and weapon systems, and in doing so will frame the analysis in terms of those tools used by the U.S. Department of Defense (DoD) to conduct these analyses.

7.2 Complex Targets Sensitive to Fragments

As in the previous chapter, the vulnerability assessment for more complex targets consists of generating the vulnerable area of the target to single fragments of specified weight traveling at a specified speed and direction when impacting the target.

Simplified methods were developed in order to compute the lethal area of a target for a given fragment; however, more advanced methods are required to deal with more complex targets. For example, subtle changes in target design may provide substantially more shielding of vulnerable components, leading to a reduction in the vulnerable area and mean area of effectiveness (MAE).

The example shown in Fig. 7.1 clearly illustrates a complex target that may be made up of thousands of individual components.

For a target such as this, it is not surprising that the specification of the target, groups, subsystems, individual components, their fragility curves, and kill tree relationships takes considerable time. The level of detail in specifying these data is always traded off against the accuracy of the resulting vulnerability assessment. Higher fidelity models produce more accurate predictions but take more resources to develop.

In addition, the availability of detailed data on newly fielded systems has to be considered. It may be that only photographs of the target are available,

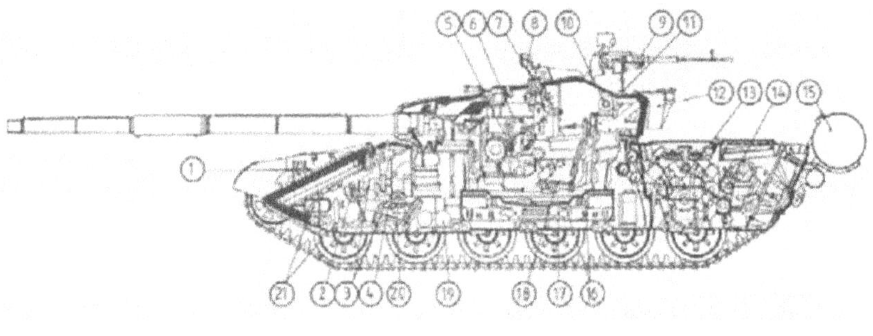

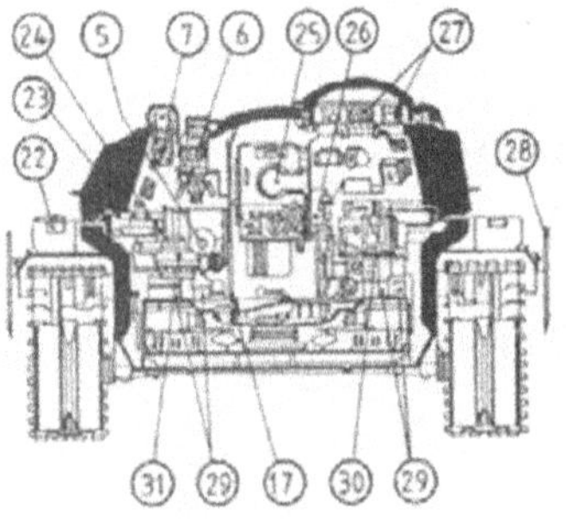

(1) driver's FG 125 headlamp (2) steering tillers (3) NBC system (4) gear lever (5) gun elevation mechanism (6) TPD-2 gunner's sight (7) TPN 1-49-23 gunner's night sight (8) searchlight for use with TKN 3 observation device (9) 12.7 mm anti-aircraft machine gun to rear (10) hoist for ammunition containers (11) antenna base (12) turret bin for deep fording equipment and cold rations (13) 780 bhp diesel engine (14) gearbox (15) long-range fuel tanks which can be jettisoned (16) charges and projectiles in containers on ammunition hoist platform (17) ammunition hoist platform (18) gunner's seat (19) NBC decontamination equipment (20) driver's adjustable seat (21) parking brake (22) external stowage for spares, tools and accessories (23) manual drive for turret traverse gear (24) turret traverse indicator (25) 125 mm gun sliding wedge breech-block (26) 7.62 mm PKT coaxial machine gun (27) tank commander's periscopes (28) skirting plates (29) 7.62 mm PKT machine gun ammunition boxes (30) radio set (31) hydraulic turret traverse gear

Fig. 7.1   Example of complex target.

providing no information on the internal design and layout. In such cases, the analyst may have to resort to surrogation methods from what is perceived to be a close match to a known target for which data are available.

7.3 OVERVIEW OF TOOLS USED FOR VULNERABILITY ANALYSIS

Although the process has become more complex, the same basic process described in Chapter 6 will be repeated; that is, a user-defined threat (fragment, projectile, shaped charge) is imposed on a target resulting in a vulnerable area for a given direction of approach. By initially considering the threat to be fragments, as in the previous chapter, the following requirements now have to be considered:

- Because the target is more complex, a formal description of the target geometric model (TGM) has to be developed in order to include all critical and noncritical components, the materials that make up the components, and the spatial relationship between them within the target itself.
- An equally complex failure analysis logic tree (FALT) analysis has to be developed in order to identify which critical components lead to different kill definitions associated with the target under study.
- Fragment impact directions have to be specified so that the vulnerability from fragments coming from any direction may be determined.
- The loss of fragment velocity and mass as it penetrates through the target and the components it encounters has to be calculated.

These requirements have existed for more than 50 years and have been met by a set of tools assisted by increasingly capable computational support. Although many of these tools may be operated in a stand-alone manner, their combined use to achieve overall vulnerable area calculations has resulted in several packages including the Computation of Vulnerable Area Tool (COVART)* and the Advanced Joint Effectiveness Model (AJEM). We will describe in detail the COVART suite of programs; for our purposes, AJEM performs similar tasks.

COVART has been developed with service-wide support to provide a consistent assessment of a specific target vulnerability to different threats. It contains several modules that address the requirements listed previously, and is shown in functional form in Fig. 7.2. Although we will describe COVART in terms of its constituent modules, a more complete description may be found in [1].

In the next few sections, we will address each of the four previously listed requirements needed to address complex target vulnerability assessment in terms of the modules in COVART that perform the equivalent task. Again, we will restrict the discussion to those details needed to understand the module function; if more details are needed, the reader is directed to [1] and [2].

---

* Also defined as Computational of Vulnerable Area and Repair Times.

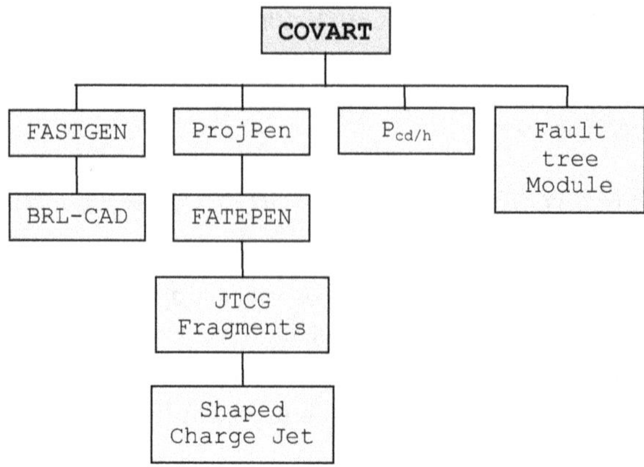

Fig. 7.2 COVART structure.

7.4 TARGET GEOMETRIC MODEL (TGM)

The formal representation of a target in terms of its constituent components is in the form of a computer-aided design (CAD) model incorporated into a programming environment known as BRL-CAD [3], as indicated in Fig. 7.2. This system is a powerful open source programming structure running on a variety of platforms. Modeling may be achieved using several commonly accepted techniques such as the following:

- Constructive solid geometry (CSG) objects based on a library of primitive shapes
- Explicit boundary representation (BREP)
- Non-uniform rational B-spline (NURBS) surfaces
- N-manifold polygonal mesh geometry (NMG)
- Triangle mesh (BoT) geometry

By way of example of the first of these [4], a set of primitive shapes, such as the ones shown in Fig. 7.3, can be used with suitable Boolean operations to form more complex shapes, such as those shown in Fig. 7.4.

Using this process, targets, subsystems, and individual components may be constructed from engineering drawings or translated from other commercial CAD formats such as CATIA, Solidworks, and Pro/Engineer, as shown in Fig. 7.5. This allows both surface models (Fig. 7.6) and interior component models (Fig. 7.7) to be generated.

The important consequence of modeling targets in this manner is that when a fragment approach direction is defined and the target model is rotated so that the direction is normal to the viewing plane (Fig. 7.7), all the internal components are also rotated. This means that if a line of sight for fragment penetration is visualized normal to the viewing plane, all components along that line

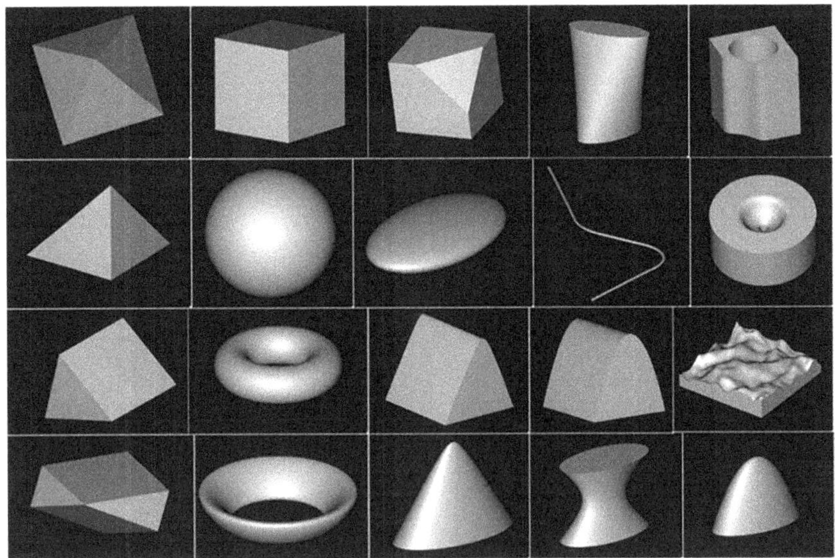

Fig. 7.3 Examples of BRL-CAD primitive shapes.

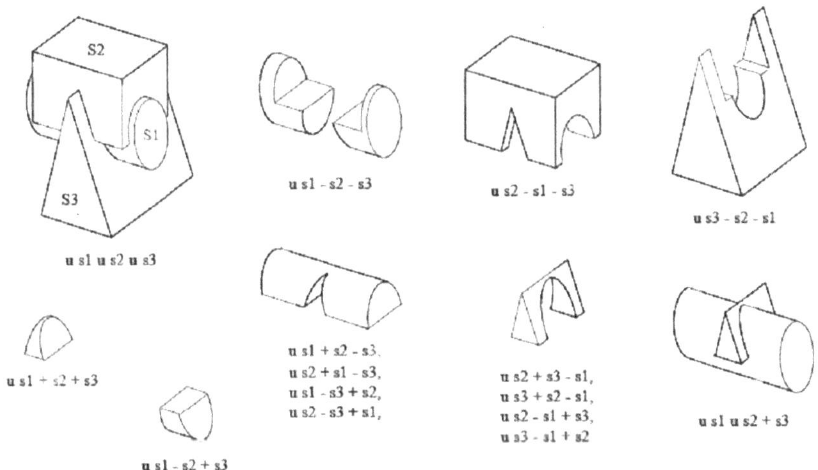

Fig. 7.4 CSG complex shape generation from primitives.

of sight may be extracted from the CAD model. This line of sight is usually called a *shotline* and will be dealt with shortly.

7.5 *FAILURE ANALYSIS LOGIC TREE (FALT) DIAGRAMS*

Just as the TGM increases in complexity when real targets are considered, so do the FALT diagrams associated with the appropriate failure modes. This is represented as the fault tree module in Fig. 7.2. Because of the complexity

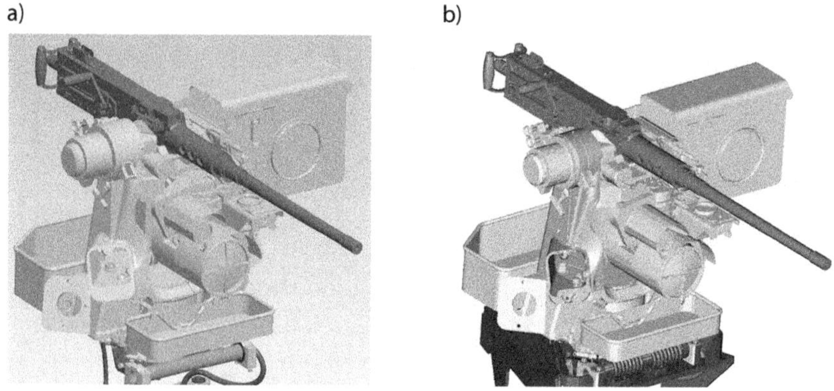

Fig. 7.5   a) Commercial and b) CAD file converted to BRL-CAD.

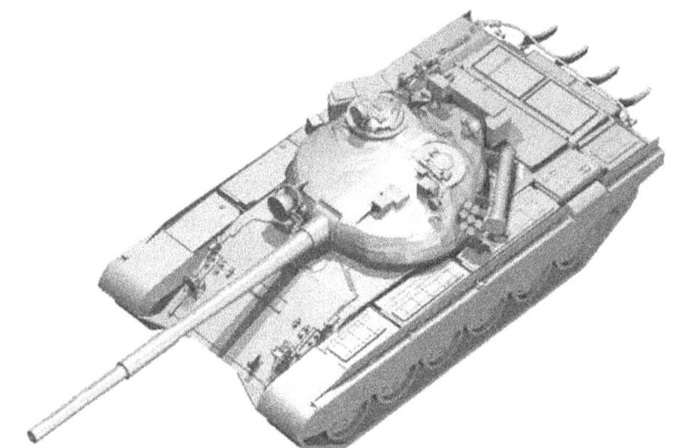

Fig. 7.6   BRL-CAD target surface model.

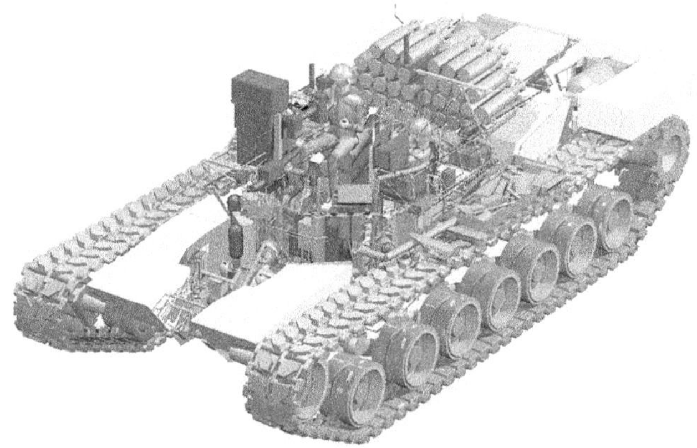

Fig. 7.7   BRL-CAD target interior model.

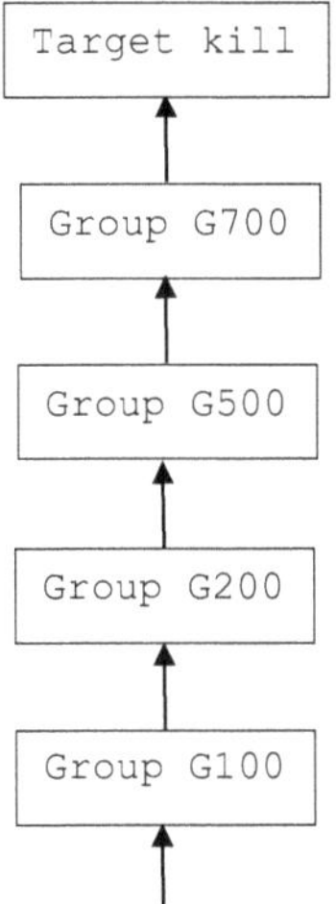

Fig. 7.8   Top-level FALT diagram for fragment kill (mobility).

of a FALT diagram for a complete system, it is usually structured in a hierarchical manner where the top level corresponds to the major systems leading to a kill, the next level addresses for each system those subsystems that would kill the parent system, and so on down to the component level.

In order to illustrate this approach and to allow unlimited distribution of this material, individual critical components will not be given their real names but defined by a four-digit numerical number such as 5213 or 0204. Consider the mobility kill definition for the target shown in Fig. 7.8. This level of damage may be achieved by either fragments, blast, penetrators, or projectiles. For fragment damage, there is a hierarchy of systems, subsystems, and components leading to a target kill. This may be written in the same logic tree form, using the convention discussed in Chapter 6, Section 6.10.

These groups might correspond to engine, tracks, driver, and so forth. This logic tree produces a mobility kill in the following manner:

Target kill if [G100 killed OR G200 killed OR G500 killed OR G700 killed]

For this particular target, each subgroup—G100, G200, G500, and G700—contains 3, 24, 138, and 5 critical components, respectively. A corresponding logic diagram for group G100 might be as shown in Fig. 7.9.

Note that in this example there are two redundant components—1003, 1004—which correspond to the following statement:

G100 kill if [Component 1002 killed OR (Component 1003 AND Component 1004) killed]

For blast there are five critical components, any of which produces a mobility kill (see Fig. 7.10).

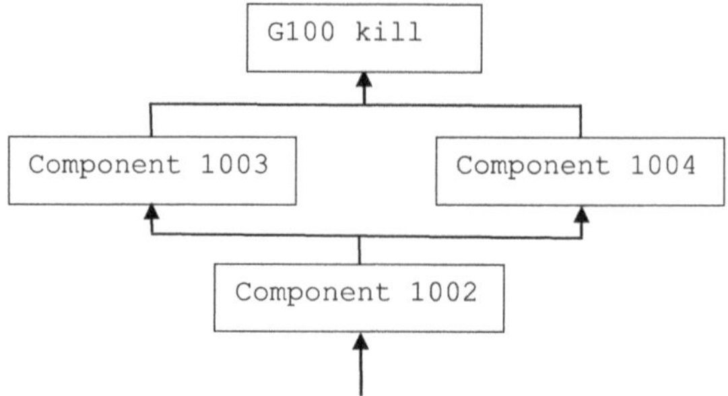

Fig. 7.9   FALT logic for group G100.

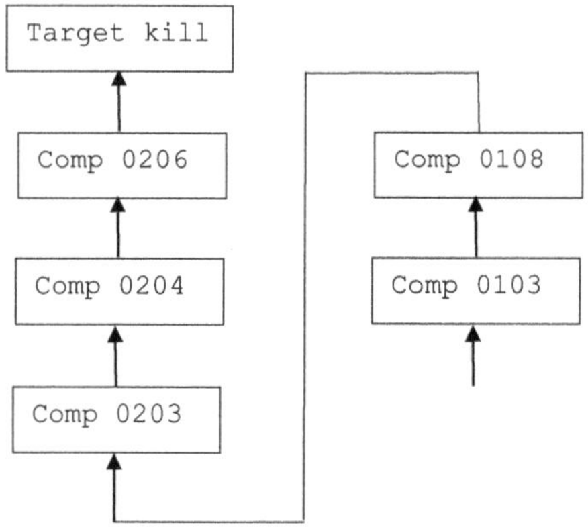

Fig. 7.10   FALT diagram for blast kill.

This corresponds to the logic statement

Target kill if [0103 killed OR 0108 killed OR … OR 0206 killed]

The complexity of the target failure modes and associated FALT diagrams will depend on the knowledge of the internal components of the target and how each contributes to a different kill definition.

7.6   DEFINITION OF FRAGMENT IMPACT ANGLES

The simple example dealt with in the previous chapter assumed all fragments impacted the target horizontally from the left side. To generalize this, we allow impacts from discrete angles in both azimuth and elevation, as shown in Fig. 7.11.

Note that a number of increments in azimuth and elevation are chosen and each combination will produce different values of A_V, $P_{K/H}$, and lethal area. It should also be noted that a target may appear externally to be symmetrical about a longitudinal axis, but internally the layout of components is not

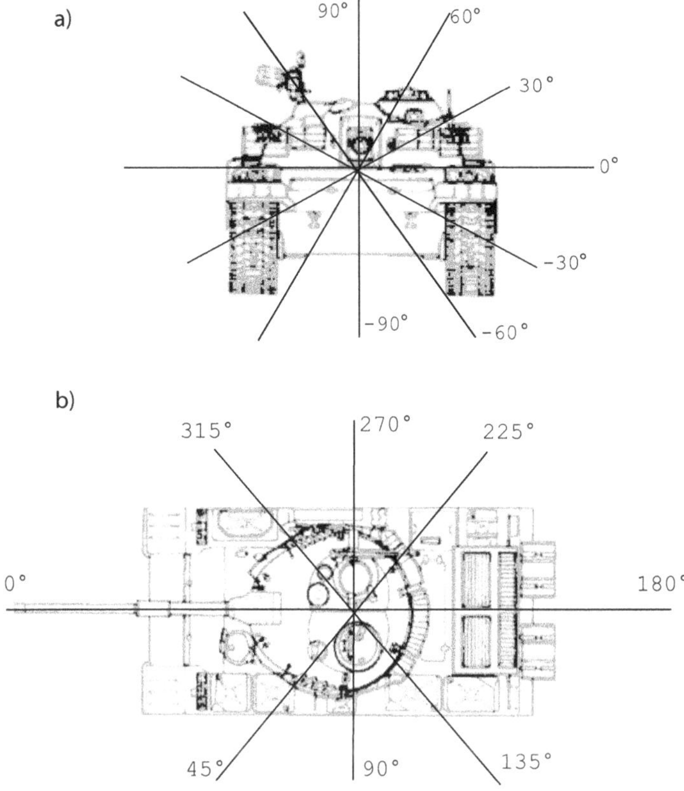

Fig. 7.11 Directions from which fragments approach target: a) elevation angles, b) azimuth angles.

symmetrical. For this reason, internal symmetry should not be assumed, and all azimuth angles should be considered. Negative elevation angles are also defined as these relate to elevated targets such as a radar dish on a tower or to evaluate the effect of mines on the bottom of targets passing over them.

7.7 SHOTLINE ANALYSIS

The last requirement needed to analyze complex targets is usually termed *shotline analysis* and deals with how a user-defined fragment traveling in a straight line through a target interacts with the various critical and noncritical components it encounters. For a given fragment approach direction (azimuth and elevation), the BRL-CAD model of the target is rotated to the resulting aspect and overlaid with a grid of cells of uniform dimensions with each cell containing one shotline, as shown in Fig. 7.12.

In rotating the target shown in Fig. 7.12, all the internal components will be rotated, too. A shotline is defined as the line of sight perpendicular to the normal viewing plane passing through the target.

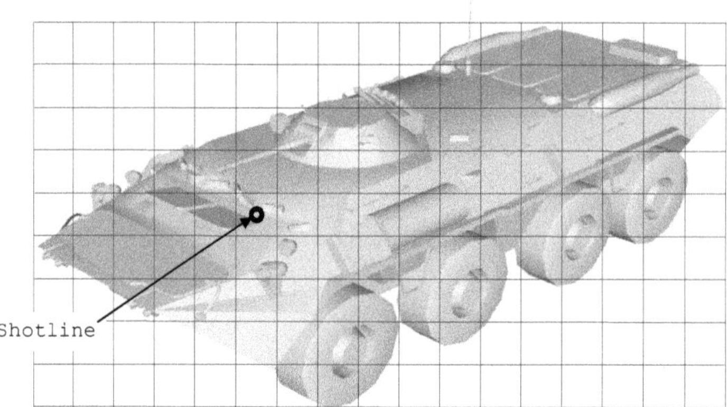

Fig. 7.12　Shotline grid overlaying target.

Grid-line spacing is user defined, but obviously more closely spaced grids will require more computational time. For many targets, a grid of 4 in. × 4 in. has been found adequate, although grids as small as 1 in. × 1 in. have been used. Once the grid has defined the shotlines, it is assumed that the weapon detonates some distance from the target so that the individual fragment paths and therefore the shotlines themselves are parallel.

The module in COVART that generates shotlines through the target as represented by the TGM is the Fast Shotline Generator (FASTGEN), which is shown in the overall COVART architecture in Fig. 7.2 earlier in the chapter.

The shotline may be located randomly within the cell or fixed at the center of each cell; components lying along each shotline are determined from the transformed TGM. This defines not only which components are impacted by the fragment, but also in which order. Shotline analysis then must make three calculations.

1. How does the fragment mass and velocity change as they propagate down the shotline and encounter a component?
2. What is the probability of killing that component?
3. What is the resulting probability of killing the target?

Note that in the vulnerability study we do not consider the reduction in fragment velocity as it moves from the point of detonation to the target, so only the velocity with which the fragment first impacts the target, the striking velocity, is of interest.

The first of these questions is probably the most difficult one to answer, because it raises two further issues: How do we represent the penetration of components located along a shotline, and how does that representation affect the mass and velocity of a fragment as it passes through a component?

7.8 COMPONENT REPRESENTATION ALONG A SHOTLINE

Once FASTGEN has been used to determine, for particular azimuth and elevation directions, those components a fragment will encounter, it is now necessary to calculate the damage done to each component due to a specified fragment mass and striking velocity. Each component is represented by two different mechanisms, a fragility curve and an equivalent plate.

The *fragility curve* was discussed in Chapter 6, Section 6.6 and expresses the probability of killing the component when it is impacted by a fragment of given mass and velocity. As before, a fragility curve is required for all critical components of the target; these data reside in the $P_{cd/h}$ ($= P_{ki/hi}$) module in COVART (see Fig. 7.2).

The *equivalent plate* is an attribute associated with both critical and non-critical components and is used to determine the loss of fragment mass and velocity as it passes through them. Most components are made up of collections of smaller physical parts, so trying to assess the interaction of a fragment with these individual parts becomes intractable, at least in terms of how much mass is removed and how the fragment slows down; therefore, these effects are embodied into a single plate.

For example, a vehicle battery might be considered a critical component for an M-kill; therefore, a fragility curve would have to be determined for it. In addition, if this battery was on a shotline of interest and a fragment passes through it, that fragment would experience a reduction in mass and velocity as it does so. Consider a fragment passing along a horizontal shotline through the battery, as shown in Fig. 7.13. It will encounter the case, a series of lead plates separated by the electrolyte, and again the case as it exits.

Instead of considering the individual parts, the battery as a whole is represented for penetration purposes by a single plate. In defining the plate, the analyst may select the dominant material, such as lead, and a thickness factor

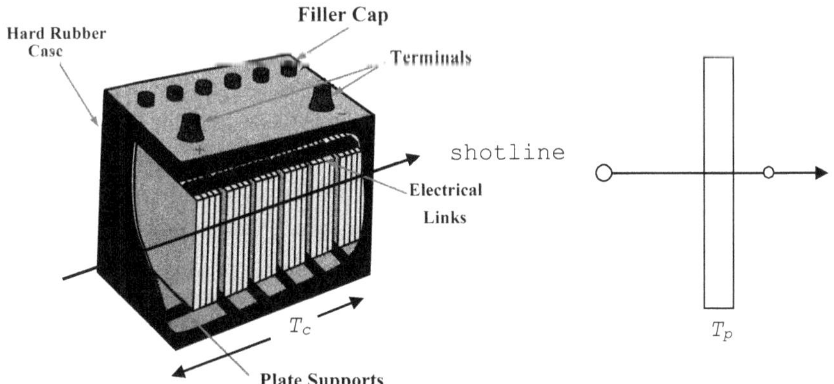

Fig. 7.13 Component represented by a single plate.

to relate the physical path length through the component T_C to the equivalent plate thickness T_P. If the width of the battery (i.e., the shotline length) is 12 in. and the thickness factor is 10% (T_P/T_C), then the battery is represented by a lead plate of thickness 1.2 in.

Some components, such as the protective armor on the side of a tank, may be accurately represented by a single plate, but others, such as the battery, require subject matter expert (SME) assessment. Assuming all critical and noncritical components are represented by equivalent plates, we can now determine the loss of mass and velocity of a penetrating fragment through a series of plates of a given material using penetration equations.

7.9 PENETRATION OF COMPONENTS ALONG A SHOTLINE

Once FASTGEN has identified the components along a particular shotline, the Fast Air Target Encounter Penetration program (FATEPEN) tracks a user-supplied fragment (mass and velocity) as it moves along it. In this process, the fragment will lose mass and velocity as it perforates each component until it reaches a component it cannot penetrate, or it can penetrate but cannot exit. If none of these conditions occurs, the fragment will exit the target at the end of the shotline.

Within the overall COVART structure (Fig. 7.2), FATEPEN deals with fragments, and ProjPen and Shaped Charged Jet do the same for different penetrating threats. Focusing on fragments, FATEPEN uses a combination of fundamental principles of mechanics supplemented with experimental data. It also allows for fracture of a component into a debris cloud, which it then tracks to determine its loading on subsequent plates.

In order to calculate mass and velocity degradation, the Joint Munitions Effectiveness Manual (JMEM) uses two methodologies for this process: THOR equations [5] and an internal FATEPEN model [6]. Results were obtained from extensive testing of regular-shaped fragments fired at plate arrays, such as those shown in Fig. 7.14. Based on results from such tests, the loss of speed and mass of a fragment may be predicted.

Although the THOR model worked well for relatively slow-moving fragments, once a material-dependent critical impact velocity is exceeded, the fragments tend to shatter rather than remain intact, and the damage mechanism changes. Sample results for residual mass and velocity for a particular combination of fragment and plate together with corresponding test data are shown in Figs. 7.15 and 7.16 [5, 6].

The FATEPEN model replaces the THOR model for faster fragments and predicts the new plate damage observed in these impacts. FATEPEN is applicable over the full range of impacts of interest to the Joint Technical Coordinating Group for Munitions Effectiveness (JTCG/ME) and provides a uniform predictive methodology over the complete range of impact speeds.

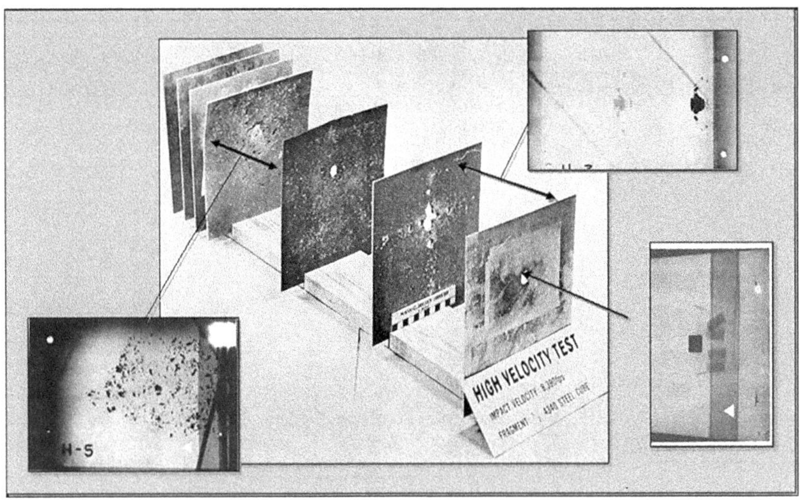

Fig. 7.14   Experimental fragment penetration test.

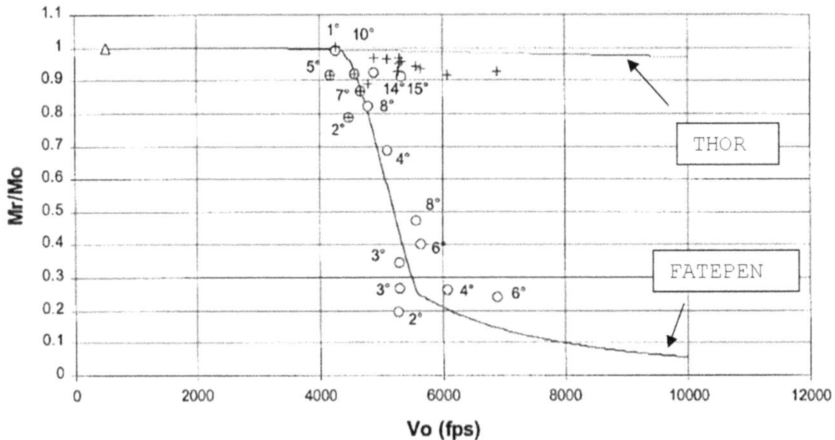

Fig. 7.15   Normalized residual fragment mass vs velocity.

This resulted in a common set of penetration equations described in the FATEPEN User and Analyst Manuals that may be used for a number of vulnerability models including COVART and AJEM.

7.10   CALCULATING THE SHOTLINE $P_{K/SH}$

Now that we have seen how a fragment can lose mass and velocity as it passes through successive components (plates), we focus attention on what target-specific components we might expect to find. It is assumed that the fragments passing along the shotlines are parallel to each other and no ricochet occurs. For example, a horizontal shotline passing through the target

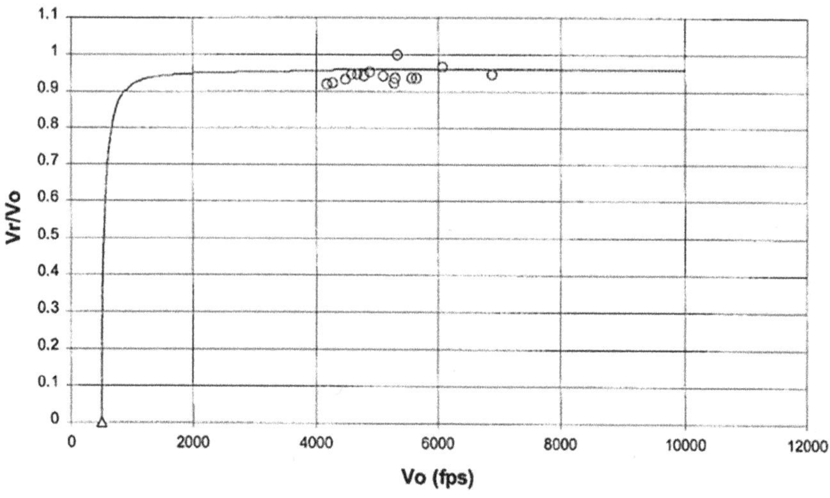

Fig. 7.16   Normalized residual fragment velocity.

shown in Fig. 7.17 might encounter the critical and noncritical components indicated. Symbolically, these components may be represented by C_1, C_2, ..., C_n, as shown in Fig. 7.18.

The fragment's mass and velocity when it impacts the first component are known, and from the TGM we know the physical composition of the first component, or plate equivalent. We may then use the penetration equations described earlier to determine the exit velocity of the fragment as well as any loss of material (mass) due to its passage through component C_1. The

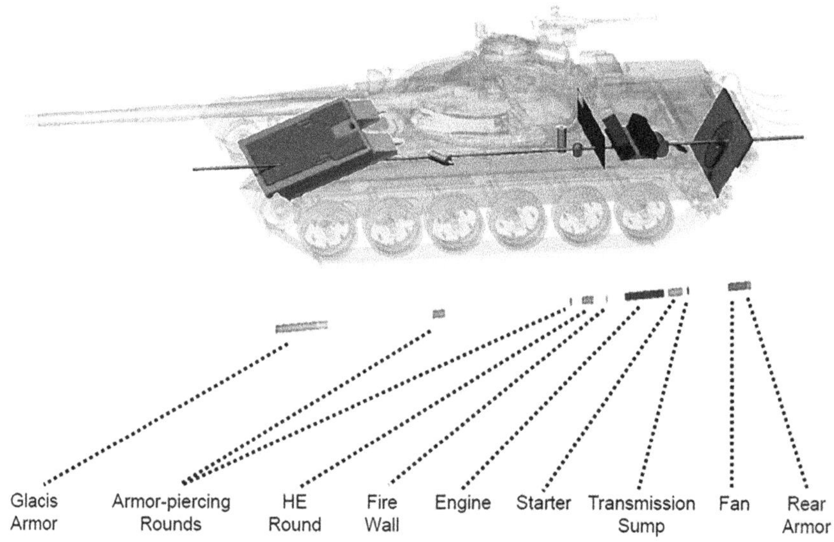

Fig. 7.17   Components encountered along a shotline.

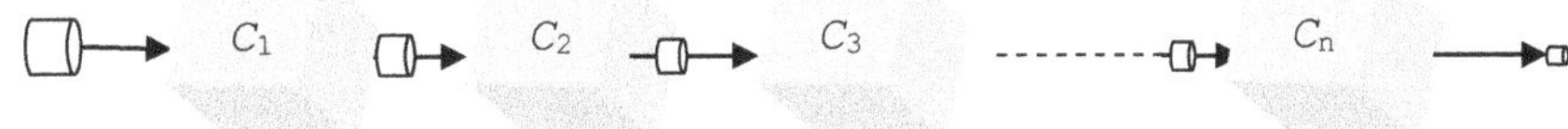

Fig. 7.18 Shotline passing through multiple components.

process is repeated until the fragment fails to enter/exit a component or it exits the target.

Fragility curves for all critical components are a required input for COVART, so knowing the fragment weight and velocity as the fragment impacts each component, we may interpolate the fragility curve to determine a $P_{ki/hi}$ value for each critical component on the shotline. Suppose the fragment passes through n critical components as it moves along the shotline. The probability that all components survive is given by

$$P_S = \left(1 - P_{k1}\right)\left(1 - P_{k2}\right)\left(1 - P_{k3}\right)\cdots\left(1 - P_{kn}\right) \tag{7.1}$$

Because all components in Eq. (7.1) are critical, the probability of killing the target if the fragment traverses this particular shotline $P_{K/sh}$ equals the probability that one or more critical components are killed.

$$P_{K/sh} = 1 - \left(1 - P_{k1}\right)\left(1 - P_{k2}\right)\left(1 - P_{k3}\right)\cdots\left(1 - P_{kn}\right) = 1 - \prod_{i=1}^{i=n}\left(1 - P_{ki}\right)$$

$$\tag{7.2}$$

In this equation, $P_{K/sh}$ is the shotline P_K and may be considered the $P_{K/H}$ if the fragment passes along that shotline. Should a component on the shotline be noncritical, the reduction of mass and velocity still has to be calculated, but such a component will not contribute to the shotline P_K.

Once the shotline $P_{K/sh}$ has been accumulated, the shotline vulnerable area of the cell is calculated from the following equation:

$$A_{v-\text{cell}} = P_{K/sh} \times A_{p-\text{cell}} \tag{7.3}$$

Here, $A_{p-\text{cell}}$ is the area of the target contained within the cell. This will usually be the same as the cell area, but there will be some cells on the edge of the target where the presented area of the target is less than that of the cell (see Fig. 7.12).

The vulnerable area of the target is then found from summing the cell A_v's over all of the p cells that contain the target.

$$A_V = \sum_{i=1}^{p} A_{v-\text{cell}} \tag{7.4}$$

Finally, the target P_K is found from the following equation:

$$P_K = \frac{A_V}{A_P} \qquad (7.5)$$

As before, A_P is the presented area of the target for this aspect. Note the difference between $P_{K/H}$ and P_K. Whenever we compute shotline probability of kill in this way, we will use the notation P_K, because it is assumed that the target is hit by a single fragment at the cell considered.

7.11 COVART COMPUTATIONAL MODEL FOR CALCULATING VULNERABLE AREAS

Now that the shotline analysis is complete, we can describe the methodology to compute vulnerable areas of a complex target for a range of user-specified aspect angles, fragment weights, and fragment velocities. This process may be defined by the following steps:

1. The model selects an azimuth and elevation angle from ranges of values input by the user to define the directions in which fragments approach the target. Typical initial values might be 90 deg and 0 deg, corresponding to the left-hand view of the target analyzed in the previous chapter.
2. The model selects a fragment weight and velocity from user inputs.
3. The target vulnerable area A_V is computed using the shotline techniques described in Section 7.10.
4. The next fragment weight is selected, and the process is repeated.
5. When all fragment weights have been processed, the fragment velocity is incremented and the process repeated again for all fragment weights.
6. When all fragment velocities have been processed, the azimuth angle is incremented by 45 deg (see Fig. 7.11), and the whole process is repeated. At this point, we have a table of vulnerable areas for user-specified ranges of fragment weights and masses for every combination of azimuth and elevation angles.
7. The vulnerable areas corresponding to a particular fragment weight and velocity, and a fixed elevation angle are averaged over the eight azimuth angles, where only the nonzero vulnerable areas are averaged.
8. For each vulnerable area, a corresponding probability of exposure (PE) is computed from

$$PE = \frac{\text{Number of azimuth angles where } A_V \neq 0}{\text{Number of azimuth angles for which } A_V \text{ was computed } (=8)} \qquad (7.6)$$

For example, consider Table 7.1, which shows a vulnerable area as a function of azimuth angle.

TABLE 7.1   SAMPLE A_V FOR ONE ELEVATION ANGLE

Azimuth (deg)	Vulnerable Area (ft²)
0	0.0
45	5.2
90	8.0
135	7.5
180	0.0
225	7.2
270	8.5
315	6.0
Sum of A_V	42.4

The average vulnerable area that we calculate is $A_V = 42.4/6 = 7.1$ (not 42.4/8), and the PE is $6/8 = 0.75$. The reason we calculate a PE is explained in the *Introductory Weaponeering* textbook.

9. The elevation angle is incremented by 30 deg, and steps 2 through 8 are repeated.

Table 7.2 shows a sample output from this process for a particular elevation angle, noting that we now have two tables, one for vulnerable areas and a corresponding one for probabilities of exposure.

In this table, the numbers 500, 1000, ... , 7000 refer to the fragment velocities in ft/s, and the tabular values of vulnerable areas are in square feet. For this case the trend is as expected—the vulnerable area increases with higher fragment weights and velocities, and below certain values of weight and velocity, the target is not vulnerable at all. As an example, the value of 7.99 ft²

TABLE 7.2   SAMPLE OUTPUT FOR ONE ELEVATION ANGLE

	Vulnerable Areas					Weight (grains)
Velocity ►	500	1000	3000	5000	7000	▼
	0.00	0.00	0.00	0.00	0.00	5
	0.00	0.00	0.00	2.85	4.65	120
	0.00	0.00	3.66	7.38	7.99	500
	0.00	0.57	8.96	10.89	11.47	5000

	Probabilities of Exposure					Weight (grains)
Velocity ►	500	1000	3000	5000	7000	▼
	0.00	0.00	0.00	0.00	0.00	5
	0.00	0.00	0.00	1.00	1.00	120
	0.00	0.00	1.00	1.00	1.00	500
	0.00	1.00	1.00	1.00	1.00	5000

would correspond to the average of the eight azimuth angles for this combination of weight and velocity. The corresponding $PE = 1$ indicates that a nonzero vulnerable area was computed for all eight azimuth angles.

A typical table of vulnerable areas may include elevation angles of -60, -30, 0, 30, and 60 deg. Together with the angles of ± 90 deg, this ensures that the target is completely defined as to its vulnerability for all fragment approach angles. Table 7.3 shows abbreviated results for user-selected elevation angles and is often referred to as an RCAS file.

Recall that the data shown in the table are for a specific target kill definition only. As described in Chapter 6, the target is now represented by the centroid of vulnerability, together with the vulnerable area and probability of exposure data given in Table 7.3. A more extensive table is shown in Table 7.4.

It is important to note that the process of calculating a shotline P_K does not directly involve the FALT diagram. A simple value is calculated for each shotline and then aggregated over the whole target for those aspect angles. The FALT tree, however, is used to define which are the critical components along the shotline corresponding to the selected damage criterion.

The RCAS file is a more generalized result than the simple 68 ft^2 of the previous chapter and takes into account a full range of fragment masses and velocities as well as the variation in fragment approach angles to the target. It should also be recognized that the A_V data generated from the shotline analysis are expressed in the normal plane. Before averaging around the azimuth, the vulnerable area was for a specific direction of fragment impact as shown, for example, in Fig. 7.12, where the normal plane is perpendicular to the line of sight (LOS). After averaging, it is normal to the elevation angle only.

This final form of vulnerability data is suitable for input to the effectiveness assessment, which will examine the target's interaction with a specific weapon. This will be discussed in detail in the next chapter.

7.12 VULNERABILITY ASSESSMENT DUE TO BLAST

It was shown in Chapter 6, Section 6.16 that the vulnerability of a target due to blast effects may be represented by critical blast radius R_B such that if the target is closer than R_B, it is damaged; otherwise, it survives. Clearly, the value of this blast radius depends on the amount of explosive fill in the weapon; however, there are complicating issues.

Not all the explosive fill is available to generate the blast wave, because some energy is used to expand and fracture the weapon case into fragments. The more energy used to do this, the less energy available for the blast wave. Consider the two Mk-84, 2000-lb bomb bodies shown in Fig. 7.19.

The upper image shows the general-purpose Mk-84 bomb; the lower image shows the BLU-109 variant with a thicker casing designed to survive penetration through concrete. Although both have the same nominal weight, they contain different amounts of explosive fill, which along with the stronger case

TABLE 7.3   OUTPUT DATA FROM COVART: THE RCAS FILE

Elevation Angle ▼	Vulnerable Areas					Weight (grains) ▼
	Fragment Velocity (ft/s)					
	500	1000	3000	5000	7000	
−90 deg	0.0000	0.0000	0.0000	0.0000	0.0000	5
	0.0000	0.0000	0.0000	2.8500	4.6500	120
	0.0000	0.0000	3.6600	7.3800	7.9900	500
	0.0000	0.5700	8.9600	10.890	11.470	5000
0 deg	0.0000	0.0000	0.0000	0.0000	0.0000	5
	0.000	0.0000	0.0000	0.0263	0.0400	120
	0.0000	0.0000	0.0526	0.0667	0.4800	500
	0.0000	0.0833	1.1034	3.6800	5.0600	5000
30 deg	0.0000	0.0000	0.0000	0.0000	0.0000	5
	0.0000	0.0833	0.0267	0.1494	0.2874	120
	0.0000	0.0833	0.1494	0.8900	1.7800	500
	0.0400	0.0345	2.9600	6.6400	8.5200	5000
60 deg	0.0000	0.0000	0.0833	0.0400	0.0400	5
	0.0000	0.0400	0.0667	1.1100	2.0800	120
	0.0000	0.0800	1.5400	4.1600	5.5100	500
	0.0800	0.0690	7.7400	12.310	14.680	5000
90 deg	0.0000	0.0000	0.0100	0.0200	0.0300	5
	0.0000	0.0300	0.0700	2.2000	3.0600	120
	0.0300	0.0600	2.2810	4.8500	6.5600	500
	0.0800	0.0900	8.6300	10.850	12.580	5000

Elevation Angle ▼	Probabilities of Exposure					Weight (grains) ▼
	Fragment Velocity (ft/s)					
	500	1000	3000	5000	7000	
−90 deg	0.0000	0.0000	0.0000	0.0000	0.0000	5
	0.0000	0.0000	0.0000	1.0000	1.0000	120
	0.0000	0.0000	1.0000	1.0000	1.0000	500
	0.0000	1.0000	1.0000	1.0000	1.0000	5000
0 deg	0.0000	0.0000	0.0000	0.0000	0.0000	5
	0.000	0.0000	0.0000	0.3750	0.5000	120
	0.0000	0.0000	0.3875	0.7500	1.0000	500
	0.0000	0.1250	0.8750	1.0000	1.0000	5000
30 deg	0.0000	0.0000	0.0000	0.0000	0.0000	5
	0.0000	0.1250	0.7500	0.8750	0.8750	120
	0.0000	0.1250	0.8750	1.0000	1.0000	500
	0.2500	0.8750	1.0000	1.0000	1.0000	5000
60 deg	0.0000	0.0000	0.1250	0.2500	0.2500	5
	0.0000	0.2500	0.7500	1.0000	1.0000	120
	0.0000	0.2500	1.0000	1.0000	1.0000	500
	0.2500	0.8750	1.0000	1.0000	1.0000	5000
90 deg	0.0000	0.0000	1.0000	1.0000	1.0000	5
	0.0000	1.0000	1.0000	1.0000	1.0000	120
	1.0000	1.0000	1.0000	1.0000	1.0000	500
	1.0000	1.0000	1.0000	1.0000	1.0000	5000

TABLE 7.4   COMPOSITE A_V AND PE TABLE

90 Degree Elevation									M-Kill (C-Category)	
Fragment Mass (grains)	Fragment Impact Velocity for $A_v=0$	Fragment Impact Velocity (FPS)								
		500		1000		2000		3000		
		$\bar{A}_v$	P_E	$\bar{A}_v$	P_E	$\bar{A}_v$	P_E	$\bar{A}_v$	P_E	
2	7000	0.00	0.00	0.00	0.00	0.00	0.00	0.00	0.00	
5	7000	0.00	0.00	0.00	0.00	0.00	0.00	0.00	0.00	
10	2000	0.00	0.00	0.00	0.00	0.00	0.00	5.56	1.00	
15	1734	0.00	0.00	0.00	0.00	1.58	1.00	5.81	1.00	
30	1203	0.00	0.00	0.00	0.00	8.01	1.00	11.04	1.00	
60	1203	0.00	0.00	0.00	0.00	11.11	1.00	13.94	1.00	
120	859	0.00	0.00	1.44	1.00	13.37	1.00	15.21	1.00	
240	641	0.00	0.00	8.82	1.00	15.14	1.00	17.49	1.00	
500	641	0.00	0.00	11.65	1.00	18.23	1.00	20.83	1.00	
1000	391	0.20	1.00	13.96	1.00	22.03	1.00	30.30	1.00	
2000	391	0.72	1.00	17.31	1.00	27.19	1.00	35.87	1.00	
5000	391	0.74	1.00	19.50	1.00	28.95	1.00	37.04	1.00	
10000	391	1.52	1.00	20.52	1.00	30.35	1.00	37.59	1.00	
Fragment Mass (grains)	Fragment Impact Velocity for $A_v=0$	Fragment Impact Velocity (FPS)								
		4000		5000		6000		7000		
		$\bar{A}_v$	P_E	$\bar{A}_v$	P_E	$\bar{A}_v$	P_E	$\bar{A}_v$	P_E	
2	7000	0.00	0.00	0.00	0.00	0.00	0.00	0.00	0.00	
5	7000	0.00	0.00	0.00	0.00	0.00	0.00	0.00	0.00	
10	2000	5.56	1.00	5.56	1.00	5.56	1.00	5.56	1.00	
15	1734	7.45	1.00	5.58	1.00	5.59	1.00	5.56	1.00	
30	1203	13.22	1.00	13.54	1.00	13.57	1.00	13.19	1.00	
60	1203	14.73	1.00	15.00	1.00	15.39	1.00	14.80	1.00	
120	859	16.66	1.00	17.44	1.00	17.34	1.00	17.15	1.00	
240	641	21.51	1.00	23.55	1.00	24.00	1.00	23.82	1.00	
500	641	24.08	1.00	27.98	1.00	28.81	1.00	28.76	1.00	
1000	391	34.59	1.00	38.24	1.00	38.83	1.00	37.76	1.00	
2000	391	40.33	1.00	43.54	1.00	43.74	1.00	44.19	1.00	
5000	391	41.23	1.00	44.49	1.00	45.07	1.00	45.29	1.00	
10000	391	41.70	1.00	44.88	1.00	45.58	1.00	46.60	1.00	

suggests the blast produced by the BLU-109 will be considerably less than the standard Mk-84. Note also that the explosive material used in each is different—tritonal vs AFX-708.

We cannot calculate a value of R_B for each warhead and explosive fill, so each is reduced to an equivalent weight of uncased (bare) TNT using a two-step process.

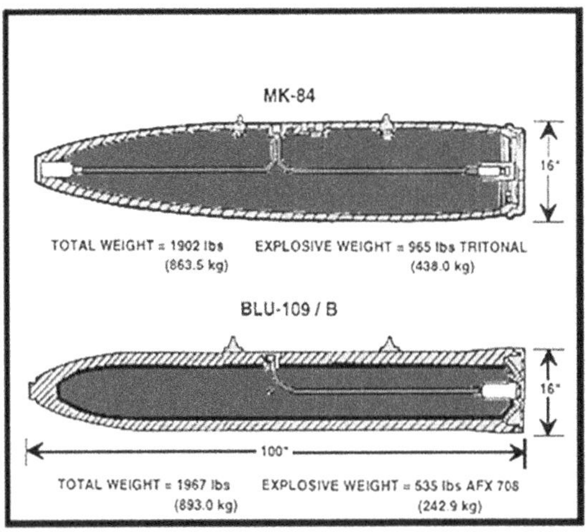

Fig. 7.19   Mk-84 bomb body variants.

1. The equivalent spherical, uncased weight of the weapon-specific explosive fill is calculated using the modified Fano equation.

$$W_U = W\left[0.6 + \frac{0.4}{1 + 2M/c}\right] \qquad (7.7)$$

where
W = total weight of explosive in the warhead
c = charge weight/unit length of cylindrical portion of the bomb
M = metal weight/unit length of cylindrical portion of the bomb

This enables all weapons to be represented by an equivalent, uncased weight of explosive fill, with the casing effects removed, as shown in Table 7.5.

Note the difference between the uncased weight and the charge weight due to the energy needed to fracture the case and provide kinetic energy to the resulting fragments.

TABLE 7.5   UNCASED CHARGE WEIGHT FOR SOME COMMON WEAPONS

Warhead	Total weight W - lb	Charge weight c - lb	Case weight M - lb	Equivalent uncased weight Wu - lb
Mk 82 warhead	500	192	308	133.4
Mk 83 warhead	1000	445	555	317.9
Mk 84 warhead	2000	945	1055	683.9
155 mm artillery round	95	15	80	9.5

TABLE 7.6   MULTIPLIER TO CONVERT TO TNT

Explosive	K
TNT	1.00
H-6	1.35
Tritonol	1.07
Comp B	1.11
Comp A3	1.07
Comp C4	1.30
Explosive D	0.92
HBX-1	1.17
HBX-3	1.14
Minol II	1.20

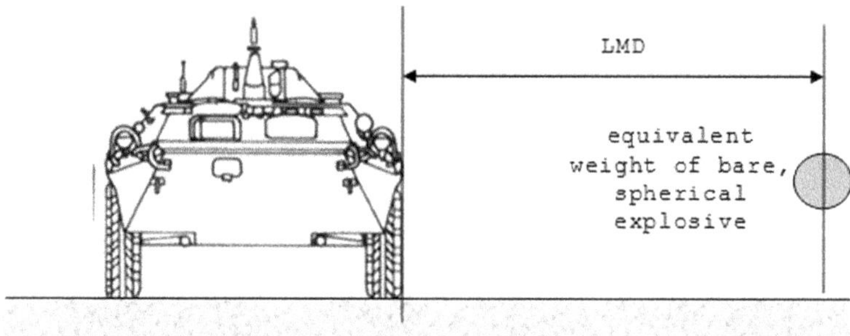

Fig. 7.20   Lethal miss distance for blast.

2. Then W_U is multiplied by a factor K that depends on the type of fill to convert it to TNT, as shown in Table 7.6.

The equivalent, spherical bare charge weight of TNT is then given by the following equation:

$$W_e = K \times W_U \tag{7.8}$$

For a given target, the vulnerability to blast is expressed by the lethal miss distance (LMD) shown as R_B in Fig. 7.20, and depends only on the calculated value of W_e, and the standoff distance.

Although relatively simple in theory, calculating the value of R_B is not straightforward due to reflections of the blast wave from the ground and the various faces of the target. The reader is referred to the *Advanced Weaponeering* textbook for further details.

7.13   THE MECHANICS OF PENETRATION

When we consider a target impacted by a penetrator (shaped charge, EFP, bullets, long rod), there are two different processes that may take place depending on the impact speed.

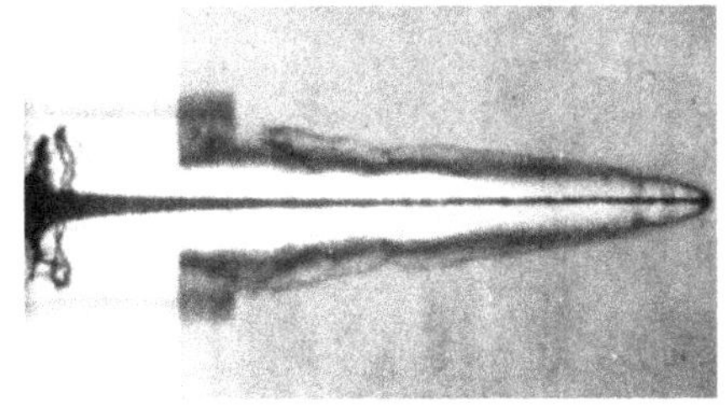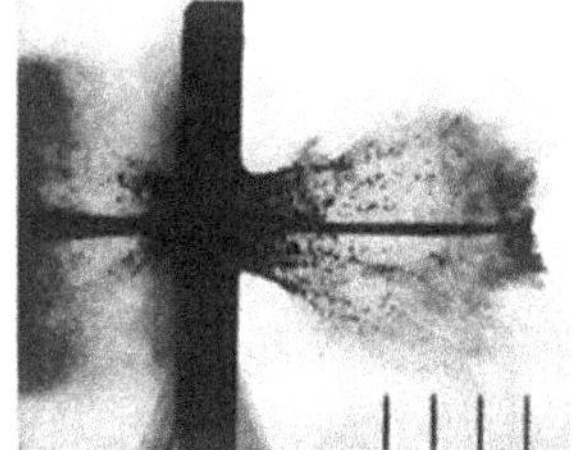

Fig. 7.21   Hydrodynamic penetration.

For high impact speed of the order of several kilometers/sec, the penetrator may be considered a fluid and for a given velocity, the penetration depth depends on the relative density of the penetrator and target materials. This is known as hydrodynamic penetration, as seen in Fig. 7.21.

For low impact speed of the order of tens of meters/sec, the penetrator maintains its rigid shape and for a given velocity, the penetration depth depends on the relative strength of the penetrator and target materials. This is known as rigid body penetration, and is illustrated in Fig. 7.22.

When considering penetration of a target by a projectile, consideration of the initial impact speed will determine which process is applicable.

7.14   VULNERABILITY ASSESSMENT FOR SHAPED CHARGE WARHEADS

A shaped charge warhead produces a high-speed jet of metal, commonly copper, which can produce a small hole in a considerable thickness of armor and is therefore an example of hydrodynamic penetration. An example of a shaped charge tank round is shown in Fig. 7.23.

It is worthwhile considering how penetration occurs because copper is a relatively soft metal whereas armor is clearly much harder. The following analysis applies to both shaped charge and rod penetrators, because they both have the same damage mechanism if the impacting velocities are sufficiently high.

Consider Fig. 7.24, which shows the jet impacting the armor plate. At the interaction surface, a considerable stagnation pressure is generated in order to bring the penetrator to a halt.

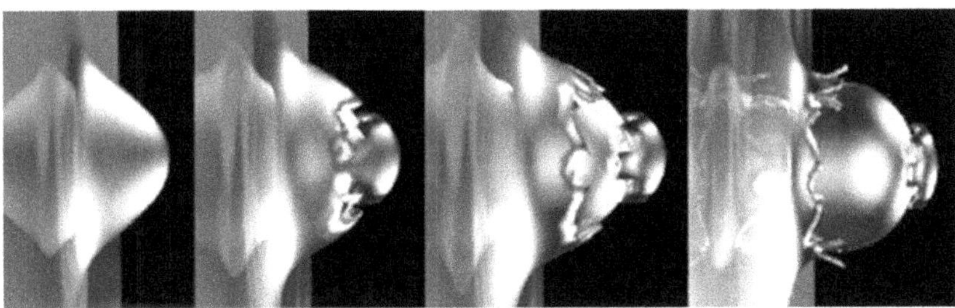

Fig. 7.22   Rigid body penetration.

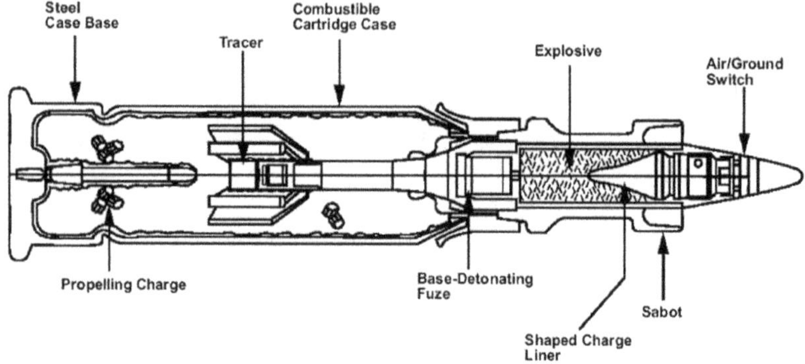

Fig. 7.23   Shaped charge 120-mm tank high-explosive antitank (HEAT) round.

Assuming for now that the jet behaves as an incompressible fluid, we can use Bernoulli's equation to relate the pressures, velocities, and elevations before and after impact as follows:

$$p_1 + \frac{1}{2}\rho_1 v_1^2 + gz_1 = p_2 + \frac{1}{2}\rho_2 v_2^2 + gz_2 \tag{7.9}$$

For copper impacting a thick armor plate, and ignoring changes in elevation, this equation may be used to estimate the stagnation pressure for some typical data:

$$p_2 = \frac{1}{2}\rho v_1^2 = 0.5 \times 8960 \times (1000)^2 = 4.48 \times 10^9 \text{ N/m}^2 \tag{7.10}$$

This is much larger than the yield strength of both the penetrator and the target armor, so the assumption of the penetrator and armor plate behaving like a fluid is justified—hydrodynamic penetration.

Although a difficult concept to visualize, this interaction is like one liquid cylinder trying to penetrate a "plate" of another liquid (armor). Whether the penetration can occur does not depend on the strength of the two materials in their solid state but in the ability of the liquid jet to push the resisting fluid out

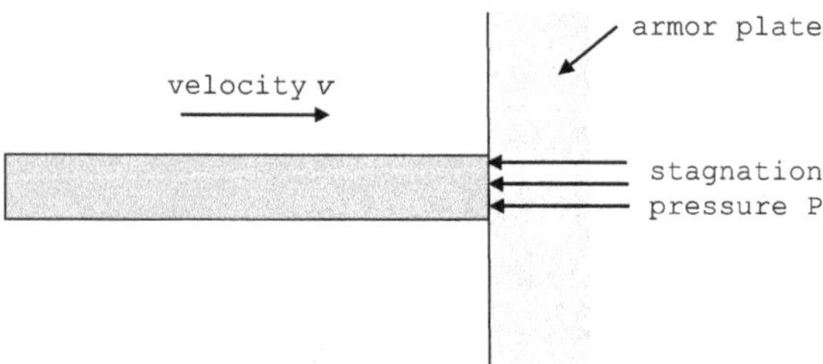

Fig. 7.24   Cylinder brought to rest.

of the way. In this case, the ability of the jet to move through the armor and the armor to resist depends more on their inertial properties than strength; therefore, material density becomes a dominant factor.

We now consider hydrodynamic penetration more closely by considering the jet penetrating into a monolithic material where the jet velocity is v and the velocity of the penetration front is u, as shown in Fig. 7.25. Note the hole formed is greater than the jet diameter.

If we now superimpose a velocity of u to the left on the whole system, the penetration front becomes stationary, the target is moving to the left at velocity u, and the jet is moving to the right at velocity $(v - u)$. This is shown in Fig. 7.26.

Because the pressures at the interface must be equal, and the system shown here is in steady state, Bernoulli's equation [Eq. (7.9)] may be applied to the jet for the left side and the target on the right side, leading to the following:

$$\frac{1}{2}\rho_j(v - u)^2 = \frac{1}{2}\rho_t u^2 \tag{7.11}$$

Note, however, from Fig. 7.25 that the jet is of finite length; therefore, the maximum penetration P will occur when the rear of the jet reaches the origin of the fixed coordinate frame. This time is given by

$$t = \frac{L}{(v - u)} \tag{7.12}$$

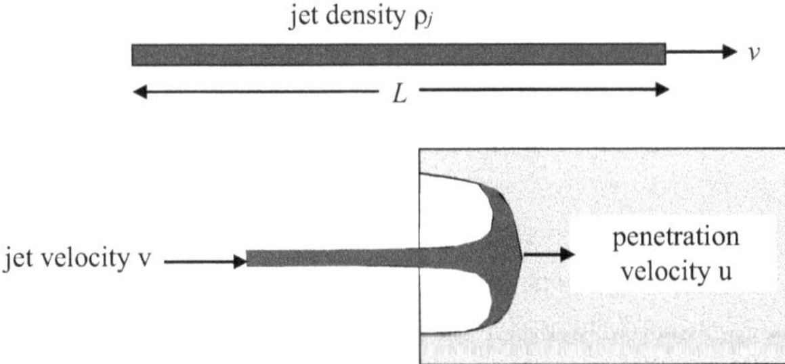

Fig. 7.25   Shaped charge and target geometry.

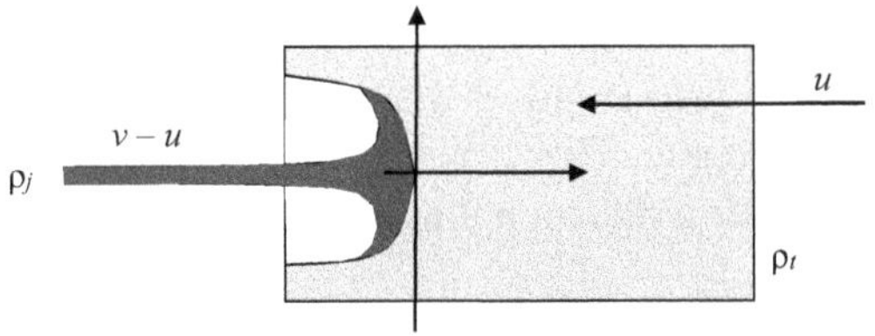

Fig. 7.26   Stationary interface coordinate frame.

**TABLE 7.7   SHAPED CHARGE AND
TARGET MATERIAL DENSITIES**

Material	Density (kg/m3)
Steel	7870
Copper	8960
Aluminum	2700
Depleted uranium	19,100
Lead	11,350

During this interval, the target material will have moved to the left a distance equal to the penetration depth P, hence for the target

$$P = u \times t = u \frac{L}{v - u} \tag{7.13}$$

Substituting Eq. (7.11) into Eq. (7.13) gives the so-called density law

$$P = L \sqrt{\frac{\rho_J}{\rho_t}} \tag{7.14}$$

Table 7.7 shows the relative densities of possible penetrator materials and a steel target plate for use in Eq. (7.14).

Based on hydrodynamic penetration theory, we see from the table why a copper shaped charge can penetrate steel armor. This is a somewhat simplistic approach but provides an estimate of the penetration expected in a single plate. A more detailed description of the development of penetration models may be found in [8].

Early methods used Eq. (7.14) together with a compartment model to determine whether a crew compartment (K-kill) or engine compartment (M-kill) in an armored vehicle was penetrated [9]. The damage mechanism is not simply penetration into the compartment, but that there is sufficient residual jet to cause damage. A rule of thumb was that about 2.5 in. of residual jet was sufficient to accomplish this.

It is also important to recognize that the penetration distance given by Eq.(7.14) is achieved only if the optimum standoff between the weapon and plate is achieved at detonation. Figure 7.27 shows typical penetration depth as a function of standoff distance for a particular penetrator/target combination.

In this figure, the built-in standoff is the distance from the target the warhead is designed to detonate. This is achieved by mounting a standoff trigger ahead of the warhead by a distance that maximizes penetration, as shown in Fig. 7.28.

The simple penetration methodology is fine for single plate penetration, but a higher fidelity method is needed if internal components are represented by

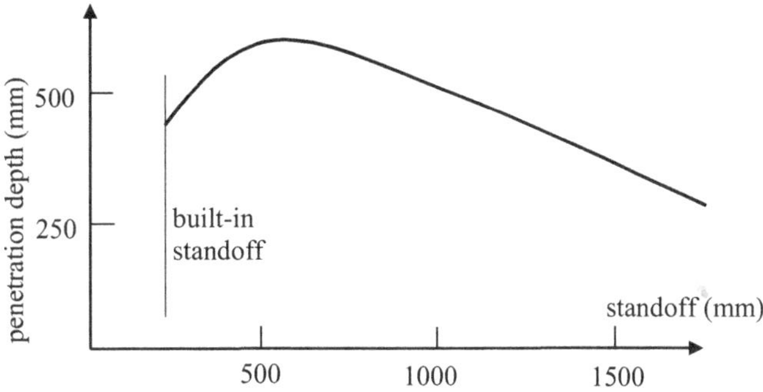

Fig. 7.27   Nominal shaped charge penetration depth vs standoff.

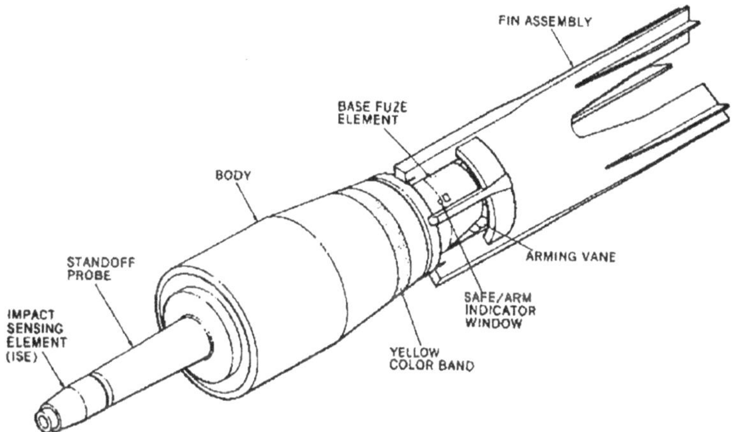

Fig. 7.28   Shaped charge warhead with standoff trigger.

successive plates, as was done by the FATEPEN program for fragment penetration. There is such a higher fidelity model known as the theory of residual penetration, and although too complex to describe in detail here, may be represented by Fig. 7.29.

As each plate is encountered, the residual penetration capacity is calculated using Eqs. (7.9) through (7.14). The jet is considered to propagate along the shotline in a manner similar to a single fragment, being degraded by each plate until it reaches a component it cannot penetrate or exits the rear of the target.

In terms of vulnerable area data for a shaped charge warhead against a specific target, typical values are shown in Table 7.8.

The weapon is characterized by a nominal diameter, and the vulnerable area of a specific target is shown for two kill definitions, each being a function of weapon elevation angle relative to the target. This table is similar to the RCAS file shown in Table 7.4, although the PE table is not needed.

TABLE 7.8   VULNERABLE AREA DATA FOR SHAPED CHARGE

Kill Definition	Diameter (in.)	Elevation Angle (deg.)			
		0	30	60	90
K-kill	1.00	1.70	1.37	1.72	4.10
	2.00	3.52	2.52	3.48	6.31
	3.00	4.70	4.18	5.51	6.94
	4.00	5.19	5.46	6.58	7.89
	5.00	5.62	6.37	7.74	8.20
	6.00	5.62	6.58	7.79	8.49
M-kill	1.00	6.70	6.17	5.99	11.01
	2.00	11.15	11.71	12.25	18.24
	3.00	17.98	21.00	23.95	31.77
	4.00	26.89	36.00	38.86	46.24
	5.00	36.08	50.65	55.71	60.40
	6.00	38.45	54.40	60.35	61.97
NOTIONAL DATA					

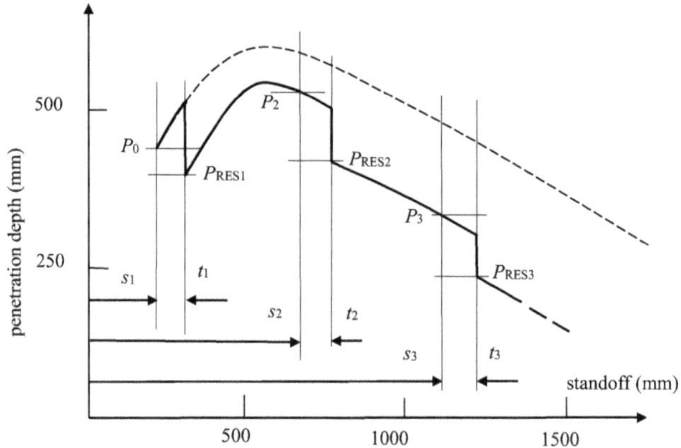

Fig. 7.29   Residual jet propagating through components.

7.15   VULNERABILITY TO PROJECTILES (RIGID BODY PENETRATION)

Bullets usually consist of a solid projectile (ball round) or one that has a case that breaks up and ignites, such as an armor-piercing incendiary (API) round. Simple bullet construction of small to medium caliber is shown in Fig. 7.30.

The cartridge or casing contains a chemical propellant that is initiated by a primer charge, usually by a mechanically generated impact from a firing pin. This propels the bullet along a barrel, which imparts a stabilizing spin though the spiral grooves machined on the inside bore. The bullet itself consists of a core (usually lead) surrounded by a casing, or jacket, commonly made of copper.

Figure 7.31 shows a larger penetrator round that may be fired from the M1A1 tank main gun comprising a discarding sabot rod that projects a solid

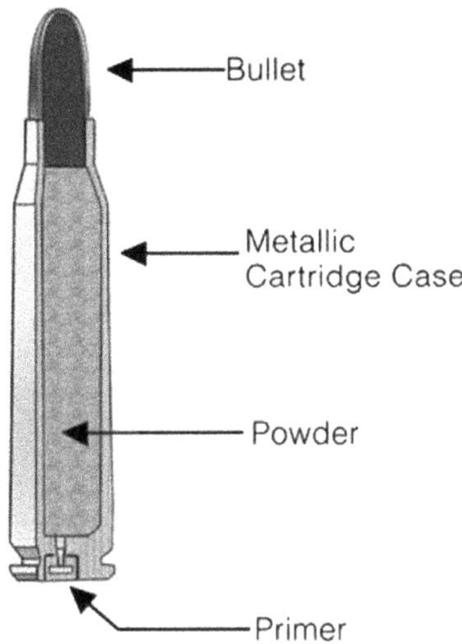

Fig. 7.30   Simple bullet construction.

depleted uranium (DU) dart. Notice that these rounds are subcaliber sabot types, and therefore neither is rifled. In this case, stabilization of flight is achieved using the fins shown to the rear of the projectile.

The penetrating object may be considered a metal cylinder that impacts the target at high speed, typically around 1 km/s. However, this velocity is below that needed to produce hydrodynamic penetration as was the case for the shaped charge; therefore, the mechanics are different. This mechanism is called rigid body penetration and is more like that for single fragments discussed earlier.

From an analytical perspective, the same tools used to generate vulnerability data due to fragments are used for single penetrators (i.e., COVART and AJEM). The TGM and FALT diagrams are essentially unchanged; however, the fragility curves will reflect the greater energy imparted to the critical components due to the larger mass of the impacting object. Recall Fig. 7.2 showing the COVART modules ProjPen and Shaped Charge Jet to accommodate these threats. Note that FATEPEN is used for both fragments and monolithic projectiles whereas ProjPen is used for nonmonolithic or segmented projectiles.

In terms of specific data to be used in an effectiveness study, tables similar to the RCAS file (Table 7.3), are generated by COVART and AJEM. An example of nominal data for small-caliber ball rounds is shown in Table 7.9.

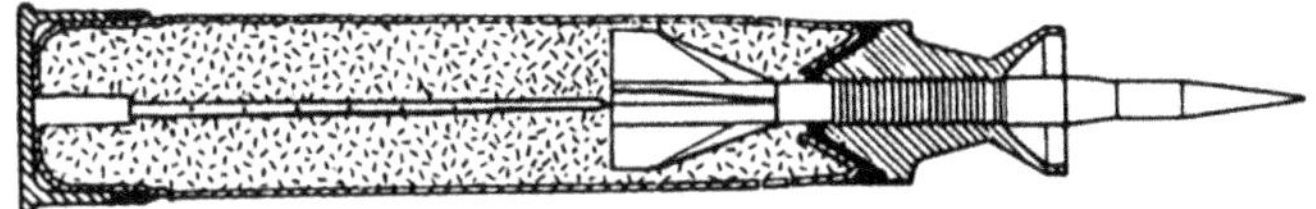

Fig. 7.31   Penetrator 120-mm tank round.

TABLE 7.9   VULNERABLE AREA DATA FOR SMALL-CALIBER
BULLETS

Caliber	Impact Velocity (ft/s)	Impact Angle (deg.)			
		0	15	45	90
0.50 in.	1000	0.00	0.03	0.14	0.22
	2000	0.40	0.54	4.34	9.72
	3000	2.84	2.88	9.02	13.07
	4000	5.90	5.07	11.93	13.54
	5000	7.78	7.10	13.12	14.24
25 mm	1000	0.58	1.22	5.77	9.99
	2000	7.69	6.67	14.60	13.69
	3000	11.23	11.14	17.36	15.51
	4000	13.17	14.21	19.06	16.08
	5000	14.51	16.11	20.29	16.64
30 mm	1000	0.36	0.53	5.70	9.46
	2000	7.80	7.17	14.17	14.40
	3000	12.42	12.31	17.49	15.40
	4000	14.30	15.36	19.05	16.41
	5000	15.61	16.98	20.47	16.84
NOTIONAL DATA					

Here, the vulnerable area in square feet depends on the impacting angle of
the bullet and the initial impacting velocity. Once more, the vulnerable areas
are averaged around the azimuth and represent normal plane data.

From Table 7.8 and Table 7.9 the trends are as expected—as the projectile
and shaped charge get bigger, so does the vulnerable area. The increasing
projectile impact velocity also increases the target's vulnerable area.

7.16   CHAPTER SUMMARY

- This chapter extended the work of the previous chapter to include more
realistic targets comprising many more components; however, the sequential process of the vulnerability study followed by the effectiveness study
is preserved. This chapter deals with only the first of these.
- Shotline analysis is crucial to calculating vulnerable areas and uses penetration equations and fragility curves to calculate vulnerable area for user-
supplied fragment characteristics.
- The TGM and FALT diagram is explained in terms of the COVART program, which uses shotline analysis to produce tables of vulnerability data,
again for user-supplied fragment characteristics.
- These data are elevation dependent but averaged around the azimuth,
introducing the associated probability of exposure (PE) for each calculated
vulnerable area.

- After dealing with fragments, the chapter described damage due to blast, shaped charge, and other penetrating rounds.

REFERENCES

[1] "Vulnerability Toolkit," DSIAC Bulletin, 10-2017. https://www.dsiac.org/resources/models/vulnerability-toolkit/

[2] Staley, T., "COVART 6: Modularization of Vulnerability Models," *Aircraft Survivability*, Fall 2009. https://ndiastorage.blob.core.usgovcloudapi.net/ndia/2011/test/11554WednesdayHornung.pdf

[3] Hornung, S., "The Impact of High Accuracy Target Geometry in Modeling and Simulation to Support Live-Fire Test and Evaluation," U.S. Army Ballistics and NBC Division, Army Research Laboratory, Survivability/Lethality Analysis Directorate, 16 March 2011.

[4] Butler, L. A., Edwards, E. W., and Kregel, D. L., "BRL-CAD Tutorial Series: Volume III—Principles of Effective Modeling," ARL Report SR-119, Sept. 2003.

[5] Ballistics Analysis Laboratory, Johns Hopkins University, "The Resistance of Various Metallic Materials to Perforation by Steel Fragments; Empirical Relations for Fragment Residual Velocity and Residual Weight," Project THOR, Technical Report 51, April 1963.

[6] Yatteau, J., Zernow, R., Recht, R., Dickinson, D., and Wasmund, T, *FATEPEN, A Model to Predict Terminal Ballistic Penetration and Damage to Military Targets*, AIAA, Reston, VA, 1999.

[7] Dehn, J., "A Unified Theory of Penetration," Technical Report BRL-TR-2770, Dec. 1986.

[8] Walthers, W. P., and Zukas, J. A., *Fundamentals of Shaped Charges*, John Wiley and Sons, New York, NY, 1989.

[9] U.S. Army Materiel Command, "Research and Development of Materiel, Engineering Design Handbook, Ammunition Series, Section 2, Design for Terminal Effects," AMCP 706-245, July 1964.

Chapter 8

Effectiveness Implementation

8.1 Process Overview

This chapter reviews the basic process of calculating the effectiveness of a weapon for a particular placement of a target resulting in the computation of a damage or lethal area matrix. Common programs to perform this task are the legacy General Full Spray Materiel (GFSM) program and its successor the Joint Mean Area of Effects (JMAE) program. As in Chapter 6, the vulnerability data of the target are combined with a warhead of known characteristics in a defined geometric relationship to the target.

We are assuming that the vulnerability study has been completed and the RCAS file for the target and kill definition is available in the form shown in Chapter 7, Table 7.3. In addition, the Z-data file for the selected weapon is available. The process to be followed is the same as that described in Chapter 6, Section 6.13, where the weapon is considered fixed and a specific engagement geometry is defined by placing the target on the ground plane to determine the $P_{K/D}$ for that location. The target is then moved over the ground plane as in Chapter 6, Section 6.15 to generate the lethal area matrix, or damage matrix. From this matrix, the MAE_F effectiveness index (EI) may be calculated as previously described.

8.2 Generating the Lethal Area Matrix

The process is best understood by considering the following example and sequential steps:

1. The specific weapon–target interaction geometry is shown in Fig. 8.1. This scenario represents an air-bursting weapon that propagates a blast wave and fragments from the detonation point. The point O directly below the warhead centroid is defined as ground zero, the origin of the ground plane coordinate frame. Assuming the velocity vector of the weapon is

195

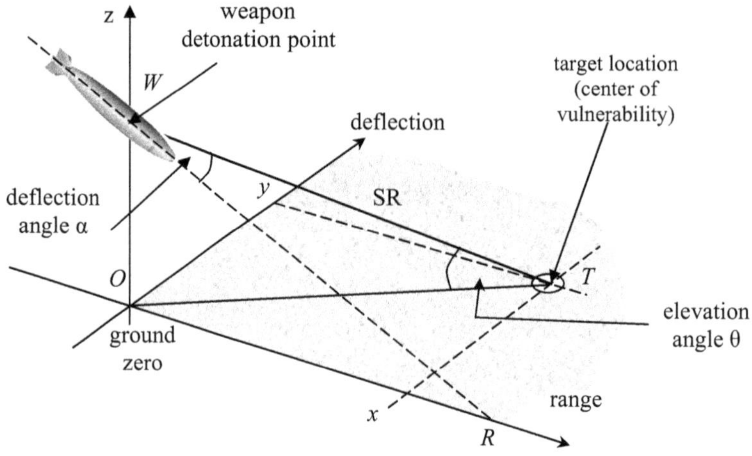

Fig. 8.1   Weapon–target interaction geometry.

coincident with the weapon axis, projecting this axis to the ground plane results in point R. The line OR defines the x, or range axis; the y, or deflection axis, is perpendicular to it with z pointing up.

2. The target, represented by the center of vulnerability, will be located on the ground plane at coordinates (x, y) relative to ground zero. It may be seen from Fig. 8.1 that the target is located at a particular deflection angle α relative to the weapon axis, and the weapon is located at a particular elevation angle θ relative to the target. This has two implications:

    •   The deflection angle determines the fragment zone in which the target is located.

    •   The elevation angle determines in which section of the RCAS file the weapon is located, thereby identifying the vulnerable area and probability of exposure tables.

3. Suppose the deflection angle happens to be 45 deg. The static fragment zone containing this angle may be extracted from the Z-data file, as shown in Table 8.1. This indicates that all fragments, although of different masses, are traveling with the same velocity of 5000 ft/s. If the deflection angle were 40 deg, the velocity would be interpolated from the 30-deg and 45-deg values.

    Note that this value 5000 ft/s is the velocity the fragments leave the warhead, not the striking velocity on the target. This has to be calculated using a

TABLE 8.1   STATIC FRAGMENT ZONE FOR 45-DEG ELEVATION

α's			Velocities			# wts
30.000	60.000	45.000	4500	6050	5000	4

frag weight ->	0.440	8.500	50.500	200.61
# of frags ->	88.96	17.79	17.790	35.590

TABLE 8.2   VULNERABLE AREA/PE FOR AN ELEVATION ANGLE OF 30 DEG

Vulnerable Areas						Weight (grains) ▼
Velocity ▶	**500**	**1000**	**3000**	**5000**	**7000**	
	0.0000	0.0000	0.0000	0.0000	0.0000	**5**
	0.0000	0.0833	0.0267	0.1494	0.2874	**120**
	0.0000	0.0833	0.1494	0.8900	1.7800	**500**
	0.0400	0.0345	2.9600	6.6400	8.5200	**5000**

Probabilities of Exposure						Weight (grains) ▼
Velocity ▶	500	1000	3000	5000	7000	
	0.0000	0.0000	0.0000	0.0000	0.0000	**5**
	0.0000	0.1250	0.7500	0.8750	0.8750	**120**
	0.0000	0.1250	0.8750	1.0000	1.0000	**500**
	0.2500	0.8750	1.0000	1.0000	1.0000	**5000**

trajectory model which includes a drag coefficient and the distance between the warhead and target, SR. Suppose the result of using the trajectory model results in a striking velocity of 3000 ft/s

4. Suppose the elevation angle is 30 deg. The vulnerable area/probability of exposure (PE) data are taken directly from the RCAS file (Chapter 7, Table 7.3) for this angle and are reproduced in Table 8.2. If the elevation angle did not correspond to one of those in the table, such as 50 deg, the data would have to be linearly interpolated from Table 7.3 between two adjacent elevation angles, 30 deg and 60 deg in this case.

5. Because all fragments are traveling with the same velocity (3000 ft/s), that column from Table 8.2 may be extracted to obtain values of A_V and PE for this velocity. If the velocity didn't coincide exactly with one listed in the table, it would again have to be interpolated.

   At this point we have found the vulnerable area for the scenario elevation angle corresponding to the common velocity of all fragments for the appropriate fragmentation zone (Table 8.3). The remaining discrepancy is that Table 8.3 does not contain A_V and PE values for the fragment weights of the weapon under consideration, and interpolation for the weight is necessary.

6. For the ith weight group of the weapon fragments, we interpolate Table 8.3 to obtain a vulnerable area A_{vi} and probability of exposure PE_i. For example, the fragment weight of 50.5 grains would be interpolated between the 5- and 120-grain fragments in Table 8.3 to get the A_V and PE values. The calculation for the vulnerable area would be

$$A_{V(50.5)} = \frac{(0.0267 - 0)}{(120 - 5)} \times 50.5 + 0 = 0.0117 \text{ ft}^2 \tag{8.1}$$

**TABLE 8.3   VULNERABLE AREA/
PE DATA FOR 3000 FT/S**

A_v	WEIGHT
0.0000	5
0.0267	120
0.1494	500
2.9600	5000
PE	
0.0000	5
0.7500	120
0.8750	500
1.0000	5000

The PE value for this fragment weight would also be interpolated from the RCAS file in a similar manner. Based on Eq. (6.21), we can write for the ith fragment weight

$$P_{K\text{-}FRAG(i)} = PE_i[1 - \exp(-\rho_i A_{Vi})] = PE_i\left[1 - \exp\left(\frac{-K_i \times A_{Vi}}{SR^2 \times \Omega}\right)\right] \quad (8.2)$$

where

  PE_i = probability of exposure for the ith weight
  K_i = # of fragments of weight i
  A_{Vi} = vulnerable area for the ith fragment weight
  SR = slant range from the weapon to the target
  Ω = fragment zone solid angle = $2\pi(\cos\varphi_1 - \cos\varphi_2)$

Notice how the probability of exposure (PE_i) is included to allow for some aspect ratios having a zero-valued vulnerable area and that the vulnerable areas were averaged over the nonzero values only. For example, if $PE = 0.5$, only half the azimuth aspects have a nonzero vulnerable area; hence, the P_K is scaled by 0.5 to reflect this.

7. The total probability of surviving all N fragment weight groups becomes

$$P_{S\text{-}FRAG} = \prod_{i=1}^{N}(1 - P_{K\text{-}FRAG(i)}) \quad (8.3)$$

leading to the probability of killing the target by one or more fragment weight groups [Eq. (6.22), Chapter 6] as

$$P_{K\text{-}FRAG} = 1 - \prod_{i=1}^{N}(1 - P_{K\text{-}FRAG(i)}) \quad (8.4)$$

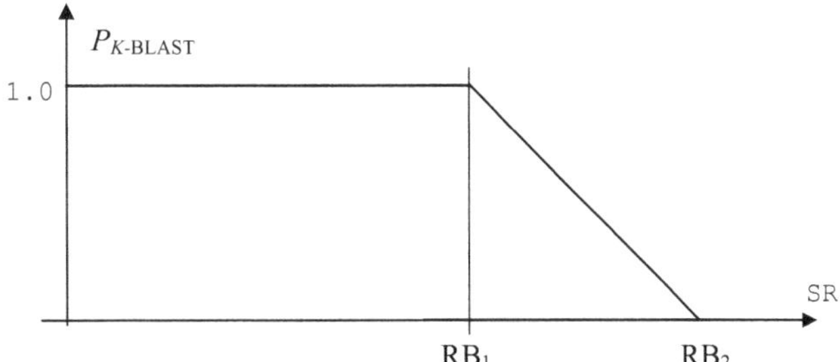

Fig. 8.2   Blast-radius characteristics.

8. This completes the damage due to fragments; now we add in the effects of blast. Although this was dealt with in a simplistic form in Chapter 6, Section 6.16 and Chapter 7, Section 7.12, the blast radius R_B is modified slightly into the form shown in Fig. 8.2, where there is a linear decrease in P_K rather than an abrupt cutoff. This damage mechanism is now represented by two blast radii, RB_1 and RB_2, as shown in Fig. 8.2.

    Given the slant range SR from Fig. 8.1, the value of $P_{K\text{-BLAST}}$ may be read directly from the graph.

9. The probability that the target survives both blast and fragments is given by

$$P_{S/D} = [1 - P_{K\text{-BLAST}}] \times [1 - P_{K\text{-FRAG}}] \tag{8.5}$$

giving the total $P_{K/D}$ due to blast and fragments as

$$P_{K/D} = 1 - \{[1 - P_{K\text{-BLAST}}][1 - P_{K\text{-FRAG}}]\} \tag{8.6}$$

10. This process is repeated for all target locations on a grid centered beneath the weapon detonation point, as long as the resulting P_K is nonzero. Observe that as the target moves into different cells relative to the weapon, it will move into different fragmentation zones and different elevation angles. A typical lethal area matrix of P_K at these grid points from GFSM is shown in Table 8.4.

    Only part of the matrix is shown here; it extends further in the deflection direction, getting smaller in size until it disappears at about 600 ft. In addition, due to the symmetry assumed for the fragment distribution about the weapon axis, the matrix is symmetrical in the deflection direction, so only half is shown here. The matrix form of P_K shown may be referred to as a P_K matrix, damage matrix, or lethal area matrix. A complete matrix is shown in Fig. 8.3, where in the computer display, the cells are color coded for the P_K rather than having numerical values; Table 8.4 represents the right half of Fig. 8.3.

11. Given the damage matrix, the lethal area specified as MAE_F is obtained from Eq. (8.7) by multiplying the P_K for each cell by the area of the cell and summing over the matrix for which there is a nonzero P_K.

TABLE 8.4 DAMAGE MATRIX: RIGHT HALF

Deflection ▶ / Range ▼	37.9	75.8	113.7	151.6	189.5	227.4	265.3	303.2	341.1	379.0
−114.4	0	0	0	0	0	0	0	0	0	0
−100.1	0	0	0	0	0	0	0	0	0	0
−85.8	0	0	0	0	0	0	0	0	0.0001	0.0001
−71.5	0.0001	0	0	0	0	0	0.0001	0.0002	0.0001	0.0001
−57.2	0.0011	0	0	0	0	0.0003	0.0004	0.0002	0.0001	0.0001
−42.9	0.0028	0	0	0	0.0009	0.0008	0.0004	0.0002	0.0001	0.0001
−28.6	0.0064	0.0001	0.0006	0.0029	0.0017	0.0009	0.0005	0.0002	0.0001	0.0001
−14.3	0.1402	0.0059	0.0099	0.0042	0.0019	0.0009	0.0005	0.0002	0.0001	0.0001
0	0.5571	0.0459	0.0127	0.0045	0.0019	0.0009	0.0005	0.0002	0.0001	0.0001
14.3	0.6794	0.0891	0.0156	0.0045	0.0019	0.0009	0.0005	0.0002	0.0001	0.0001
28.6	0.1741	0.0927	0.0325	0.0116	0.0041	0.0012	0.0005	0.0002	0.0001	0.0001
42.9	0.0060	0.0186	0.0258	0.0128	0.0063	0.0034	0.0016	0.0006	0.0002	0.0001
57.2	0.0007	0.0050	0.0105	0.0118	0.0061	0.0032	0.0017	0.0010	0.0006	0.0003
71.5	0	0.0024	0.0015	0.0072	0.0056	0.0031	0.0017	0.0010	0.0006	0.0004
85.8	0	0.0010	0.0012	0.0011	0.0045	0.0028	0.0017	0.0009	0.0005	0.0003
100.1	0	0.0003	0.0009	0.0005	0.0012	0.0025	0.0015	0.0009	0.0005	0.0003
114.4	0	0	0.0006	0.0004	0.0002	0.0011	0.0014	0.0009	0.0005	0.0003
128.7	0	0	0.0003	0.0003	0.0002	0.0001	0.0009	0.0007	0.0004	0.0003
143.0	0	0	0.0001	0.0003	0.0001	0.0001	0.0001	0.0006	0.0004	0.0003
157.3	0	0	0	0.0002	0.0001	0.0001	0	0.0002	0.0004	0.0002
171.6	0	0	0	0.0001	0.0001	0.0001	0	0	0.0002	0.0002
185.9	0	0	0	0.0001	0.0001	0.0001	0	0	0	0.0001
200.2	0	0	0	0	0.0001	0	0	0	0	0
214.5	0	0	0	0	0.0001	0	0	0	0	0
228.8	0	0	0	0	0	0	0	0	0	0
243.1	0	0	0	0	0	0	0	0	0	0

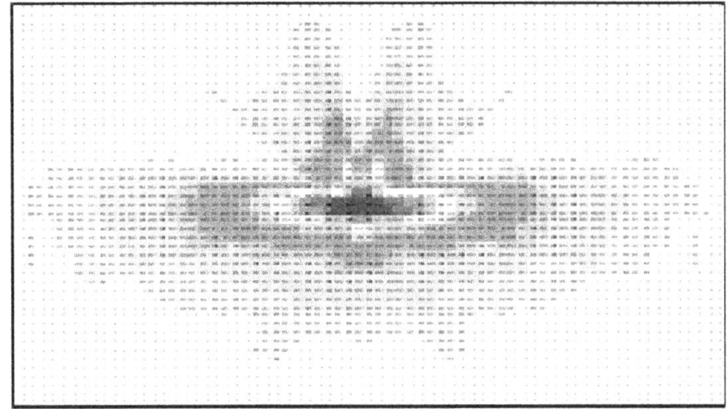

Fig. 8.3 Complete lethal area matrix.

$$A_L = \text{MAE}_F = \sum_{i=1}^{n} P_K(i) \times A_{\text{cell}} \tag{8.7}$$

For example, for the partial damage matrix shown in Table 8.4, applying Eq. (8.7) gives $\text{MAE}_F = 1135$ ft^2; however, this is only half of the matrix, so the total EI is $\text{MAE}_F = 2270$ ft^2.

Note that the damage matrix will be a function of the weapon impact angle due to varying fragment spray patterns, as indicated in Fig. 8.1, and that although the effectiveness index is written as MAE_F, it contains the effect of blast, too. This is because most general purpose warheads produce both damage mechanisms simultaneously.

One other feature of GFSM is that it includes the velocity of the weapon at detonation to determine the total velocity of fragments. Recall the Z-data file lists the fragment velocity as the value of the fragment at the warhead obtained from an arena test where the weapon is stationary. In a real attack, the weapon terminal velocity may be quite large, and this has to be added to the static velocities given in the Z-data file, as shown in Fig. 8.4.

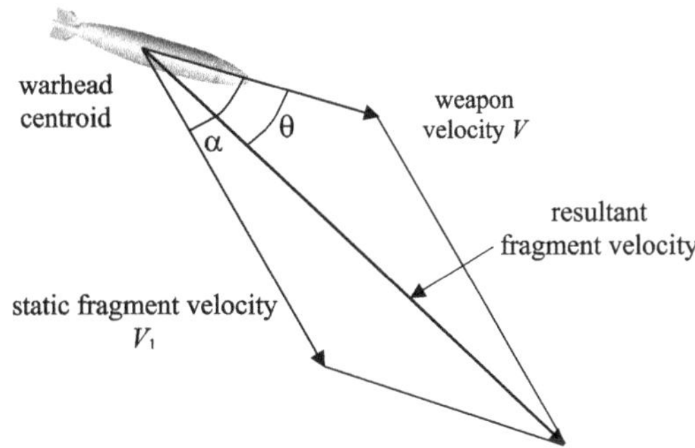

Fig. 8.4 Vector addition of weapon and fragment velocities.

It may be observed that the maximum P_K is 0.6794, and yet we might expect 1.0 due to blast at the detonation point. A possible explanation is that if the scenario involved an air-bursting weapon, a target at ground zero might be further away than RB_2. Also, due to the relatively large grid size, a target located at the center of the cell closest to ground zero might again be further away than RB_2.

It may also be noticed that the cell dimensions (37.9 ft × 14.3 ft) are an unusual choice; in fact, these are not chosen but are determined by the total number of cells GFSM can process, which is 40 × 40. Recall the program dates back to the 1970s when computer memory and processing speed were both very inferior to contemporary levels, so the total cells processed were limited to these fixed values, and the cell size adjusted so the total area of effects would be covered by the same number of cells.

Given the area covered by the lethal area matrix in Table 8.4, we can surmise it corresponds to a relatively large weapon extending several hundred feet in both range and deflection. If a small warhead weapon were processed into the same 40 × 40 matrix, the dimensions of each cell would be correspondingly smaller. When JMAE replaced GFSM, there were no restrictions on cell size or numbers.

We may represent the lethal area matrix as a surface above the horizontal plane where the height is proportional to the P_K in each cell. Such a surface is shown in Fig. 8.5.

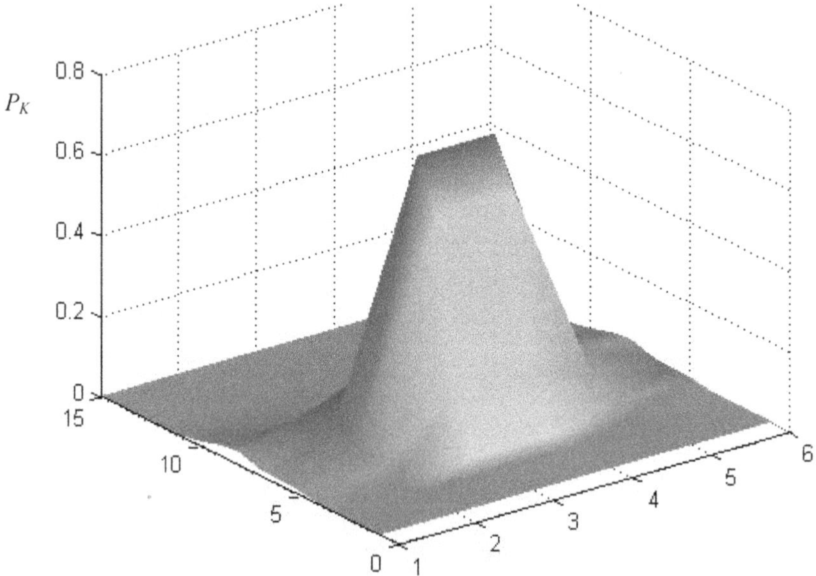

Fig. 8.5   Lethal area matrix as a 3-D surface.

With some imagination, this figure resembles a bivariate probability distribution discussed in Chapter 2; this will be discussed further later in this chapter.

8.3 JOINT MEAN AREA OF EFFECTS (JMAE) MODEL

JMAE is similar to GFSM in that it can produce a lethal area matrix, or surface, from which a MAE may be calculated; however, it has more capability than GFSM and somewhat higher fidelity in how the target is represented. We will not discuss the operation of JMAE in detail; if students wish to obtain more information, it can be found in the *Introductory Weaponeering* textbook.

One of the more significant improvements in the JMAE is that it is not limited to the 40 × 40 lethal area matrix that GFSM is constrained by. JMAE users can specify the number and dimensions of the matrix and also obtain a matrix with logarithmically scaled cells, as shown in Fig. 8.6.

Note how the cells are now user defined and, in this example, logarithmically distributed about the impact point, allowing better resolution near where the weapon detonates and coarser resolution far from it.

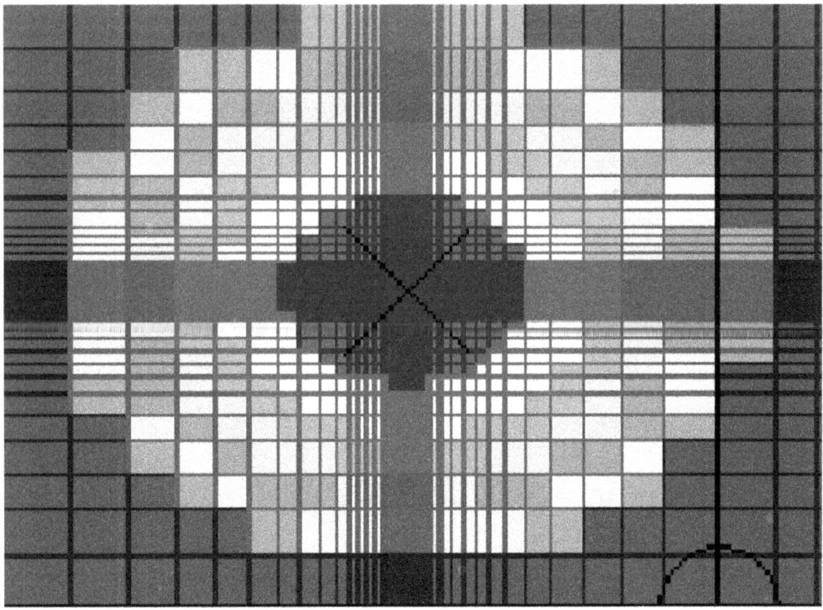

Fig. 8.6 JMAE-generated lethal area matrix cells.

8.4 SIMPLIFICATION OF DAMAGE MATRIX

Although the damage matrix shown in Table 8.4 represents the P_K values in cells around the weapon, we shall see later in Chapter 9 that it is too cumbersome to use in fast-running weaponeering methodologies, so approximations have to be used to describe it. From Table 8.4 we can plot contours of constant values of P_K around the weapon detonation point, as shown in Fig. 8.7. Sometimes this representation is termed the "bug-splat." Note that these contours appear approximately elliptical, and represent the 3-D surface shown in Fig. 8.5.

Also shown in Fig. 8.7 are Gaussian-like approximations to the actual contour profiles when viewed in the range and deflection directions. Note how the parameters describing the spread of these distributions are different. This observation leads to an analytical expression for an approximation to the tabulated damage function (P_K's) and the bug splat contour, which is referred to as the Carlton damage function and takes the form

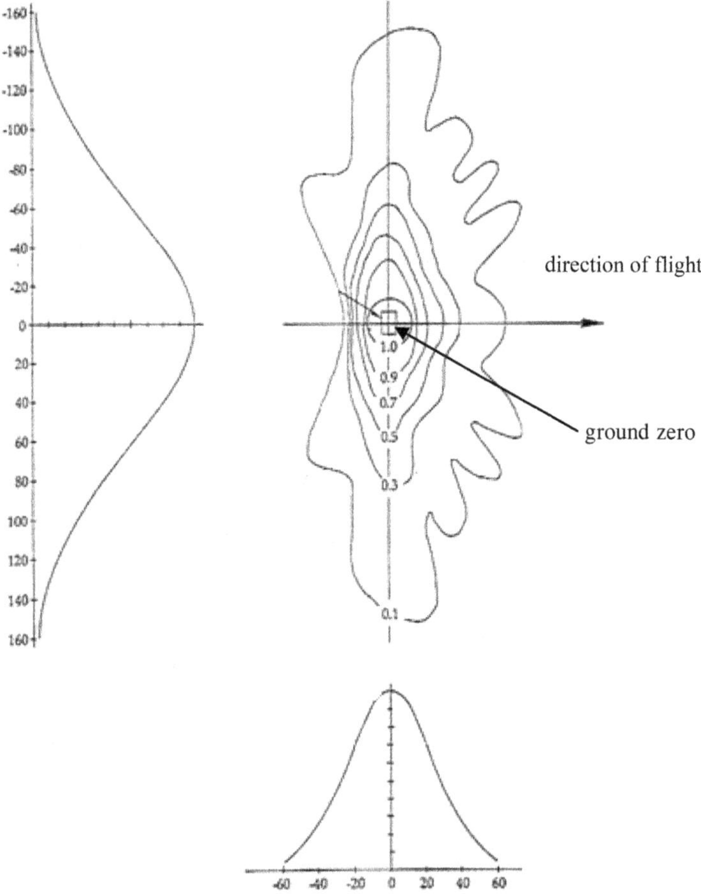

Fig. 8.7 P_K **contours with Gaussian approximations.**

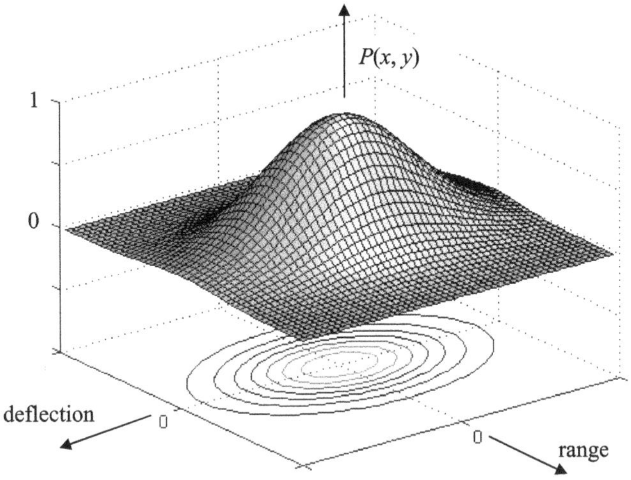

Fig. 8.8 Carlton damage function.

$$P(x,y) = \exp - \left[\frac{x^2}{\text{WR}_r{}^2} + \frac{y^2}{\text{WR}_d{}^2} \right] \tag{8.8}$$

where WR_r and WR_d are defined as the weapon radius in the range and deflection directions, respectively. Note that the Carlton damage function is similar in shape to the bivariate Gaussian, or normal distribution, shown in Chapter 2, Section 2.3, Figure 2.7, except that it has a value of unity at the point (0, 0), which the Gaussian distribution does not.

Figure 8.8 shows a three-dimensional view of the Carlton damage function and the elliptical nature of horizontal sections (constant P_K contours) through it. Consider how Fig. 8.8 is a smoothed version of Fig. 8.5.

We do not know the values of WR_r and WR_d because they depend on the rate at which P_K decreases, but note that in general they will be different. The lethal area may be calculated from the Carlton damage function by summing the products of cell area and cell P_K values, but if we have an analytical form for the damage function, this summation process is equivalent to an integral.

$$A_L = \text{MAE}_F = \sum_{i=1}^{N} P_K(i) \times A_{\text{cell}} = \int_A P_K \, \mathrm{d}A \tag{8.9}$$

Substituting the Carlton damage function, we get

$$A_L = \text{MAE}_F = \int_{-\infty}^{\infty} \int_{-\infty}^{\infty} \exp - \left[\frac{x^2}{\text{WR}_r{}^2} + \frac{y^2}{\text{WR}_d{}^2} \right] \mathrm{d}x\mathrm{d}y \tag{8.10}$$

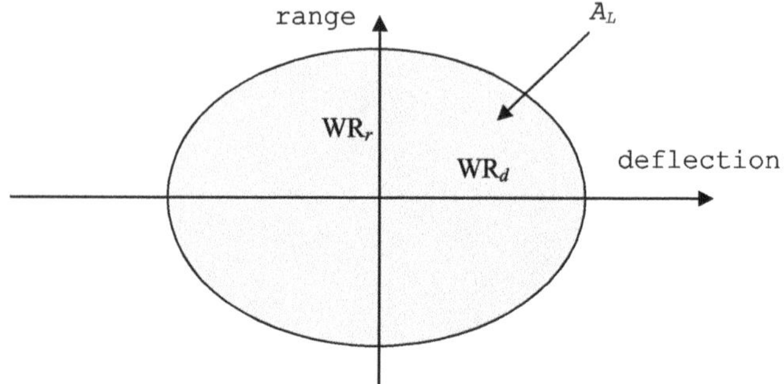

Fig. 8.9   Elliptical representation of the lethal area.

It may be shown that the integral evaluates to

$$A_L = \mathrm{MAE}_F = \pi \times \mathrm{WR}_r \times \mathrm{WR}_d \tag{8.11}$$

which is the equation for the area of an ellipse with radii WR_R and WR_D. The lethal area may therefore be equated to the area of the ellipse shown in Fig. 8.9.

This damage function is different from those discussed previously in that the lethality is concentrated inside a specified boundary and is zero outside the boundary. This type of damage function is known as a "cookie cutter" for obvious reasons. Hence, the damage matrix may be represented by either the Carlton damage function extending over the whole of the ground plane or a cookie-cutter elliptical damage function where the P_K inside is unity and outside is zero.

Sometimes the Carlton damage function is described in terms of the weapon diameters in range and deflection L'_{ET} and W'_{ET}, respectively, rather than the weapon radii, where $L'_{ET} = 2\mathrm{WR}_r$ and $W'_{ET} = 2\mathrm{WR}_d$. The corresponding Carlton damage function takes the form

$$P(x,y) = \exp -\left[\frac{4x^2}{L'^{2}_{ET}} + \frac{4y^2}{W'^{2}_{ET}} \right] \tag{8.12}$$

Unfortunately for some applications, even the Carlton damage function or the elliptical cookie cutter damage function shown in Fig. 8.9 is not sufficiently simple to allow a mathematically tractable solution, so further modifications are necessary. To do this, a rectangular approximation is made to the elliptical lethal area, as shown in Fig. 8.10.

This damage function is another cookie cutter function because the value of P_K inside the function is unity and outside is zero. Therefore, the lethal area in this representation is

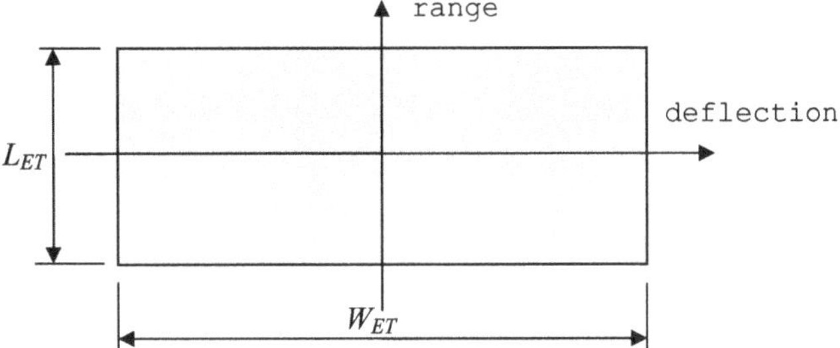

Fig. 8.10   Rectangular approximation to the lethal area.

$$A_L = \mathrm{MAE}_F = L_{ET} \times W_{ET} \tag{8.13}$$

Although confusing, it should be remembered that L'_{ET} and W'_{ET} are associated with the Gaussian or Carlton damage function, whereas L_{ET} and W_{ET} are the effective target length and width, respectively, and define an equivalent rectangular cookie cutter damage function.

8.5   CALCULATING DIFFERENT DAMAGE FUNCTIONS

So far we have simplified the lethal area matrix into the Carlton damage function, then to the elliptical cookie cutter, and finally into the rectangular cookie cutter, allowing us to write the following:

$$\begin{aligned}
\mathrm{MAE}_F &= \pi \times \mathrm{WR}_r \times \mathrm{WR}_d \\
&= \frac{\pi \times L'_{ET} \times W'_{ET}}{4} \\
&= L_{ET} \times W_{ET}
\end{aligned} \tag{8.14}$$

Suppose we have a lethal area matrix such as the one shown in Table 8.4. We can calculate the lethal area using Eq. (8.9) by multiplying each cell P_K by the area of the cell and summing over the ground plane for all cells with a non-zero P_K. The result is $\mathrm{MAE}_F = 2270 \ \mathrm{ft}^2$. If we want to represent the lethal area matrix by an equivalent Carlton damage function, we have to select values of WR_r and WR_d such that lethality is conserved, that is,

$$\pi \times \mathrm{WR}_r \times \mathrm{WR}_d = 2270 \tag{8.15}$$

By itself, Eq. (8.15) does not allow WR_r and WR_d to be calculated because there are an infinite number of combinations of weapon radii that would give us the correct answer. Which combination is correct?

To answer this, consider the shape of the damage function, particularly the aspect ratio, which is the ratio of the "length" to the "width." The aspect ratio measures whether the lethal area matrix and Carlton damage function is circular or stretched out in one direction or another. For the Carlton damage function, the aspect ratio is defined by the following:

$$a = \frac{\mathrm{WR}_r}{\mathrm{WR}_d} \tag{8.16}$$

Due to the way fragments are projected from the warhead, we know the shape, hence, the aspect ratio is affected by the impact angle of the weapon. In the 1970s, a study linking the aspect ratio of the lethal area matrix (LAM) to the impact angle I resulted in the following empirical equation:

$$a = \mathrm{MAX}[1 - 0.8\cos(I), 0.3] \tag{8.17}$$

where I is measured in degrees. This second equation enables us to find unique values for the weapon radii if the impact angle and MAE_F are known, by substituting WR_d from Eq. (8.16) into Eq. (8.15) to get

$$\mathrm{MAE}_F = \pi \times \mathrm{WR}_R \times \frac{\mathrm{WR}_R}{a} \tag{8.18}$$

Therefore

$$\mathrm{WR}_r = \sqrt{\mathrm{MAE}_F \times a/\pi} \tag{8.19}$$

and

$$\mathrm{WR}_d = \mathrm{WR}_r/a \tag{8.20}$$

This allows the Carlton damage function to be defined as follows. Suppose the LAM was generated knowing the impact angle was 65 deg. Substituting $\mathrm{MAE}_F = 2270$ and $I = 65$ yields

$$a = 0.662 \tag{8.21}$$

$$\mathrm{WR}_r = \sqrt{2270 \times 0.662/\pi} = 21.87 \text{ ft} \tag{8.22}$$

$$\mathrm{WR}_d = 21.87/0.662 = 33.04 \text{ ft} \tag{8.23}$$

The resulting Carlton damage function becomes

$$P(x,y) = \exp-\left[\frac{x^2}{21.87^2} + \frac{y^2}{33.04^2}\right] \tag{8.24}$$

If the Carlton damage function uses the weapon diameters instead of the weapon radii, the corresponding equations are

$$L'_{ET} = 2 \times WR_r = 1.128\sqrt{MAE_F \times a} \qquad (8.25)$$

$$W'_{ET} = 2 \times WR_d = \frac{L'_{ET}}{a} \qquad (8.26)$$

Note that the same aspect ratio parameter a applies to the elliptical and rectangular cookie cutters, too, allowing Eq. (8.16) to be extended to the following variables:

$$a = \frac{WR_r}{WR_d} = \frac{L'_{ET}}{W'_{ET}} = \frac{L_{ET}}{W_{ET}} \qquad (8.27)$$

Representing the damage function as an elliptical cookie cutter as shown in Fig. 8.9 would result in an ellipse of major radius 33.04 ft and minor radius 21.87 ft. Because the probability of damage inside the cookie cutter is unity and is zero outside, the lethal area is simply the area of the ellipse, or

$$MAE_F = \pi \times 21.87 \times 3.04 = 2270 \qquad (8.28)$$

For the rectangular cookie cutter we have

$$a = \frac{L_{ET}}{W_{ET}} \qquad (8.29)$$

$$MAE_F = L_{ET} \times W_{ET} \qquad (8.30)$$

In this case, the effective target length and width are given by

$$L_{ET} = \sqrt{MAE_F \times a}, \quad W_{ET} = L_{ET}/a \qquad (8.31)$$

For the data given

$$L_{ET} = \sqrt{2270 \times 0.662} = 38.76 \text{ ft},$$
$$W_{ET} = 38.76/0.662 = 58.56 \text{ ft} \qquad (8.32)$$

The lethal area is again $L_{ET} \times W_{ET} = 2270$ ft^2, as expected. This analysis shows how by modifying the damage function from the most detailed (lethal area matrix) to the simplest (rectangular cookie cutter), the lethality of the weapon/target/damage criteria is preserved.

In this section, we have simplified the lethal area matrix into the Carlton damage function and then into the elliptical and rectangular cookie cutter damage functions and shown that with the correct selection of parameters, they all represent the same MAE_F. The question then arises: If we use an MAE_F to calculate weapon effectiveness, which representation should we use? We will see in the next chapter how to calculate target damage and how we might use a deterministic or Monte Carlo solution for this calculation. Unfortunately, the preferred faster deterministic approach is not always possible using all three representations of the damage functions, so the answer to which one should we use is: the highest fidelity model possible that allows a solution to be found.

8.6 LETHAL AREA CALCULATION FOR TARGETS SENSITIVE TO BLAST

So far, only fragment-sensitive targets have been discussed, although it is important to remember that the P_K matrices shown in Table 8.4 include the effect of blast as well as fragments. In some cases, however, the target may be vulnerable only to blast; damage functions for this mechanism have to be represented in mathematical form for calculating effects.

This is accomplished by defining the mean area of effects for blast MAE_B and then representing that number by a damage function. The usual representation is an area on the ground plane within which the P_K is unity and outside of which P_K is zero, as discussed in Chapter 6, Section 6.16.

This is, therefore, a cookie cutter EI similar to the one described in the previous section. The EI may be defined in terms of the weapon blast radii shown in Fig. 8.2 and takes the form of a truncated cone, as shown in Fig. 8.11.

The $P_{K\text{-BLAST}}$ could be determined for each cell beneath the cone and the MAE_B determined as for the fragmentation case. Any cell closer to ground zero than RB_1 would have a $P_K = 1$, and any cell further away than RB_2 would

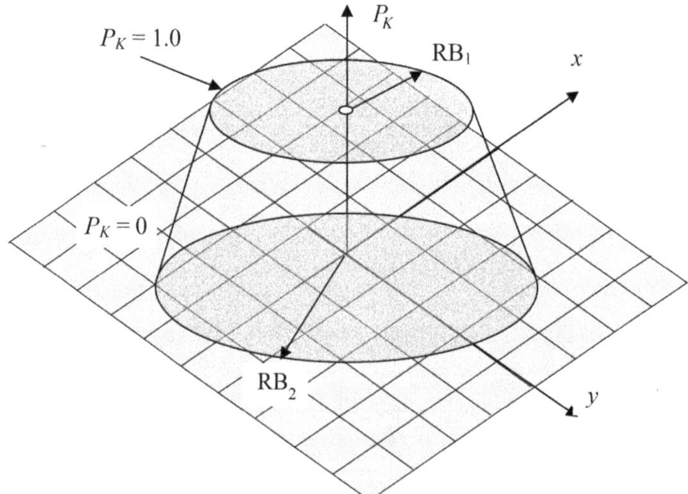

Fig. 8.11   Blast envelope placed over ground plane cells.

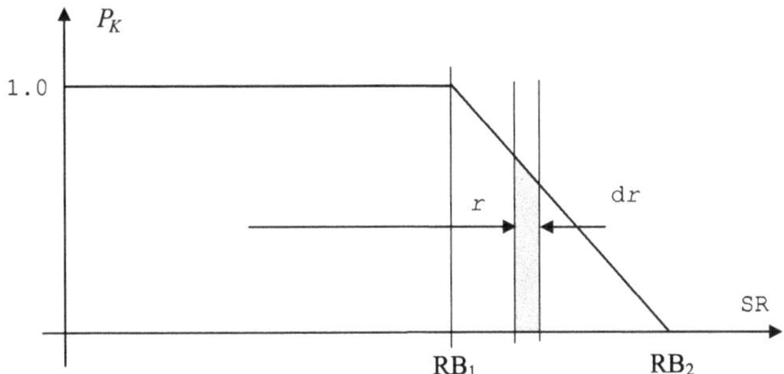

Fig. 8.12   Variation in $P_{K/H}$ for blast.

have a $P_K = 0$. Cells between RB$_1$ and RB$_2$ would have the P_K determined from Fig. 8.12. The blast equivalent to the Carlton damage function for MAE$_F$ would be the truncated cone centered over ground zero shown in Fig. 8.11.

The symmetry of the blast envelope, however, lends itself to a simpler calculation for the MAE$_B$ using the general expression for the lethal area in Eq. (8.9)

$$\text{MAE}_B = \int P_K \, dA \qquad (8.33)$$

The variation in $P_{K\text{-BLAST}}$ with radius is shown in Fig. 8.12 as a symmetrical function of radius. Consider an annulus of radius r and width dr where it is assumed that P_K is constant within the narrow annulus.

Equation (8.33) may be integrated directly if we put it in the polar form

$$\text{MAE}_B = \int_{r=0}^{r=\infty} P_K(r) \times 2\pi r \, dr \qquad (8.34)$$

After some manipulation, this reduces to

$$\text{MAE}_B = \frac{\pi}{3}[\text{RB}_1{}^2 + \text{RB}_1\text{RB}_2 + \text{RB}_2{}^2] \qquad (8.35)$$

For weaponeering purposes, it is more convenient if the damage function is represented in the ground plane, this time as a square defined again by L_{ET} and W_{ET}, where

$$L_{ET} = W_{ET} = \sqrt{\text{MAE}_B} \qquad (8.36)$$

Graphically, this cookie cutter damage function may be represented in the ground plane as shown in Fig. 8.13.

Note the similarity between Eqs. (8.31) and (8.36). The difference is the absence of the aspect ratio parameter a, because blast propagates radially from

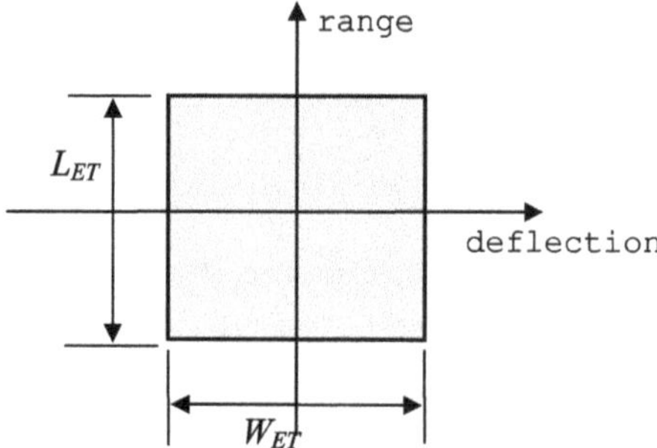

Fig. 8.13   Blast cookie cutter damage function.

the point of detonation and will therefore be independent of the weapon impact angle.

In summary, the highest fidelity representation of the interaction between a weapon and a target has been derived in the form of the lethal area matrix. Although this may be conveniently represented by a single number, the MAE_F, the distribution of lethality around the weapon is lost when this is done.

8.7   PERSONNEL TARGETS

Although GFSM and JMAE have been used to describe the effectiveness of a weapon against materiel targets, they may also be used to assess injuries to personnel. We will discuss an associated tool called the General Full Spray Personnel (GFSP) program that operates in much the same way as GFSM. As for materiel targets, the effect of fragments and blast on personnel will be dealt with separately.

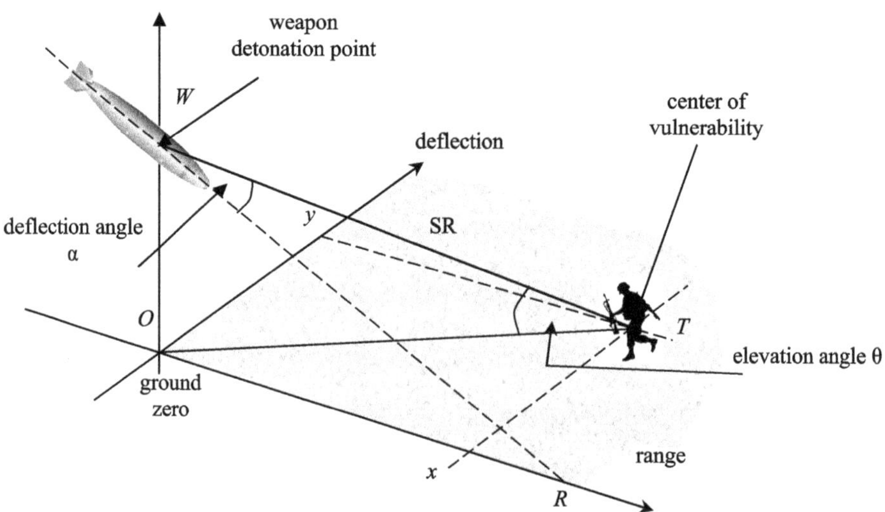

Fig. 8.14   Scenario for GFSP program.

8.7.1 FRAGMENT EFFECTS

GFSP establishes the weapon/target geometry in the same way as GFSM, as indicated in Fig. 8.14. The warhead is represented by the same Z-data file as before in order to calculate the effects of fragments.

GFSP uses the same process as GFSM to produce a lethal area for a single personnel target by placing the target at the center of successive cells covering the ground plane; however, there are a few modifications to calculating the "damage" inflicted on the target. Recall the expression used for materiel targets for the effects of a single (*i*th) fragment weight group in the Z-data file, Eq. (8.2).

$$P_{I\text{-FRAG}(i)} = PE_i[1 - \exp(-\rho_i A_{Vi})] \tag{8.37}$$

It is assumed that the target is equally vulnerable from all azimuth directions; therefore, the probability of exposure is unity.

$$P_{I\text{-FRAG}(i)} = 1 - \exp(-\rho_i A_{Vi}) \tag{8.38}$$

The effect of all fragment weight groups was then combined using Eq. (8.4).

$$P_{I\text{-FRAG}} = 1 - \prod_{i=1}^{N}(1 - P_{I\text{-FRAG}(i)}) \tag{8.39}$$

Before the total effect of fragments can be assessed, each of the terms in Eq. (8.38) has to be determined. Recalling from the definition of vulnerable area

$$A_{Vi} = A_P \times P_{K/H(i)} \tag{8.40}$$

Equation (8.38) may be written as

$$P_{I\text{-FRAG}(i)} = 1 - \exp(-\rho_i A_P P_{K/H(i)}) = 1 - \exp A_P(-\rho_i P_{I/H(i)}) \tag{8.41}$$

Note that $P_{K/H(i)}$ has been changed to the probability of incapacitation given a hit $P_{I/H(i)}$, which is more appropriate for personnel targets. We now focus on calculating the three terms in Eq. (8.41).

Based on the work of Sperrazza and Kokinakis [1], $P_{I/H(i)}$ is given by the following equation:

$$P_{I/H(i)} = 1 - \exp[-a(m_i V^{3/2} - b)^n] \tag{8.42}$$

where
   m = fragment mass (grains)
   V = fragment striking velocity (ft/s)
   a, b, n = parameters that depend on the casualty criteria

TABLE 8.5   VALUES OF A, B, AND N FOR CASUALTY CRITERIA

Casualty Criterion	a	b	n
30-s defense	$8.8771.10^{-4}$	31400	0.45106
30-s assault	$7.6442.10^{-4}$	31000	0.49570
5-min assault	$1.0454.10^{-3}$	31000	0.48781
12-h supply	$2.1973.10^{-3}$	29000	0.44350

Note the parameter b represents a threshold energy level below which a fragment will not penetrate; this will depend on what the person is wearing—civilian clothes, heavy uniform, body armor, helmet, and so forth. The values of m_i and V are obtained from the Z-data file for each weight group, and a, b, and n are given in Table 8.5.

For a given casualty criterion, therefore, a, b, and n can be selected from Table 8.5 and substituted into Eq. (8.42); the resulting $P_{I/H}$ can then be substituted into Eq. (8.41). The fragment density ρ_i is calculated for each fragment weight group where the number of fragments also comes from the Z-data file, leaving only the value of A_P to find.

The problem of finding the presented area of a personnel target can be approached in several ways. The simplest assumption is that a standing man may be represented by a cylinder of about height 2 m and diameter 0.5 m; therefore, from any horizontal angle this would be the presented area of 1 m². However, the U.S. Army conducted considerable work on estimating soldier presented areas, which included posture (standing, prone, foxhole), the effect of air-bursting weapons, and the effect of partial defilade due to terrain. The interested reader is referred to the *Advanced Weaponeering* textbook for more information.

To summarize, the target presented area, fragment density, and $P_{I/H}$ may all be calculated for a given weight group defined by the Z-data file, allowing Eq. (8.41) to determine the effects of each fragment weight group. These are then combined using the survivor rule using Eq. (8.39) to obtain the effect due to all fragments projected from the warhead.

8.7.2 BLAST EFFECTS

As far as blast is concerned, the origins of its representation in GFSP is somewhat obscure. It appears to use the same RB_1 and RB_2 method described earlier; however, it is expressed in the following form:

$$RB_1 = a_1 + b_1 \times (W_G)^{0.35} + c_1 \times (W_G)^{0.7} \tag{8.43}$$

$$RB_2 = a_2 + b_2 \times (W_G)^{0.35} + c_2 \times (W_G)^{0.7} \tag{8.44}$$

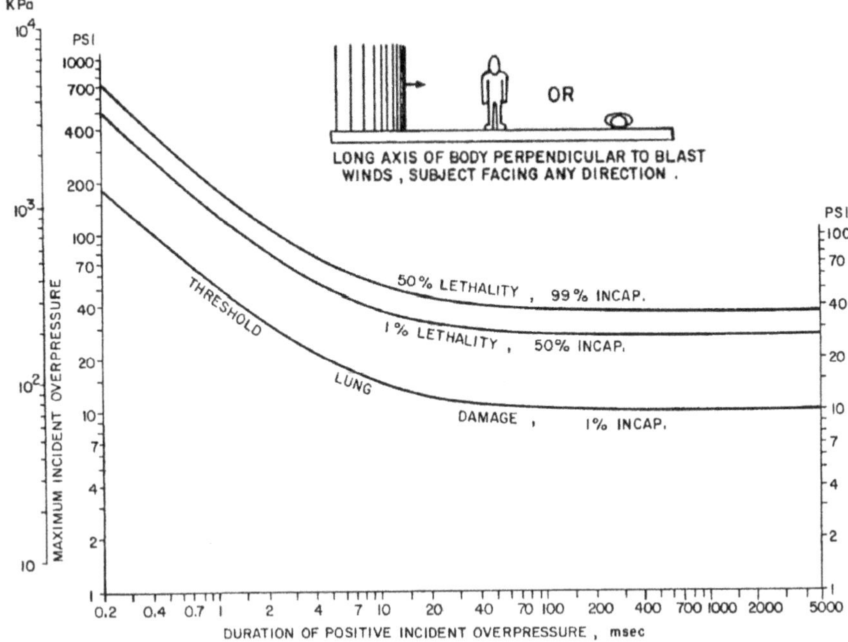

Fig. 8.15   Bowen curves for blast injury.

where W_G is the bare weight TNT equivalent of the warhead charge, and a, b, and c are constants that depend on the posture and orientation of the target relative to the oncoming blast wave.

Perhaps a more widely used method attributable to Bowen [2] can be used to give a simpler approach. This is summarized in Fig. 8.15, which shows the probability of incapacitation in terms of the peak overpressure of the blast wave and its duration.

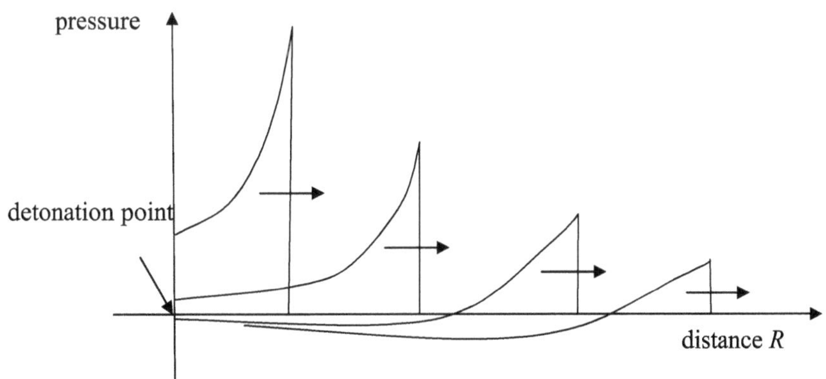

Fig. 8.16   Changes in peak pressure and blast duration.

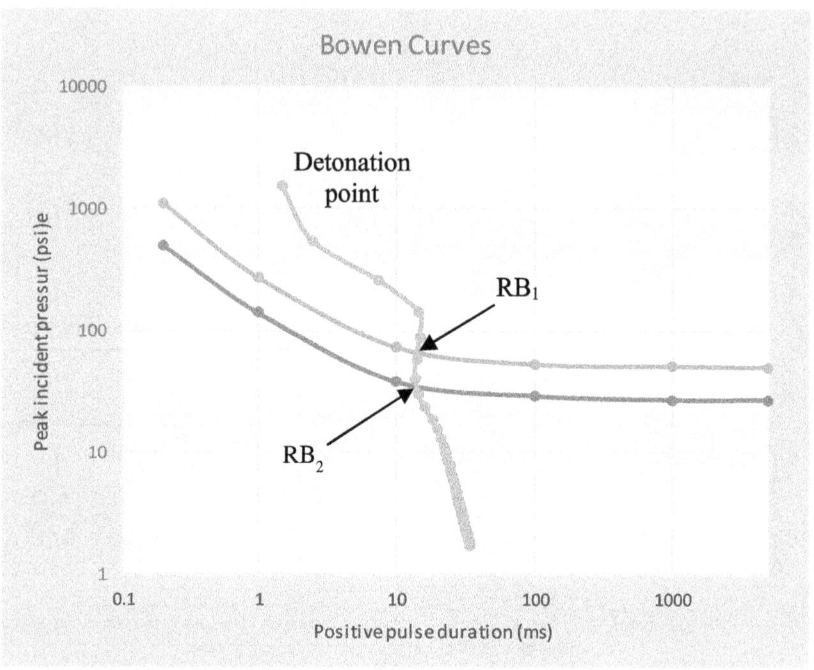

Fig. 8.17   Blast wave superimposed on Bowen curves.

Both of these parameters can be seen in Figure 1.9 in Chapter 1; however, the way they change may best be observed in Fig. 1.11, which is reproduced here in Figure 8.16.

As the blast wave moves away from the detonation point, the peak pressure decreases and the duration of the positive pulse increases. This path may be superimposed onto the Bowen curve as shown in Fig. 8.17.

Where the blast wave contour crosses the 99% and 1% incapacitation curves determines RB_1 and RB_2, respectively. A more detailed example of calculating RB_1 and RB_2 is provided in Chapter 17, Section 17.8, but for now assume the probability of incapacitation due to blast $P_{I\text{-BLAST}}$ is available from Fig. 8.17.

The effects of both blast and fragments are combined using the survivor rule to produce the following equation, which is analogous to Eq. (8.6) for materiel targets:

$$P_{I/D} = 1 - (1 - P_{I\text{-FRAG}})(1 - P_{I\text{-BLAST}}) \qquad (8.45)$$

This completes the calculation of $P_{I/D}$ for the specific location of the target shown in Fig. 8.1. By moving it to different cells on the ground plane, the lethal area matrix is generated in the usual way. As for materiel targets, this lethal area may be represented by a single number, the MAE_F, and simplified into successive simpler versions such as the Carlton damage function or the rectangular cookie cutter.

Recall that the lethal area matrix depends on the injury criterion selected, and this is obtained from Table 8.5; therefore, as the injury level changes, the size of the lethal area matrix will also change.

8.8 CHAPTER SUMMARY

- The chapter examined the effectiveness study for materiel targets by extending the process described in Chapter 6 of fixing the weapon and placing the target in a succession of ground plane cells to calculate the $P_{K/D}$ caused by both fragments and blast to determine the lethal area matrix.
- The MAE_F was calculated by summing the product of cell area and $P_{K/D}$ over the whole of the ground plane.
- This was done using the GFSM program as the implementation method, but JMAE was also described and compared to GFSM.
- Anticipating possible mathematical difficulties using the lethal area matrix directly, it was simplified into the Carlton damage function and then into a rectangular cookie cutter, all with equivalent lethality.
- The methodology used in GFSM was extended to personnel targets, as implemented in the GFSP program.

REFERENCES

[1] Sperrazza, J., and Kokinakis, W., "Criteria for Incapacitating Soldiers with Fragments and Flechettes," Report 1269, Ballistics Research Laboratories, Aberdeen Proving Ground, MD, 1965.

[2] Bowen, I. G., Fletcher, E. R., and Richmond, D. R., "Estimate of Man's Tolerance to the Direct Effects of Air Blast," DASA 2113, Lovelace Foundation for Medical Education and Research, Albuquerque, NM, 1968.

Chapter 9

Calculating Damage Probabilities

9.1 Introduction

The preceding chapters have developed the basic data needed to calculate the effectiveness of a weapon directed toward a target. In this chapter, to begin with, the simplest possible case will be considered (i.e., a single or unitary weapon directed toward a single target). We are interested in the value of the probability of damage caused by a single weapon, defined as PD_1, using

- Measures of delivery accuracy
- Weapon-specific warhead effects such as fragments, blast, and shaped charge
- Target vulnerability to these effects

This assumes knowledge of range error probable (REP), deflection error probable (DEP), or circular error probable (CEP), and a damage function represented by either a lethal area matrix, a Carlton damage function, or a cookie cutter of specified size and shape. The exact method of determining PD_1 will vary somewhat depending on the specific data supplied, but the basic methodology will be the same.

We are dealing with single weapon releases, so the two principle error sources—mean point of impact (MPI) and precision—may be combined into a single term, as shown in Chapter 4, Eq. (4.21). Therefore, as far as delivery accuracy is concerned, we have single values of σ_x and σ_y or their error-probable equivalents.

9.2 Basic Approach to Effectiveness Calculation

Consider the scenario shown in Fig. 9.1 showing someone throwing a Frisbee (with a hole in it) and trying to hit the center of a cross (target) drawn on the ground.

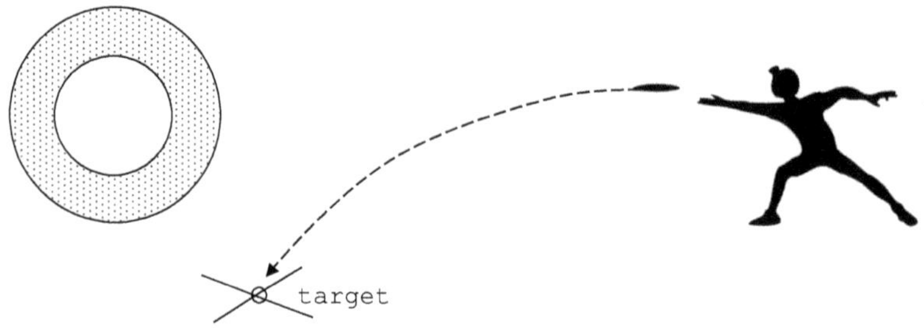

Fig. 9.1   Simplified effectiveness methodology.

In this experiment we see the following analogies:

- The delivery accuracy is how accurately the Frisbee is thrown.
- The weapon lethality/target vulnerability/damage criterion is represented by the size of the hole in the Frisbee, corresponding to the lethal area. The appropriately scaled radius of the hole R would be given by

$$\mathrm{MAE}_F = \pi R^2 \tag{9.1}$$

- Here the cookie cutter damage function is unusually circular; however, throwing a rectangular cookie cutter with a rectangular hole in it presents challenges, particularly if the orientation has to be preserved. We therefore use a conventional circular Frisbee.

If the Frisbee is thrown such that the center of the cross is inside the circular hole, we define the target as killed; otherwise, it survives. The thrower positions themself, say, 15 ft from the target, aims at the target, and throws the Frisbee. The result of whether the target is killed or not is recorded. This process is repeated, say, 100 ($=N$) times. If the number of kills recorded is N_K, then the probability of kill is calculated from the following equation:

$$P_K = \frac{N_K}{N} \tag{9.2}$$

The value of P_K is called the *expected result of the test*; in other words, it is the likelihood of killing the target if the mission were executed a large number of trials. Another interpretation of the result is the probability a single throw will be successful. It will not tell us if one particular trial will succeed or fail, only the expectation of success.

The manner in which this result was obtained is known as a *Monte Carlo* or *stochastic* methodology because it mimics stylistically what would happen in practice—a lethal area is delivered against a target with some defined accuracy. To summarize, this result may be interpreted in two ways.

1. It is the proportion of a large number of trials that will result in success.
2. It is the probability a single trial will result in success.

We now adopt a more formal approach to this intuitive example where the damage function is a rectangular cookie cutter, and single weapon delivery error parameters REP and DEP are known. The target is again a point located at the origin of a coordinate frame on the ground plane and a weapon is directed at the target, as shown in Fig. 9.2. If we were to write a computer program to implement the Monte Carlo simulation, it would consist of the following steps:

1. Initialize a counter, the P_K counter, to zero.
2. Using REP and DEP and a normal random number generator, calculate the weapon impact point on the ground plane (see Fig. 9.2).
3. Draw the lethal area represented by the rectangular cookie cutter centered on the impact point.
4. Determine if the target, represented as a point, is enclosed within the cookie cutter.
5. If it is, then the target is killed, so increment the P_K counter by one,
6. Repeat from step 2.
7. After a sufficiently large number of steps, terminate the simulation.
8. Calculate the PD_1 from Eq. (9.2).

$$\mathrm{PD}_1 = \frac{P_K \text{ counter}}{\text{Number of iterations}} \tag{9.3}$$

A sample iteration of this simulation where the target is not killed is shown in Fig. 9.2.

So, if we perform 100,000 iterations, and 47,396 of them resulted in the point target being enclosed by the lethal area, Eq. (9.3) would give the following:

$$\mathrm{PD}_1 = \frac{47{,}396}{100{,}000} = 0.474 \tag{9.4}$$

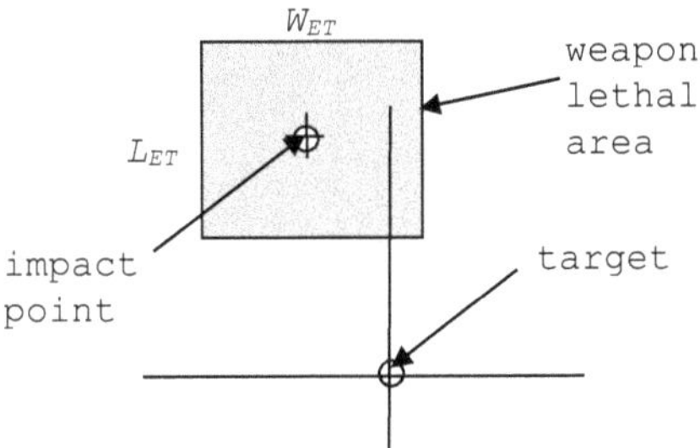

Fig. 9.2   One iteration of a Monte Carlo simulation.

Note that this method would work for a blast/fragment warhead MAE_F damage function represented by a rectangular cookie cutter or a blast warhead MAE_B represented by a square cookie cutter.

9.3 INTERPRETATION OF THE WEAPON LETHAL AREA

The relationship between the target and weapon lethal area may be viewed in two ways and often leads to some confusion. For example, in the previous chapter the rectangular cookie cutter damage function was derived from the effectiveness of the weapon and yet is given the dimensions L_{ET} by W_{ET}, the effective target length and width. So, does the rectangle represent the weapon or the target? The answer is both, because the lethal area represents both the damage effects of the warhead and the target's vulnerability to those effects.

Calculating effectiveness can take advantage of this duality. Consider the situation shown in Fig. 9.2 (reproduced in Fig. 9.3a) in which the point target is fixed at the origin, and the weapon lethal area is centered on the randomly derived impact point. Figure 9.3b shows an alternative view of the interaction in which the target is represented by the lethal area rectangle fixed at the origin, and the weapon, represented as a point, impacts at some randomly determined point relative to it.

This again begs the question: Which representation is correct? Is the target a point and the weapon lethal area a rectangle (Fig. 9.3a) or is the target a rectangle and the weapon represented by a point (Fig. 9.3b)? As before, the answer is that both are correct interpretations of the interaction, and in calculating the effectiveness it is seen that the probability of placing the lethal area rectangle on a fixed point target is the same as the probability of hitting a fixed rectangle with a single impact point.

Historically, weaponeers have used both representations interchangeably when discussing effectiveness, so care must be exercised when determining which representation is being analyzed. In implementing computer solutions for effectiveness, however, the method shown in Fig. 9.3b is almost always used.

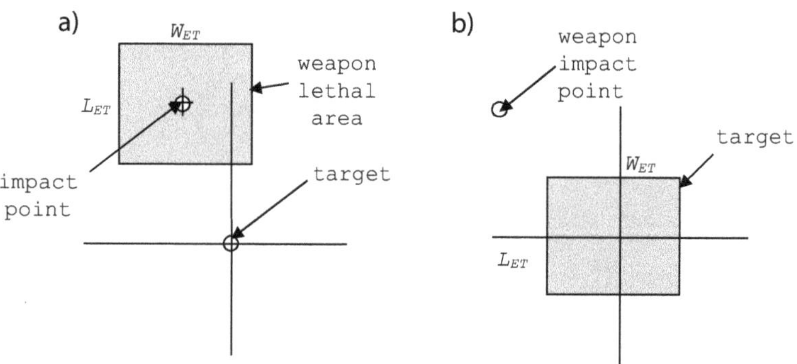

Fig. 9.3 Alternate representation of weapon and target.

9.4 DAMAGE CALCULATION FOR A SIMPLE CASE

Using this alternative interpretation makes it easier to solve the example given in the previous section using a Monte Carlo simulation by fixing the weapon lethal area at the origin of the coordinate frame and determining the probability the impact point lies within it.

Consider the example where the target has a lethal area that is a 15 ft × 15 ft rectangle. We are dealing with single weapon deliveries, so the combined dependent and independent accuracies in range and deflection are equal and represented by a zero mean circular normal distribution with a CEP of 10 ft. A computer program may be written to simulate the process shown in Fig. 9.3b resulting in the typical impact points shown in Fig. 9.4.

Running this program for the data shown produces the result shown in Table 9.1.

We can modify this approach to make it more "weaponeer friendly" by specifying the MAE_F, REP, DEP, and impact angle in order to compute L_{ET} and W_{ET} and allow for REP and DEP rather than a CEP. We can then use the following equations derived in Chapter 8, Section 8.5:

$$a = \max[(1 - 0.8 \times \cos I), 0.3] \tag{9.5}$$

$$L_{ET} = \sqrt{\mathrm{MAE}_F \times a} \tag{9.6}$$

$$W_{ET} = \frac{L_{ET}}{a} \tag{9.7}$$

$$\sigma_x = \frac{\mathrm{REP}}{0.6745} \tag{9.8}$$

$$\sigma_y = \frac{\mathrm{DEP}}{0.6745} \tag{9.9}$$

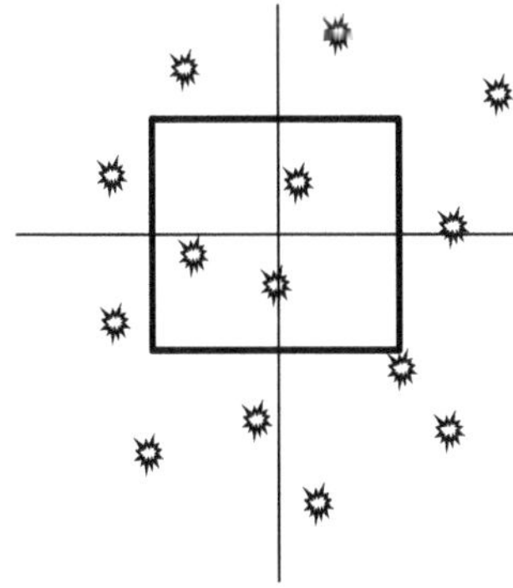

Fig. 9.4 Independent, single weapon deliveries.

TABLE 9.1   PROGRAM BASED ON MONTE CARLO SIMULATION

INPUTS	
Target area length	15
Target area width	15
CEP	10
Number of iterations	500000
OUTPUTS	
PD1	0.3878
Compute	

These equations may be incorporated into the spreadsheet program shown in Table 9.2, which for the inputs shown produce the result $PD_1 = 0.4636$.

Note the result here (0.4636) is not comparable with the previous result (0.3878) because in one case L_{ET} and W_{ET} are defined, whereas in the other case they are calculated from the MAE_F and the impact angle.

Instead of using a Monte Carlo–based method we can also use a *deterministic* approach using an equation to compute the result, again using the representation shown in Fig. 9.3b. Consider the probability that a round lands inside the target in the range direction. This is the same as the probability a random sample of the delivery accuracy distribution lies between $-L_{ET}/2$ and $+L_{ET}/2$, as shown in Fig. 9.5.

This problem is the same as that in Chapter 2, Example 2.2 and is calculated as

$$PD_{1x} = (ND(L_{ET}/2,0,\sigma_x,1) - ND(-L_{ET}/2,0,\sigma_x,1)) \qquad (9.10)$$

TABLE 9.2   IMPROVED SPREADSHEET BASED ON
MONTE CARLO SIMULATION

PD1 calculator using Monte Carlo methods	
INPUTS	
MAE_F	2270
Impact angle	65
REP	20
DEP	10
Number of iterations	500000
OUTPUTS	
PD1	0.4636
Compute	

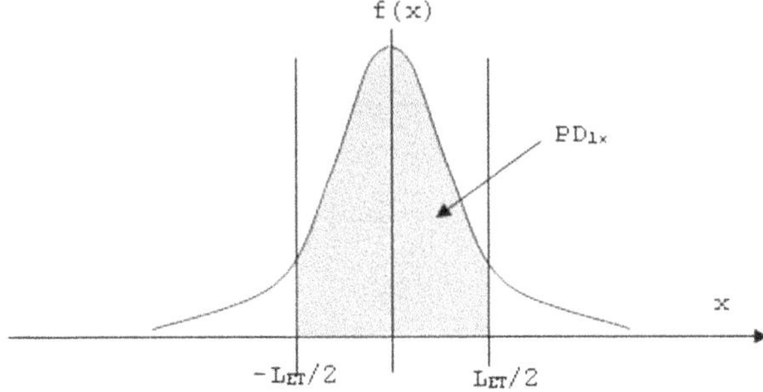

Fig. 9.5   Probability round lands inside lethal area in range.

Similarly, the probability it lands inside the rectangle in the deflection direction is

$$PD_{1y} = (ND(W_{ET}/2,0,\sigma_y,1) - ND(-W_{ET}/2,0,\sigma_y,1)) \qquad (9.11)$$

Therefore, the probability the impact point is inside the rectangle is

$$PD_1 = PD_{1x} \times PD_{1y} \qquad (9.12)$$

Table 9.3 shows the Excel implementation of this deterministic or analytical method.

Comparing the Monte Carlo result of 0.4636 with the analytical result of 0.4632, we see they are very close, and in general this will be the case so long as a sufficient number of Monte Carlo iterations are executed.

Unfortunately, the choice of either Monte Carlo or analytical solutions is not always possible, usually because in some cases the deterministic approach cannot be solved mathematically. In these cases, the Monte Carlo approach is the only option.

In the preceding analysis, the lethal area was represented by a rectangular cookie cutter damage function, and the effectiveness was calculated using both Monte Carlo and deterministic methods. Recall, however, that in Chapter 8 we saw the same lethal area also could be represented by a Carleton damage function in Eq. (8.12). It is reasonable to ask, therefore: Can damage be calculated using this damage function rather than the more simplistic rectangular cookie cutter?

The answer is, yes it can; however, that analysis will not be presented here, only the solutions, and in fact, only the analytical result. For a Carleton damage function represented by Eq. (8.12) and for a given delivery accuracy, the result is given by the following equation:

$$PD_1 = \frac{L'_{ET} \times W'_{ET}}{\sqrt{(17.6REP^2 + L'_{ET}{}^2)(17.6DEP^2 + W'_{ET}{}^2)}} \qquad (9.13)$$

TABLE 9.3   CALCULATING PD_I FOR A RECTANGULAR
DAMAGE FUNCTION

PD1 calculator using cookie cutter damage function	
INPUTS	
MAEF	2270
Impact angle	65
REP	20
DEP	10
DERIVED	
a	0.6619
Let	38.7624
Wet	58.5619
sigma_x	29.6516
sigma_y	14.8258
RESULTS	
PD1-range	0.4867
PD1-deflection	0.9517
PD1-total	0.4632

where L'_{ET} and W'_{ET} are given by Eqs. (8.25) and (8.26). Readers interested in the details of this result are directed to Chapter 10 of the *Introduction to Weaponeering* textbook.

So, in summary we have available three methods for calculating damage inflicted by a general-purpose blast/fragmentation warhead.

1. A Monte Carlo method using the rectangular cookie cutter damage function (Table 9.2)
2. A deterministic method also using the rectangular cookie cutter damage function [Eq. (9.12)]
3. A deterministic method using the Carlton damage function [Eq. (9.13)]

Which one should we use? The answer is that we use option 3 for the following reasons:

- It is a deterministic methodology; therefore, it will always run faster than a Monte Carlo–based method.
- The Carleton damage function is a more realistic approximation than the rectangular cookie cutter to the lethal area matrix.

For targets that are sensitive to blast, the EI is an MAE_B lethal area and is represented by a square in the ground plane, as shown in Fig. 9.6.

The analytical method used for the MAE_F EI can be used here also because blast is independent of the impact angle, and with reference to Chapter 8, Section 8.6 we get

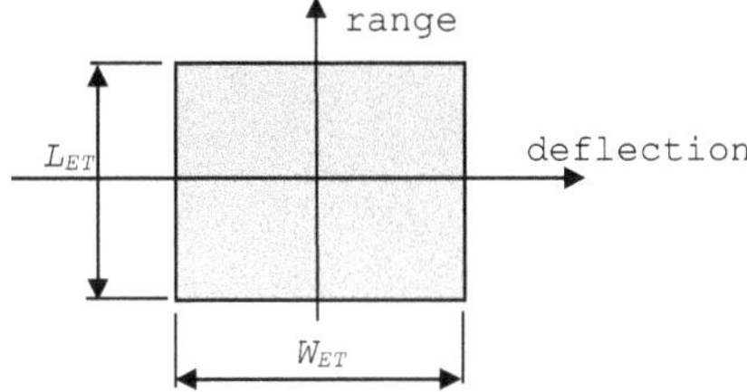

Fig. 9.6   Blast cookie cutter damage function.

$$L_{ET} = W_{ET} = \sqrt{\text{MAE}_B} \qquad (9.14)$$

Again, a simple spreadsheet implementation for representative data is shown in Table 9.4.

Similarly, the Monte Carlo code discussed previously can be modified for the blast EI, producing a value of 0.2899.

9.5 DAMAGE CALCULATION FOR PROJECTILES IN THE NORMAL PLANE

Previously we have considered the effectiveness indices to be either MAE_F or MAE_B, which are defined in the ground plane and as such are used for weapons such as bombs and artillery shells. What if we have a direct fire case such as that shown in Fig. 9.7, in which an armored vehicle is engaging another vehicle with a gun?

TABLE 9.4   CALCULATING PD_1 FOR BLAST
DAMAGE FUNCTION

PD1 calculator for MAEB	
INPUTS	
MAEB	1000
REP	20
DEP	10
DERIVED	
sigma_x	29.6516
sigma_y	14.8258
Let	31.6228
Wet	31.6228
RESULTS	
PD1-range	0.4061
PD1-deflection	0.7138
PD1-total	0.2899

Fig. 9.7   Example of direct fire in the normal plane.

Because a direct hit on the target is required to cause damage, the relevant effectiveness index is the vulnerable area A_V expressed in the normal plane. The methodology and vulnerability data were discussed in Chapter 7, Section 7.15.

The probability of damage is considered to be the probability a single round hits the target vulnerable area[1] in the normal plane. The first step to calculating this is to assume that if we are using a vulnerable area EI, then the A_V is represented as a square of dimensions L_{ET} by W_{ET} such that

$$L_{ET} = W_{ET} = \sqrt{A_V} \qquad (9.15)$$

This situation may be represented in the normal plane by Fig. 9.8, which also shows a rectangle of sides σ_x and σ_y representing the combined weapon accuracy. Note in this representation, range is defined as up and deflection perpendicular to it.

Again, the probability that one round will fall inside the square is calculated in the usual way.

$$PD_{1x} = (ND(L_{ET}/2,0,\sigma_x,1) - ND(-L_{ET}/2,0,\sigma_x,1)) \qquad (9.16)$$

$$PD_{1y} = (ND(W_{ET}/2,0,\sigma_y,1) - ND(-W_{ET}/2,0,\sigma_y,1)) \qquad (9.17)$$

$$PD_1 = PD_{1x} \times PD_{1y} \qquad (9.18)$$

---

[1] Recall that if you hit the target you may not kill it, but it will be killed if you hit the vulnerable area.

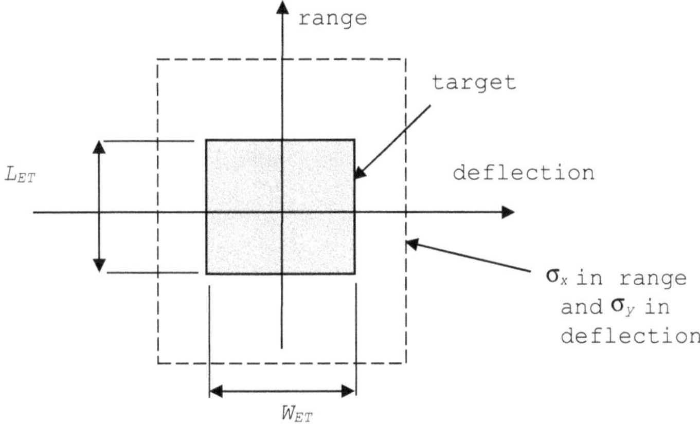

Fig. 9.8 A_V in the normal plane.

In these calculations, and assuming only one round is fired, the MPI and precision errors may be combined using the root sum square process.

$$\sigma^2_x = \sqrt{\sigma^2_{\text{MPI-}x} + \sigma^2_{\text{prec-}x}} \tag{9.19}$$

$$\sigma^2_y = \sqrt{\sigma^2_{\text{MPI-}y} + \sigma^2_{\text{prec-}y}} \tag{9.20}$$

In general, however, such weapons may be fired in a burst where the MPI and precision errors have to be kept separate; this will be dealt with in the next section.

9.6 MULTIPLE WEAPON ATTACKS

In the preceding section it was assumed that we were directing a single weapon, round, or shot, but clearly for some weapons, burst fire or multiple weapon delivery is more common, as shown by the examples in Fig. 9.9.

It was pointed out in Chapter 4, Section 4.8 that when multiple weapons are considered, the dependent error may have to be kept separate from the independent error when calculating effectiveness; otherwise, incorrect results will be obtained. This issue is now addressed in more detail.

The next question to address is: If multiple weapons are used in an attack, how is the effectiveness calculated? The answer to this question depends on whether the attacks are dependent or independent; that is, does the impact point of the second and subsequent weapons depend on the impact point of the first? If the answer is yes, then we are dealing with dependent attacks; if the answer is no, then it is a case of independent attacks, and each has a different methodology to calculate effectiveness.

Fig. 9.9   Multiple weapon deliveries.

9.6.1 MULTIPLE INDEPENDENT ATTACKS

Examples of this type of attack include the following:

- An aircraft flies over the target and drops a bomb. It then circles around and drops another bomb, or another aircraft attacks the same target with the same weapon. The impact point of the second bomb is independent of the first, because each was independently aimed.
- A moving tank shoots at an enemy tank, drives on, re-aims the cannon, and fires another shot at the same target. Each round is independently aimed.

For independent attacks, the methodology for calculating the value of PD is straightforward. We calculate the probability of damage due to a single attack with a single weapon, PD_1, and use the powering up formula to calculate PD, the damage caused by n independent, identical attacks.

$$PD = 1 - (1 - PD_1)^n \tag{9.21}$$

Recall this formula is derived from the binomial distribution, which is valid only for independent events.

9.6.2 MULTIPLE DEPENDENT ATTACKS

Examples of this type of attack include the following:

- A machine gun is aimed at a target on the wall and the trigger is pressed, releasing a burst of rounds.
- An artillery cannon fires several rounds at a target without re-aiming between rounds. Each round has the same aiming error (denoted by the MPI), and the impact point is therefore not independent for each round.

Fig. 9.10   Automatic weapon firing at a target vulnerable area.

*   An aircraft drops several global positioning system/inertial navigation system (GPS/INS) guided weapons at a set of coordinates believed to be those of the intended target. Each weapon is flying to the same set of coordinates, so the target location error (TLE) is the same for each weapon, and each weapon delivery is therefore not independent.

To illustrate the effect of treating these errors separately, or otherwise, let us review the scenario shown in Figure 9.10.

Here we have an automatic weapon firing a three-round burst at a target represented by a vulnerable area. After the first burst, the soldier examines where the rounds landed, re-aims the weapon, fires another three rounds, and so on. The appropriate EI for this case would be the vulnerable area A_V in the normal plane, and we associate this with the target. Impacts in the normal plane for this scenario are shown in Fig. 9.11.

Again, in this representation, the lethal area A_V is associated with the target, and each weapon is a dart (thrown three at one time) where the target is killed if one or more rounds lands inside it. Suppose the target is represented by a 20 ft × 20 ft vulnerable area and is fired on by a gun with a three-round burst. Assuming the MPI error at the target has a CEP of 25 ft and the precision error is a CEP of 2 ft also at the target. What is the expected probability of damaging the target (i.e., hitting the vulnerable area)?

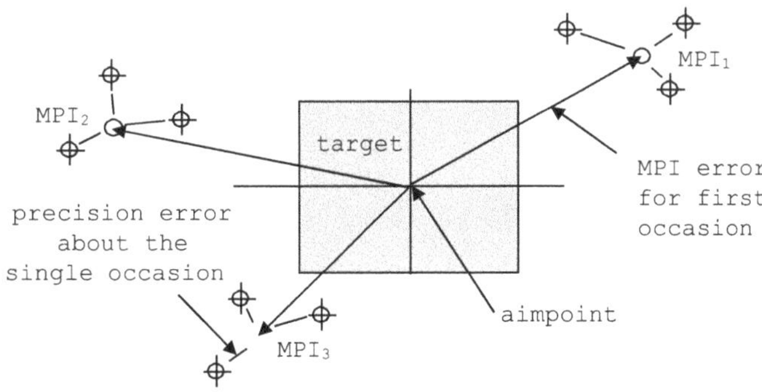

Fig. 9.11   Monte Carlo simulation example.

There is no deterministic solution to this problem, so a Monte Carlo simulation is used. Conceptually, in order to reproduce the pattern of shot shown in Fig. 9.11, we would have a Monte Carlo simulation based on the following sequence of repetitive events:

1. Take a random draw from the MPI error dispersion in range and deflection to determine the MPI for this iteration.
2. For each round in the burst, take a random draw from the precision error dispersion to determine the impact point for each round relative to the MPI.
3. If at least one round impacts inside the vulnerable area, the target is killed; otherwise, it is not.

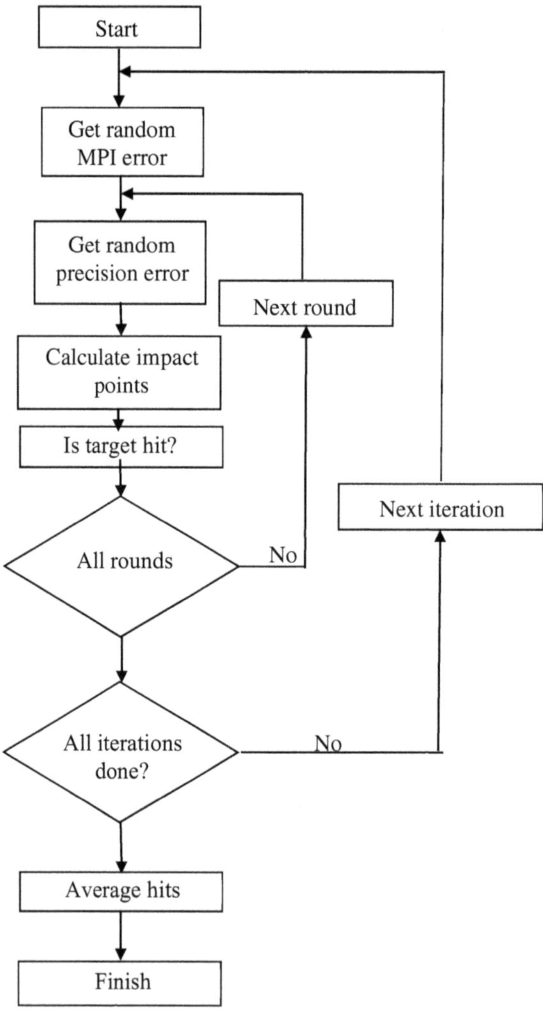

Fig. 9.12   Double loop code for Monte Carlo simulation.

4. Repeat for a sufficient number of iterations.
5. Divide the number of iterations resulting in a kill by the total number of iterations.

To implement this strategy, a computer code is written for the double loop Monte Carlo program illustrated functionally in Fig. 9.12.

This algorithm, which is easily implemented as a Visual Basic program, has a double loop—one for the MPI error and one for the precision error. Suppose the input data are as shown in Table 9.5.

Running the program for the specified 2×10^4 iterations produces the result PD = 0.199. For this case, sample impacts in the normal plane are shown in Fig. 9.13. Readers should compare the single-round impacts in Fig. 9.4 with the burst fire impacts in Fig. 9.13.

TABLE 9.5   INPUTS TO DOUBLE LOOP
MONTE CARLO PROGRAM

INPUTS

Target dim - x	20
Target dim - y	20
MPI CEP	25
Precision CEP	5
Rounds/ burst	3
Number of iterations	20000

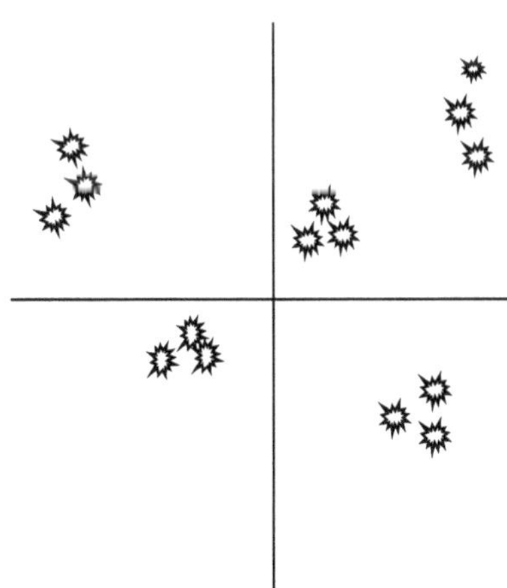

Fig. 9.13   Normal plane burst fire impact points.

If we make the common mistake of assuming both the MPI and precision errors are independent, we would combine them by a root sum squared (RSS) process. If this is done for the example shown in Fig. 9.13, we get a single, combined error term $CEP = 25.5$. Running the same program for a three-round burst but with a single error term produces the result $PD_1 = 0.336$. Clearly, by comparing the two resulting PD values, the effect of assuming independence of error sources introduces considerable error into the calculation of $PD = 0.199$ vs $PD = 0.336$.

9.7 MULTIPLE WEAPON ATTACKS: DETERMINISTIC WORKAROUND

We saw in the previous section that for multiple weapon attacks we keep the MPI and precision errors separate, and this was handled in a straightforward manner by constructing the Monte Carlo simulation with a double loop. We also stated there was no analytical solution to this problem. Although this is true, some Joint Munitions Effectiveness Manual (JMEM) methodologies use a process that provides approximate answers and, like all deterministic methods, provides a fast-running solution. This methodology attempts to reduce the number of error terms from two (MPI and precision) to one (MPI) and has the following objectives:

- Combine the precision error with the damage function, thereby eliminating it from the analysis.
- This leaves only one error term, the MPI, so the methods described in Section 9.4 are applicable.

Consider a single weapon such as an artillery shell fired from a fixed cannon, and the lethal area represented by a simple rectangle. If the cannon fires multiple rounds without re-aiming, we will expect some variability in the location of the rectangular lethal areas on the ground plane around the desired point of impact (DPI) due to precision errors alone, as shown in Fig. 9.14.

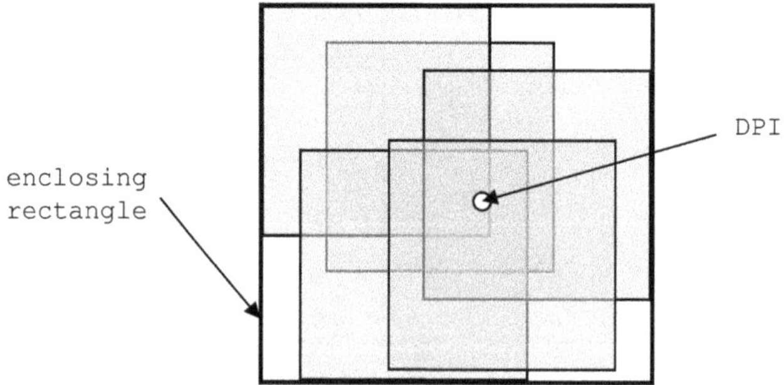

Fig. 9.14   Enclosing lethal areas due to precision error.

Consider the following question: Can we draw a rectangle that encloses all of the weapon lethal areas? If we can, then we could argue this enlarged lethal area includes all variability associated with precision error. Essentially we have enlarged the single weapon lethal area to account for the precision error; in doing so, this error is removed from further consideration.

Unfortunately, the answer to the question is no, because if we draw such a rectangle around 1000 randomly placed lethal areas and then continue the process, we will eventually find one that extends beyond the enclosing rectangle we have drawn. Assuming that precision error is normally distributed in range and deflection, we see there is a small but finite probability the lethal area could land a considerable distance from the aimpoint.

The best we can do is to define such an enclosing rectangle that will contain a certain proportion of lethal areas. For example, suppose σ_{br} is the standard deviation of the precision error in range, L_{ET} is the length of the rectangular damage function, σ_{bd} is the standard deviation of the precision error in deflection, and W_{ET} is the width of the rectangular damage function. In the range direction, we see from Fig. 2.5 in Chapter 2 that an interval $L_{ET} \pm \sigma_{br}$ will contain 68% of such rectangles, and in the deflection direction the interval $W_{ET} \pm \sigma_{bd}$ will also contain 68% of the rectangles. The following statements therefore apply:

- A rectangle $(L_{ET} \pm \sigma_{br}) \times (W_{ET} \pm \sigma_{bd})$ will enclose about 46% (68% × 68%) of the lethal areas.
- A rectangle $(L_{ET} \pm 2\sigma_{br}) \times (W_{ET} \pm 2\sigma_{bd})$ will enclose about 90% of the lethal areas.
- A rectangle $(L_{ET} \pm 3\sigma_{br}) \times (W_{ET} \pm 3\sigma_{bd})$ will enclose about 98% of the lethal areas.

Whichever one we choose, the process is to expand the single weapon rectangular lethal area by an amount that depends on the precision error, as shown in Fig. 9.15.

The size of the enlarged cookie cutter is given by the following:

$$L_B = L_{ET} + k \times \sigma_{br} \tag{9.22}$$

$$W_B = W_{ET} + k \times \sigma_{bd} \tag{9.23}$$

where $k = 2$, 4 or 6 depending on the model or analyst's choice. Note that in doing this enlargement, we cannot leave the damage probability inside the cookie cutter at 1.0; otherwise, we have a weapon with increased lethality compared to the original one. The new damage probability inside the expanded lethal area is defined as the conditional probability of damage P_{CD1} and its value is determined as the ratio of lethal areas. This ensures the lethality of both rectangles are the same.

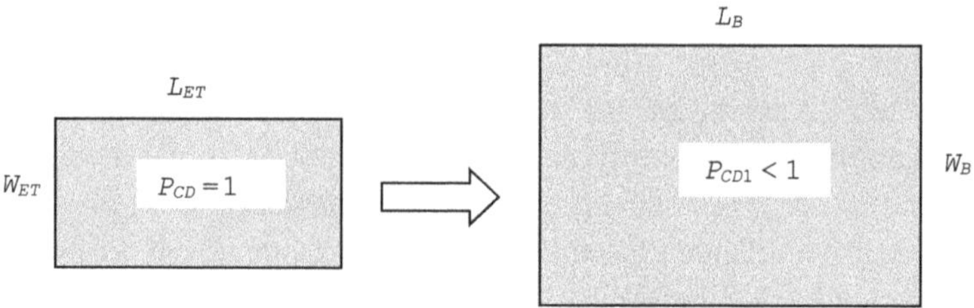

Fig. 9.15　Expanding the single weapon lethal area.

$$P_{CD1} = P_{CD}\frac{A_{ET}}{A_B} = P_{CD}\frac{L_{ET}W_{ET}}{L_B W_B} \tag{9.24}$$

The rectangle of area A_B now becomes the new weapon lethal area, and precision error has been removed from the effectiveness calculation. This leaves only the single, random MPI error, which may be used directly in a deterministic solution.

If we apply this approach to the case considered previously in Table 9.5, where the effectiveness was calculated using a Monte Carlo simulation keeping the error terms separate, what will the result be if we use the approach just described? We note first that for the precision error

$$\sigma_{br} = \sigma_{bd} = \frac{\text{CEP}_{\text{prec}}}{1.1774} = 4.247 \tag{9.25}$$

We expand the lethal area using $k = 6$ and the precision error of 5 ft to get

$$L_B = 20 + 6 \times 4.247 = 45.48\,\text{ft} \tag{9.26}$$

$$W_B = 20 + 6 \times 4.247 = 45.48\,\text{ft} \tag{9.27}$$

Equation (9.24) gives the probability of damage inside the enlarged lethal area as

$$P_{CD1} = P_{CD}\frac{20 \times 20}{45.48 \times 45.48} = 0.193 \tag{9.28}$$

A single weapon is now represented by a lethal area $L_B \times W_B$ delivered with an accuracy of the MPI (i.e., a 25-ft CEP). The resulting damage probability is evaluated using Eqs. (9.10) and (9.11) written in the form

$$\text{PD}_{1x} = (\text{ND}(L_B/2,0,\sigma_x,1) - \text{ND}(-L_B/2,0,\sigma_x,1)) \tag{9.29}$$

$$\text{PD}_{1y} = (\text{ND}(W_B/2,0,\sigma_y,1) - \text{ND}(-W_B/2,0,\sigma_y,1)) \tag{9.30}$$

In calculating PD_1, we have to modify Eq. (9.12) to allow for the reduced conditional probability of damage inside the enlarged lethal area.

$$PD_1 = P_{CD1} \times PD_{1x} \times PD_{1y} \tag{9.31}$$

First, we convert the MPI CEP to sigma values with

$$\sigma_x = \sigma_y = \frac{CEP_{MPI}}{1.1774} = 21.23 \text{ ft} \tag{9.32}$$

Then, evaluating Eqs. (9.29) through (9.31), we get $PD_1 = 0.099$. Because we have eliminated the precision error, we have only the MPI error left as the single random error, so we use the powering up formula for the three rounds in the burst.

$$PD = 1 - (1 - PD_1)^n = 1 - (1 - 0.099)^3 = 0.269 \tag{9.33}$$

Clearly this is closer to the Monte Carlo value (0.199) than the result where we incorrectly RSS the MPI and precision errors (0.326) and illustrates the utility of this approach; however, variations from this result may occur for different values of k.

9.8 CALCULATING DAMAGE FOR NON-NORMAL ACCURACY DISTRIBUTIONS

This calculation is usually associated with guided weapons because their accuracy distributions may be non-Gaussian, as discussed in Chapter 4, Section 4.7, where the P_{HIT}/P_{NM} method was described. The calculation of PD_1 is a little different than for weapons described by a Gaussian distributed miss distance.

Consider Fig. 4.12 in Chapter 4 again. Suppose $P_{HIT} = 1$ and $P_{NM} = 0$. This would indicate that the weapon always hits the target, and there is no Gaussian distribution of miss distances. If the weapon is well matched against the target, this would always give $PD_1 = 1$. Now consider $P_{HIT} = 0$ and $P_{NM} = 1$. This implies the weapon behaves with a Gaussian distribution of miss distances, and the methodology based on Gaussian delivery errors described earlier would work just fine in these circumstances. For intermediate values of P_{HIT} and P_{NM}, we need an estimate of PD_1 that takes both factors into account. To achieve this, P_{HIT} and P_{NM} are used as weighting factors to combine the direct hit and Gaussian miss distribution values of PD_1. The weighted summation function takes the following form:

$$PD_1 = [PD1_1 \times P_{NM} + PD1_2 \times P_{HIT}] \tag{9.34}$$

In this equation, $PD1_1$ is the value for PD_1 computed using the Gaussian-based methodology described in the first part of this chapter using REP and DEP or perhaps CEP. $PD1_2$ is the same as $PD1_1$ except that a value of $CEP = 0$ is used (or $REP = DEP = 0$), to simulate a weapon that always hits the target. It is seen that Eq. (9.34) is indeed a weighted combination of the pure Gaussian and direct hit PD_1 values.

Equation (9.34) gives the single weapon PD_1, but if more than one guided weapon is used, the resulting PD is obtained by powering up the single weapon PD_1 in the usual manner,

$$PD = 1 - [1 - PD_1]^n \qquad (9.35)$$

where n is the number of weapons launched. Note that an implicit assumption here is that each weapon launched is individually aimed at the target.

9.9 BOMB BURIAL

When a single weapon impacts a soft material such as soil or sand, it will generally become partially buried before it detonates, resulting in a degraded value for the PD_1. Based on test data and high fidelity computer models of the resulting blast and fragment patterns, the following empirical equations were found to represent the degradation of PD_1 for 10%, 25%, and 50% burial in terms of the unburied PD_1 value:

$$PD1_{10} = 0.3897PD_1^4 - 0.1047PD_1^3 - 0.2210PD_1^2 + 0.8443PD_1 \qquad (9.36)$$

$$PD1_{25} = 0.2034PD_1^4 - 0.1051PD_1^3 - 0.1606PD_1^2 + 0.6321PD_1 \qquad (9.37)$$

$$PD1_{50} = 0.3148PD_1^4 - 0.2279PD_1^3 - 0.3190PD_1^2 + 0.4013PD_1 \qquad (9.38)$$

These equations form the legacy JMEM methodology for accounting for mitigated blast and fragment effects for partially buried warheads. These effects are shown in Fig. 9.16.

Clearly the percent burial has to be estimated based on bomb size, impact velocity, and ground material. More information may be found in the *Advanced Weaponeering* textbook.

9.10 CHAPTER SUMMARY

- There are two ways to interpret the lethal area: In the first case, the target is represented by a fixed point at the aimpoint with the weapon represented by a randomly placed lethal area. Alternatively, the target is represented by

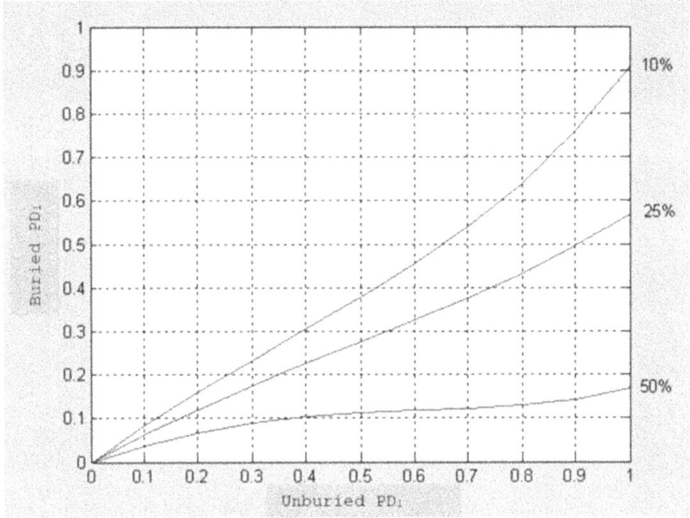

Fig. 9.16   Effect of bomb burial on PD₁.

the lethal area centered on the aimpoint and the weapon is a randomly placed point. Mathematically, both produce the same result.

- The probability of damage can be calculated using either a Monte Carlo simulation or a deterministic calculation. In some cases, either may be used; however, for more complicated deliveries, a deterministic solution may not be available.
- When single weapon deliveries are used, the MPI and precision errors may be combined (RSS), but for multiple weapon deliveries they have to be kept apart; otherwise, the results will be incorrect.
- The damage caused by guided weapons having non-Gaussian distributions of miss distance may be calculated as a weighted sum of two PD_1 calculations using the P_{HIT}/P_{NM} methodology.

Chapter 10

SINGLE WEAPONS DIRECTED AGAINST AN AREA OF TARGETS

10.1 INTRODUCTION

Up to this point, all of our targets have been considered unitary (i.e., one tank, one truck, one soldier, etc.). Sometimes, however, we want to attack a collection of targets grouped together, perhaps a staging area for vehicles or a group of soldiers advancing in the open. This chapter will extend the previous analysis to include targets that contain multiple elements inside a defined area.

Although an area of targets is sometimes referred to as simply an area target, there is an important point to keep in mind. A target such as a liquid storage tank may be considered to occupy a large area on the ground; however, because there is only one storage tank, it is a unitary target, and the methods developed in the previous chapter will apply. On the other hand, an area of targets as defined here are understood to contain multiple, identical target elements, perhaps in this example a collection of storage tanks.

This concept is consistent with the previous discussion of weapon/target representation in Chapter 9, Section 9.3, where a single target may be represented by a point, whatever its size, or by an area, although this area represents the lethal area and not the physical size of one target element.

In analyzing targets of this type, the following assumptions are made:

- All target elements are unitary and identical.
- They are contained in the smallest possible rectangular area of dimensions L_A and W_A.
- The target elements are uniformly distributed within this rectangle.

Figures 10.1 and 10.2 show how real-world target scenarios are stylized into compliance with the definition of an area of target elements.

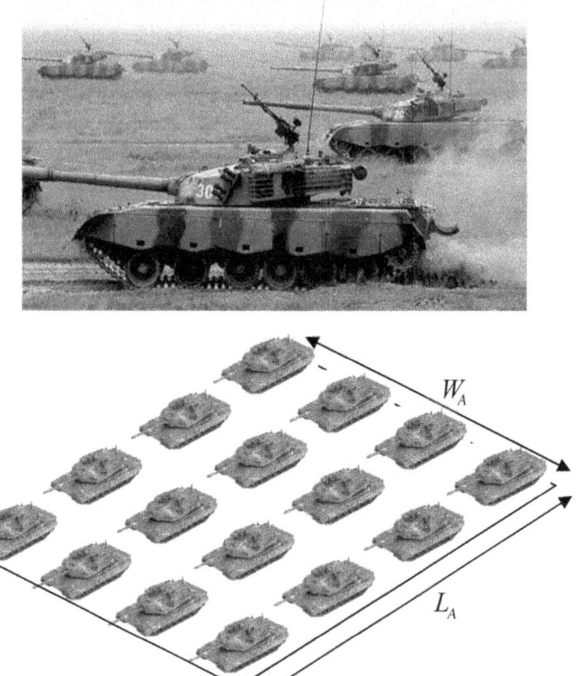

Fig. 10.1   Area of vehicle elements of dimensions L_A by W_A.

Fig. 10.2   Area of personnel elements, also L_A by W_A.

10.2   MEASUREMENT OF DAMAGE

For targets of this type, we will use the rectangular cookie cutter lethal area for the damage function. In dealing with areas of targets, we do not describe the damage done by a single weapon in terms of a PD_1, as was done for a unitary target. Instead, the damage metric is the fractional damage FD_1. This is because the target area contains a fixed number of target elements, and of primary interest is the fraction of those elements damaged. Consider the situation shown in Fig. 10.3, where an area of targets is 50% covered by a rectangular weapon lethal area.

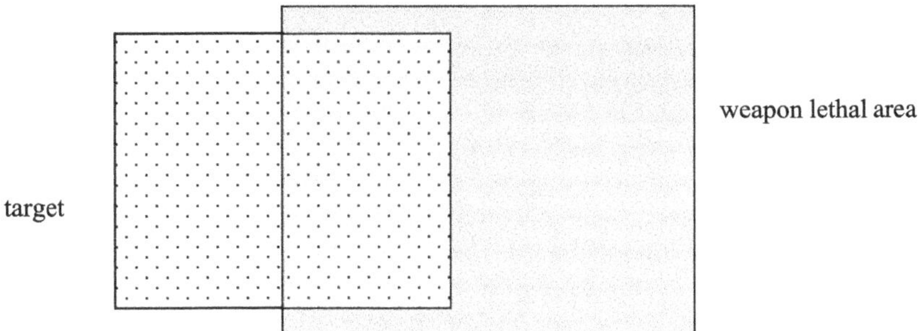

Fig. 10.3   Target half covered by weapon lethal area.

For this particular case, if there were 100 target elements, 50 would be covered by the weapon lethal area; if the probability of damage inside the damage function were 1.0, those 50 elements would be damaged to the level for which the lethal area is defined. Suppose we consider the general case where perhaps the probability of damage inside the lethal area is 0.8; in this case, 40 elements would be damaged. We can therefore write

$$\mathrm{FD}_1 = F_C \times P_{CD} \tag{10.1}$$

where FD_1 is the fractional damage due to a single weapon; F_C is the fractional coverage, 50% in this example; and P_{CD} is the conditional damage probability inside the rectangular lethal area. Usually $P_{CD}=1$ for a single weapon cookie cutter lethal area, but we saw in Chapter 9, Section 9.7 how this is not always true. Therefore, because P_{CD} is fixed and assumed known, the problem of finding FD_1 is the same as finding the fractional coverage F_C.

Of course, we have assumed that the weapon lands in the location shown such that it covers exactly half of the target. Because the impact point will be random, we should now ask the following: In a large number of trials, what is the average fractional coverage of the target by the weapon lethal area? This situation is shown in Fig. 10.4, where randomly placed lethal areas cover the target area by different amounts.

Figure 10.4 suggests how FD_1 might be calculated in a Monte Carlo simulation by calculating the fractional area of the target covered by the weapon lethal area for a sufficient number of iterations and then averaging. We will see that this task is not as daunting as it might appear.

10.3   FRACTIONAL COVERAGE FUNCTION

Although the fractional coverage may be thought of in terms of a random placement of the weapon lethal area relative to the area of target elements, we apply a more methodical approach here by considering placement of the lethal area in the x and y directions separately. Consider first what happens when the

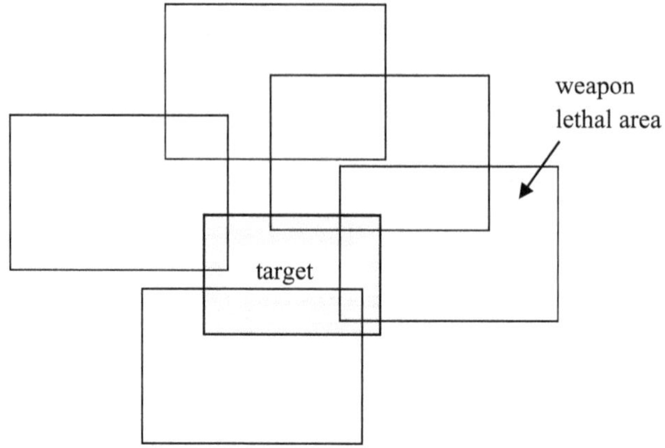

Fig. 10.4   Fractional coverage of area of targets.

target area is fixed at the origin and the weapon lethal area, assumed larger than the target, is far to the left of it, as shown in Fig. 10.5. Here the position of the weapon is defined as the value of x (range) from the origin to the center of the lethal area.

We may simply write from observation that the fractional coverage in range for this case is

$$F_R = 0 \qquad (10.2)$$

Now consider the weapon moving to the right, that is, x increasing in a positive sense. Equation (10.2) will still apply until the condition shown in Fig. 10.6 occurs.

At this point we see that

$$x = \frac{L_{ET}}{2} + \frac{L_A}{2} = \frac{L_{ET} + L_A}{2} \qquad (10.3)$$

Further movement of the weapon area to the right will cause a partial coverage of the left side of the target area, as shown in Fig. 10.7.

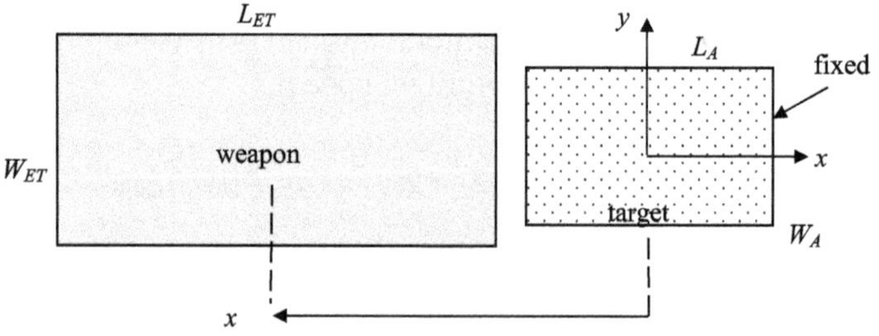

Fig. 10.5   Weapon misses the target to the left.

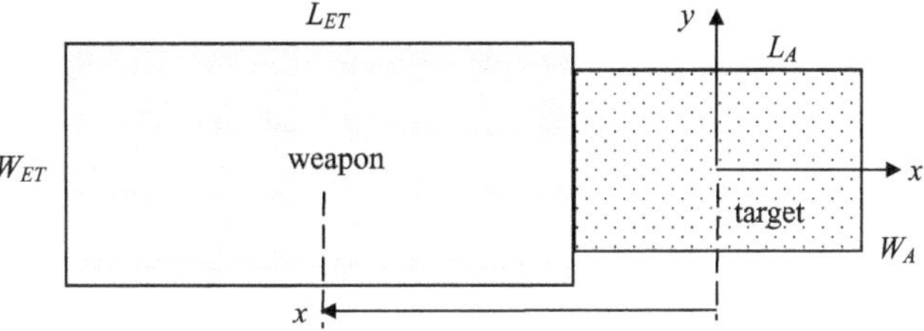

Fig. 10.6 Weapon abuts the target on the left.

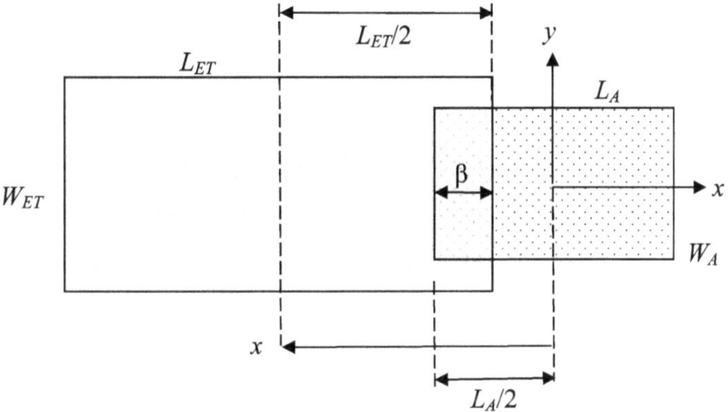

Fig. 10.7 Partial left overlap of target by weapon.

We can see that the value of F_R is simply the fraction of the target area overlapped by the lethal area (β) and will vary linearly from zero to unity depending on x. It may be shown that the fractional coverage in range is given by the following:

$$F_R = \frac{L_A + L_{ET}}{2L_A} + \frac{x}{L_A} \tag{10.4}$$

This is simply the equation of a straight line

$$y = mx + c \tag{10.5}$$

By setting $F_R = 0$, we get the value of x when partial covering begins as

$$x = \frac{L_{ET} + L_A}{2} \tag{10.6}$$

As x increases further and the lethal area moves to the right, at some value of x the target is completely covered ($F_R = 1$), as shown in Fig. 10.8.

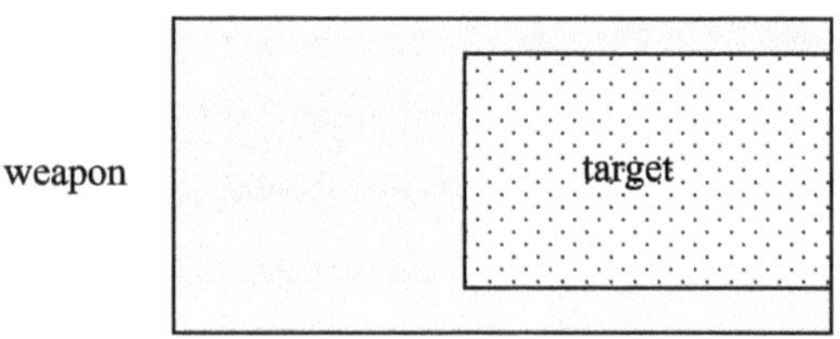

Fig. 10.8   Weapon covering target from the left.

The value of x when this happens is obtained from Eq. (10.4) by setting $F_R = 1$, giving

$$x = \frac{L_A - L_{ET}}{2} \tag{10.7}$$

At this point, the weapon area completely covers the target area, in which case

$$F_R = 1 \tag{10.8}$$

Increasing x further, the fractional coverage remains at this value until the weapon area is on the point of uncovering the target area on the left side, as shown in Fig. 10.9.

Moving the weapon lethal area further to the right begins to uncover the target area on the left side, as shown in Fig. 10.10.

Here the fractional coverage in the range direction is given by

$$F_R = \frac{L_{ET} + L_A}{2L_A} - \frac{x}{L_A} \tag{10.9}$$

which is again the equation of another straight line, this time with negative slope. Partial uncovering of the target begins by putting $F_R = 1$ in Eq. (10.9), giving

$$x = \frac{L_{ET} - L_A}{2} \tag{10.10}$$

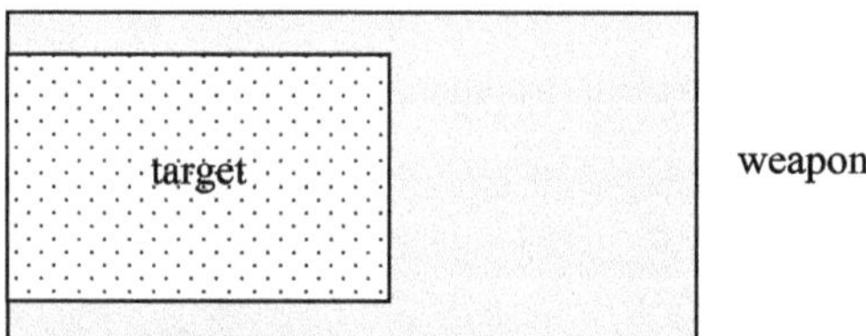

Fig. 10.9   Weapon covering target from the right.

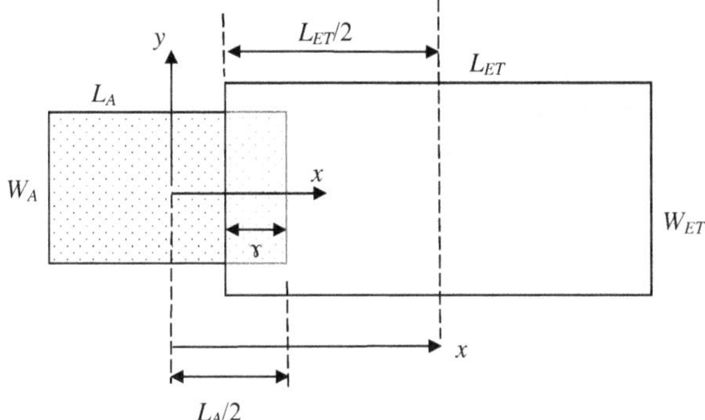

Fig. 10.10 Partial right overlap of target by weapon.

Increasing x further reveals more of the target until the situation shown in Fig. 10.11 occurs.

The value of x when this occurs is obtained by setting $F_R=0$ in Eq. (10.9), yielding

$$x_{\max} \frac{L_{ET} + L_A}{2} \tag{10.11}$$

Now use the shorthand notation

$$s = \frac{L_{ET} + L_A}{2} \qquad t = \frac{L_{ET} - L_A}{2} \tag{10.12}$$

Further movement of the weapon lethal area to the right results in $F_R=0$. Figure 10.12 summarizes fractional coverage as a function of x, the coordinate in the range direction measured from the center of the fixed target to the center of the weapon lethal area. This fractional coverage function is known as a piecewise, continuous, trapezoidal function.

A Monte Carlo approach to calculating the fractional coverage in the range direction would use this function and consist of the following steps:

1. Calculate a random placement of the weapon lethal area x from a random draw from a normally distributed, zero mean distribution defined by range error probable (REP).

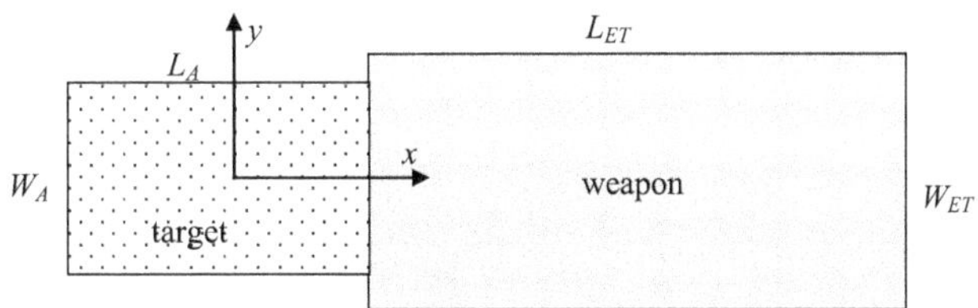

Fig. 10.11 Weapon abuts target on the right.

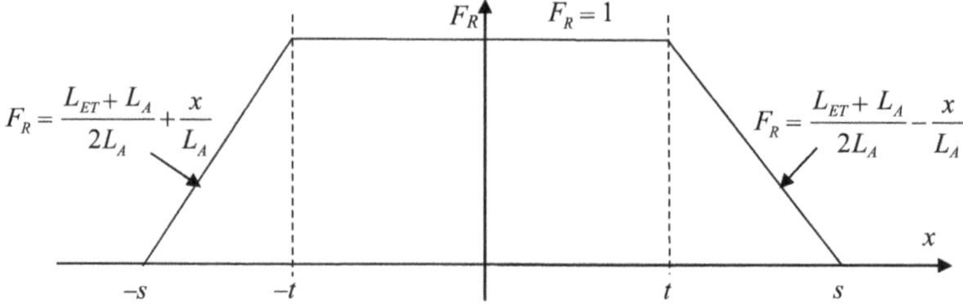

Fig. 10.12   Fractional coverage function in range.

2. For that value of x, from Fig. 10.12 calculate the fractional coverage in the range direction.
3. Repeat the process and average over a sufficient number of iterations.

Figure 10.13 shows how the fractional coverage function is sampled for a few random values of x.

What about the fractional coverage F_D in the deflection direction? Exactly the same analysis may be done in deflection, with the same results except the inputs will be W_A, W_{ET}, and deflection error probable (DEP). The final result for fractional coverage is obtained from

$$F_C = F_R \times F_D \tag{10.13}$$

The fractional damage representing the proportion of target elements damaged is obtained from Eq. (11.1).

$$\mathrm{FD}_1 = F_C \times P_{CD} \times R \tag{10.14}$$

where R is introduced as the weapon reliability and P_{CD} is the conditional probability of damage inside the rectangular damage function, usually unity. Table 10.1 shows a spreadsheet implementation of this method.

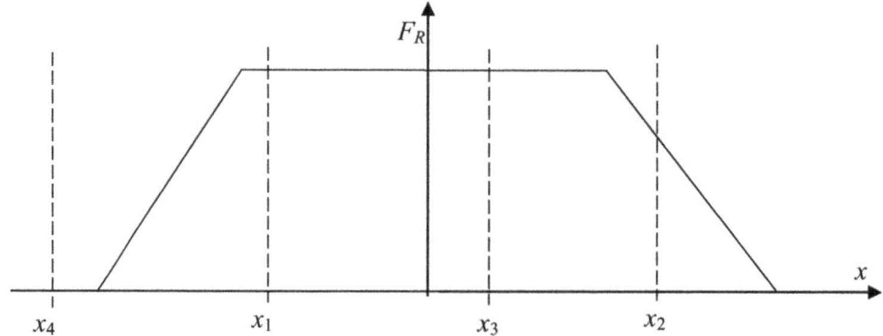

Fig. 10.13   Sampling the fractional coverage function.

TABLE 10.1   SPREADSHEET TO CALCULATE FRACTIONAL DAMAGE

INPUTS	
MAE_F	2270
Impact angle	65
Target area length	25
Target area width	15
REP	20
DEP	10
Number of iterations	100000
OUTPUTS	
Fractional coverage range	0.4737
Fractional coverage defl.	0.9420
Fractional Damage	0.4463

Compute

All the preceding analysis assumes that the weapon lethal area is larger than the target area. What if the lethal area is smaller than the target? Figure 10.14 shows the overlay of the target for this case where the weapon is shown for several critical positions.

Figure 10.12 is changed into Fig. 10.15 because the maximum fractional coverage is no longer unity.

In addition, the parameter t becomes

$$t = \frac{L_A - L_{ET}}{2} \tag{10.15}$$

With these adjustments, the spreadsheet shown in Table 10.1 can be modified to handle both cases. Given FD_1, if the number of target elements in the area target is m, the number of elements damaged N_D is given by

$$N_D = FD_1 \times m \tag{10.16}$$

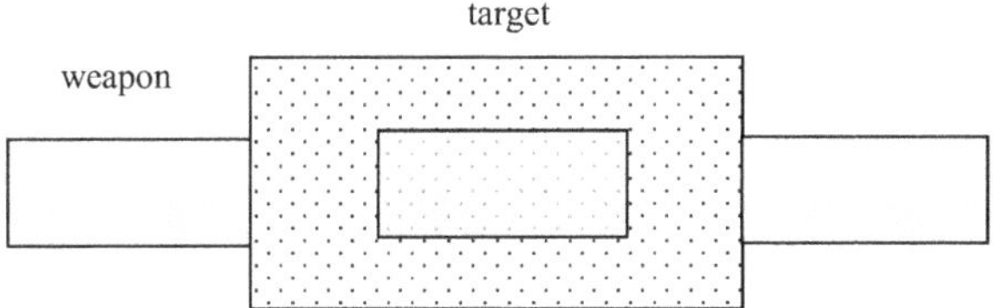

Fig. 10.14   Weapon smaller than the target.

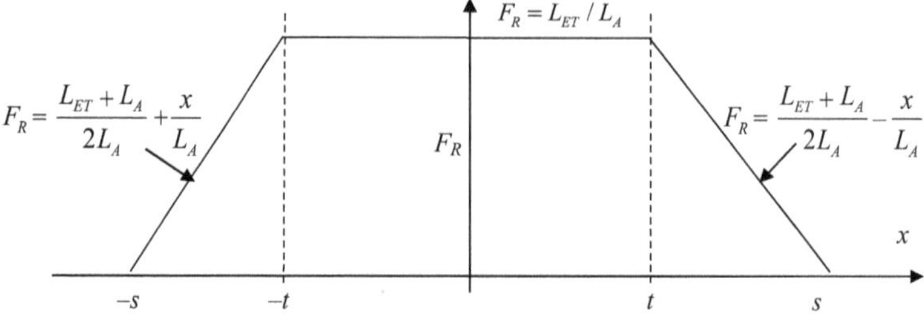

Fig. 10.15 Fractional coverage function in range.

Equation (10.14) gives the expected fractional damage due to a single sortie. If there are n independent sorties, the total fractional damage is given by

$$FD = 1 - (1 - FD_1)^n \tag{10.17}$$

Once again, the number of target elements damaged in n sorties is given by

$$N_D = FD \times m \tag{10.18}$$

Remember, this will not tell us which elements are damaged and which survive, only the number damaged.

10.4 CALCULATING FD_1 FOR GUIDED WEAPONS

The calculation of FD_1 for guided weapons is similar to that described in Chapter 9 for unitary targets [i.e., using Eq. (9.34) adapted for area targets].

$$FD_1 = \left[FD1_1 \times P_{NM} + FD1_2 \times P_{HIT} \right] \tag{10.19}$$

In this equation, $FD1_1$ is the value for FD_1 computed using the methodology described in the first part of this chapter [i.e., assuming a Gaussian miss distance characterized by REP and DEP, or circular error probable (CEP)]. $FD1_2$ is the same as $FD1_1$ except that a value of $CEP = 0$ is used (or $REP = DEP = 0$) to simulate a weapon that always hits the target. Again it is seen that Eq. (10.19) is indeed a weighted combination of the pure Gaussian and direct hit FD_1 values.

10.5 FRACTIONAL DAMAGE USING DISCRETE TARGET ELEMENTS

Now that computers are relatively fast, it is possible to develop a Monte Carlo–based methodology to calculate the fractional damage for a single weapon directed against an area of target elements. The basic approach is to calculate an impact point for the rectangular cookie cutter damage function

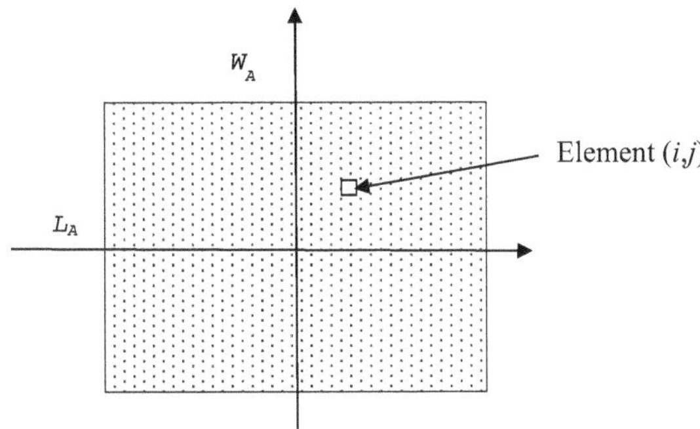

Fig. 10.16   Area of discrete target elements.

and center the rectangular cookie cutter on it. The area of target elements is inspected to see how many are killed. In detail, the method consists of the following steps:

1. The area containing the identical, uniformly distributed target elements is placed at the origin. Suppose for simplicity it is assumed there are 50 elements along each side of this rectangle; therefore, there are 2500 elements total. These elements are represented by a 2-D "kill matrix" K where if the element value is unity, it is killed; otherwise, it survives and has the value of zero. The array is initialized to all zero values and is shown in Fig. 10.16.
2. Because we are only dealing with a single weapon, the mean point of impact (MPI) and precision errors may be combined into a single accuracy. A random draw is taken from this combined delivery error distribution and indicated on the ground, as shown in Fig. 10.17.
3. Taking each target element in turn, we determine if it is enclosed by the lethal area in its current location; if it is, the value of that element in the kill matrix is set to unity.

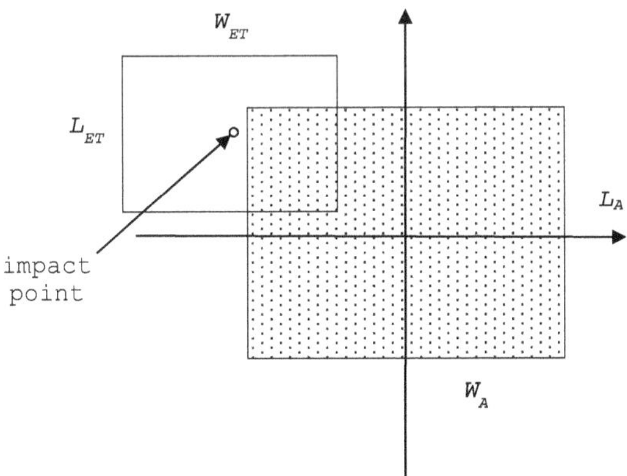

Fig. 10.17   Lethal area centered on sample MPI.

4. The fractional damage may now be calculated by summing all the elements killed in this iteration divided by the total number of elements.

$$FD_1 = \frac{\sum_{i=1}^{i=50} \sum_{j=1}^{j=50} K(i,j)}{2500} \tag{10.20}$$

5. The whole process is repeated for another random draw from the delivery error distribution over a user-specified number of Monte Carlo iterations, and the resulting values of fractional damage are averaged.

The spreadsheet shown in Table 10.1 can be expanded to include this discrete target element method and compared to the one using the random sampling of the trapezoidal damage function. This is shown in Table 10.2.

Notice that there is a small difference in results due to the discrete nature of the target elements and the number of Monte Carlo iterations. Given a choice, the trapezoidal function should be used to get the most accurate result.

TABLE 10.2   COMPARISON OF METHODS TO CALCULATE
FRACTIONAL DAMAGE

INPUTS	
MAE_F	2270
Impact angle	65
Target area length	25
Target area width	15
REP	20
DEP	10
Number of iterations	5000
OUTPUTS - TRAPEZOIDAL FUNCTION	
Fractional coverage range	0.477
Fractional coverage defl.	0.944
Fractional coverage	0.451
OUTPUTS - DISCRETE ELEMENTS	
Fractional coverage	0.416
Compute	

Students should also note that the trapezoidal fractional coverage function shown in Fig. 10.15 can be used to develop a deterministic solution to calculating fractional damage; however, for reasons of simplicity, it is not presented here. The interested reader is referred to the *Introductory Weaponeering* textbook.

10.6   CHAPTER SUMMARY

- A new target type was introduced known as an area of target elements. It consists of a collection of identical unitary targets uniformly distributed within a defined rectangular area on the ground.
- For this target type, the damage metric is fractional damage and depends on the fractional coverage of the single weapon rectangular cookie cutter damage function over the rectangle containing the target elements.
- By determining a trapezoidal fractional coverage function for both the range and deflection directions, and by considering only single weapon attacks, the fractional damage may be estimated using Monte Carlo methods.
- Another method using discrete target elements was also described.

Chapter 11

STICKS, VOLLEYS, AND CLUSTER MUNITIONS

11.1 INTRODUCTION

So far, we have considered multiple weapon deliveries to originate from one source (i.e., a single artillery cannon firing several rounds or a machine gun firing a burst of bullets). In both these cases all weapons are directed toward a common aimpoint. In many cases, multiple weapons are directed towards a target with differing aimpoints, resulting in a greater spread of the individual impacts and hence lethality.

Figure 11.1 shows an aircraft releasing a stick of unguided bombs with the intent to give a larger lethal area on the ground in which damage can occur.

A stick release involves more than one weapon being released, not at the same instant, but over a short period of time with a specified delay between them. Similarly, a battery of artillery cannons may deliver a volley where all

Fig. 11.1   Stick delivery.

Fig. 11.2   Artillery volley fire.

guns fire one round, each directed toward a different aimpoint, as shown in Fig. 11.2.

In both cases, the area on the ground covered by the weapons is called the *pattern*, and because the attacks are usually used against area targets, the effectiveness measure is fractional damage (FD). Assuming for now that each weapon is represented by a rectangular cookie cutter, a typical pattern is shown in Fig. 11.3, where the shaded areas are individual weapon lethal areas.

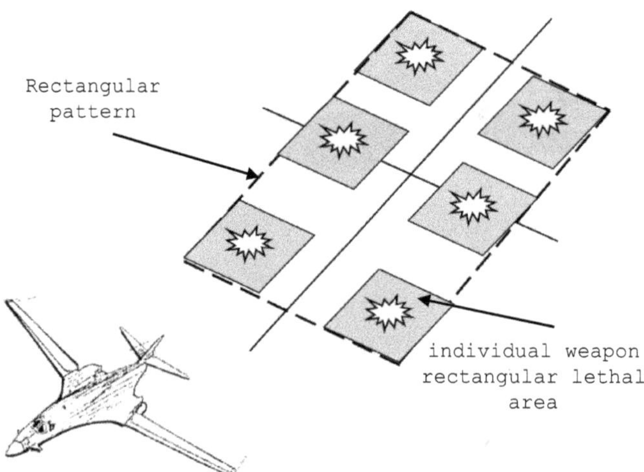

Fig. 11.3   Individual weapon impacts forming the pattern.

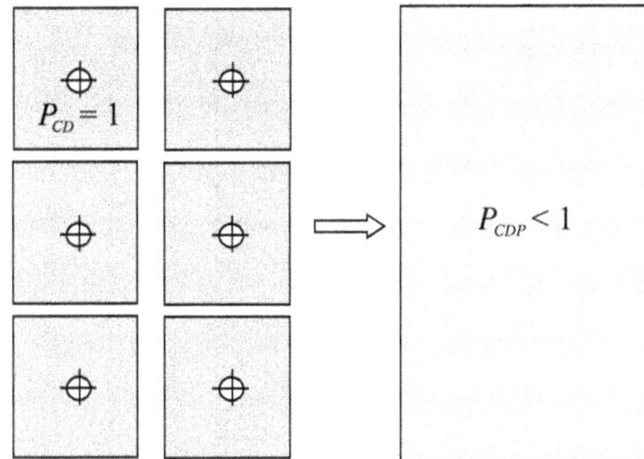

Fig. 11.4 Combining individual lethal areas into a pattern.

This pattern could be produced by a single aircraft dropping six unguided bombs in a stick as shown, or a six-gun artillery battery firing a single volley. The basic approach in dealing with these deliveries is as follows:

1. Determine the impact points on the ground where each weapon in the stick or volley lands.
2. Determine the dimensions of the lethal area from the weapon trajectory and effectiveness index.
3. Place a rectangular damage function on the ground, centered on the impact points.
4. Determine the rectangular pattern dimensions $L_P \times W_P$ by enclosing all the weapon lethal areas with a single rectangle.
5. Calculate the probability of damage P_{CDP} averaged within the pattern based on the number and spacing of the lethal areas.
6. For an area of targets, calculate FD_1 as described in Chapter 10 by considering the stick or volley to be equivalent to a single effectiveness area with the dimensions and $P_{CD} = P_{CDP}$ defined earlier.

The aggregation of individual weapon lethal areas into a single rectangle with a probability of damage P_{CDP} inside it is shown in Fig. 11.4.

The stick is then treated as a single weapon represented by a rectangular lethal area of known dimensions and a probability of damage inside it. The main tasks, therefore, are determining the pattern size and its associated conditional probability of damage P_{CDP}. Although Fig. 11.4 shows the lethal areas completely separated, they may overlap in range, deflection, or both. The next sections deal with each of the items listed previously; however, before moving forward, we first differentiate between the stick and the pattern.

11.2 DETERMINING THE STICK DIMENSIONS

Consider the six impact points and associated lethal areas shown in Fig. 11.5.

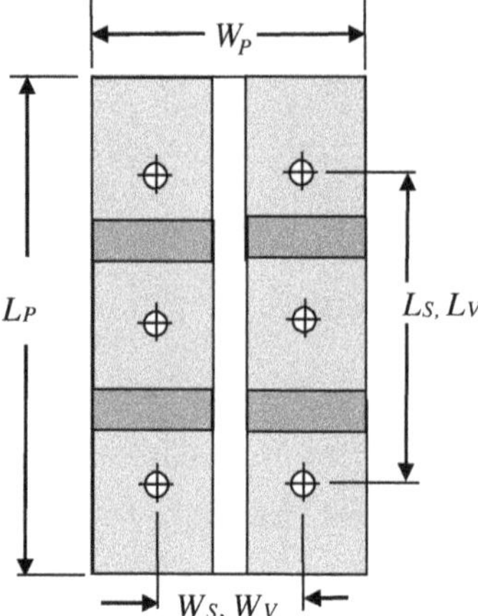

Fig. 11.5   Definition of stick, volley, and pattern dimensions.

- The stick and volley dimensions are those of the smallest rectangle enclosing the impact points.
- The pattern dimensions are those of the smallest rectangle enclosing the lethal areas.

The first task, therefore, is to determine the stick or volley dimensions. This is done differently for unguided, air-launched bombs and unguided artillery projectiles.

11.2.1   AIR-LAUNCHED WEAPONS: STICK DIMENSIONS

In the range direction, the stick dimension L_S is determined by the aircraft intervalometer setting. An intervalometer is a timer that sends electrical pulses to the bomb racks to release the weapons. For each pulse, one or more weapons may be released. The pilot is able to set the intervalometer to fixed values; for example, the A-10 intervalometer may be set between 0.010 and 0.999 s in 0.01-s steps.

Aircraft with more advanced avionics allow the pilot to enter a desired stick length on the ground, and sensors provide data regarding the aircraft state in order to calculate the correct intervalometer delays. The number of intervalometer pulses is denoted by the variable n_r, and the number of weapons released for each pulse is n_p. Figure 11.6 illustrates some weapon lethal area patterns for various n_r and n_p values.

11.2.1.1 STICK WIDTH

Bombs may be carried on hard points on the aircraft wings and fuselage or internally for larger aircraft. In addition, bomb racks capable of holding more than one weapon can be attached to the hard points to increase the payload of the aircraft. Figure 11.7 shows bomb racks capable of holding six weapons on a single aircraft hard point.

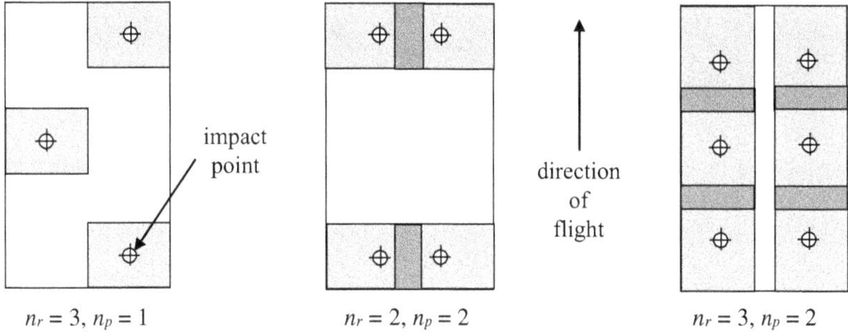

Fig. 11.6   Sample stick patterns for different n_r and n_p.

Fig. 11.7   Multiple ejection bomb rack.

A TER carries three weapons at a single station, whereas a multiple ejection rack (MER) consists of two TERs in tandem. Each weapon is ejected in the manner indicated in Fig. 11.8.

This figure shows that the weapons will spread in the deflection direction when ejected and will therefore influence the pattern width on the ground. A main rack release is defined as one where the weapons drop vertically from the aircraft attachment points; therefore, the stick width is set to the interstation distance on the aircraft. As shown in Fig. 11.9, this will vary for each aircraft.

The stick width depends on each airframe, so data for a particular aircraft must be obtained, as indicated in Table 11.1.

For MER releases where a lateral velocity is present, the stick width is computed by determining the value of this velocity. Then a transverse trajectory model is used to determine, for the altitude at which the bomb is

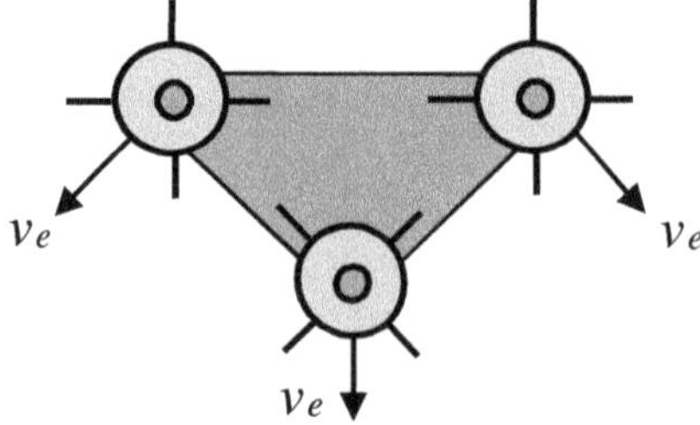

Fig. 11.8   Bomb ejection from a MER/TER.

Fig. 11.9   Weapon station locations.

TABLE 11.1   DISTANCE BETWEEN STATIONS (FT)

Aircraft	Centerline	Inboard	Intermediate	Outboard
A-10	0, 3.8	12	None	20
AV-8B	0	7.2	9.3	11.1
F-14	0, 4	None	None	None
F-15	0	None	None	20
F-16	0	12	None	20
B-52	None	None	None	36

released, how far in the deflection direction an unguided weapon will move as it falls to the ground. A simple zero drag trajectory model as discussed in Chapter 3 may be used for this calculation.

11.2.1.2 STICK LENGTH

Although the stick width (hence the pattern width) may be estimated from the locations of the hard points on the airframe, the stick length (and hence the pattern length) is calculated from the aircraft release conditions and the intervalometer settings. If the aircraft is flying straight and level, we have the situation shown in Fig. 11.10.

For this case, the problem becomes somewhat simplified because the stick length between the last release pulse and the first one is the same distance the aircraft travels horizontally; therefore,

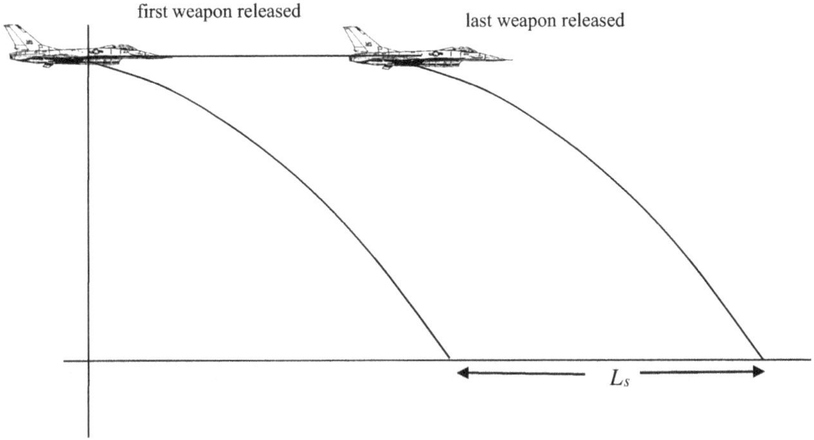

Fig. 11.10   Level deliveries of air-launched weapons.

$$L_s = v_a \left(n_r - 1 \right) \Delta t \tag{11.1}$$

In this equation, v_a is the aircraft speed in the horizontal direction, and Δt is the intervalometer setting. This equation is based on the assumption that all weapon trajectories are identical, resulting in a uniform distribution of impact points in the ground plane. Dive releases are more complicated and will not be dealt with here.

11.2.2   SURFACE-LAUNCHED WEAPONS: VOLLEY DIMENSIONS

For an artillery battery comprising several cannons, determining the impact points for a single volley is easier than the air-launched case, because each cannon will be aimed at a different point than the others. Because these weapon systems are usually employed against a rectangular area of targets, the aimpoints are spread out in a uniform manner, as shown in Fig. 11.11.

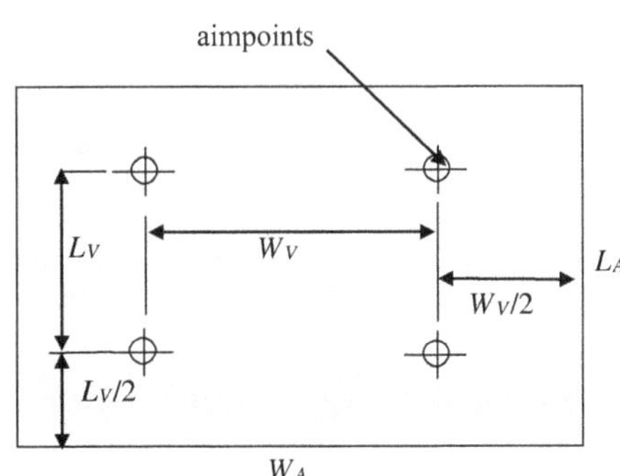

Fig. 11.11   Aimpoints for artillery on a rectangular target.

The volley length and width for the four artillery cannons distributed uniformly within a rectangular target area will give the following:

$$L_V = L_A / 2 \tag{11.2}$$

$$W_V = W_A / 2 \tag{11.3}$$

A similar strategy is employed if the battery has more than four guns. This completes the calculation of the stick and volley dimensions.

11.3 DETERMINING THE WEAPON LETHAL AREA

Both systems represent the weapon lethal areas with a rectangular cookie cutter damage function, so the methods for air-launched and surface-launched weapons are essentially the same. It begins by using the methods described previously where the damage function is described by the MAE_F lethal area and the dimensions depend on the impact angle. We saw before that the impact angle aspect ratio is given by

$$a = \max\left(1 - 0.8\cos(I), 0.3\right) \tag{11.4}$$

This is applicable to air-launched bombs, but a different expression is used for artillery projectiles

$$a = 1 - 0.8\cos(I) \tag{11.5}$$

The reason is that artillery shells may impact at a shallower angle, so the constraining value of $a = 0.3$ is removed. The equations developed in Chapter 9, Section 9.4 are used to find the damage function dimensions.

$$L_{ET} = \sqrt{\mathrm{MAE}_F \times a} \tag{11.6}$$

$$W_{ET} = \frac{L_{ET}}{a} \tag{11.7}$$

This results in the lethal area shown in Fig. 11.12, which could then be centered over each impact point for either the air-launched stick or surface-launched volley.

There is, however, a problem. Recall Chapter 4, Section 4.8, where it was pointed out how there are two accuracy terms, the dependent and independent errors, and for multiple weapons they have to be kept separate. Clearly for sticks and volleys we are dealing with multiple weapons, so how is this to be handled?

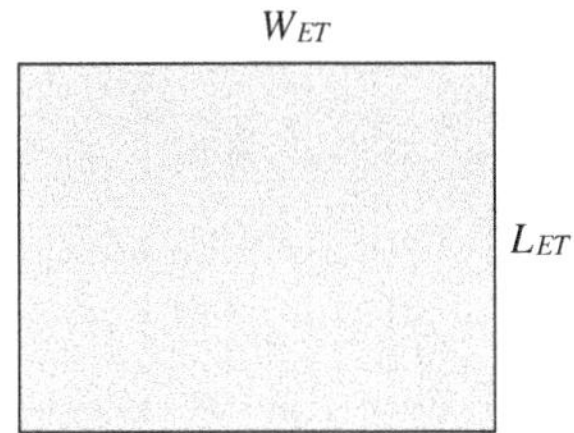

Fig. 11.12   Single weapon lethal area.

11.4   STICKS AND VOLLEYS USING DISCRETE TARGET ELEMENTS

Recall the discussion in Chapter 10, Section 10.5 where fractional damage due to a single warhead was calculated by assuming the target area contained 50×50 elements distributed uniformly within it. This process may be extended by replacing the single weapon lethal area shown in Fig. 10.17 by a stick or volley.

Because we are dealing with multiple weapon deliveries, the mean point of impact (MPI) and precision errors have to be kept separate, unlike the single weapon case where they were combined. Therefore, in Fig. 10.17 the random draw for the impact point now refers to the random draw for the MPI error for the stick. Suppose the volley or stick comprises four weapons at the corner of a rectangle, as indicated by the four ideal weapon impact points shown in Fig. 11.13.

The actual impact points will not be exactly at the indicated locations due to the precision error, so four samples are now drawn from this distribution to determine the impact points of each weapon lethal area. The individual weapon lethal areas are drawn centered on the actual impact points of each round, as shown in Fig. 11.14.

As before, a kill matrix of 50×50 elements corresponds to the status of each element in the area of target elements, where a value of zero represents survived and unity represents killed. Taking each target element in turn, we determine if

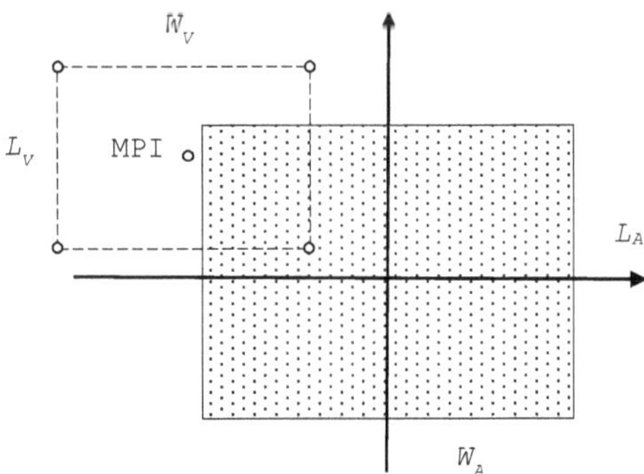

Fig. 11.13   Stick/volley centered on sample MPI.

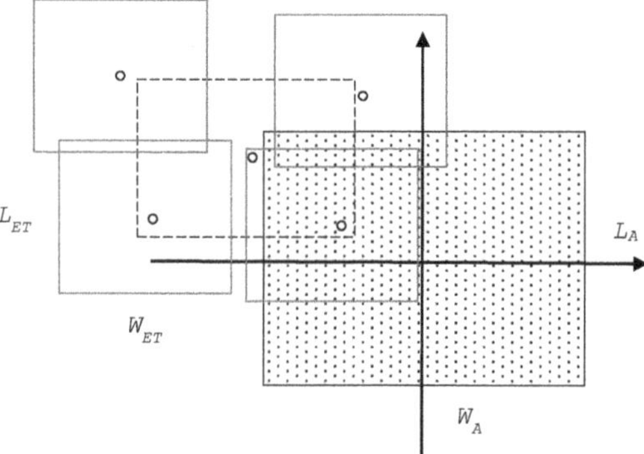

Fig. 11.14 Lethal areas placed on actual impact points.

it is enclosed by any of the round lethal areas comprising the volley; if it is, the value of that element in the kill matrix is set to unity. As seen in Fig. 11.14, in the top left of the target area some elements fall inside two weapon lethal areas, so if it is killed by the first weapon its value is already set to unity, and because the second weapon cannot kill it again, it remains set to unity. This determines which target elements are killed by each round in this volley.

Keeping the same MPI error and the kill matrix unchanged, the process is repeated for the second volley by taking another four draws from the precision distribution so the individual round impact points will have different locations relative to the corners of the volley rectangle. The individual element coverage process is repeated to see if any additional target elements have been killed for this iteration and, if so, more elements of the kill matrix are set to unity. This is repeated n times and results in an n-volley kill matrix for a single iteration of the Monte Carlo simulation. From this, the fractional damage may be calculated by summing all the elements in the kill matrix and dividing by the total number of elements present, which for the example case of 2500 elements is given by the following:

$$\text{FD}_n = \frac{\sum_{i=1}^{i=50} \sum_{j=1}^{j=50} K(i,j)}{2500} \tag{11.8}$$

Equation (11.8) is for one value of MPI error. The whole process is repeated for random draws from the MPI error distribution and the n-volley results averaged for final result.

The results for this approach are shown in Fig. 11.15, where the fractional damage is plotted against the number of volleys.

Students familiar with the *Advanced Weaponeering* textbook may recall there is a deterministic method to calculate fractional damage involving expansion of the lethal area to accommodate the precision error; however, this introduces anomalies that the stochastic method just outlined does not exhibit.

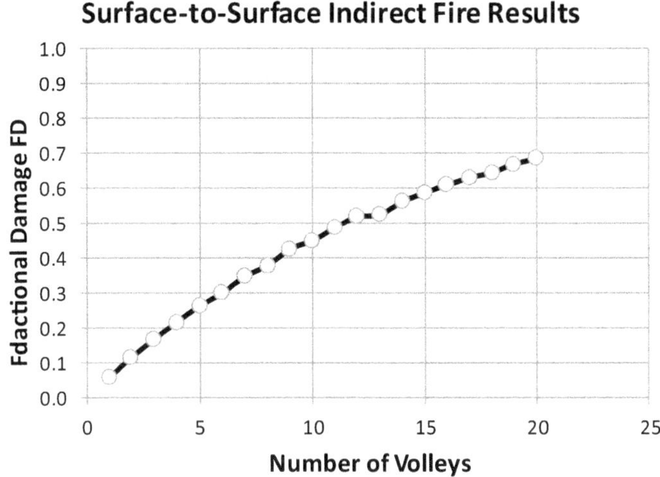

Fig. 11.15   Fractional damage for n volleys.

11.5   CLUSTER MUNITIONS

To attack large area targets, large warhead unitary weapons or sticks of smaller unguided weapons may be used; however, better coverage is often obtained with the use of one or more cluster weapons. A cluster weapon comprises a single container that is released like a unitary weapon from an aircraft or fired from a cannon, but at a particular point in the trajectory it opens up to release a large number of smaller submunitions. These then disperse over a large area on the ground before detonation, as indicated in Fig. 11.16. An air-dropped dispenser is often referred to as a tactical munitions dispenser (TMD), and an artillery round carrying smaller submunitions is called either a cargo round or an improved conventional munition (ICM).

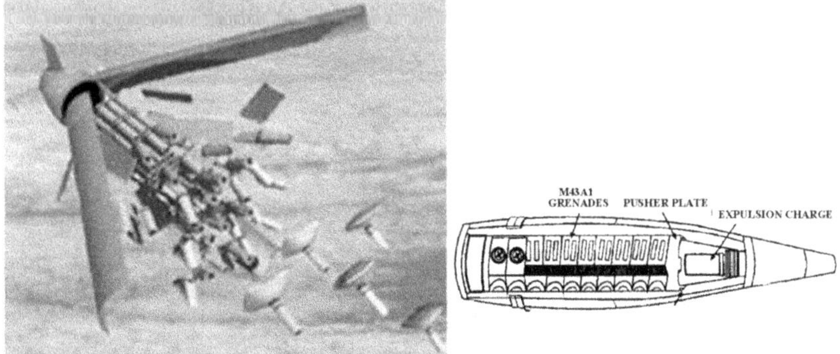

Fig. 11.16   Air-launched and artillery cluster munitions.

The submunitions may project either small fragments designed to incapacitate personnel or larger fragments more suited to lightly armored materiel targets such as trucks. Many cluster munitions carry a combination of such submunitions including shaped charge warheads intended specifically for armored vehicles. A single weapon may contain a few dozen large, sophisticated submunitions or several hundred grenade-size antipersonnel types.

Most weapons are unguided; however, the guided multiple launch rocket system (GMLRS) and some cruise missiles have cluster munition warheads.

The basic approach in dealing with cluster munition deliveries is similar to that used for sticks and is shown in Fig. 11.17.

1. Assume the pattern on the ground is rectangular in shape and there is a uniform distribution of submunitions within it.
2. Determine the dimensions of the rectangle from the release conditions, weapon trajectory, and formulae for the submunition spread.
3. Calculate the probability of damage P_{CDP}, averaged within the pattern based on the density of weapons.
4. For an area of targets, calculate fractional damage by considering the stick to be equivalent to a single effectiveness area with these calculated characteristics, as described in Chapter 10.

Figure 11.17 shows the uniform distribution of submunitions in the pattern, and Fig. 11.18 indicates the equivalent rectangular cookie cutter representation.

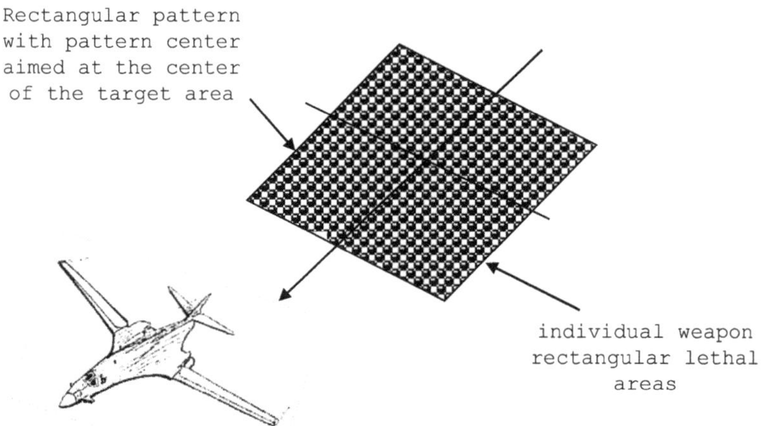

Fig. 11.17   Individual submunition impacts in the pattern.

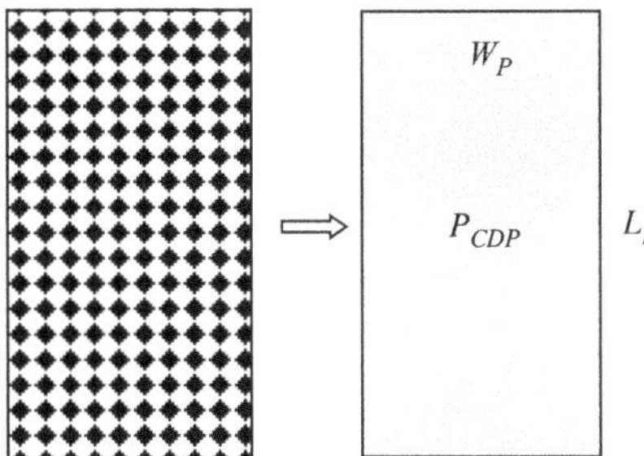

Fig. 11.18   Combining individual lethal areas into a pattern.

If this can be achieved, the cluster munition may be analyzed using techniques applicable to single weapons, and fractional damage values may be calculated in the usual manner.

Typical accuracy however is poor since the submunitions may be influenced by unknown, local wind conditions if descent is controlled by a parachute or ribbon.

11.6   GENERAL ANALYTICAL TREATMENT OF CLUSTER MUNITIONS

The operation of cluster munitions is somewhat complex and varies from one type to another. The computation of the fractional damage is also more complicated than for unitary weapons, and some assumptions have to be made to ensure a tractable solution can be computed in a short period of time. In general, the process may be considered in two parts, corresponding to the two stages of the weapon trajectory.

1. The trajectory of the dispenser is calculated from the weapon release point to the dispense point.
2. The trajectory of a single "mean" submunition from the dispense point to ground impact is computed. This locates the center of the pattern formed by the submunitions for the purposes of computing slant range and precision error effects. In addition, the mean submunition impact angle can be obtained to determine the shape of the lethal area. The size of the pattern is determined separately.

The goal is to determine the dimensions of a rectangular area of effectiveness on the ground corresponding to a lethal area for a unitary weapon, and a probability of damage within it. The first step is to investigate the trajectories of both the dispenser and submunition. Figure 11.19 shows the two phases of cluster munition weapon delivery.

These trajectories will be used to determine the pattern dimensions.

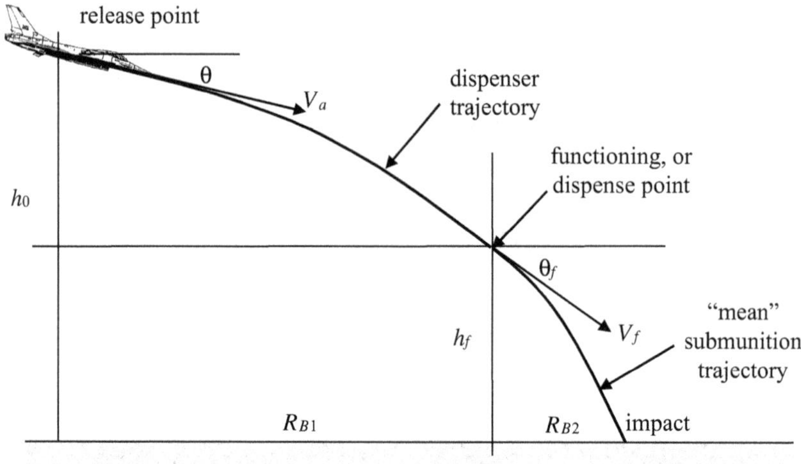

Fig. 11.19   Cluster munition and submunition trajectory.

11.7   CALCULATING PATTERN DIMENSIONS

Historically, the Joint Munitions Effectiveness Manuals (JMEMs) have used different methods to calculate pattern dimensions for air-launched and surface-launched weapons. Clearly it is not possible in real time to determine where each of perhaps several hundred submunitions will land, so some simplifying approximations are made.

11.7.1   AIR-LAUNCHED SUBMUNITIONS

This method is based on some high fidelity simulations performed in the 1970s in which individual trajectories of submunitions were estimated and the resulting pattern on the ground was predicted. It was found that the pattern size depended on three critical parameters corresponding to the submunition release, or functioning, conditions: the altitude, angle, and velocity shown in Fig. 11.19.

Comparing the pattern dimensions for a range of functioning parameters resulted in a set of empirical equations relating the two. As an example of this process, the functioning altitude and functioning velocity are adjusted to allow for the functioning delay and are given by

$$h_1 = h_f - 38 \tag{11.9}$$

$$v_1 = v_f - 84 \tag{11.10}$$

Then two intermediate variables are then computed.

$$v_t = 0.041 \times R_{\text{spin}} + 13 \qquad (11.11)$$

where R_{spin} is the dispenser spin rate in rev/min. The second intermediate variable is

$$T = \frac{v_t}{\sqrt{v_t^2 + v_1^2}} \qquad (11.12)$$

If

$$v_t \cos\theta_f > v_1 \sin\theta_f \qquad (11.13)$$

then the pattern length is computed using

$$L = 553 + 0.2h_1 \sin\theta_f - 10\sqrt{h_1/T} + 1.4v_t \cos\theta_f \qquad (11.14)$$

otherwise the pattern length is given by

$$L = 581 + 0.5h_1 \sin\theta_f - 6.5\sqrt{h_1/T} - 583\sqrt{T} + 1.8v_t \cos\theta_f \qquad (11.15)$$

The exact equations are not important here, but observe the final result where the pattern length is simply a function of the submunition release conditions. The same process was repeated for the pattern width and extended to both altitude and time-fuzed munitions. Given the trajectory of the dispenser, the functioning parameters can be calculated; therefore, so can the pattern dimensions.

11.7.2 SURFACE-LAUNCHED SUBMUNITIONS (ICM)

As for air-launched dispenser weapons, ICM warheads burst at a prescribed altitude or time from firing and dispense a fixed number of submunitions over an area on the ground. Different warheads contain different numbers of submunitions and different mixes of antimateriel and antipersonnel types. A single ICM will produce a pattern on the ground that is assumed to be elliptical in shape and have dimensions that are functions of the range between the howitzer and the target. Table 11.2 shows typical precomputed pattern sizes for an artillery round.

These pattern dimensions were also computed from high fidelity simulations, verified by limited testing; however, the inputs to the calculations are different from the air-launched weapon systems. Because elliptical patterns are difficult to use when calculating effects, they are changed to a rectangle with equal area and aspect.

TABLE 11.2   ELLIPTICAL ICM PATTERN DIMENSIONS

Range (km)	Average Height of Burst (m)	Ellipse Axis L_E (range, m)	Ellipse Axis W_E (deflection, m)
4	290	57	55
7	340	67	59
12	425	103	60
14	425	111	73

$$\pi \times L_E \times W_E = L_P \times W_P, \qquad \frac{L_E}{W_E} = \frac{L_P}{W_P} \qquad (11.16)$$

11.8 PROBABILITY OF DAMAGE WITHIN THE PATTERN

At this point we consider the coverage within the pattern by the submunitions. The probability of damage to a target element within the pattern due to a single submunition of lethal area A'_{ET} is given by

$$P_S = \frac{A'_{ET} R_S}{L_P \times W_P} \qquad (11.17)$$

Here we are expanding the submunition lethal area over the whole of the pattern. Assuming the submunition impacts are uniformly distributed within the pattern and there are n_b submunitions, the probability of damage to the target is obtained by the survivor rule.

$$P_{CDP} = 1 - \left(1 - \frac{A'_{ET} R_S}{L_P \times W_P}\right)^{n_b} \qquad (11.18)$$

The end result is a rectangular cookie cutter lethal area in the ground plane of known dimensions and known probability of damage inside (see Fig. 11.20).

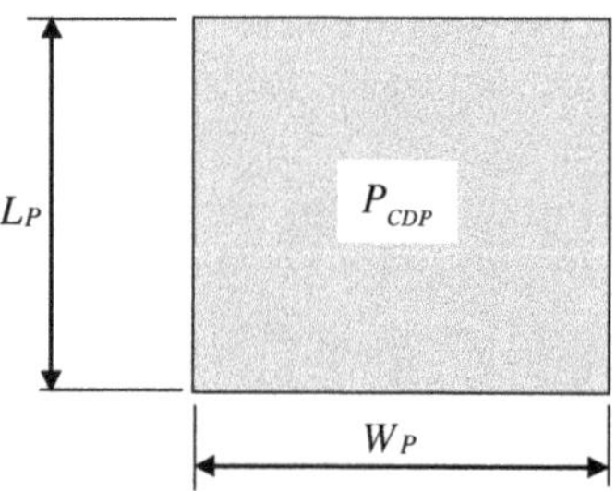

Fig. 11.20   Final pattern dimensions and damage probability.

It represents a single weapon lethal area, and the methods described in Chapter 10 against a unitary target, or areas of target elements, may be used.

EXAMPLE 11.1

Suppose we have an ICM containing 100 grenades (see Fig. 11.16) and a pattern corresponding to the second line of Table 11.2. Calculate the equivalent rectangular cookie cutter damage function.

From Eq. (11.16) we get

$$\pi \times 67 \times 59 = L_P \times W_P = 12420, \qquad \frac{67}{59} = \frac{L_P}{W_P} = 1.13 \qquad (11.19)$$

This gives $L_P = 118.5$m and $W_P = 104.8$m. Assume each grenade has a lethal area corresponding to a radius of 10 m and a reliability of 95%. Equation (11.17) gives

$$P_S = \frac{\pi \times (10)^2 \times 0.95}{12420} = 0.024 \qquad (11.20)$$

From Eq. (11.18)

$$P_{CDP} = 1 - (1 - 0.024)^{100} = 0.912 \qquad (11.21)$$

Notice how the individual grenade lethal area spread over the pattern is small, however once powered up by the large number of grenades, the whole weapon is quite effective over the pattern.

11.9 CHAPTER SUMMARY

- For stick and volley fire deliveries, the process of calculating the effectiveness resulted in aggregating all the weapons into a single rectangular cookie cutter lethal area.
- For clusters, the calculation was different but still resulted in a single rectangular lethal area.
- If the dimensions of the lethal area and the conditional probability of damage inside it are known, the fractional damage may be calculated for an area of target elements using the same techniques as for a single weapon described in Chapter 10.

Chapter 12

BRIDGES, LINEAR TARGETS, EFFECTIVE MISS DISTANCE, AND TUNNELS

12.1 INTRODUCTION

The effectiveness indexes (EIs) we have developed thus far in the book have been MAE_F, MAE_B, and vulnerable area A_V (VAN), and in some sense are generic in nature; for example, an MAE_F could reference a 1000-lb bomb warhead or a 155-mm artillery round used against a common target. Many weapon/target combinations may be accurately modeled using these EI types.

Early Joint Munitions Effectiveness Manual (JMEM) developers, however, realized that some target types are not well represented by either an MAE or A_V, and some modification of the methodology was needed. This led to some different EI types being defined that were more target specific, including the following discussed in this chapter:

- Bridge effectiveness index (BEI)
- Disruptive crater diameter ($\mathrm{D_C}$)
- Effective miss distance (EMD)
- Mean area of effectiveness for buildings ($\mathrm{MAE_{BLDG}}$)

The basic process for addressing these three new EI types is the same: each will be reduced to an equivalent rectangle in the ground plane having dimensions L_{ET} and W_{ET}. By representing the weapon as a dart (point), the probability of damaging the target is the probability of hitting this rectangle with the dart, as shown in Fig. 12.1.

This calculation assumes that the probability of damage inside the rectangle is 1.0; however, we generalize the situation by considering that this is not always the case. It is assigned the probability of damage given a hit and is denoted by P_{HD} to allow this to have a value other than unity. P_{HD} is not a separate EI but is involved in the effectiveness calculations leading to the following equation:

$$\text{Prob\{damage\}} = \text{Prob\{hit\}} \times \text{Prob\{damage given a hit\}} \qquad (12.1)$$

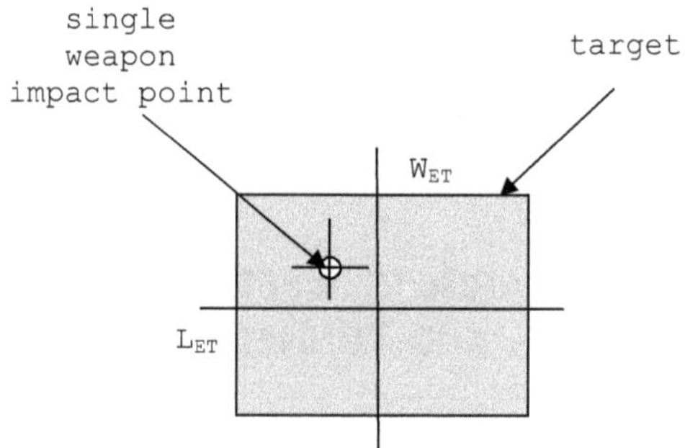

Fig. 12.1   Targets represented by a ground plane rectangle.

Or using conventional terminology

$$\mathrm{PD}_1 = P_{\mathrm{HIT}} \times P_{HD} \times R \tag{12.2}$$

We have seen that P_{HIT}[1] may be calculated using stochastic or deterministic methods, as long as the delivery accuracy data for the weapon are available. For example, from Chapter 9, Eqs. (9.10) through (9.12), a deterministic solution would take the following form:

$$P_{\mathrm{HIT}-x} = (\mathrm{ND}(\mathrm{L}_{ET}/2,0,\sigma_x,1) - \mathrm{ND}(-\mathrm{L}_{ET}/2,0,\sigma_x,1)) \tag{12.3}$$

$$P_{\mathrm{HIT}-y} = (\mathrm{ND}(\mathrm{W}_{ET}/2,0,\sigma_y,1) - \mathrm{ND}(-\mathrm{W}_{ET}/2,0,\sigma_y,1)) \tag{12.4}$$

$$P_{\mathrm{HIT}} = P_{\mathrm{HIT}-x} \times P_{\mathrm{HIT}-y} \tag{12.5}$$

In most cases considered here, P_{HD} is set to unity; the exception is in analyzing bridges, which are considered next.

12.2   BRIDGES

Bridges are constructed in many different forms and materials, and their vulnerability to air- or surface-launched weapons depends on many factors. Some examples of construction methods, terminology, and features are shown in Fig. 12.2.

Clearly, it will be difficult to generalize an overall damage function for all bridges, but data obtained from bomb damage assessment (BDA) studies have provided some guidance on the issue.

Tactically, for bridge attacks the question arises: Where should we aim the weapon or which bridge components should be damaged? Bridge abutments are usually large earthen or concrete structures that are resilient to conven-

---

[1] Here P_{HIT} is the probability of hitting the target and not the proportion of excess direct hits as described in Chapter 4.7.

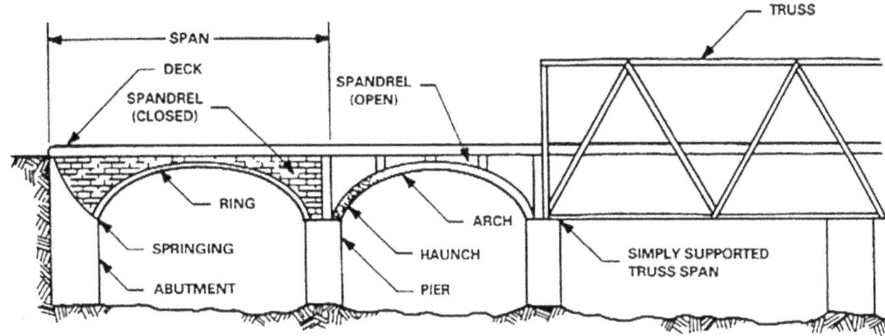

Fig. 12.2   Bridge construction elements.

tional weapons, and it is difficult to detonate a weapon close to a supporting pier. Therefore, the damage criteria for a bridge are to drop a span (see Appendix), so weapons are aimed at, and designed to damage, the span or deck of the bridge.

The bridge deck will have a physical length L_T and width W_T, but the effective target length L_{ET} and width W_{ET} are defined relative to the attack direction, as shown in Fig. 12.3.

In order to calculate P_{HIT}, the values of L_{ET} and W_{ET} are set equal to the bridge dimensions so for Fig. 12.3a, we have

$$L_{ET} = W_T \tag{12.6}$$

$$W_{ET} = L_T \tag{12.7}$$

whereas for Fig. 12.3b these are reversed.

$$L_{ET} = L_T \tag{12.8}$$

$$W_{ET} = W_T \tag{12.9}$$

Note that only orthogonal attacks are considered. Equations (12.3) to (12.5) may now be used to calculate P_{HIT}.

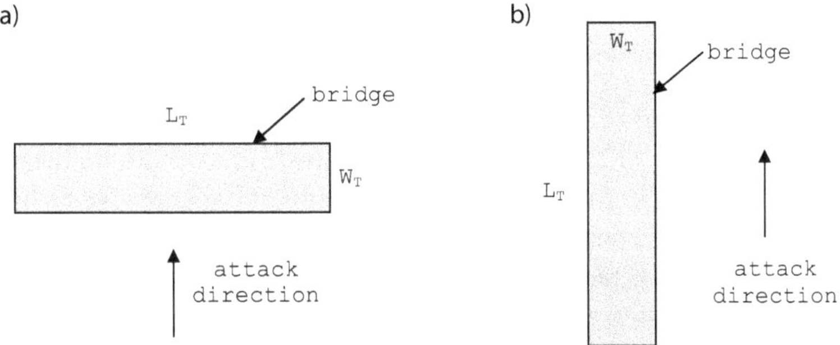

Fig. 12.3   Different attack directions for bridge target.

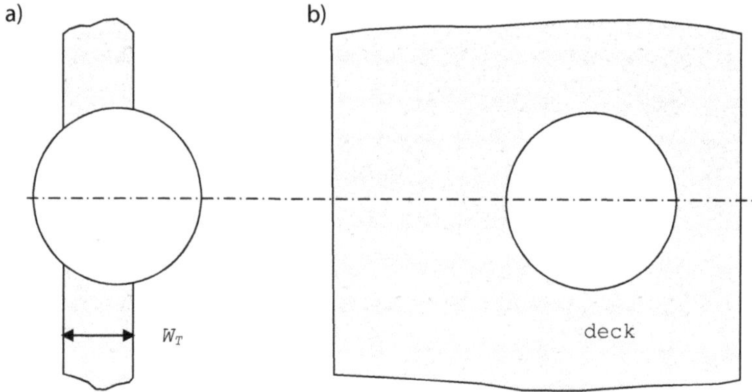

Fig. 12.4   Crater damage on the bridge deck.

If the probability of dropping a span were simply equal to the probability of hitting the span, then misleading results would be obtained. For example, suppose for a particular bridge type and weapon used to attack it, the weapon produces a clear crater of 40-ft diameter in the deck. If the bridge width is 20 ft and the weapon hits it, clearly the span is severed, and the span will drop as shown in Fig. 12.4a. If, however, the bridge width is 100 ft and the weapon hits it, the span is not severed and may not drop, as shown in Fig. 12.4b.

For the same bridge length L_T, the probability of hitting a wider bridge is higher than for a narrow bridge, but the probability of dropping the span appears to go down. Therefore, the probability of hitting the bridge cannot be the only criterion used to determine if the span fails. This paradox is resolved by introducing the variable P_{HD}, which intuitively has to decrease as the bridge width increases and takes the following form:

$$P_{HD} = 1 - \exp\left[-\frac{\text{BEI}}{W_b}\right] \tag{12.10}$$

where BEI is a constant that is determined by the bridge construction method, and W_b is the bridge width (in feet) defined by

$$W_b = \text{MIN}(L_T, W_T) \tag{12.11}$$

P_{HD} as a function of bridge width and a specific BEI is shown in Fig. 12.5.

Reconsidering Eq. (12.2), it is seen that as the bridge width increases, the probability of hitting it goes up but the probability of damage given a hit goes down, which appears consistent with operational experience.

The JMEMs have a pictorial library of bridge types from which the user selects the one closest to the target of interest. Then, by selecting the weapon for the attack, a value of BEI is obtained from a classified database.

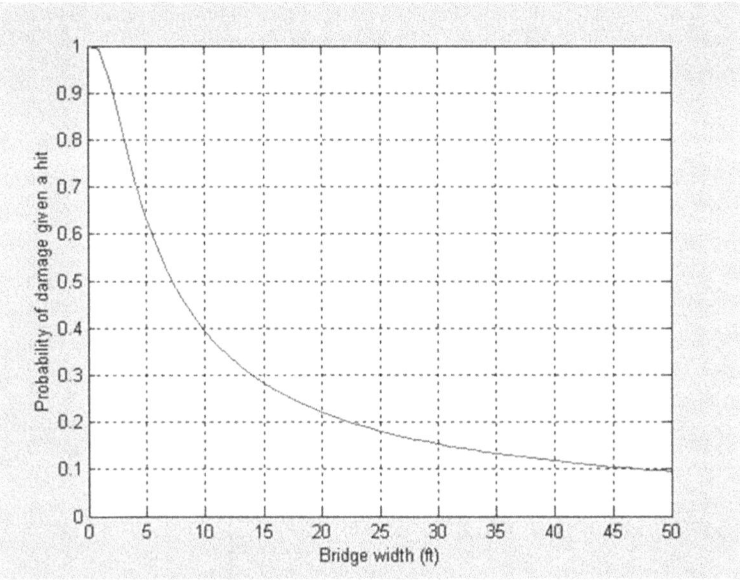

Fig. 12.5 P_{HD} **as a function of bridge width for BEI = 5 ft.**

EXAMPLE 12.1

A bridge has dimensions 30 ft wide by 500 ft long and is attacked by a single unguided weapon aimed at the center. The aircraft attacks in a direction parallel to the 500-ft length of the bridge. Calculate the probability of dropping the span of the bridge, given that for this type of construction and the weapon used, the value of the effectiveness index is BEI = 5 ft, the delivery accuracy has a CEP of 75 ft, and the weapon is 95% reliable. First, we calculate P_{HD} from

$$P_{HD} = 1 - \exp\left[-\frac{\text{BEI}}{W_b}\right] = 1 - \exp\left[-\frac{5}{30}\right] = 0.154 \qquad (12.12)$$

Then the values of L_{ET} and W_{ET} are calculated from the bridge dimensions, taking into account the direction of attack.

$$L_{ET} = 500 \qquad W_{ET} = 30 \qquad (12.13)$$

The probability of damage may be calculated using the spreadsheet shown in Table 12.1. This table is representative of bridge attacks using unguided weapons and indicates the difficulty of a successful single mission, or the large number of missions needed for a reasonable PD. It was

TABLE 12.1   WEAPON EFFECTIVENESS FOR BRIDGES

Bridge Attacks	
Inputs	
Bridge length	500
Bridge width	30
CEP	75
BEI	5
Reliability	0.95
Calculations	
PHD	0.154
sigma_x, sigma_y	63.700
P_hit-x	1.000
P_hit-y	0.186
P_hit	0.186
Output	
PD1	0.027

results such as these that prompted the development of laser-guided bombs, which were originally used against bridges in Vietnam.

12.3   *LINEAR TARGET METHOD*

In this section we investigate the chances of a single weapon represented by a circular damage function partially or totally cutting a linear, rectangular target. Typical applications of this scenario involve a weapon creating a circular crater of diameter D_C impacting a long and narrow rectangular target of interest such as an aircraft runway, road, or railway.

There are some common elements between this method and that dealing with bridges—after all, a bridge may be considered a linear target; however, there are some differences.

- We will allow for nonorthogonal attacks.
- The damage criterion for a bridge is to drop a span, whereas here we are interested in restricting the width of the linear target so that traffic may be prevented.
- The BEI method is deterministic whereas the linear target method will be Monte Carlo based.

Figure 12.6 shows the effect of multiple-bomb crater damage to a runway, and although such a target is clearly unusable, it should be kept in mind that some damage may be repaired more quickly than others. For example, roads

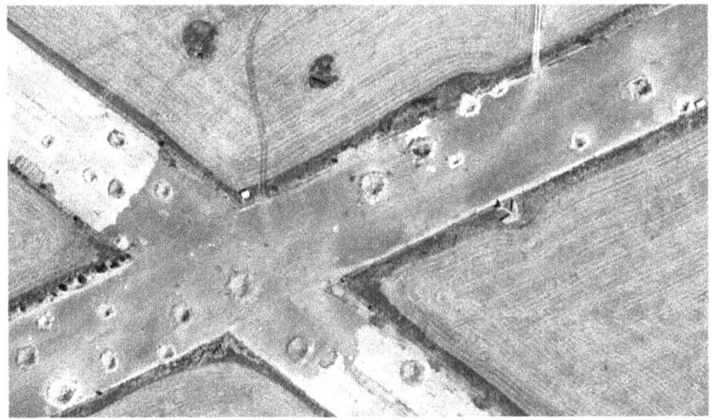

Fig. 12.6   Bomb crater damage to runways.

and runways may be temporarily repaired by filling in the holes with rubble and placing a metal plate over the patch. Other linear targets such as railway lines may take longer to bring back into use.

It should also be noted that a crater does not necessarily have to damage the whole width of the target in order to make it inoperable. The damage done by a single crater in Fig. 12.6 may prevent any aircraft from using the runway for takeoffs or landings.

We limit the analysis to a single weapon aimed at the center of a linear target, which is assumed long enough to ignore damage to its ends; therefore, only damage to the target width is of interest. Early JMEM methods attempted to derive an analytical solution to this problem, but they made many assumptions in the process; therefore, the solution presented here is based on a Monte Carlo simulation and is sufficiently simple that it runs in less than a second on an ordinary personal computer.

Consider the scenario shown in Fig. 12.7, which shows a random impact point relative to the linear target. It is assumed that the attack direction is at an angle θ to the longest direction, and the impact point in weapon coordinates (x_w, y_w) is a random draw from the range and deflection accuracies range error probable (REP) and deflection error probable (DEP) (or σ_x, σ_y). The impact point in target coordinates (x, y) is given by

$$x_i = x_w \cos\theta - y_w \sin\theta \tag{12.14}$$

$$y_i = x_w \sin\theta - y_w \cos\theta \tag{12.15}$$

If the crater radius is r_c, we need to calculate what fraction of the target is cut. The simplest way to do this is to consider the analogous case in Chapter 10 where we calculated the fractional coverage of a weapon lethal area over an area of target elements. Consider the case shown in Fig. 12.7, where the attack angle θ is 90 deg, so the range axis is perpendicular to the target axis.

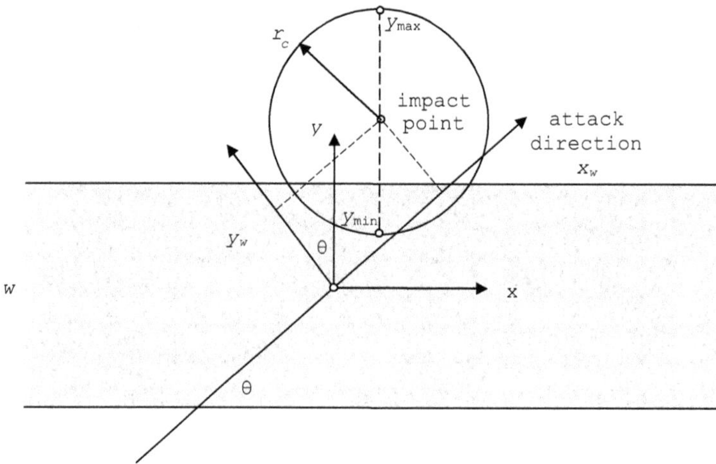

Fig. 12.7   Geometry of impact point and target.

This situation is shown in Fig. 12.8, where the geometry has been rotated 90 deg for compatibility with Fig. 10.5.

We see that the crater moving to the right and cutting the target in Fig. 12.8 is the same as the rectangular weapon lethal area moving to the right in Figs. 10.5 to 10.11 in Chapter 10 and covering the rectangular target area. This implies we can define a similar trapezoidal damage function like the one shown in Fig. 10.12 for the crater cutting the target shown Fig. 12.7. Therefore, the spreadsheet shown in Table 10.1 in Chapter 10 may be modified for the linear target method, with the following considerations:

1. For the target area coverage, the dimensions were L_A by W_A; however, for the linear target it is semi-infinite; that is, its length is unspecified. The target is therefore defined as $L_A = W$ and $W_A = 500$ ft.

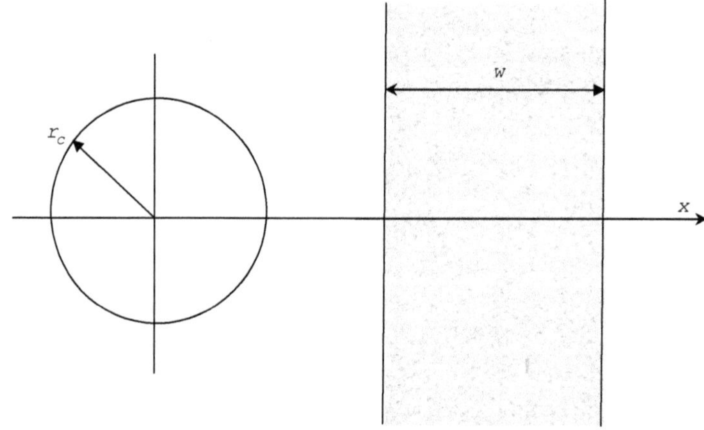

Fig. 12.8   Crater moving left to right perpendicular to target.

TABLE 12.2 LINEAR TARGET SPREADSHEET

Linear Target Method	
Inputs	
Target width	50
Crater radius	30
Attack angle	45
REP	20
DEP	20
# iterations	100000
Output	
Percentage cut	0.634

Calculate

2. The weapon lethal area in this case is the circular crater; therefore, $L_{ET} = 2r_c$.
3. The location of the crater in the deflection direction does not increase the cut, so we are only interested in the cut in the range direction.
4. Although Fig. 12.8 shows an attack perpendicular to the target for simplification purposes, an oblique attack such as the one shown in Fig. 12.7 is accomplished by using Eqs. (12.14) and (12.15) to determine the impact point.

Substitution of these variables in the fractional coverage spreadsheet produces the version for the linear target method shown in Table 12.2.

Running the program produces the following observations:

- If the accuracy in range and direction is the same, the damage to the target is independent of the attack angle.
- If the accuracy in range is different from that in deflection, maximum damage is achieved with an attack angle that puts the smallest dispersion across the target width. This means that if REP < DEP, theta should be set to 90 deg; however, if DEP < REP, theta should be zero.

A further refinement would be to consider a stick of bombs; however, this case is beyond the scope of this book.

12.4 EFFECTIVE MISS DISTANCE (EMD) DAMAGE FUNCTION

If we consider a typical MAE_F for a particular weapon and target combination, such as an Mk-84 unguided bomb against an armored vehicle, a value such as 500 ft^2 might be appropriate. Obviously, the size of the vehicle is fixed, so there is little ambiguity in defining a single EI.

There are targets, however, that can exist in many different sizes; for example, cylindrical liquid petroleum, oil, and lubricant (POL) storage tanks

Fig. 12.9   Construction method for POL storage tanks.

can vary from a few feet to several hundred feet in diameter. These storage tanks might have the same construction method (i.e., 1/2-in. rolled steel plates welded along the edges) as shown in Fig. 12.9.

Consider that the damage criterion for attacking such a tank is to puncture it so that it drains within 30 min. For a given weapon, say the same Mk-84 2000-lb unguided bomb, if we hit the tank, we would clearly achieve the damage required. Suppose we miss it by 5 ft—is the target damaged sufficiently? The answer is yes, surely. What if we miss by 10 ft? Again, the answer is probably yes. As we increase the miss distance, we will get to a point where the required damage is not met. This distance is a new EI called the effective miss distance (EMD), and it depends on the target type, weapon, and damage criterion.

The principle feature of this EI is that because it is not dependent on the target size, we do not need to compute different EI values for each possible POL tank size, but a single EMD will apply to all tanks.

For the POL tank example, the EMD might be 50 ft, which allows us to define a value of effective target size for any given physical-size tank. The area surrounding the target and enclosed by the EMD is treated exactly as if it is part of the target, as shown in Fig. 12.10.

For the rectangular target shown, L_{ET} and W_{ET} are computed using

$$L_{ET} = L_T + 2 \times \text{EMD} \tag{12.16}$$

$$W_{ET} = W_T + 2 \times \text{EMD} \tag{12.17}$$

Where, as usual, L_T is the dimension of the target parallel to the attack direction, and W_T is the dimension of the target element perpendicular to it. Now consider the two-dimensional circular target shown in Fig. 12.11.

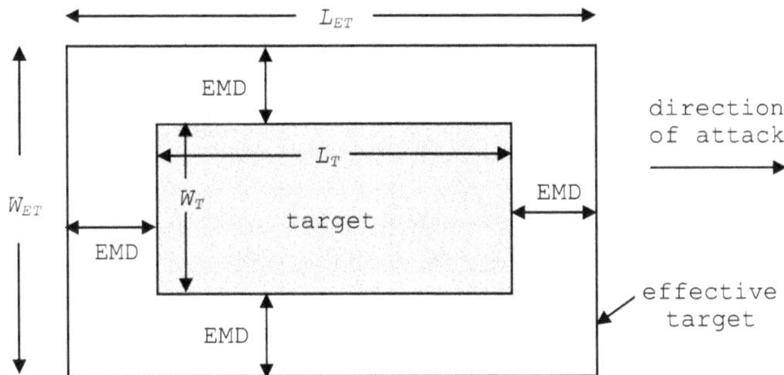

Fig. 12.10   EMD for 2-D rectangular target in the ground plane.

It is approximated by a square of equal area having dimensions L_{ET} and W_{ET} in the ground plane, where L_{ET} and W_{ET} are defined by

$$L_{ET} = W_{ET} = \sqrt{\pi}\,(R_T + \text{EMD}) \qquad (12.18)$$

Here, R_T is the radius of the target (in feet). We can extend the concept of EMD to targets having three dimensions, such as a building, as shown in Fig. 12.12. If we consider only the target footprint ($L_T \times W_T$), then certainly hitting this rectangle would mean the target is hit; however, it would not include the probability of hitting the roof of the building as shown in Fig. 12.12. From this figure we see that the target may be hit on the roof while the projected ground plane impact point relative to the building footprint would be outside the footprint, implying that the building was not hit.

What we have to do is to generate an equivalent rectangle in the ground plane that, if hit, would correspond to hitting the building, taking into account hits in the footprint and on the roof. This is done by defining the building "shadow."

For three-dimensional targets such as the one shown in Fig. 12.12, the shadow length L_{SH} is defined from Fig. 12.13, where the upper-right corner of the target is projected down to the ground plane by the impact angle of the

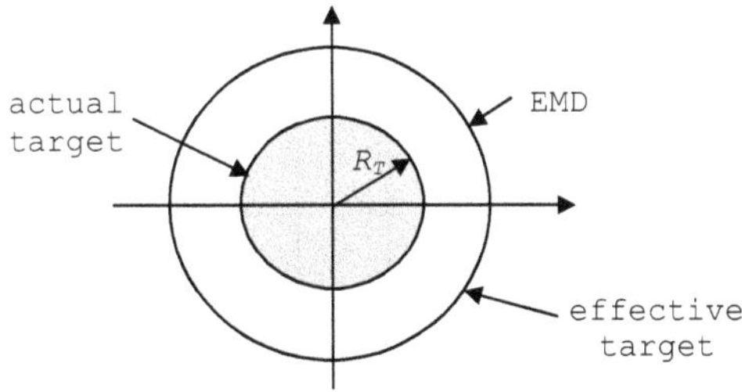

Fig. 12.11   Circular target.

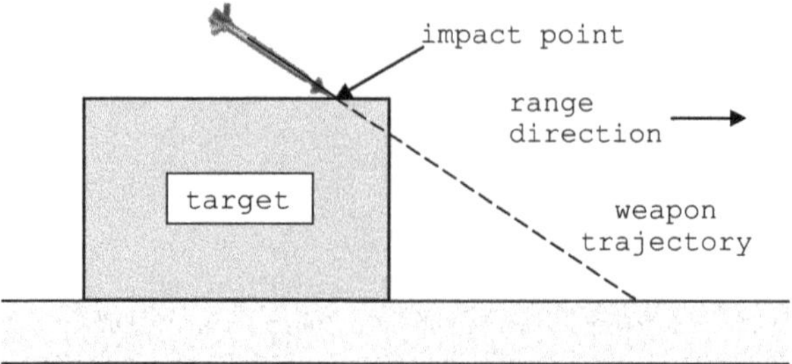

Fig. 12.12   Weapon striking the top of a 3-D target.

weapon to the point P. For this example, let us also consider there is a 2-D ground plane EMD around the rectangular target.

The shadow length L_S is defined as the distance beyond the downrange EMD to point P and is computed using the expression

$$L_{SH} = \frac{H_T}{\tan(I)} - \text{EMD} \qquad (12.19)$$

where H_T is the target element height and I is the impact angle of the weapon. If L_{SH} is less than or equal to zero (i.e., the shadow does not extend beyond the EMD), L_{ET} and W_{ET} are computed using Eqs. (12.16) and (12.17), respectively, for rectangular targets, or Eq. (12.18) for circular targets.

We can see from Fig. 12.13 that an impact anywhere inside the shaded area corresponds to hitting the building side or roof. Unfortunately, this is not a simple rectangle in the ground plane as required by Eqs. (12.3) through (12.5);

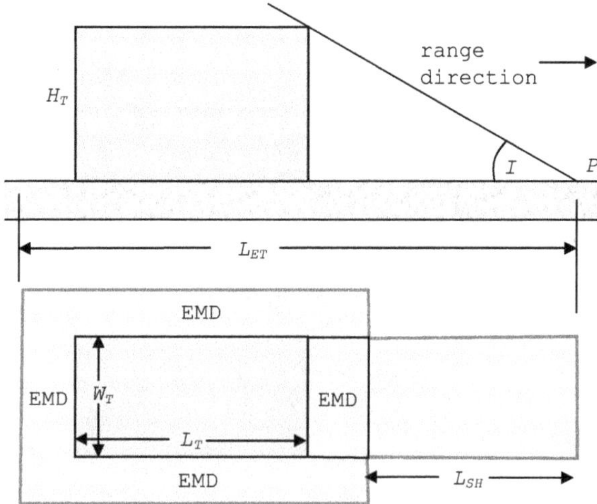

Fig. 12.13   Three-dimensional target with EMD.

therefore, we have to convert this area to a simple, equivalent rectangle. If L_{SH} is greater than zero as indicated in Fig. 12.13, L_{ET} is computed using

$$L_{ET} = L_T + 2 \times \text{EMD} + L_{SH} \tag{12.20}$$

for rectangular targets, or

$$L_{ET} = \sqrt{\pi}(R_T + \text{EMD}) + L_{SH} \tag{12.21}$$

for cylindrical targets. The effective target area in the ground plane A_T is then computed using

$$A_T = [L_T + 2\text{EMD}][W_T + 2\text{EMD}] + L_{SH}W_T \tag{12.22}$$

for rectangular targets, or for cylindrical targets from

$$A_T = \pi(R_T + \text{EMD})^2 + 2R_T L_{SH} \tag{12.23}$$

The value of W_{ET} is then adjusted so that the same effective target area in the ground plane will be produced, that is,

$$W_{ET} = \frac{A_T}{L_{ET}} \tag{12.24}$$

Vertical targets such as tunnel entrances, caves, and shelter doors, shown in Fig. 12.14, can be weaponeered by entering a target length L_T of zero, the door width as the target width W_T, and the door height as the target height H_T. Typical damage definitions for these targets include damaging any protective doors, if fitted, or the production of rubble at the entrance to prevent passage of materiel or personnel. As such, damage may be produced either by trying to impact the weapon inside the tunnel or by hitting an area around the portal

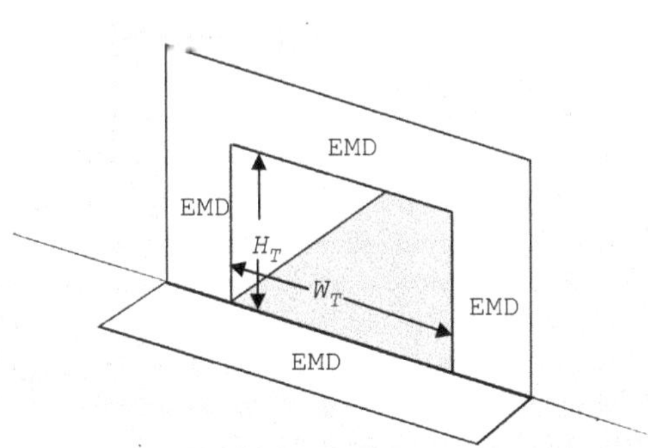

Fig. 12.14   Vertical target with EMD.

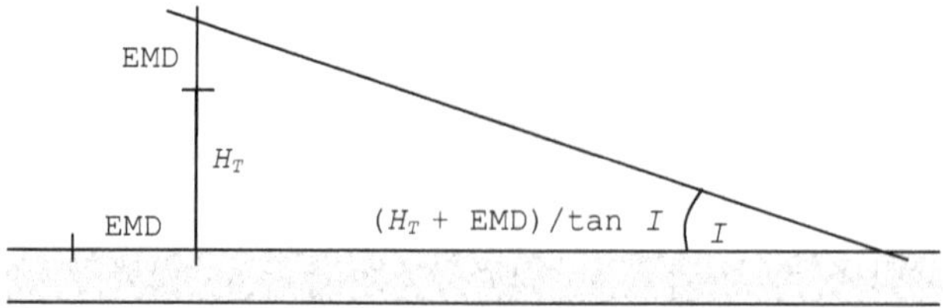

Fig. 12.15   Side view of tunnel entrance.

or the apron in front of the portal. Therefore, as shown in Fig. 12.14, there will be an EMD in the vertical plane and in front of the target.

How can this target be reduced to a single rectangle in the ground plane? Consider the side view of the tunnel shown in Fig. 12.15. If the impact angle of the weapon is known, we can project a point at the top of the EMD above the portal onto the ground plane for the shadow length given by

$$L_{SH} = \frac{H_T + \text{EMD}}{\tan(I)} \tag{12.25}$$

Because the target length is zero, the effective target length is the length of the shadow plus the EMD in front of the target, or

$$L_{ET} = \frac{H_T + \text{EMD}}{\tan(I)} + EMD \tag{12.26}$$

W_{ET} is computed as before from a frontal view of the tunnel and using Eq. (12.17).

$$W_{ET} = W_T + 2 \times \text{EMD} \tag{12.27}$$

The vertical tunnel target has been transformed into a ground plane rectangle of dimensions $L_{ET} \times W_{ET}$. For all these EMD targets, once L_{ET} and W_{ET} have been determined, the PD_1 can be calculated using Eqs. (12.3) through (12.5). Table 12.3 shows a spreadsheet implementation for a rectangular 2-D or 3-D EMD effectiveness index.

Some modifications will be required to handle circular targets.

12.5   TUNNELS

Early JMEM tools such as the JMEM Air-to-Surface Weaponeering System (JAWS) did not include methodologies for assessing damage to tunnels. However, reference to tunnel damage criteria may be found in the Appendix together with recommended weapons designed to create sufficient rubble to block access to the tunnel.

TABLE 12.3　WEAPON EFFECTIVENESS FOR RECTANGULAR EMD

Effective Miss Distance (EMD)	
Inputs	
Target length LT	50
Target width WT	30
Target height HT	20
EMD	5
Impact angle I	65
REP	20
DEP	10
Reliability	1
Calculations	
Shadow length Lsh	4.326
L_et	64.326
A_et	2529.785
W_et	39.327
P_hit-x	0.722
P_hit-y	0.815
Output	
PD1	0.589

In more recent times, certain high value targets are increasingly concealed inside tunnel structures, which may be quite large and complex in design. By their nature, such targets may be difficult to defeat; however, methods are needed to assess the effect of conventional weapons against them, and more careful analysis has to be developed regarding ways to defeat them.

The scope of tunnel target types is remarkably wide, ranging from single-bore tubes to permit the passage of troops to complex multibranch designs intended to protect whole airfields including aircraft, munitions, personnel, and fuel supplies (see Fig. 12.16 and [1]). Such tunnel systems may accommodate dozens of aircraft and support equipment and hundreds of people.

Because most tunnels are buried beneath massive geological structures, their weakest element is the portal, and this is the usual aimpoint. Figure 12.17 shows the typical elements of a tunnel portal.

Another approach for attacking tunnel targets is where the intent is to penetrate the side of the mountain and detonate in the vicinity of the tunnel close to the portal. Collapsing the tunnel will block any entrance or exit. To perform this task, the U.S. Air Force has a 30,000-lb weapon—the GBU-57/A massive ordnance penetrator (MOP), which is shown in Fig. 12.18.

Fig. 12.16   Underground airfield facility.

Even for this weapon, which has more penetration than more commonly used bombs, a limit of about 60-ft penetration into rock is the best that can be expected (see Fig. 12.19).

Another tactic that was frequently employed to penetrate considerable thickness of rock above a tunnel was to use two or more blast/fragmentation warheads to incrementally enlarge the crater formed by the previous weapon. Although this may work conceptually, the difficulty of detonating several weapons in exactly the same spot is challenging.

Fig. 12.17   Typical tunnel portal features.

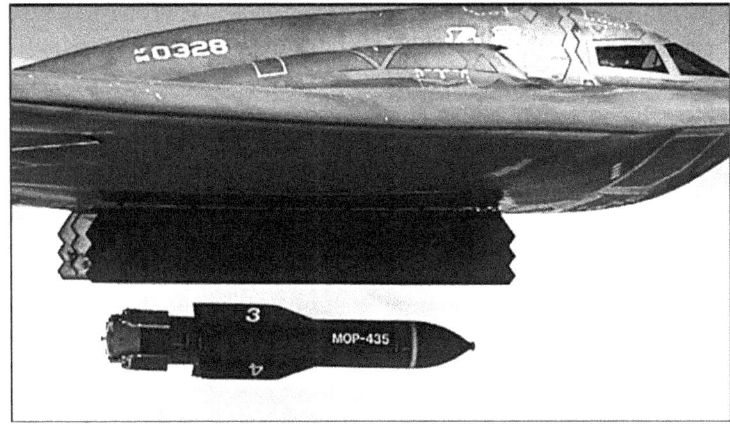

Fig. 12.18　GBU-57A massive ordnance penetrator.

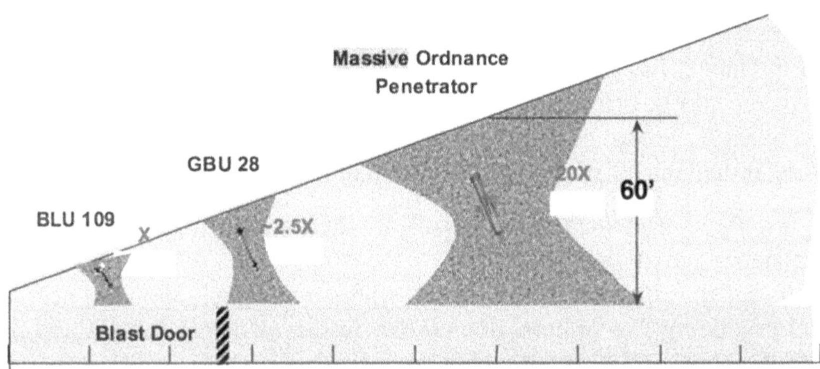

Fig. 12.19　Typical weapon performance for tunnel attacks.

Instead of using multiple weapons or a single, large warhead, a dual or tandem warhead device may be effective against tunnel targets. The JSOW-154C is an example of this and is shown in Fig. 12.20.

This shows a large shaped charge broach warhead (cylinder) designed to produce a hole in the rock above the tunnel, allowing the follow-up blast/fragmentation warhead to continue the penetration process, hopefully into the tunnel. This ensures that at least two warheads will detonate in the same location.

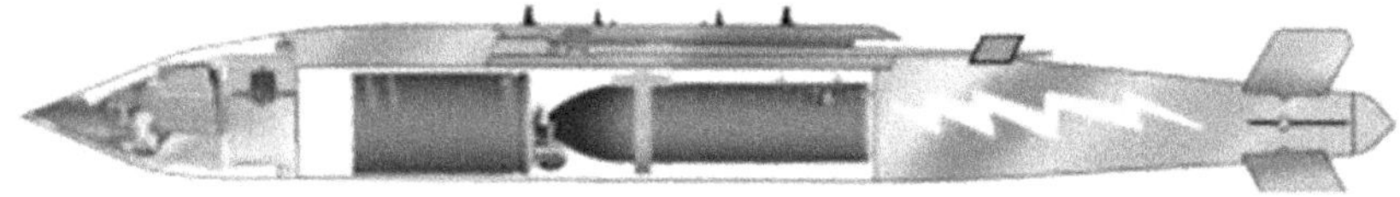

Fig. 12.20　Tandem warhead design.

Because tunnels deeply buried in rock are very difficult to attack successfully, in practice, the following strategies have been the most productive:

- Detonating a warhead close to the portal so that rubble will at least temporarily prevent traffic in or out. This technique was discussed earlier in this chapter using the effective miss distance (EMD) methodology.
- Detonating a weapon inside the rock surrounding the tunnel passageway close to the tunnel walls or roof to produce a crater in the surrounding rock. If the crater is close enough to the tunnel wall, it will vent into the passageway and project rubble into the tunnel itself. This detonation will damage equipment and/or personnel inside the tunnel but is difficult to plan unless the internal layout is known.
- Probably the most damage that can be caused to a tunnel complex is if the weapon penetrates into the tunnel itself and then detonates. If that occurs, then the effects of blast, and to some extent fragments, are enhanced by the constraining walls, roof, and floor, producing more damage than would be expected in a free-air explosion.
- If the tunnel has a blast door designed to protect the tunnel interior, then passage may be effectively blocked by jamming the door. These doors tend to be large, strong, and heavy (see Fig. 12.21), so the repair time may be longer than that of simply clearing rubble.

The last three strategies listed above are expanded on in the Advanced Weaponeering textbook.

Consider the situation in Fig. 12.22, where a warhead detonates near the portal producing two craters, one on the outside of the mountain and one inside the tunnel. The crater rubble from the lower crater will fall into the

Fig. 12.21   Blast doors at the tunnel portal.

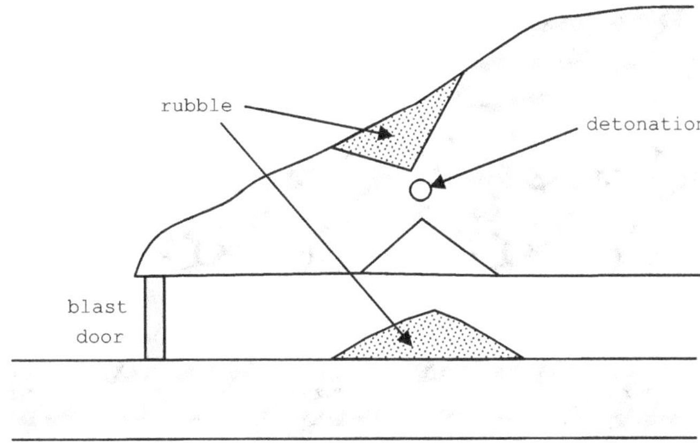

Fig. 12.22   Single or multiple warheads producing two craters.

tunnel, causing partial blockage, while the rubble from the upper crater will fall back into the crater. Suppose a second warhead (Fig. 12.20) detonates in exactly the same location. It will enlarge the crater size and perhaps connect it with the lower crater. As a result, the rubble volume produced by the original and expanded upper crater may fall into the tunnel, considerably increasing the probability of blockage. This phenomenon is termed a *coal chute*. The addition of crater volumes involves some complex geometric calculations. The interested reader is referred to [2].

12.6   CHAPTER SUMMARY

- Three new EI types were discussed dealing with bridges, disruptive craters, and effective miss distances. This completes all EI types used in the weaponeering program.
- The basic approach has been to represent each target type as a rectangle in the ground plane and calculate the probability of damage as that of a random impact point landing inside this rectangle.
- This approach was modified for bridges using the P_{HD} value and for buildings by calculating a fractional damage.
- For completeness, tunnel targets were also discussed, but apart from portal damage using an EMD, internal cratering and air blast are not supported in the weaponeering program.

REFERENCES

[1] O'Connor, S., and Kopp, C., *Assessing PLA Underground Airbasing Capability*, Air Power Australia Analysis 2011, February 2011.
[2] Applied Research Associates, *FIST 1.0 Software Analysts Manual*, Vol. 2, Tunnels, JTCG/ME Program Office, Sept. 2011.

Chapter 13

GROUND PENETRATION AND CRATERING

13.1 INTRODUCTION

The first part of this chapter deals with the penetration of weapons into a variety of materials—soil, concrete, and rock—including layers of these materials. If a weapon impacts a hard surface at certain angles, penetration may not occur, and the weapon could ricochet or broach. The conditions under which these phenomena happen will also be discussed.

As far as penetration models are concerned, two have been used in the Joint Munitions Effectiveness Manual (JMEM). The simpler one, which is discussed in this chapter, is PC Effects, and it uses a straight-line penetration path. The other one, PENCRV3D, models curvilinear trajectories through a resistive medium. Before moving ahead, however, two terms associated with penetrating weapons need to be defined.

- *Penetration* occurs when a weapon moves into a material and comes to rest in it; for example, a bomb may penetrate a 20-ft-thick layer of soil to a depth of 10 ft before detonating.
- *Perforation* occurs when a weapon passes completely through a finite thickness of material; for example, a weapon may perforate a concrete runway before detonating in the soil beneath it.

The second part of this chapter deals with the production of craters in the ground for soil, concrete, and a concrete slab covering soil, as in the case of a runway or a floor slab in a building. These scenarios are shown in Fig. 13.1.

In particular, we are concerned with calculating the crater dimensions produced by a specific warhead based on the depth at which detonation occurs. These methods can be found as the basis for many weaponeering tools for which the study of penetration and cratering is required, such as Integrated Munitions Effects Assessment (IMEA), "Fundamentals of Protective Design for Conventional Weapons" (CONWEP), and JMEM Weaponeering System (JWS).

293

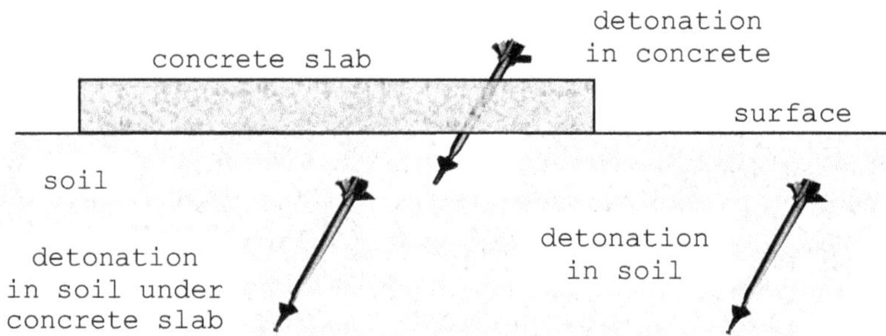

Fig. 13.1　Weapon detonation locations.

13.2　RICOCHET AND BROACH

As mentioned before, penetration of a resistive medium does not always occur; for some ranges of impact angle and velocity, the weapon may ricochet or broach. Ricochet is when the projectile does not penetrate the medium to a depth greater than its diameter. In a broach trajectory, penetration to more than a diameter does occur, but the weapon totally or partially re-emerges from the medium. Representative trajectories are shown in Fig. 13.2.

The angles at which either of these two phenomena occur depend on the hardness of the material impacted, the impact velocity, the impact angle, and the angle of attack of the weapon. The critical ricochet angle is defined as the

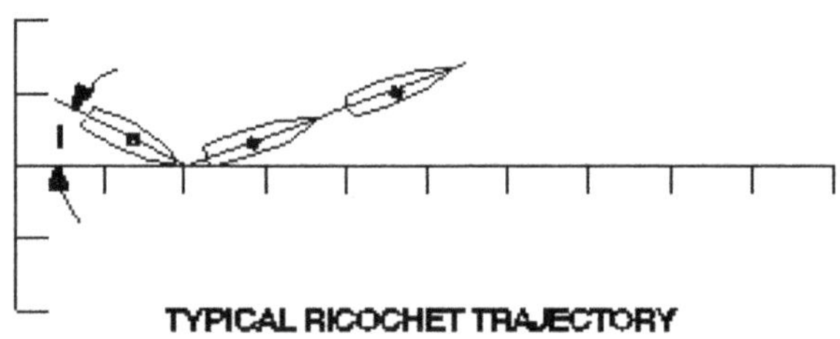

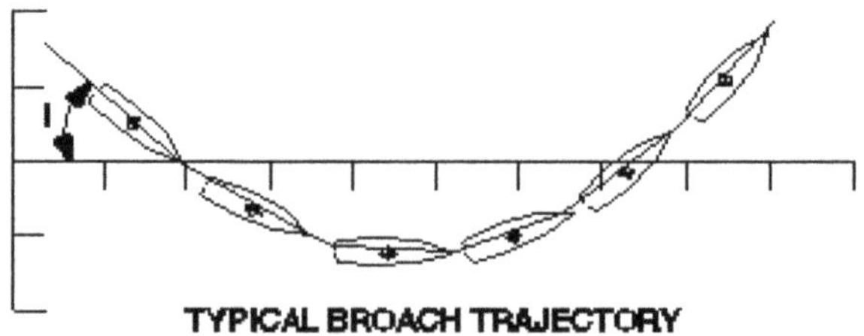

Fig. 13.2　Ricochet and broach trajectories.

TABLE 13.1   RICOCHET AND BROACH ANGLES FOR SOIL

Weapon	Impact Velocity (ft/s)	Impact angle (deg)					
		Hard soil $S=4$		Medium soil $S=11$		Soft soil $S=35$	
		Broach	Ricochet	Broach	Ricochet	Broach	Ricochet
Small GP bomb	300	40	40	36	31	45	19
	700	38	35	45	25	45	15
	900	40	30	45	22	43	15
Medium GP bomb	300	41	41	37	30	50	22
	700	41	34	37	25	50	16
	900	39	32	42	42	46	15
Large GP bomb	300	40	40	36	33	47	23
	700	40	35	36	25	46	19
	900	38	33	40	23	42	19

GP = general purpose.

impact angle for which 50% of the weapons will ricochet and 50% will not, and is usually determined experimentally; however, some simulations have been performed for general purpose bombs impacting soil. These results are shown in Table 13.1. The soil strength parameter S is the same one that will be shown in Table 13.2 later in this chapter. These are not critical angles, nor do they take into account the effect of impact velocity and angle of attack; however, the data do give an indication of impact angles to avoid in order to prevent lack of penetration.

13.3   PENETRATION MODELING USING THE PC EFFECTS PROGRAM

In this model we are considering discrete layers of material (soil, concrete, rock) through which a munition passes, either penetrating or perforating. In this idealized representation, each material is considered separately.

13.3.1   PENETRATION IN SOIL

Consider the situation shown in Fig. 13.3 in which a weapon falls through air to impact and penetrates a layer of soil thickness L. The weapon is assumed to pass through the soil from point A to point B in a straight line, and while doing so decelerates from the impact velocity at point A to the exit velocity at point B.

The weapon is modeled as a point mass that has initial impact conditions (velocity and angle) and experiences a resistive force (F) from the soil. The soil is modeled such that the force resisting motion is a function of the static soil resistance and a function of the velocity of the penetrator. This formulation is known as a Poncelet model.

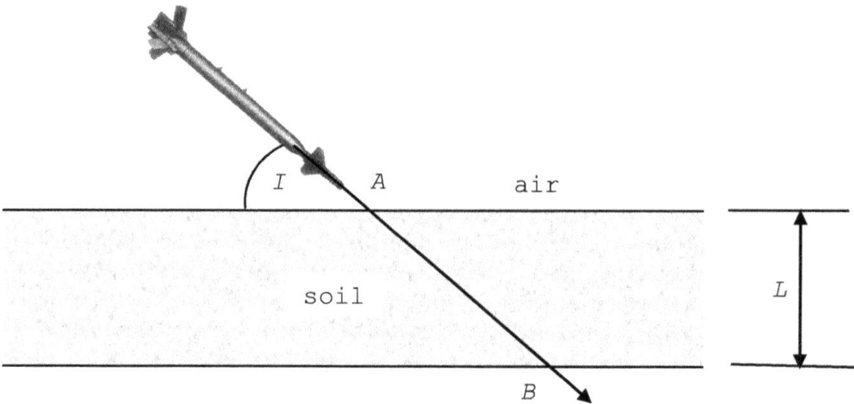

Fig. 13.3   Penetration of soil from air.

Applying Newton's second law of motion to the weapon in the direction of motion yields

$$F = ma = m\frac{\mathrm{d}v}{\mathrm{d}t} \tag{13.1}$$

where

$$F = -\pi r^2 (\gamma_{SL} + \beta_{SL} v^2) \tag{13.2}$$

In these equations

a = acceleration
v = velocity
r = radius of weapon
γ_{SL} = static soil constant
β_{SL} = dynamic soil constant

Because the equation of motion is nonlinear (v^2), it has to be solved numerically. We begin by writing

$$a = \frac{F}{m} = -\frac{\pi r^2}{m}(\gamma_{SL} + \beta_{SL} v^2) \tag{13.3}$$

A finite difference approximation to the acceleration is made in the following form:

$$a = \frac{\mathrm{d}v}{\mathrm{d}t} = \frac{v(t + \mathrm{d}t) - v(t)}{\mathrm{d}t} \tag{13.4}$$

Combining Eqs. (13.3) and (13.4) gives

$$v(t + dt) = v(t) + a\,dt = v(t) - \frac{\pi r^2}{m}\left(\gamma_{SL} + \beta_{SL}v(t)^2\right) \qquad (13.5)$$

The solution begins at time $t=0$ by calculating acceleration a from Eq. (13.3), where $v=v(t)$ is the impact velocity. The velocity at the end of a small time step $v(t+dt)$ is calculated from Eq. (13.5). The distance traveled down the straight line penetration path is obtained from the average velocity over the time step.

$$dx = \frac{v(t + dt) + v(t)}{2} \times dt \qquad (13.6)$$

Substituting $v(t)=v(t+dt)$ allows the recursive solution for weapon velocity, as well as the total penetration depth, to proceed by summing the incremental penetrations for each time step. The weapon exits the layer when

$$\sum dx \geq \frac{L}{\sin I} \qquad (13.7)$$

where L is the layer thickness and I is the impact angle. To investigate fuzing options, the total weapon penetration time may be obtained by summing the time steps from initial contact with the layer to exit.

This model may have several layers of soil, each with different static and dynamic characteristics. Sometimes the soil characteristics are specified in terms of a strength number S that is related to the static and dynamic parameters by Table 13.2.

TABLE 13.2 SOIL PENETRABILITY CHARACTERISTICS

Soil Type	Examples	γ_{SL} (lbf/in.^{2})	β_{SL} (lbf/s^2/in.^{4})
Hard ($S<6$)	Dry, silty sand Permafrost Glacier ice Hard, dense, clayey silt	750	0.0000824
Medium ($6<S<12$)	Loess or loam Loose, dry, sandy soil Moist, sandy, silty clay Loose, moist sand	350	0.0000689
Soft ($S>12$)	Partially saturated clay Wet, soft, tide-flat mud Wet, soft clay	160	0.0000592

13.3.2 PENETRATION IN CONCRETE

Concrete penetration equations are known as the Sandia equations [1] and are empirical in nature. They take the form

$$D = 0.00178 \times S \times N \times (v - 100) \times \left(\frac{W}{A}\right)^{0.7} \tag{13.8}$$

where

D = distance penetrated (in.)
S = concrete penetration resistance factor (lbf/in.^{2})
N = weapon nose shape constant
W = bomb weight at penetration (lb)
A = bomb cross-sectional area (in.^{2})

The bomb nose shape factor for an ogive-shaped nose is given by

$$N = 0.183R + 0.56 \tag{13.9}$$

whereas for a conical shape it is

$$N = 0.25R + 0.56 \tag{13.10}$$

Here, R is the weapon nose ratio, which is the weapon nose length divided by the diameter. The concrete penetration resistance factor is calculated from

$$S = 0.085K_e \left(11 - P\right)\left(t_c T_c\right)^{-0.06} \left(\frac{5000}{f_c}\right)^{0.3} \tag{13.11}$$

In this equation,

P = volumetric percent of rebar (typically 1–2%)
t_c = concrete cure time in years, max value of 1
T_c = target thickness in penetrator calibers (usually a maximum of 6)
f_c = concrete compressive strength (lbf/in.^{2})
K_e = 1 for most cases

The concrete compressive strength is supplied by the user, but representative data for low/medium/high strength would be 2500/4000/6000 lbf/in.^{2}. The penetration time for a semi-infinite concrete layer is given by

$$T_{\text{PEN}} = \frac{D}{12V}\left(\frac{n}{n-1}\right) \tag{13.12}$$

where

$$n = 9.957 f_c^{0.25} \tag{13.13}$$

If the layer is perforated, the exit velocity is given by

$$v_E = V\left[1 - (0.75T - D)\right]^{1/n} \tag{13.14}$$

In this equation, V is the initial impact velocity, T is the path length of the weapon through the concrete layer. The time taken to perforate the layer is

$$T_{\text{PER}} = \frac{D}{12V}\frac{n}{n-1}\left[1 - \left(1 - \frac{0.75T}{D}\right)^{\frac{n-1}{n}}\right] \tag{13.15}$$

13.3.3 PENETRATION IN ROCK

Rock is treated in a manner similar to concrete except that there may be a difference in laboratory strength testing and in situ values. This difference is measured by the rock mass rating (RMR), which takes into account joints, cracks, or fissures that exist at the weapon impact site. The rock penetrability factor S is obtained from the Sandia rock model

$$S = 12(\sigma_u Q)^{-0.3} \tag{13.16}$$

where

σ_u = unconfined compressive strength (lbf/in.^{2})
Q = rock quality

The rock quality is calculated from the RMR as follows:

$$Q = \left[1 - \left(1 - \exp\left(\frac{\text{RMR} - 100}{18}\right)\right)^{1.5}\right]^{\frac{1}{1.5}} \tag{13.17}$$

Once calculated, the factor S is used in the concrete model given in Eq. (13.8).

13.3.4 PENETRATION IN CONCRETE OVER SOIL

This situation relates to targets such as roads or runways and is analyzed by a combination of the methods already outlined. For example, a 1-ft-thick runway laid over soil uses the Sandia equations to determine penetration and perforation of the concrete, and Poncelet equations to determine subsequent penetration of the soil. The exit conditions from the concrete become the entry conditions for the soil. The PC Effects program allows custom layering of air, soil, concrete, and rock to simulate buried targets such as bunkers.

Although these equations may seem somewhat complex, they may be programmed into a spreadsheet to investigate multilayer penetration by an impacting weapon, as shown in Fig. 13.4.

In this figure, the user selects the appropriate weapon and impact conditions and can specify a total of six layers of soil, concrete, air, or rock. The curved line in the right-hand part of the figure shows the penetration path through the various materials with the times to reach the material interfaces indicated. This allows estimates of fuze settings so, for example, if the figure represented a bunker with a concrete roof and 10-ft soil overburden, a fuze setting between 14.2 ms and 35.3 ms would detonate the bomb in the bunker.

13.4 Treatment of Cylindrical Charges for Cratering

In interactions between a weapon and different materials or layers of material, it is important to consider the finite size of the weapon itself. As shown in Fig. 13.5, a weapon may penetrate a sequence of material layers (concrete and soil in this case) before detonating, but in doing so may be embedded in more than one layer. In general, these layers may be combinations of air, soil, concrete, or other materials.

Clearly the effects of cratering in both soil and concrete would have to be considered and aggregated in some way. This is accomplished by considering the weapon to consist of a fixed number of charge elements, some of which detonate in soil while some detonate in concrete, depending on the overall penetration of the weapon. Each element will produce its own crater, the dimensions of which will be aggregated to find the total crater dimensions. This process for a distributed charge warhead partially buried in soil is shown in Fig. 13.6.

In the sections that follow, therefore, we assume that when we refer to a "weapon detonating in the medium" we are referring to those elements of the warhead that detonate in the medium, recognizing there may be other elements detonating in a different medium.

13.5 Crater Geometry and Definitions

Figure 13.7 shows some of the terminology and definitions associated with crater geometry [2].

It may be seen that damage may be caused by more than just the hole produced by the crater itself. The upthrust, or heave, may render an aircraft runway unusable beyond the crater dimensions, and the material ejected from the crater formation and not falling back into the hole (ejecta) may damage personnel or unprotected materiel targets.

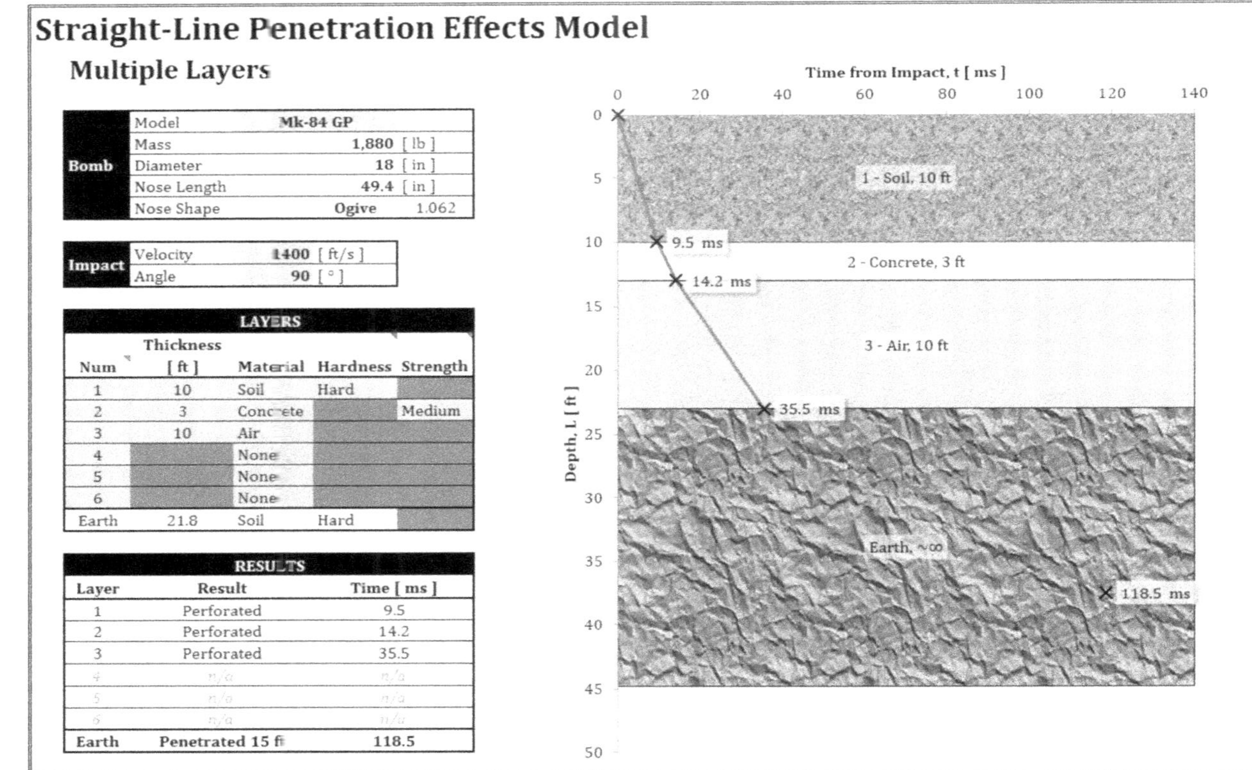

Fig. 13.4 Straight line penetration of layered materials.

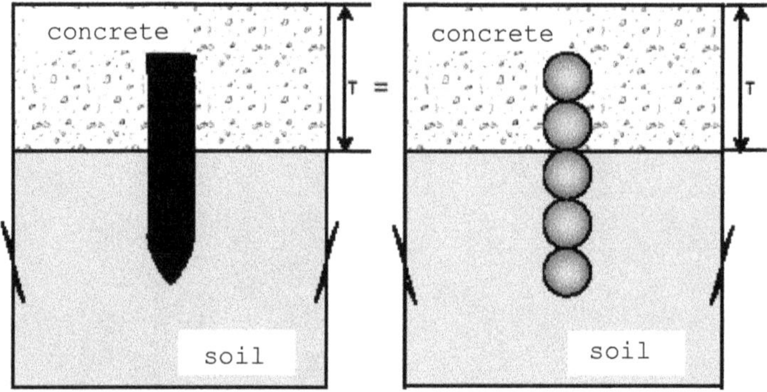

Fig. 13.5   Representing weapon by charge elements.

The crater dimensions as defined previously are functions of several primary and secondary parameters.

* Weight of charge at detonation
* Depth of burial (DOB) below the pre-explosion surface
* Material between the detonation and the surface
* Moisture content and cohesiveness of the material forming the crater (soil only)

13.6   CRATERS IN SOIL

This section deals with calculating surface crater dimensions (depth and diameter) caused by a weapon detonating in soil. The methodology used stems from "Fundamentals of Protective Design for Conventional Weapons," usually abbreviated to CONWEP. This methodology is based on calculating the crater dimensions from the explosive weight of the weapon and its depth of burial. Programs such as PC Effects may be used to determine the latter. Although the CONWEP technical manual has limited distribution, the documents used in its compilation, at least for cratering, are not restricted; therefore, in the following descriptions, references are made to these earlier publications.

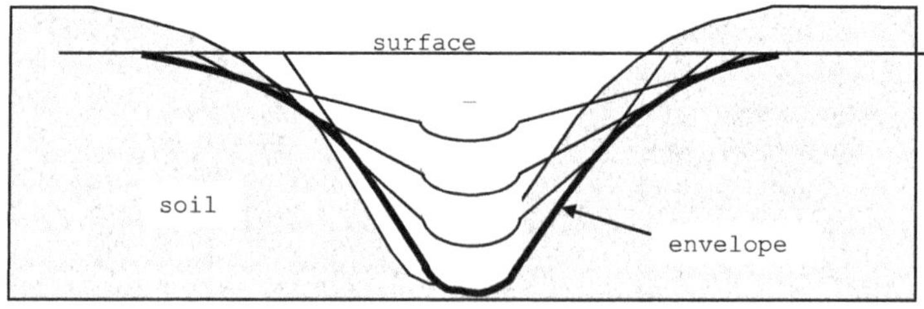

Fig. 13.6   Individual craters forming a single crater.

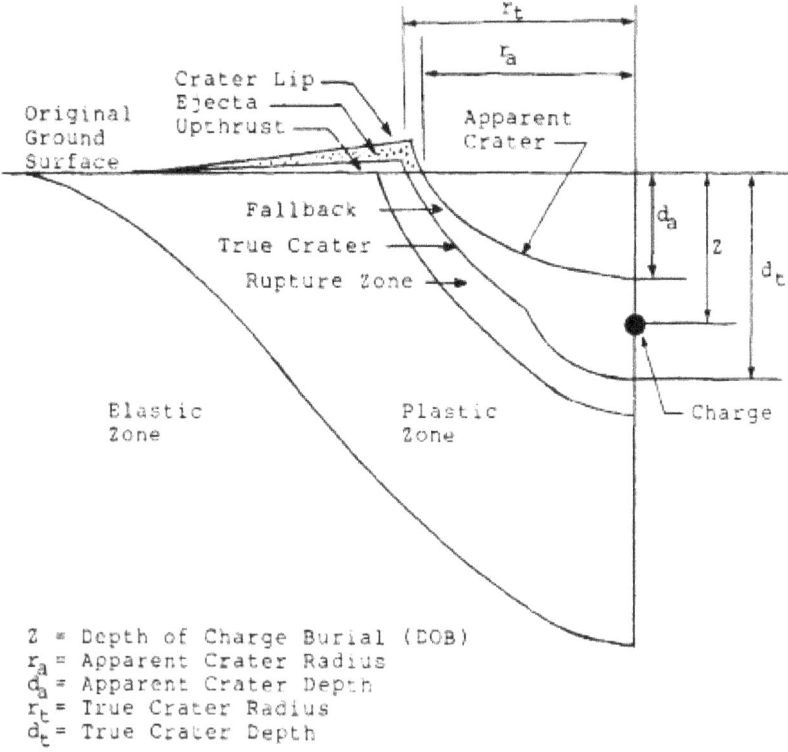

Fig. 13.7 Crater (half shown) features and definitions.

In these studies, the explosive weight used in the calculations is the TNT equivalent weight of explosive considered as a spherical charge with no casing. Specific warheads have a casing, and the thickness of the casing will vary from weapon to weapon. The first step, therefore, is to calculate the effective bare explosive weight W_e from the modified Fano equation converted to TNT equivalent as given in Chapter 7, Eqs. (7.7) and (7.8):

$$W_e = K \times W \left[0.6 + \frac{0.4}{1 + 2M/c} \right] \qquad (13.18)$$

where

W = total weight of explosive in the bomb
c = charge weight/unit length of cylindrical portion of the bomb
M = metal weight/unit length of cylindrical portion of the bomb
K = factor converting explosive fill to TNT

Clearly, the size of a crater will depend on the amount of explosive used; however, it has been observed that the crater dimensions scale with the cube root (Hopkinson scaling) of the explosive equivalent. Therefore, all linear dimensions will be scaled by dividing by this quantity. Here the scaled diameter and radius are given by

$$d = \frac{d_c}{W_e^{1/3}} \tag{13.19}$$

$$r = \frac{r_c}{W_e^{1/3}} \tag{13.20}$$

We will also use the depth of burst (DOB) of the warhead, and this is scaled in a similar manner.

$$\text{DOB} = \frac{\text{DOB}_c}{W_e^{1/3}} \tag{13.21}$$

where the subscript c implies the actual crater dimension. The shape of the crater formed is dependent on the depth of the warhead when it detonates. For airburst or near-surface explosions the crater is shallow, and the crater debris, or rubble, is thrown almost horizontally a considerable distance from the point of detonation (see Fig. 13.8a–c). As the detonation depth increases, the rubble is ejected in a more vertical direction (see Fig. 13.8d) and eventually begins to fall back into the crater, partially filling it. At a particular depth, the pressures produced by the detonation gasses are too small to eject the material above it, and a condition known as a camouflet may be produced (Fig. 13.8f). This is the generation of an underground cavity with just a small earth mound directly above it at the surface.

The crater dimensions, defined in terms of the diameter and depth, are shown in Fig. 13.9 for different materials and moisture content [3, 5].

The diagram shows that to maximize the crater dimensions, the scaled depth of detonation should be about 1.5 ft/lb$^{1/3}$. Some programs curve fit a sixth-order polynomial, sometimes referred to as the CONWEP equations, to these data in the following form:

$$z_c = a_0 + a_1 z + a_2 z^2 + a_3 z^3 + a_4 z^4 + a_5 z^5 + a_6 z^6 \tag{13.22}$$

$$d_c = b_0 + b_1 z + b_2 z^2 + b_3 z^3 + b_4 z^4 + b_5 z^5 + b_6 z^6 \tag{13.23}$$

Here, z is the scaled DOB, z_c is the scaled crater depth, d_c is the scaled crater diameter, and the coefficients a_i and b_i are tabulated for common soil types. Taking the medium soil data points shown in Fig. 13.9 and fitting a sixth order polynomial produces the results shown in Table 13.3.

An example of Eqs. (13.22) and (13.23) for this material and the coefficients given in Table 13.3 is shown in Fig. 13.10, where the data points have been taken from Fig. 13.9.

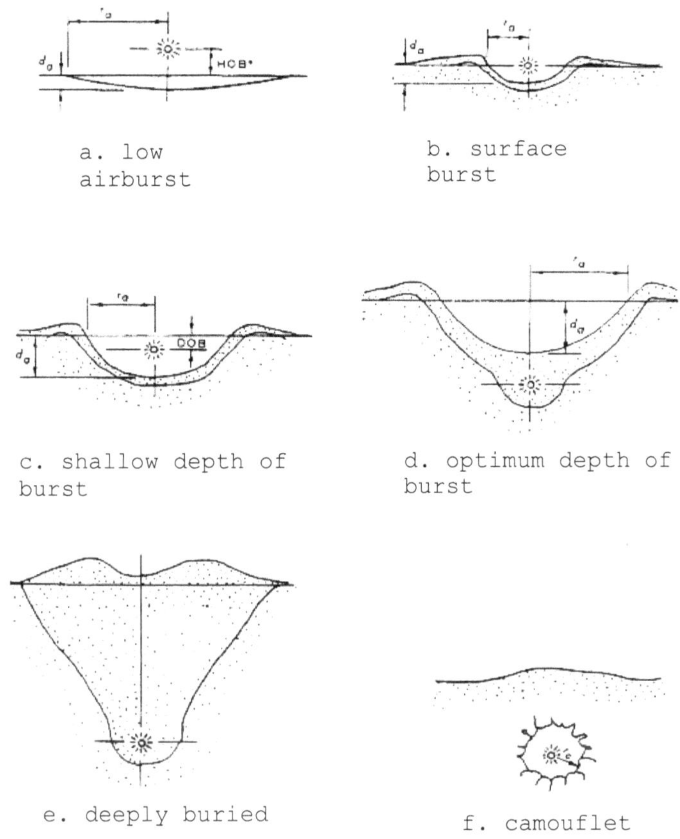

Fig. 13.8   Crater types in soil.

TABLE 13.3   CONWEP COEFFICIENTS FOR CRATER DIMENSIONS

	0	1	2	3	4	5	6
a	0.8971	1.2242	−0.1702	−0.7877	0.7885	−0.3062	0.0400
b	2.9014	1.6387	0.1173	−0.5520	0.2697	−0.0769	0.0089

This approximation to an average material property allows effectiveness computer programs to quickly generate crater dimensions without storing excessive data.

13.7   SURFACE CRATERS IN CONCRETE

These crater dimensions are from [5] and are shown in Fig. 13.11, where data are for concrete with typical strengths in the 2000–5000 lbf/in.^{2} range. The graphs also apply to medium strength rock with similar strength. Within this range, weaker materials will result in the larger crater dimensions.

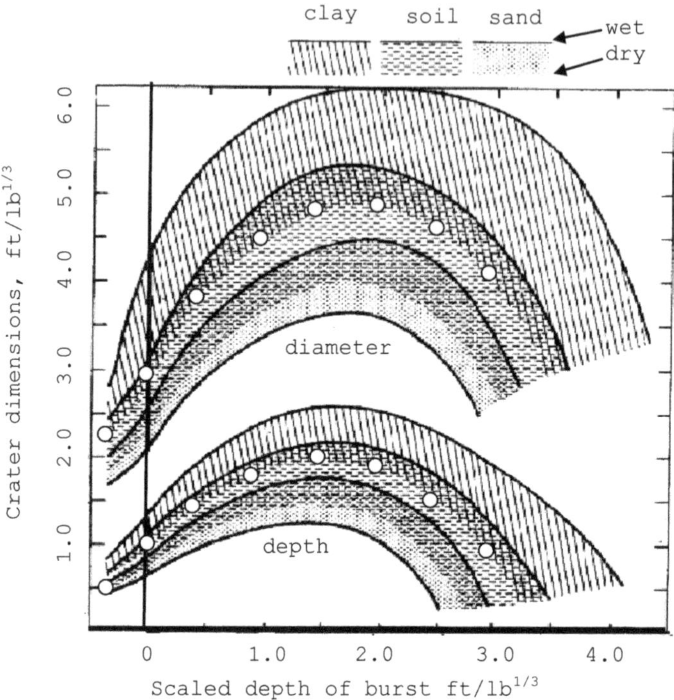

Fig. 13.9   Apparent crater dimensions in soil.

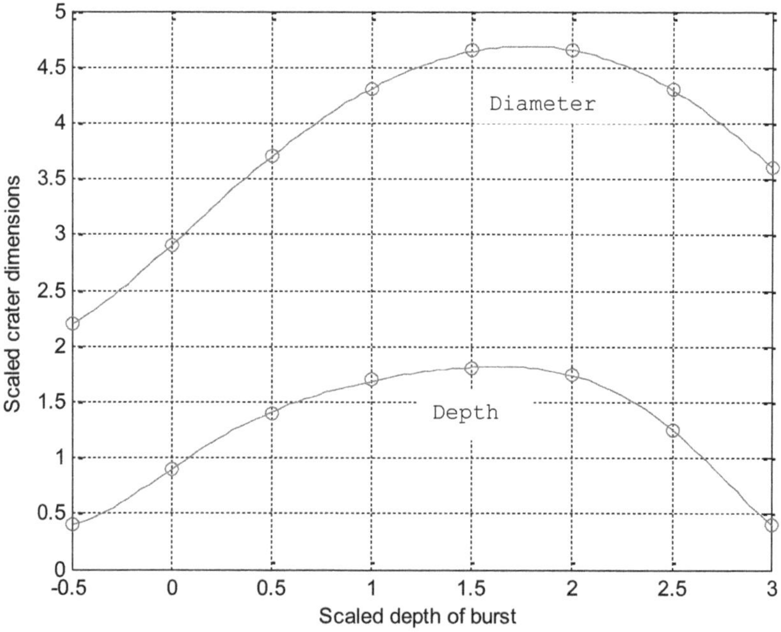

Fig. 13.10   Predicted apparent crater dimensions.

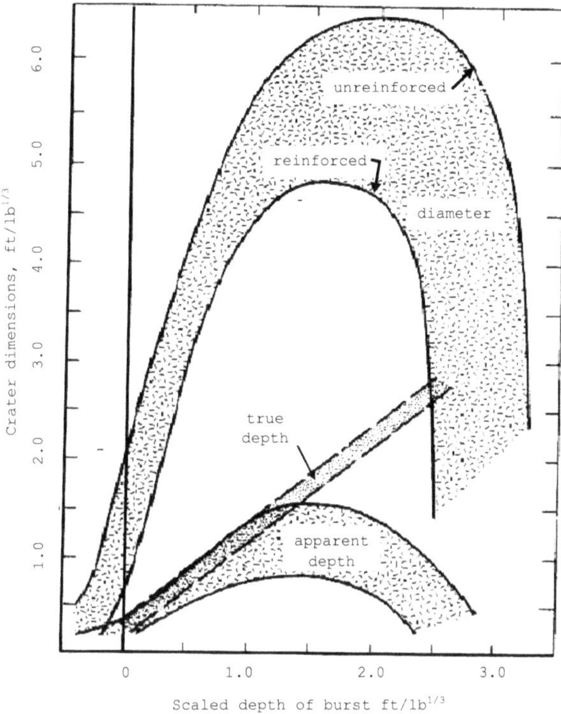

Fig. 13.11   Apparent crater dimensions in concrete and rock [5].

In this figure both true and apparent crater depths are shown. Although not developed here, a single curve fit to an "average" concrete or rock material, as was done for soil, allows a simple, computationally rapid way to calculate crater dimensions.

13.8   SURFACE CRATERS IN CONCRETE OVER SOIL

It is assumed in this scenario that the weapon impacts the concrete at a sufficiently steep angle as to perforate the concrete and penetrate the under- lying soil to some depth before detonating. The depth of burial may be determined using the PC Effects program as was suggested for a simple soil layer.

A layer of concrete backed by soil or air is relevant to several target types such as airfield runways, the floor slab in a building, the deck of a bridge, or the exterior wall of a buried bunker. As for simple soil craters, concrete- covered soil craters may form different shapes depending on the detonation depth, as shown in Fig. 13.12.

In assessing damage to this type of target, we will be interested in calculat- ing the crater dimensions as well as any heaving of the surrounding concrete slab surrounding the hole. The analysis that follows is based on the work of

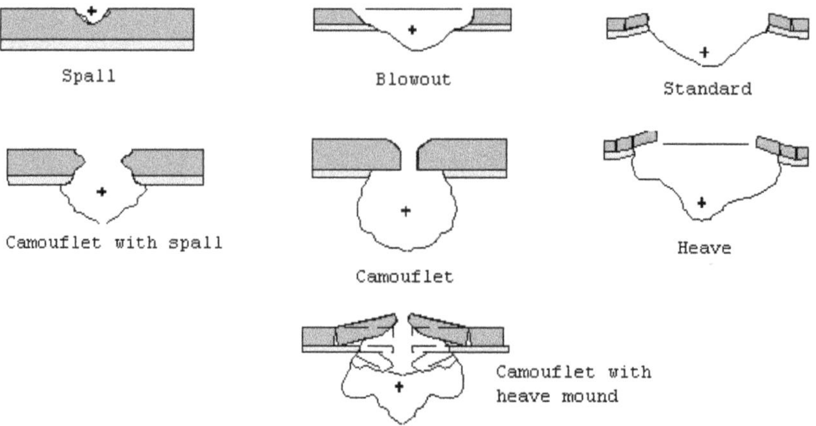

Fig. 13.12   Crater types in concrete-covered earth.

Ross and Rosengren [6] and forms the basis of the IMEA model for craters in concrete over soil.

In this model, the objective is to determine a relationship among the depth of burst (DOB), the resulting crater radius, and the amount of heave in the surrounding slab. It begins with the scenario shown in Fig. 13.13.

The detonation of the warhead will produce a pressure wave in the soil that radiates out in a spherical manner, decreasing in magnitude with distance from the detonation point, until it encounters the bottom of the slab, at which point it produces a load distribution under the slab. This load will be a maximum directly above the charge, decreasing with radius r from the vertical line.

This geometrically and time-dependent load will transmit an impulse to the concrete slab that may cause it to fail due to localized shear stress exceeding that of the yield strength of the concrete. Because this stress will be a maximum directly above the detonation at the instant the pressure wave arrives, and it decreases both with distance from the vertical and with

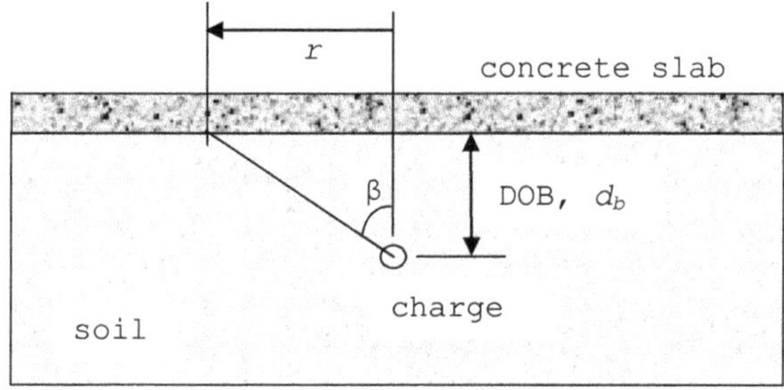

Fig. 13.13   Detonation of charge in soil beneath concrete slab.

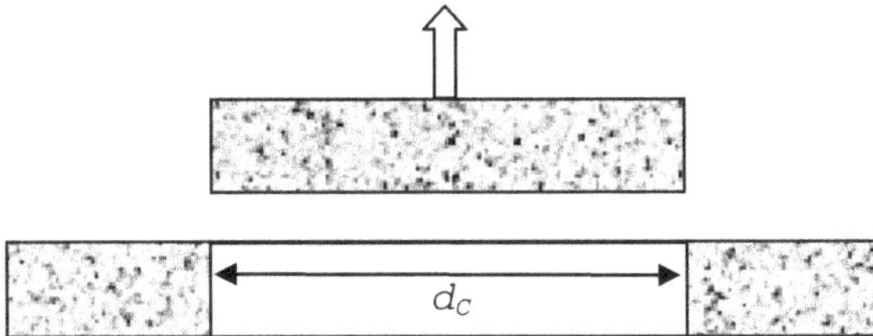

Fig. 13.14   Blowout plug from slab.

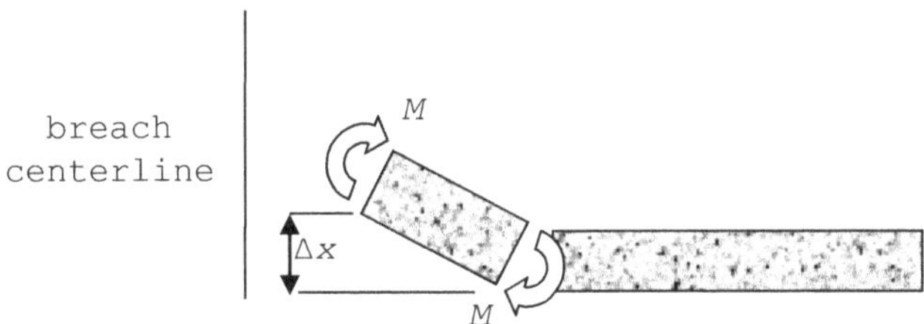

Fig. 13.15   Slab heave estimation.

time, if failure occurs it will eject a circular plug from the slab, as shown in Fig. 13.14.

The mathematical analysis is somewhat complex but is dealt with in the *Advanced Weaponeering* book. The end result is to determine, for a given weight of explosive, the size of the breach plug as a function of detonation depth. Apart from the breach or blowout plug, damage may occur to the concrete slab due to fracturing around the perimeter of the hole, resulting in heave mounding. This may impair the functionality of the slab, as in the case of an aircraft runway, so the amount of heave is of interest.

To determine the amount of heave in the concrete surrounding the breach hole, the plug is removed and the remaining load on the bottom of the slab around the hole is treated as producing a bending moment. Integrating this bending moment over the remaining slab area produces estimates of the deflection of the surrounding material, as shown in Fig. 13.15.

Reference [6] computes an example detonation beneath a slab where the depth of burst is varied, the results of which are shown in Fig. 13.16.

These results come from a simulation where the slab is 20 ft × 20 ft × 1 ft thick is resting on sandy soil. The explosive was 6.7 lb of TNT located directly beneath the center of the slab, where the DOB is measured relative to the midpoint of the slab's thickness. It is seen that at about 34-in. burial, the breach radius goes to zero, which is about the same DOB for maximum heave.

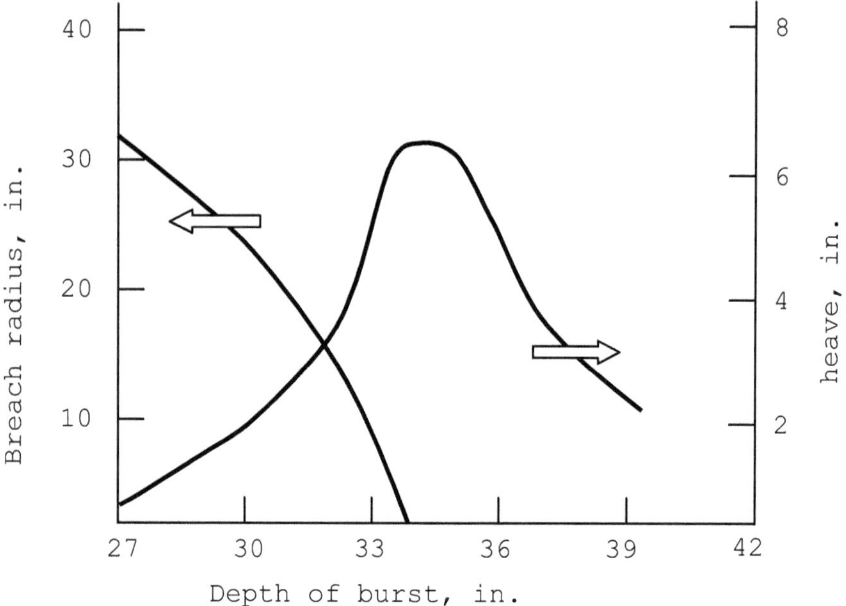

Fig. 13.16   Sample results for heave crater.

13.9 EJECTA

Because the material ejected from the crater that doesn't fall back into the crater (ejecta) may cause damage to some target types, it is useful to characterize it in terms of the amount (mass, volume) of material, size, and area affected. In terms of damage, two different quantities may be considered:

- The volume of material ejected can block access to something such as a road, railroad, or tunnel entrance.
- Solid ejecta associated with rock or cohesive soil may form projectiles that can injure personnel or some fragile materiel targets. If sufficient material is deposited on a structure, it may collapse under the static load.

In estimating the amount of material permanently ejected from the crater, it may be seen that for a given charge weight, Figs. 13.9 and 13.10 would provide depth and radius of the apparent crater. The volume of the apparent crater equals the volume of ejecta.

In addition, it will be of interest to know what area around the crater is affected by ejecta (i.e., how far does the debris field extend?). Figure 13.17 from [5] provides a guide to these distances for both rock and soil, although it might be expected there is a dependence on depth of charge burial that is not indicated.

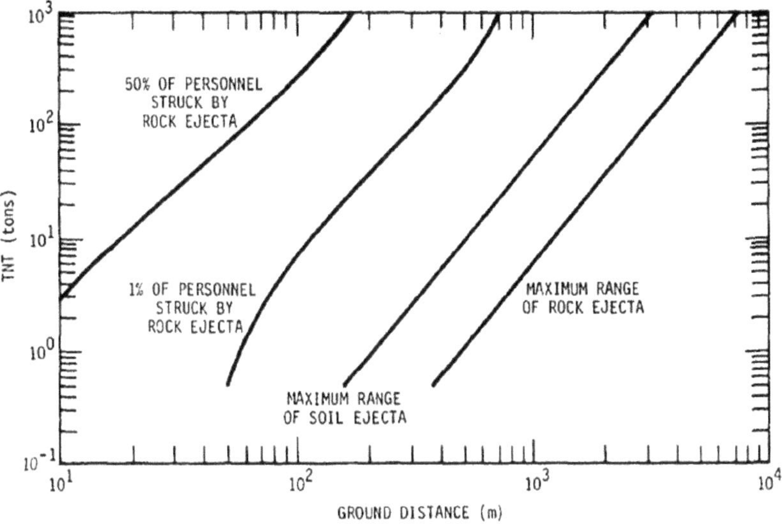

Fig. 13.17 Damage-distance criteria for ejecta [5].

13.10 CHAPTER SUMMARY

- The chapter first looked at penetration and perforation of different layered materials to determine for a given projectile impact velocity the exit velocity and time of transit, if applicable. This allowed a spreadsheet to be developed to investigate the trajectory of a penetrator through layers of user-defined materials and thicknesses.

- Assuming the projectile comes to rest in a particular material and then detonates, the resulting crater geometry has been determined.

- For the special case of a concrete layer over soil, the dimensions of the crater and surrounding heave have been described. Finally, estimates of material ejected from the crater that does not fall back have been made. These methods can be found as the basis for many weaponeering tools where the study of cratering is required, such as IMEA, CONWEP, and the JWS crater methodology.

REFERENCES

[1] Young, C. W., "Penetration Equations," Sandia National Laboratory report SAND97-2426, Oct. 1997.

[2] Brownell, K. C., and Charlie, W. A., "Centrifuge Modeling of Explosion Induced Craters in Unsaturated Sand," Colorado State University, Fort Collins, Nov. 1992.

[3] Strange, J. N., Denzel, C. W., and McLane, T. I., "Cratering from High Explosive Charges, Analysis of Crater Data," Technical Report 2-547, U.S. Army Engineers Waterways Experiment Station, Vicksburg, MS, June 1961.

[4] Joachim, C. E., "Ejecta Hazard Ranges from Underground Munitions Storage Magazines," U.S. Army Engineers Waterways Experiment Station, Vicksburg, MS, Aug. 1990.

[5] Kiger, S. A., and Balsara, J. P., "A Review of the 1983 Revision of TM 5-855-1—Fundamentals of Protective Design (Nonnuclear)," U.S. Army Engineers Waterways Experiment Station, Vicksburg, MS, May 1983.

[6] Ross, C. A., and Rosengren, P. L., "Expedient Nonlinear Analysis of Reinforced Concrete Structures," Second International Symposium on the Interaction of Non-nuclear Munitions with Structures, Panama City Beach, FL, April 1985.

Chapter 14

AIR BLAST

14.1   INTRODUCTION

In order to study the effects that the detonation of an explosive has on target elements located in close proximity to it, some basic understanding of the mechanism of air blast is needed. This is not intended to be a comprehensive explanation of this subject; such a treatment may be found in [1]. Air blast can affect many target types including buildings, bridges, and personnel, and is particularly effective if a detonation occurs inside a confined area.

14.2   EXPLOSION CHARACTERISTICS

When an explosive material detonates, a large amount of energy is released in a short period of time that rapidly increases the temperature and pressure of the air around the detonation point. This increased pressure will propagate in a radial manner to the adjacent undisturbed air in the form of a shock or blast wave. Damage is caused to surrounding objects by their exposure to this wave, causing the collapse or rupture of structural elements or injury to personnel.

In the vicinity of the explosion, the near field, the pressure waves are often assumed to be propagating in a spherical manner, as indicated in Fig. 14.1. At much larger distances from the detonation point, the radius of curvature of the spherical blast front is sufficiently large where it is often assumed be flat (i.e., planar). This characterizes the blast wave in the far field.

Following detonation, the pressure measured at a fixed location some distance (far field) from the detonation point is shown in Fig. 14.2.

The figure shows that when the wave arrives, a large positive overpressure is generated that falls in an approximately exponential manner until a negative pressure is experienced, followed by a gradual return to atmospheric pressure. The effect on a structural target of the passage of this blast wave moving from right to left is shown in Fig. 14.3 [1].

313

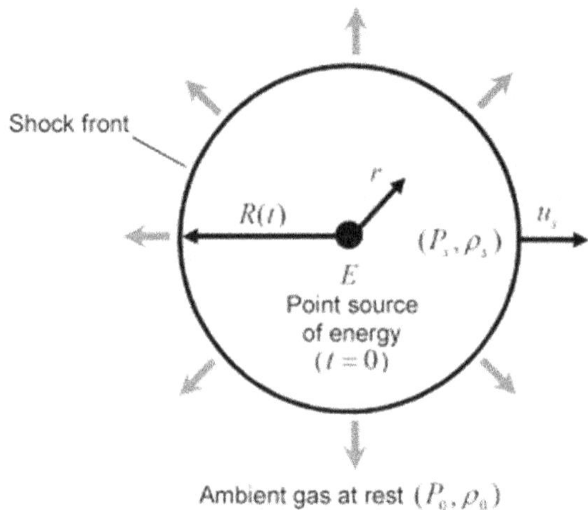

Fig. 14.1   Blast wave propagation.

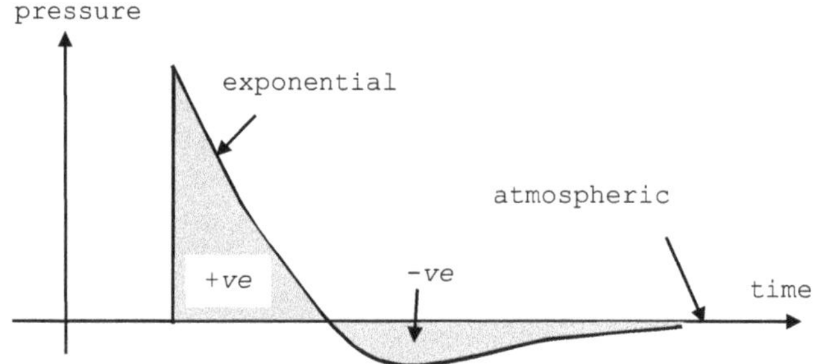

Fig. 14.2   Pressure-time for one-dimensional propagation.

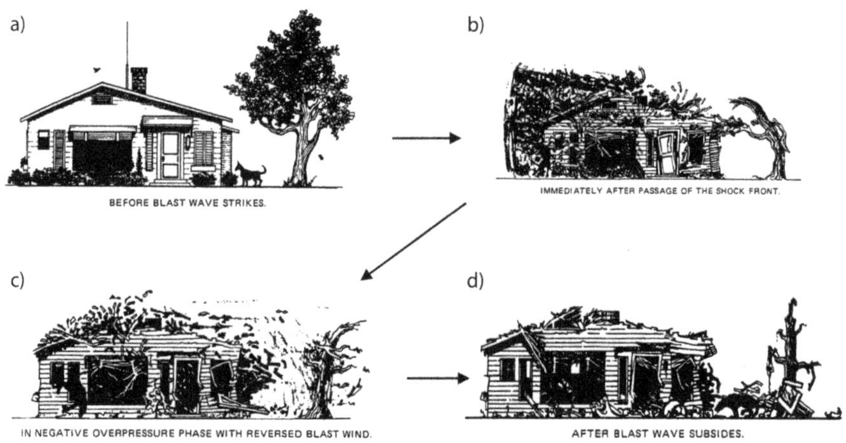

Fig. 14.3   Building response to blast wave moving right to left [1].

Initially the blast wave travels at supersonic speed, but gradually it slows down until it has the characteristics of a sound wave traveling at the local speed of sound.

14.3 THE SHOCK FRONT

We will start this section with a discussion of terminology. The phenomenon of interest is sometimes referred to as a *blast wave* or a *shock wave*. Usually a blast wave describes the result of a real weapon detonation, whereas a shock wave is associated with a theoretical study of gas dynamics, such as that presented in this section. Often these terms are used interchangeably.

It is convenient to initially consider a blast wave in terms of its planar shock front traveling at supersonic speed through still air. This situation, known as normal shock, provides a convenient basis for analytical treatment of how air properties change due to the passage of the shock. A complete analysis of the passage of a shock front through air is beyond the scope of this book; however, we will provide an overview, together with the more important, useful equations.

Consider the situation in Fig. 14.4 where the shock wave (dotted line) is moving from left to right with velocity u_x. Air properties upstream of the shock have subscript x whereas those of the air after the shock has passed have the subscript y.

It may be shown that by assuming air to be an ideal gas, the properties of the air across the shock front are given by the following equations:

Overpressure:

$$p = P_Y - P_X = \frac{7\left(M_X^2 - 1\right)}{6} P_X \tag{14.1}$$

Temperature:

$$\frac{T_Y}{T_X} = \frac{\left(5 + M_X^2\right)\left(7M_X^2 - 1\right)}{36M_X^2} \tag{14.2}$$

Velocity:

$$u_Y = u_X - \frac{5a_X\left[M_X^2 - 1\right]}{6M_X^2} \tag{14.3}$$

where M_X is the Mach number of the shock wave. The last of these equations, for velocity, defines what is known as the blast wind u_p, which is the air particle velocity immediately after the shock wave. The blast wind is in the direction of shock movement, and its speed is calculated from

$$u_P = u_X - u_Y \tag{14.4}$$

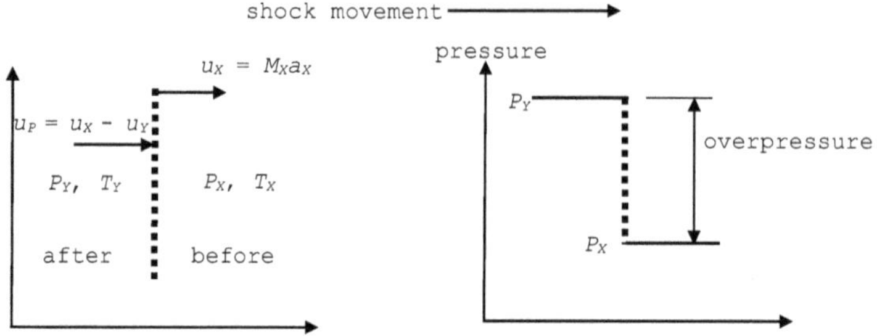

Fig. 14.4   Shock front parameters.

It follows from Eqs. (14.3) and (14.4) that

$$u_P = \frac{5a_X\left[M_X^2 - 1\right]}{6M_X^2} \tag{14.5}$$

The reader need not be overly concerned with the mathematics of these equations, because they may be programmed into a simple spreadsheet to observe their importance. The change in air conditions across the shock front depends on the Mach number, and as we have previously mentioned, this decreases with distance from the point of detonation. Therefore, assuming that at a particular location the Mach number is known (e.g., 2.0), the overpressure and blast wind observed at that point may be calculated as shown in Table 14.1.

In summary, as the shock moves through still air at standard pressure and temperature, the pressure, temperature, and wind speed all increase by an amount dependent on the Mach number. Recall from Fig. 8.15 in Chapter 8 that these output values are sufficient to be fatal to personnel targets, and buildings will be totally destroyed.

TABLE 14.1   SHOCK WAVE PARAMETER CALCULATIONS

Normal Shock Parameters		
Inputs		
Mach number	2.0	
Acoustic speed	761	mph
Atmospheric pressure	14.7	psi
Outputs		
Overpressure	51.5	psi
Blast wind	476	mph

14.4 SCALING LAWS

Perhaps the most critical parameters in determining damage are the amount of explosive used and the distance of the detonation point from the target. Although it will be expected that different warheads will produce different effects at a fixed distance from the detonation point, it is possible to relate these effects using what is known as *scaled distance*. This is sometimes referred to as Hopkinson or cube root scaling, and is defined through the parameter Z as follows:

$$Z = \frac{R}{W^{1/3}} \tag{14.6}$$

where W is the weight of the explosive and R is the distance to the detonation point. It was found that blast parameters such as pressure, temperature, and the like are the same for identical values of Z. A simple example will explain. Suppose that 100 lb of explosive produces an overpressure P a distance 50 ft from the detonation point. This gives the value of $Z = 21.5$. At what distance would 500 lb of the same explosive produce the same overpressure? This is solved by recognizing that for the two cases, the value of Z will be the same.

$$Z = \frac{50}{100^{1/3}} = 21.5 = \frac{R}{500^{1/3}} \tag{14.7}$$

This gives the result $R = 171$ ft. The same process also allows us to compare the effects on a target located a fixed distance from the detonation point if warheads of different weights are detonated.

14.5 EXPLOSIVE YIELD

As previously mentioned, the amount of damage inflicted on a target will depend on the amount of explosive contained in a particular warhead; however, this is not the only consideration in determining the strength of the blast wave. When the explosive detonates, it releases a fixed amount of energy that is used to perform the following functions:

- Expanding the metal case of the warhead prior to it fracturing
- After fracturing the case, propelling the resulting fragments away from the point of detonation
- Generating the blast wave

The total weight of the warhead is therefore not the sole criterion for determining the blast wave. For example, consider the 2000-lb bomb variants shown in Fig. 14.5. Although the nominal all-up weight is the same, the

WARHEAD COMPARISONS

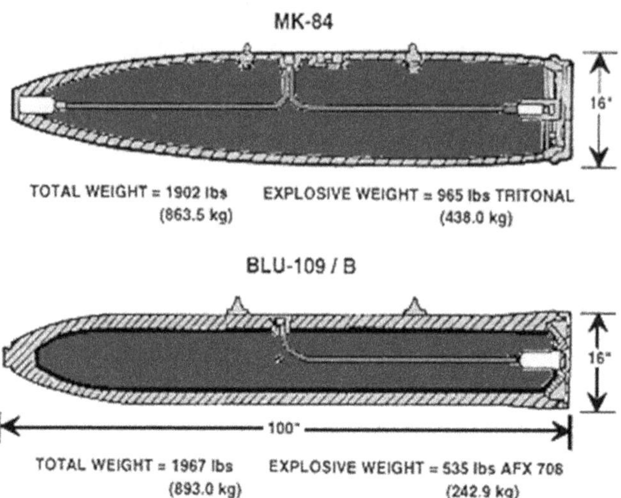

Fig. 14.5   Mk-84 and BLU-109 bomb variants.

BLU-109 version is designed to penetrate hard targets, so it has a much thicker case than the standard Mk-84. This has two consequences:

1. The additional case weight results in a smaller weight of explosive fill.
2. Because the case is stronger, more energy is needed to break it up into fragments; therefore, less energy is available to generate a blast wave.

Note also that the type of fill is different; therefore, some account must be taken of the amount of energy released per pound of these two different fills.

The solution to these inconsistencies is to represent each warhead as an equivalent, uncased, spherical charge of TNT. This was dealt with previously in Chapter 7, Section 7.12 by using the modified Fano equation and converting the explosive fill type by multiplying by the factor K, as shown here:

$$W_e = K \times W \left[0.6 + \frac{0.4}{1 + 2M/c} \right] \quad (14.8)$$

Also, the presence of a rigid reflective surface may enhance the explosive yield due to reflection of part of the blast wave. The resultant equivalent, uncased charge weight is then given by the equation

$$W = W_e \times 2^n \quad (14.9)$$

where n is the number of reflective surfaces ($n = 0$, 1, 2, 3) adjacent to the detonation point. This enhancement is shown in Fig. 14.6.

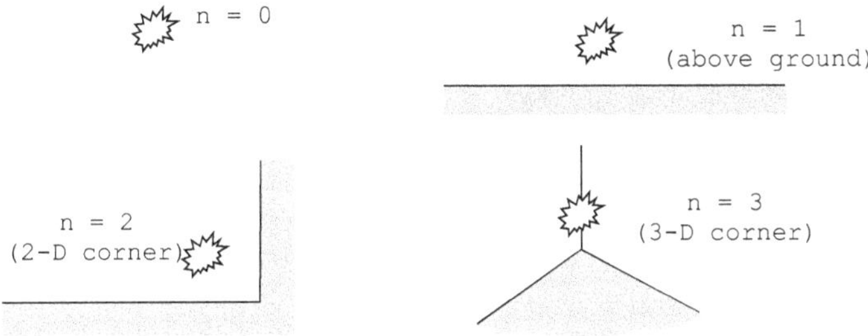

Fig. 14.6   Enhancement of yield due to reflective surfaces.

These effects are considered only if the target is "close" to the reflecting surfaces, usually when the standoff distance r satisfies the following equation:

$$r \leq 1.5 W_e^{\frac{1}{3}} \tag{14.10}$$

Another way of viewing this is to consider the presence of one or more rigid reflective surfaces as enhancing the explosive yield by constraining the propagation of the air blast energy upon a portion of the free field spherical volume. For example, a ground burst produces a hemispherical volume containing the same energy, hence the pressure is doubled; an explosion in a corner produces a quarter-sphere, hence the pressure is multiplied by four, and so on.

These three techniques—cased charges, non-TNT fill, and reflective surfaces—are all means to reduce a real-world detonation to an equivalent, free field, spherical charge of TNT by which different warheads may be compared.

14.6   MACH STEM

The Mach stem is a blast wave that propagates radially outward from the detonation point and results from the interaction of the incident and ground-reflected blast waves for an air-bursting weapon, as shown in Fig. 14.7.

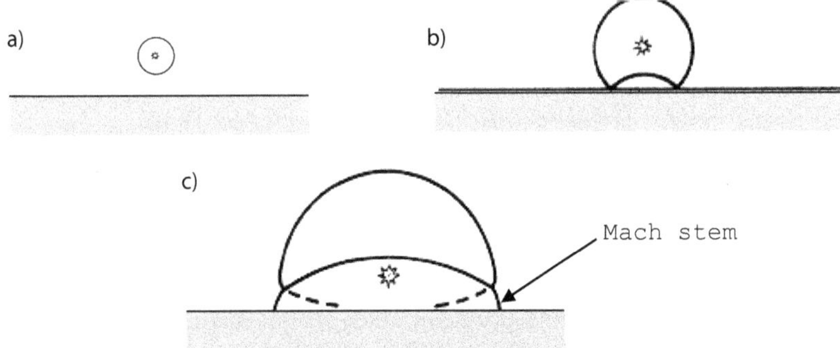

Fig. 14.7   Formation of Mach stem.

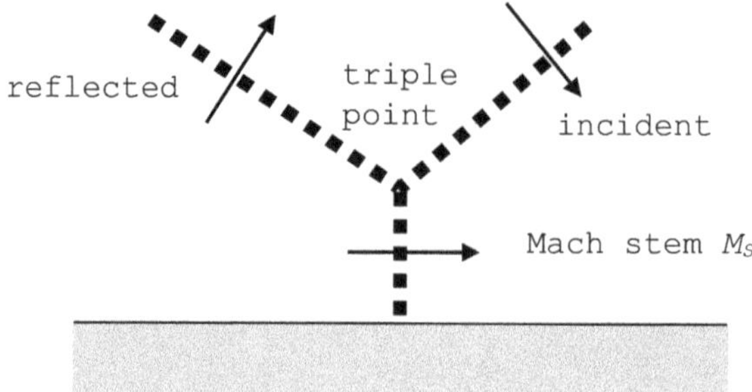

Fig. 14.8   Interaction of incident and reflected waves.

This figure shows

Fig. 14.7a: The spherical blast wave propagating from the air burst
Fig. 14.7b: The reflected blast wave from the ground
Fig. 14.7c: The Mach stem forming at the interaction of the incident and
    reflected waves.

A more detailed view of the Mach stem region is shown in Fig. 14.8. As can
be seen from the figure, the Mach stem blast wave is planar and propagates
radially out from ground zero.

The Mach stem has important implications for air-bursting weapons
because the pressure behind it is higher than the incident wave. Consider a
detonation at a variable height. At zero burst height there is no Mach stem, and
an object on the ground such as a building experiences just the incident pressure of the blast wave.

As the height of burst (HOB) increases, a Mach stem is formed, and because
Fig. 14.6 indicates a doubling of pressure ($n = 1$) due to ground reflection, a
building would experience twice the pressure of that due to the blast wave by
itself. For greater HOB, the detonation is so far away from the ground that
although a Mach stem exists, it is relatively weak. It follows, therefore, that for
a given charge weight, there is an optimum HOB for which the blast wave has
maximum pressure.

14.7   CHARACTERISTICS OF THE BLAST WAVE

The difference between shock and blast waves was defined earlier in this
chapter; however, there are other differences—for example, the shock wave
is assumed planar whereas the blast wave includes some elements of the
spherical nature of real detonations. As a result, the equations given in Section
14.3 provide a suitable starting point for analysis but have to be extended and
supplemented with experimental data. This section, therefore, transitions the
discussion from theoretical shock to more practical blast.

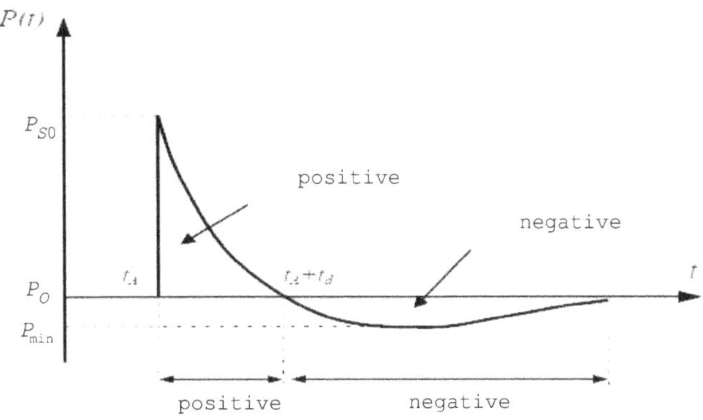

Fig. 14.9   Characteristics of blast wave.

Figure 14.9 shows a more detailed view of the blast wave with some of the more significant parameters indicated, including the following.

- *Peak overpressure P_{S0}:* This is the peak pressure above atmospheric generated by the blast wave.
- *Arrival time t_A:* This is the time between when the warhead detonates and when the blast wave arrives at a specified location.
- *Duration t_d:* This is the duration of the positive portion of the shock wave. There is another duration time, that of the negative portion, but this is usually ignored.
- *Impulse per unit area I/A:* Mathematically, the impulse is defined as the integral over time of the pressure wave, and therefore is a measure of how long the pressure is exerted over a target area.
- *Dynamic pressure q:* When the air mass moving at the blast wind speed is brought to rest by, for example, the side of a building, the kinetic energy of the air is converted to an additional pressure—the dynamic pressure.

Mathematical equations for these quantities are difficult to develop, so experimental data are used to represent them.

14.8   BLAST CHARACTERISTICS: KINGERY-BULMASH DATA

Although theoretical shock-generated pressures, temperatures, and wind speeds have been predicted in Section 14.3, the presence of the ground and the spherical nature of the blast wave make these predictions unrealistic for real explosions. The most widely accepted way of representing blast characteristics is known as the Kingery or Kingery-Bulmash (K-B) method [3, 4]. These are based on extensive tests of TNT air burst and surface explosions. Of principle interest is the manner in which the blast wave characteristics vary with charge weight and distance from the point of detonation. K-B data are presented in graphical form and plot the various blast wave parameters as a function of distance from the point of detonation and use two methods discussed earlier.

1. The charge used is the equivalent TNT, bare spherical charge defined by Eq. (14.8).
2. The distance used in the graph is the scaled distance defined earlier in Eq. (14.6), making the graphs independent of charge weight.

These graphs are shown in Figs. 14.10 and 14.11 for surface burst and air burst, respectively, where a free air burst is defined as a detonation without any reflection from the ground.

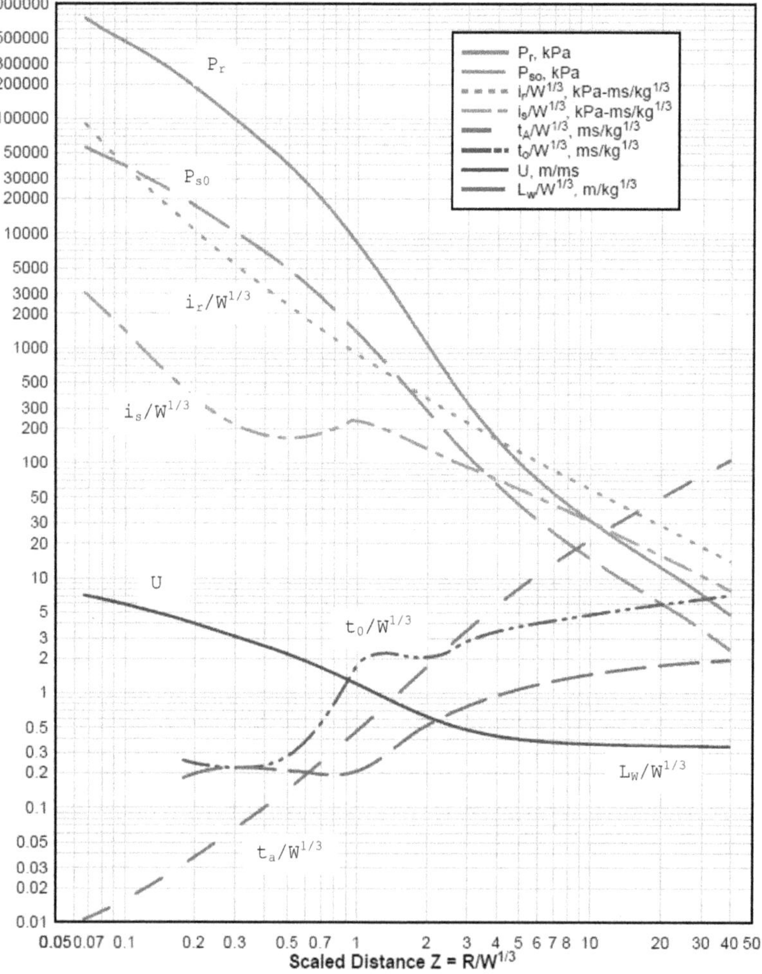

Fig. 14.10   Kingery-Bulmash blast wave chart for surface burst.

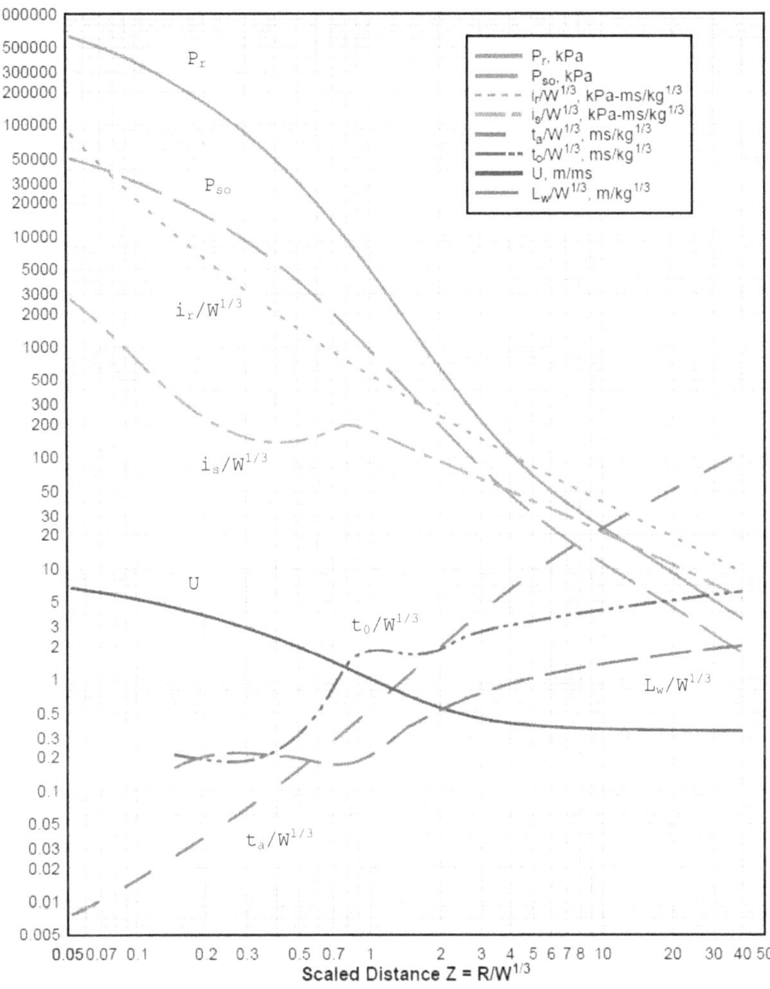

Fig. 14.11   Kingery-Bulmash blast wave chart for a free air burst.

The parameters shown in the charts may differ from those used earlier in this chapter, so some redefinitions are needed.

- P_r = peak total pressure at reflected surface
- P_{SO} = peak pressure in incident wave
- i_r = impulse applied to reflected surface
- i_{SO} = impulse of incident wave
- t_0 = wave duration
- L_W = wavelength of incident wave (not used here)
- t_a = arrival time (right axis)

The difference between incident pressure/impulse and reflected pressure/impulse is that the former would be that measured by an observer as the blast

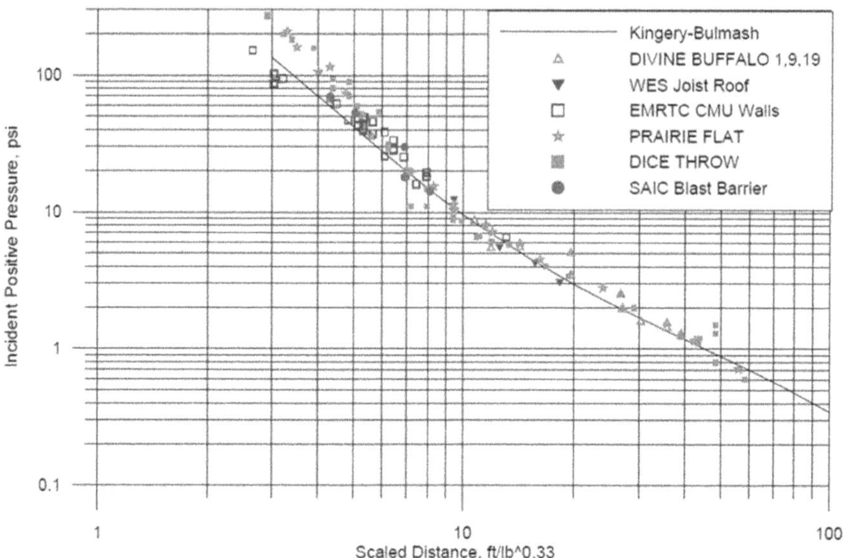

Fig. 14.12   Comparing Kingery-Bulmash P_{S0} with test data [5].

wave passes, whereas the latter is that measured at a reflecting surface such as a wall. The original report by Kingery does not include the peak total pressure because for the test data reported, there were no reflecting surfaces, so a reflection coefficient appears to have been used to calculate these values.

Figure 14.12 compares one of the curves shown in the previous figures— incident positive pressure P_{SO}—to some test data and is taken from [5].

Although useful, these graphs are cumbersome to use, so further work by others such as Swisdak [6] fit equations to them as an aid to any computational tool that requires the data.

If the scaled range is given by

$$Z = \frac{X}{W^{1/3}} \tag{14.11}$$

and we define

$$T = \log_n(Z) \tag{14.12}$$

then the general function F is given by the following:

$$F = \exp\left[A + BT + CT^2 + DT^3 + ET^4 + FT^5 + GT^6\right] \tag{14.13}$$

F takes each of the blast wave characteristics (e.g., time of arrival, incident pressure, etc.), and the coefficients for that feature are given in Table 14.2, where the coefficients A–G depend somewhat on the scaled distance to the detonation point.

TABLE 14.2　KINGERY-BULMASH AIR BLAST POLYNOMIAL COEFFICIENTS

Positive Phase Duration, T ($ms/kg^{1/3}$)						
Range, Z ($m/kg^{1/3}$)	A	B	C	D	E	F	G
0.2–1.02	0.5426	3.2299	−1.5931	−5.9667	−4.0815	−0.9149	0
1.02–2.80	0.5440	2.7082	−9.7354	14.3425	−9.7791	2.8535	0
2.80–40	−2.4608	7.1639	−5.6215	2.2711	−0.44994	0.03486	0

Positive Phase Duration, T ($ms/lb^{1/3}$)						
Range, Z ($ft/lb^{1/3}$)	A	B	C	D	E	F	G
0.5–2.5	−1.7221	0.45	1.3552	1.1249	−0.05773	−0.608	0
2.5–7	−18.7701	55.0513	−60.4348	32.0236	−8.3256	0.8817	0
7–100	−13.0597	19.7805	−11.2975	3.2552	−0.4647	0.02624	0

Incident Impulse, II ($kPa\text{-}ms/kg^{1/3}$)						
Range, Z ($m/kg^{1/3}$)	A	B	C	D	E	F	G
0.2–0.96	5.522	1.117	0.6	−0.292	−0.087	0	0
0.96–2.38	5.465	−0.308	−1.464	1.362	−0.432	0	0
2.38–33.7	5.2749	−0.4677	−0.2499	0.0588	−0.00554	0	0
33.7–158.7	5.9825	−1.062	0	0	0	0	0

(Continued)

TABLE 14.2 KINGERY AIR BLAST POLYNOMIAL COEFFICIENTS (*Continued*)

Incident Impulse, II (psi-ms/lb$^{1/3}$)						
Range, Z (ft/lb$^{1/3}$)	A	B	C	D	E	F	G
0.5–2.41	2.975	−0.466	0.963	0.03	−0.087	0	0
2.41–6.0	0.911	7.26	−7.459	2.960	−0.432	0	0
6.0–85	3.2484	0.1633	−0.4416	0.0793	−0.00554	0	0
85–400	4.7702	−1.062	0	0	0	0	0

Time of Arrival, TOA (ms/kg$^{1/3}$)						
Range, Z (m/kg$^{1/3}$)	A	B	C	D	E	F	G
0.06–1.50	−0.7604	1.8058	0.1257	−0.0437	−0.0310	−0.00669	0
1.50–40	−0.7137	1.5732	0.5561	−0.4213	0.1054	−0.00929	0

Time of Arrival, TOA (ms/lb$^{1/3}$)						
Range, Z (ft/lb$^{1/3}$)	A	B	C	D	E	F	G
0.2–4.5	−2.5671	1.5348	0.1313	0.01825	0.003656	−0.008615	0
4.5–100	−1.79097	−0.44021	2.01409	−0.78101	0.13045	−0.0081529	0

Incident Pressure, PI (kPa)						
Range, Z ($m/kg^{1/3}$)	A	B	C	D	E	F	G
0.2–2.9	7.2106	−2.1069	−0.3229	0.1117	0.0685	0	0
2.9–23.8	7.5938	−3.0523	0.40977	0.0261	−0.01267	0	0
23.8–198.5	6.0536	−1.4066	0	0	0	0	0

Incident Pressure, PI (psi)						
Range, Z ($ft/lb^{1/3}$)	A	B	C	D	E	F	G
0.5–7.25	6.9137	−1.4398	−0.2815	−0.1416	0.0685	0	0
7.25–60	8.8035	−3.7001	0.2709	0.0733	−0.0127	0	0
60–500	5.4233	−1.4066	0	0	0	0	0

Reflected Pressure, PR (kPa)						
Range, Z ($m/kg^{1/3}$)	A	B	C	D	E	F	G
0.06–2.00	9.006	−2.6893	−0.6295	0.1011	0.29255	0.13505	0.019736
2.00–40	8.8396	−1.733	−2.64	2.293	−0.8232	0.14247	−0.0099

Reflected Pressure, PR (psi)						
Range, Z ($ft/lb^{1/3}$)	A	B	C	D	E	F	G
0.3–4.0	9.0795	−1.7511	−0.2877	−0.2199	−0.0128	0.0696	−0.0118
4–100	5.1515	9.15826	−11.85735	5.56754	−1.33455	0.16333	−0.008181

TABLE 14.3 CALCULATING EQUIVALENT, SPHERICAL BARE CHARGE OF TNT

Warhead	Total Weight (W-lb)	Charge Weight (c-lb)	Case Weight (M-lb)	Equivalent Uncased Weight (Wu-lb)	Scale Factor for TNT	Spherical Bare Weight of TNT (lb)	Weight/ Charge Ratio
Mk 82 warhead	500	192	308	**133.4**	**1.35**	**180.2**	2.60
Mk 83 warhead	1000	445	555	**317.9**	**1.35**	**429.2**	2.25
Mk 84 warhead	2000	945	1055	**683.9**	**1.35**	**923.3**	2.12
155 mm artillery round	95	15	80	**9.5**	**1.00**	**9.5**	6.33
GMLRS HE (M31)	400	200	200	**146.7**	**1.00**	**146.7**	2.00
ATACMS HE	1000	500	500	**366.7**	**1.00**	**366.7**	2.00
User Supplied	1500	750	750	**550.0**	**1.00**	**550.0**	2.00

These tables of coefficients allow polynomial equations to be used to express how blast wave parameters vary with scaled distance. A spreadsheet has been developed to implement this polynomial representation and begins by using open source data to compute the TNT equivalent bare charge for some common warheads using the modified Fano equation, as shown in Table 14.3.

With this, or a user-supplied value of W_U, the individual graphs comprising the K-B charts may be plotted; one sample plot is shown in Fig. 14.13.

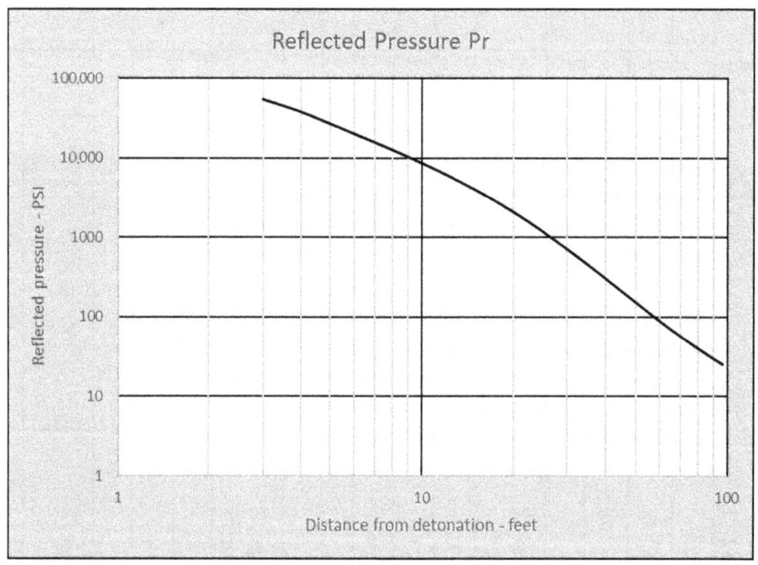

Fig. 14.13   K-B plot for reflected pressure.

14.9 EXTERNAL BLAST LOADING ON STRUCTURES

The interaction of an external blast wave on a structure such as a building may cause collapse because structures are usually designed to support vertical loads, not the sideways forces imparted by a blast wave. In addition, the building should be regarded as a dynamic structure, so the impact of the blast wave may produce sideways motion that deforms the structural elements. If these deformations are sufficiently large, the offset vertical loading may cause the building to collapse due to buckling of the supporting structure. The following approach may be found in many references such as [7], although the following analysis is based on [1].

All of the analysis so far has been "free field" in that there have been no impediments to the movement of the blast except for reflection off a rigid wall. Even in this case, the situation is not realistic because the reflecting surface is assumed to extend to infinity in all directions normal to the blast direction.

Consider the more realistic case shown in Fig. 14.14, where we have an above-ground structure with finite dimensions subjected to a planar blast wave.

The roof, side walls, front wall, and rear wall will all experience different pressure, impulse, and blast wind effects from the passing wave, but the front wall will probably be subject to the greatest values.

If the distance between the point of detonation and the building structure is known, and assuming the building is normal to the direction of the blast wave, the Kingery-Bulmash charts allow us to determine the forces acting on the structural elements from which the building is constructed—beams, columns, floor slabs, walls, and so forth. The question now is: Will these structural elements fail?

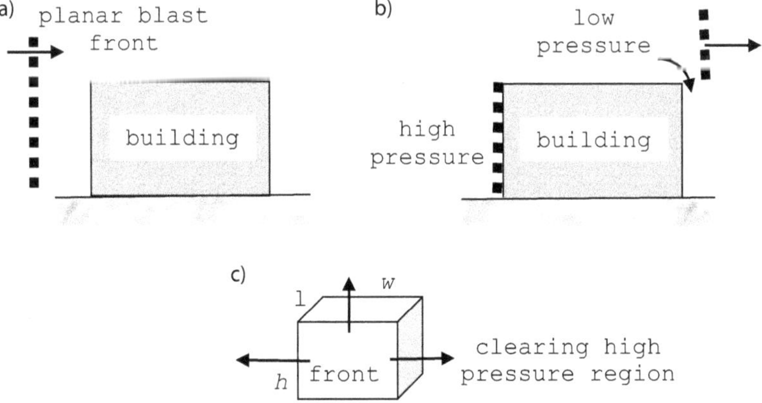

Fig. 14.14 Interaction of a structure with a shock wave.

14.10 THE FACEDAP PROGRAM

A widely used method to predict damage to building components and complete building structures is the Facility and Component Explosive Damage Assessment Program (FACEDAP) developed by the U.S. Army Corps of Engineers [8]. It is a PC-based program, and its utility lies in the relatively simple metrics that are used to predict a scale of damage based on pressure and impulse loading resulting from an adjacent explosion.

The program predicts damage at four different damage levels, defined as follows:

- *0% damage:* High level of protection
- *30% damage:* Medium level of protection
- *60% damage:* Low level of protection
- *100% damage:* Collapse

For a particular structural component, the results are presented as a set of pressure-impulse (P-I) diagrams such as the one shown in Fig. 14.15, where the scaling of pressure and impulse is different for each structural component. The interpretation of the diagram is that, for example, for any loading falling between the lines bounding the 30% range, the damage is assumed constant at 30%.

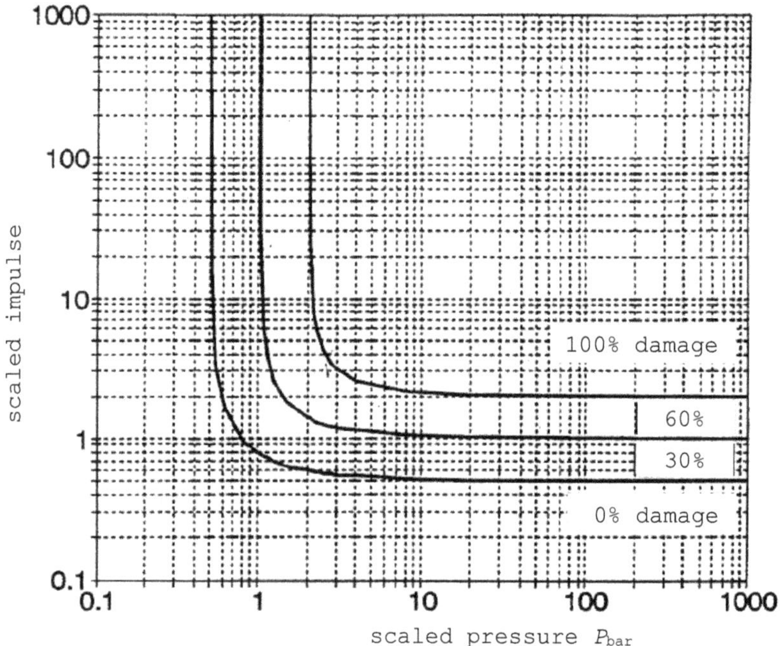

Fig. 14.15   Pressure-impulse diagram for a structural component.

These diagrams provide a rapid assessment of damage to a variety of structural components if the applied pressure and impulsive loading from an explosion can be determined.

The FACEDAP program provides P-I diagrams for 24 structural components and is based on a simple dynamic structural response. Some test data are used for a few components, but they are very limited. The component materials considered are concrete, steel, masonry, and wood, and include beams, unreinforced slabs, columns (interior and exterior), and slabs with unidirectional and bidirectional reinforcing where appropriate.

The derivation of a typical P-I diagram is somewhat complex, but is based on work–energy balances; for more information, readers are directed to [8]. The end result is a diagram such as the one shown in Fig. 14.15, where the damage level is shown graphically as a function of scaled pressure and scaled impulse. Once the P-I curves have been developed, they are used in the following manner:

- Given the location of the detonation relative to the building, use the K-B polynomials to determine the pressure and impulse loading on each structural member.
- Given the P-I chart, determine the amount of damage sustained and whether failure of that component occurs.

14.11 ENGINEERING MODELS FOR PREDICTING BLAST DAMAGE

From a weaponeering perspective, the process of estimating damage due to blast consists of two parts.

1. Determine the blast field at the target.
2. Calculate the target response to that field.

Considering buildings as an example, the study of the interaction of explosions with structures has been extensive, with an excellent overview being given by [9], from which a representative example is shown in Fig. 14.16.

In such a scenario, the most accurate prediction of damage to the structure would involve modeling the blast using computational fluid dynamics (CFD) simulation and modeling the target response using computational finite element analysis (FEA) of the building structure. Using these techniques, the two models are then coupled. Unfortunately, such coupled models take too long to run for the type of rapid results needed by the warfighter. To remedy this, engineering-level models have been produced that provide reasonable results in an acceptable time.

The following sections provide a brief overview of commonly used techniques and codes that may be characterized as an example of an engineering-level model. Again, a more extensive description and comparison may be found in [9].

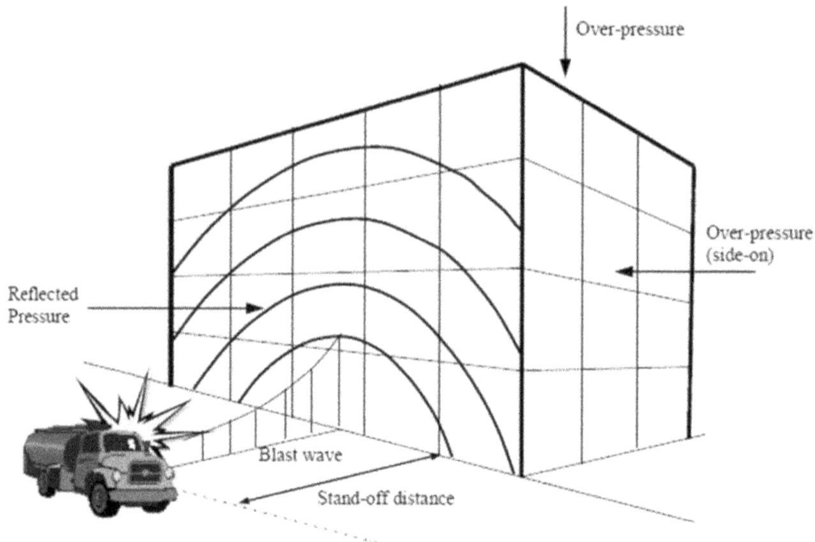

Fig. 14.16   External blast loads on building [9].

14.11.1   *FACEDAP*

This program was described in the previous section and is a good example of an engineering-level model that not only predicts the loading on a structural element, but also estimates the likelihood of failure. Not all targets may be represented by the components addressed by FACEDAP, so a need exists for simplified engineering-level models that predict the blast environment and loading produced by an explosion, which then may be coupled to different target types to estimate damage.

14.11.2   KINGERY-BULMASH

This also was discussed in Section 14.8 in terms of graphical representations of test data for spherical (airburst) and hemispherical (ground burst) graphs (see Fig. 14.10). For many models, these data represent the benchmark for shock characteristics and are used extensively to predict shock and impulse loading on a structure or equipment. The curves assume there are no reflective waves present (free field), and the negative portion of the incident shock is usually ignored. It also ignores the effect the structure has on the incident shock (i.e., they are decoupled). The original K-B paper [3] presented the data in tabular and graphical form, which is not well suited to coding into a computer program.

14.11.3   *CONWEP*

The CONWEP code is a collection of conventional weapons effects calculations based on the data from [7], including blast, fragments, projectile penetration, shaped charge, cratering, and ground shock. For the air blast portion, CONWEP uses the K-B data shown in Fig. 14.10 and fits a polynomial equation to the graphical data, so it is expected to give very similar results to it. It addresses external shock on structures and ignores the negative portion of the incident blast, but does take into account reflected blast by assuming the reflecting surface is infinite (i.e., no clearing of the incident wave and normal incidence). The CONWEP program models both airburst and surface detonations and, for the latter, includes the effect of Mach stem formation.

14.11.4 BLASTX

BlastX is a program developed by the U.S. Army Engineering Research and Development Center (ERDC) to predict the air blast environment inside or outside of structures [10]. This fast running code has at its core a "tabular explosive source" model. In this approach, a hydrocode calculation is performed for a specific explosive material, charge shape (sphere, cylinder, 3-D form), and charge orientation. The outputs of the calculation are the peak pressure, density, time of arrival, and particle velocity at a set of distances (ranges) from the charge surface. The simulation is in free air with no reflecting surfaces. Outside these ranges, the blast parameters are extrapolated; between the discrete ranges, the data are interpolated. Although the tabular explosive model is preferred, the user may use the Kingery-Bulmash (K-B) TNT model if they choose.

The user will select the explosive material of interest, as well as its weight, charge shape, and orientation, and BlastX will choose the appropriate table of blast parameter data vs range from its database. This core function of BlastX in producing the free field blast environment is then coupled with the capability to introduce reflecting surfaces to produce a complex blast loading on those surfaces. The specification of these reflecting surfaces will depend on the target being studied. In this context, BlastX has been used to simulate the loading on buildings (for both internal and external detonations), tunnels, bridges, hardened aircraft shelters, and materiel targets such as vehicles. BlastX will predict the reflected blast environment but will not estimate damage to a target; that must be done separately.

Comparison with the K-B predictions shows very close agreement for pressure and impulse values, except when very close to the detonation, as indicated in Fig. 14.17.

Here, it is argued the BlastX results are more accurate due to the use of a specific shape and material of the explosive charge, and the fidelity of modeling in this region.

14.12 INTERNAL BLAST LOADING ON STRUCTURES

The detonation of an explosive charge inside a building is different from an external detonation in that, for the latter case, the gas produced

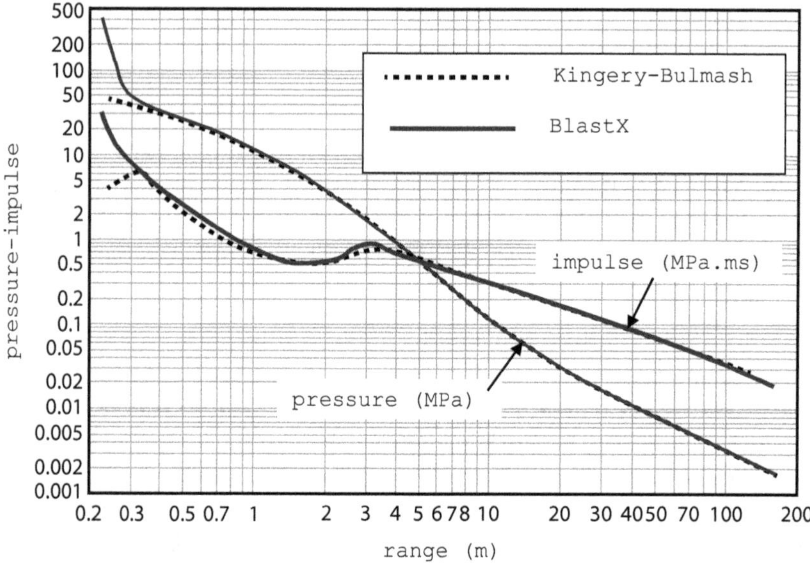

Fig. 14.17 Comparison between BlastX and Kingery-Bulmash.

by the explosion expands spherically from the detonation point. An internal detonation constrains this expansion and results in an additional loading on the interior walls. The magnitude of this loading depends on the amount of explosive, the size of the room in which the detonation takes place, and whether the room is vented to the outside. The loading from an explosive charge detonated within a vented or unvented structure is shown in Fig. 14.18.

Initially, blast waves reflect back and forth across the room, assuming the walls are not damaged. If they are, the blast wave moves past them. As the blast waves reflect and re-reflect within the structure, a quasi-static or gas pressure rise occurs, and the second phase of loading takes place. In a perfectly sealed room, this would produce an elevated, constant pressure loading on all internal surfaces; however, any venting through doors, windows, or damaged walls would return the quasi-static pressure to atmospheric.

The initial and reflected shocks are shown to occur within the first 2 ms and have the decreasing amplitudes described in Fig. 14.18. The quasi-static pressure is seen as the exponential decay back down to the ambient pressure and lasts about 15 ms. Therefore, the process is characterized as a short duration, high pressure shock loading in addition to a lower pressure, longer duration quasi-static pressure.

Although the peak value of the quasi-static pressure is difficult to predict, Fig. 14.18 shows qualitatively how this is estimated by extending the exponential decay of the secondary loading back to the arrival time of the first

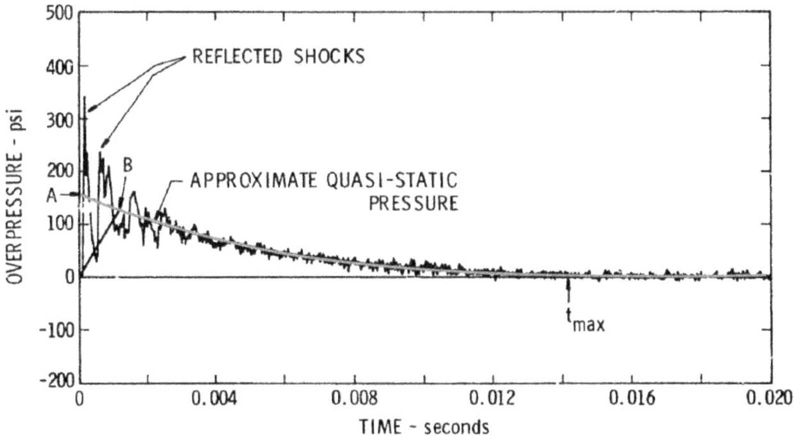

Fig. 14.18 Pressure-time history for internal detonation.

shock, point A in the figure. The degree of venting may be expressed as the ratio of the ventilation area to a measure of room surface area as follows:

$$R_{\text{VENT}} = \frac{A_{\text{VENT}}}{V^{2/3}} \tag{14.15}$$

Here, A_{VENT} is the total area of ventilation to the outside, and V is the room volume. Reference [11] then shows how the peak pressure varies with the ratio of explosive weight W in accordance with Fig. 14.19, at least for a given range of R_{VENT}.

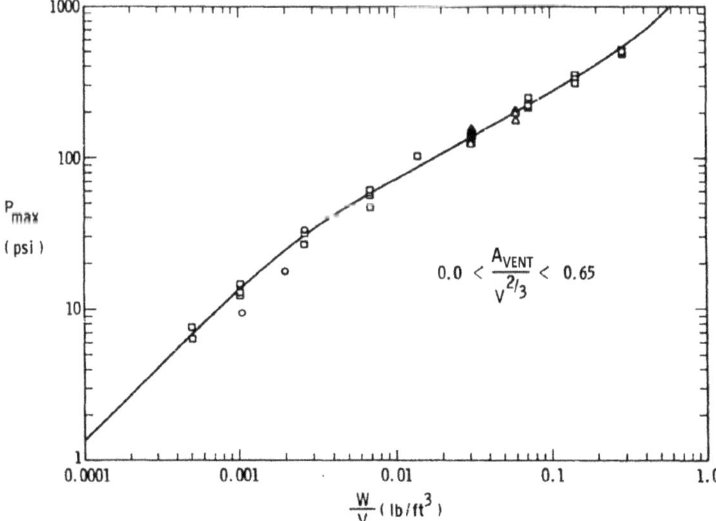

Fig. 14.19 Peak quasi-static pressure vs charge weight.

Although this completes the study of external and internal blast loading on structures such as buildings, the topic of damage, collapse, and loss of function will be deferred until the next chapter.

14.13 CHAPTER SUMMARY

- From a weaponeering perspective, air blast analysis determines the pressure, impulse, and blast wind at various distances from the point of detonation to determine damage and injury to structures and personnel.
- Estimates of these parameters were obtained from theoretical gas dynamics, and modifications were made to represent practical blast waves. This resulted in the widely used Kingery-Bulmash charts.
- Several engineering models were discussed dealing primarily with blast damage to buildings.
- A brief discussion dealt with blast inside a building and the resulting quasi-static pressure.

REFERENCES

[1]   Kinney, G., and Graham K., *Explosive Shocks in Air,* Springer-Verlag, Berlin, 1962.

[2]   Needham, C. E., *Blast Waves,* Springer, New York, NY, 2010.

[3]   Kingery, C. N., "Air Blast Parameters versus Distance for Hemispherical TNT Surface Bursts," Ballistics Research Report 1344, Aberdeen Proving Ground, MD, Sept. 1966.

[4]   Kingery, C. N., and Bulmash, G., "Air Blast Parameters from TNT Spherical Airburst and Hemispherical Surface Burst," ARBRL-TR-02555, Ballistics Research Laboratory, Aberdeen Proving Ground, MD, 1984.

[5]   Bogosian, D., Ferritto, J., and Shi, Y., "Measuring Uncertainty and Conservatism in Simplified Blast Models," *Proceedings of the 30th Explosives Safety Seminar*, Atlanta, GA, Aug. 2007.

[6]   Swisdak, M. M., "Simplified Kingery Air Blast Calculations," Technical Report, Naval Surface Warfare Center Indian Head Division, Aug. 1994.

[7]   "Fundamentals of Protective Design for Conventional Weapons," Department of the Army Technical Manual TM-5-855-1, 3 Nov. 1986.

[8]   "Facility and Component Explosive Damage Assessment Program (FACEDAP)—Theory Manual," Dept. of the Army Corps of Engineers, Omaha, NE, May 1994.

[9]   Ngo, T., Mendis, P, Gupte, A., and Ramsay, J, "Blast Loading and Blast Effects on Structures—an Overview," *Electronic Journal of Structural Engineering,* 7, 76–91, 2007.

[10] Britt, J., and McMahon, G., "Tabular Explosive Source Models Used in BlastX," *Proceedings of the 20th International Symposium on Military Aspects of Blast and Shock,* Oslo, Norway, Sept. 2008.

[11] Esparza, E., Baker, W., and Oldham, G., "Blast Pressures Inside and Outside Suppressive Structures," Edgewood Arsenal Contractors Report EM-CR-76042, report #8, Aberdeen Proving Ground, MD, Dec. 1975.

Chapter 15

ABOVE-GROUND BUILDINGS AND BUNKERS

15.1 INTRODUCTION

Above-ground buildings and buried bunkers have been targets of increasing significance in recent conflicts. This chapter will deal with damage to both of these structures caused by detonations inside or outside the structure, above and beneath the ground. Typical detonation points are shown in Fig. 15.1,

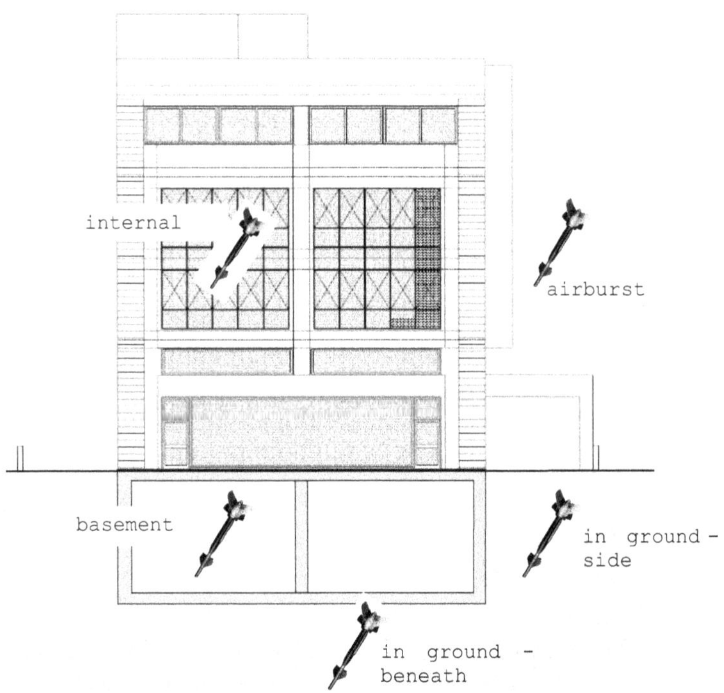

Fig. 15.1 Weapon detonation inside a building.

where a weapon may penetrate the roof, walls, and floor slabs before detonating inside a room. Alternatively, the warhead may detonate external to the building in air or after penetrating the ground. Finally, the weapon may penetrate through the building and detonate beneath the basement floor. The resulting damage may be to the building structure or to materiel and/or personnel inside the building.

The structure of interest may be the above-ground portion of the building, the basement of an otherwise above-ground building, or a bunker that may have no above-ground elements at all. Bunkers may also have features not found in above-ground buildings such as a slab of concrete, sometimes referred to as a burster slab, between the top of the bunker and ground level designed to detonate weapons penetrating the soil.

Also, although most buildings of interest to the weaponeer comply with some form of civilian building codes, bunkers are intended to resist attack and may have substantial interior and exterior walls designed to mitigate the effects of penetration and detonation effects.

Target elements inside buildings and bunkers are usually personnel and equipment that are vulnerable to blast and fragments. Some bunkers may not be completely buried beneath ground level and may have a protective mound, or berm, on top and surrounding it. Typical bunker construction and features as implemented in the JMEM Weaponeering System (JWS) are shown in Fig. 15.2.

Weapons commonly used against such targets include the GBU-28/BLU-113 (5000 lb) and GBU-27/BLU-109 (2000 lb) class penetrator weapons. A comparison between the BLU-109 and conventional Mk-84 warhead is shown in Fig. 15.3. The thicker casing and absence of a nose fuze will be noted, as will the reduced weight of explosive fill in the penetrator.

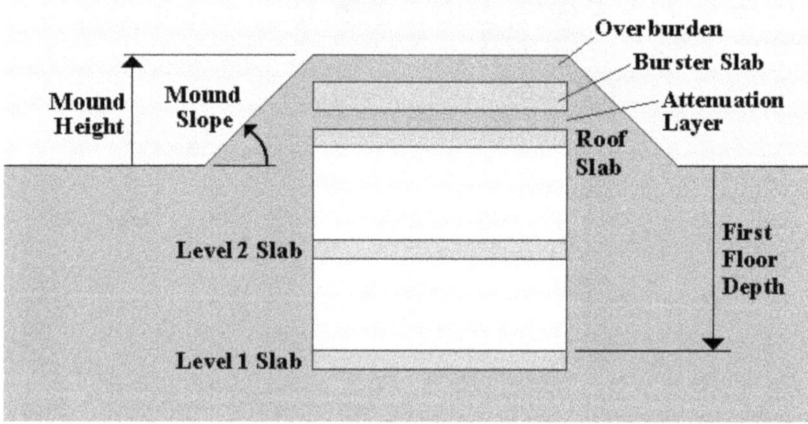

Fig. 15.2   Typical bunker features.

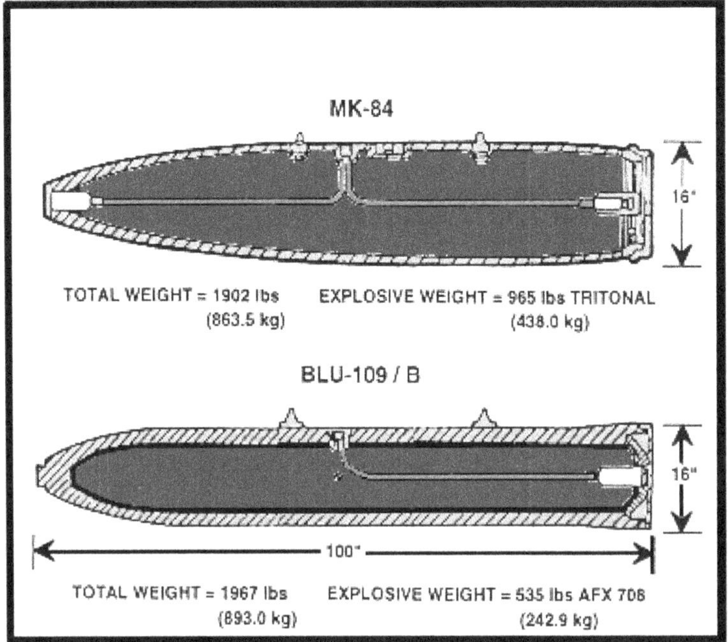

Fig. 15.3   Conventional and penetrator warhead comparison.

In general, targets of this type may be damaged in two ways: structural damage and functional damage. It should be noted that in many cases, it is only necessary to kill the function or activity performed inside the building rather than destroy the building itself. Several operational and analytical tools are available to the weaponeer to analyze the effect of conventional weapons on structures of different design and function. Such tools include

- Joint Munitions Effectiveness Manual (JMEM) legacy method based on effectiveness indices (EIs)
- JWS Fast Integrated Structural Tool (FIST)
- Integrated Munitions Effects Assessment (IMEA)
- Modular Effectiveness Vulnerability Assessments (MEVA)

These tools usually contain combinations of common methodologies for specific weapon–structure interactions, and the more complex are usually Monte Carlo in nature, with respect to the delivery accuracy of the weapon. Although this chapter will address only the first two of these methods, the following are common factors that need to be quantified before structural and functional damage may be assessed:

- A structural model of the building or bunker being attacked
- The impact conditions of the weapon on the resisting structure

- The penetration of each structural member the weapon encounters, if appropriate
- Possible modification of the weapon trajectory caused by penetration of a structural member or the material outside the structure
- The location of the weapon inside/outside the building when detonation occurs
- The effect of blast and fragments on the structure and its contents, including personnel

This chapter will develop methods for dealing with buildings in the following sequence.

1. The JMEM legacy method based on effectiveness index (EI) values
2. An overview of the FIST program
3. A simplified, unclassified version of FIST functionality

Following a discussion of buildings, the focus will move to bunkers and other hardened structures. In many cases the damage effects will be similar, but areas where they differ will be addressed in detail.

15.2 JMEM Effectiveness Index-Based Method

Building construction materials and methods vary considerably, and subtle changes in construction techniques can influence considerably the effect of a weapon on such a target. The early JMEM approach essentially correlates building damage to building type for a specific weapon based on battle damage assessment. This correlation is in the form of an effectiveness index called the MAE-Building, which is abbreviated MAE_{BLDG}. It extends sparse data for a particular building type to a range of building sizes and weapons. In this sense, the method is similar to that described earlier for bridges. The process begins by classifying buildings according to their building construction code, as indicated in Table 15.1.

Some examples of some building types from the JMEM pictorial library are shown in Fig. 15.4.

In the MAE_{BLDG} method, buildings are treated as three-dimensional targets with an effective miss distance (EMD). Because buildings are damaged

TABLE 15.1   CLASSIFICATION OF SAMPLE BUILDING TYPES

Building Type	JMEM Code
Multistory masonry load bearing wall	1c
Multistory framed—steel or concrete reinforced	1a
Single-story load-bearing masonry wall	4b
Single-story framed, column and slab construction	3b

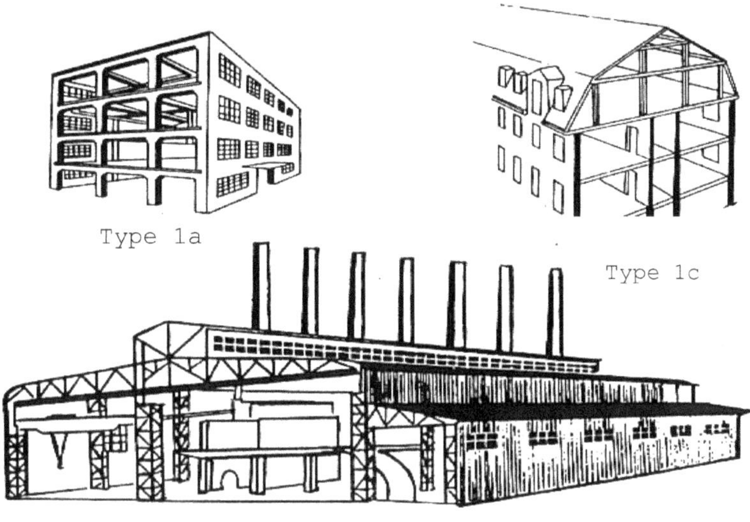

Fig. 15.4   Sample JMEM building types.

primarily by blast, the methodology for evaluating damage using MAE_{BLDG} is a modification of the EMD routines discussed in Chapter 12.

The effectiveness methodology is somewhat different for buildings compared to other target types dealt with in earlier chapters in that although the building is a unitary target, a fractional coverage calculation is made on the building footprint.

When MAE_{BLDG} is used to describe the damage function for buildings, the damage calculation represents the total floor area damaged, averaged over all the floors. In this method, P_{HD} is set to 1.0. Again, the numerical values of MAE_{BLDG} are classified and are a function of the building type and weapon. All MAE_{BLDG} values are assumed to be circular in the ground plane and are approximated by squares, regardless of the impact angle, which recognizes that damage is caused principally by blast.

The first stage of analysis is to represent the 3-D building as a 2-D rectangle in the ground plane by using the 3-D EMD approach of Chapter 12, reproduced in Fig. 15.5.

The method begins by comparing the MAE_{BLDG} EI value to the physical size of the building. Intuitively, if the EI is larger than the building footprint, it may be damaged by an external detonation, whereas if not, damage will occur only for an internal detonation.

Therefore, we start by calculating an initial value of L_{ET} from the following equation:

$$L_{ET} = \sqrt{MAE_{BLDG}}$$

$$(15.1)$$

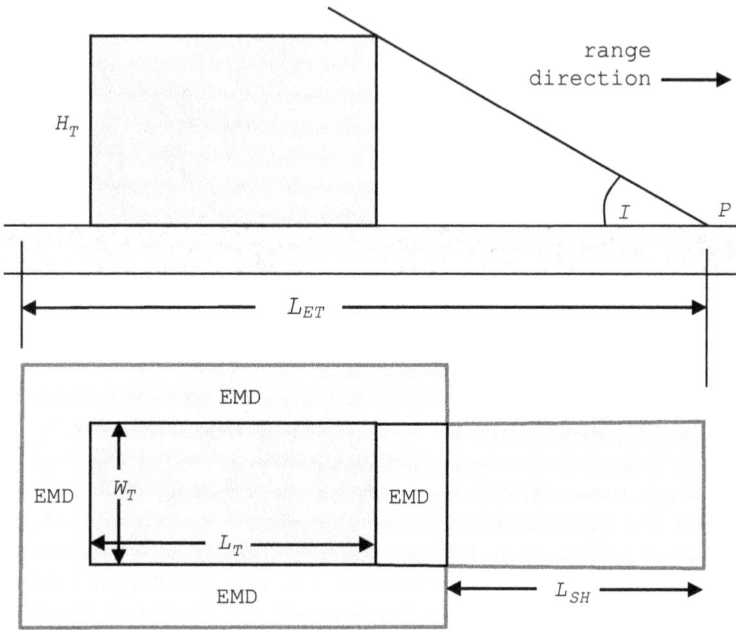

Fig. 15.5   Three-dimensional EMD geometry.

Then, if L_{ET} is greater than the length of the building, the equivalent miss distance (EMD) is computed using

$$\text{EMD} = \frac{L_{ET} - L_T}{2} \tag{15.2}$$

otherwise the EMD is set to zero. If the impact angle is 90 deg, no target shadow of the building height exists, and the shadow length and width (L_{SH} and W_{SH}) are set to zero. For all other impact angles, L_{SH} is computed as before using the following equation, where any negative values are set to zero:

$$L_{SH} = \frac{H_T}{\tan(I)} - \text{EMD} \tag{15.3}$$

Proceeding as for the EMD damage function, the new effective target length is now calculated from

$$L_{ET} = L_T + 2 \times \text{EMD} + L_{SH} \tag{15.4}$$

The effective target area in the ground plane A_{ET} shown in Fig. 15.5 is then computed using

$$A_{ET} = \left[L_T + 2\text{EMD}\right]\left[W_T + 2\text{EMD}\right] + L_{SH}W_T \tag{15.5}$$

and W_{ET} is computed from

$$W_{ET} = \frac{A_{ET}}{L_{ET}} \qquad (15.6)$$

Before calculating effectiveness from L_{ET} and W_{ET}, we demonstrate the process so far with an example.

EXAMPLE 15.1

Consider the building target shown in Fig. 15.6, which has dimensions of 100 ft (L_T) × 100 ft (W_T) and is 50 ft (H_T) high. Calculate L_{ET} and W_{ET} for the building if MAE_{BLDG} is 25,000 ft² and a weapon impact angle of 45 deg is specified. An initial calculation reveals that

$$L_{ET} = \sqrt{\text{MAE}_{\text{BLDG}}} = \sqrt{25,000} = 158.1 \text{ ft} \qquad (15.7)$$

The EMD value is

$$\text{EMD} = \frac{L_{ET} - L_T}{2} = \frac{158.1 - 100}{2} = 29.1 \text{ ft} \qquad (15.8)$$

Calculating the shadow length as

$$L_{SH} = \frac{H_T}{\tan(I)} - \text{EMD} = \frac{50}{\tan 45} - 29.1 = 20.9 \text{ ft} \qquad (15.9)$$

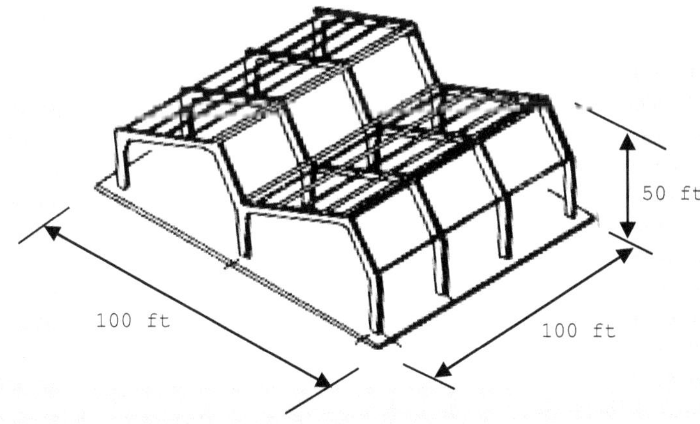

Fig. 15.6 Building target.

The effective target length is recomputed as

$$L_{ET} = L_T + 2 \times \text{EMD} + L_{SH} = 100 + 2 \times 29.1 + 20.9 = 179.1 \, \text{ft} \quad (15.10)$$

This makes the effective target area

$$A_{ET} = [L_T + 2\text{EMD}][W_T + 2\text{EMD}] + L_{SH}W_T = 27{,}094 \, \text{ft}^2 \quad (15.11)$$

From which we obtain

$$W_{ET} = \frac{A_{ET}}{L_{ET}} = \frac{27{,}094}{179.1} = 151.3 \, \text{ft} \quad (15.12)$$

Based on the bridge effectiveness index BEI and EMD effectiveness indices discussed earlier in this book, we might expect to calculate PD_1 as the probability of hitting this rectangle; however, the methodology for calculating building damage is somewhat different and will be a fractional damage calculation.

Given that L_{ET} and W_{ET} have been calculated, the process of determining the damage caused by a single weapon delivery begins by calculating the fractional coverage of the lethal area rectangle $L_{ET} \times W_{ET}$ over the target footprint $L_T \times W_T$. This may be accomplished using a modified version of the program in Table 10.1 in Chapter 10, which is shown in Table 15.2.

In this table, the items labeled as DPI offsets are not used, nor have they been defined; they will be described in more detail in Chapter 16. Recall this is for a single weapon attack. If multiple, independently aimed weapons are used, the single fractional damage FD_1 will be powered up for the number of weapons used.

$$FD = 1 - (1 - FD_1)^n \quad (15.13)$$

If we are considering a guided weapon where P_{HIT} is not zero, the previous calculations are performed twice for two different accuracies, one for the P_{NM} value and one where $\text{CEP} = 0$, and then combined in the usual way.

$$FD = P_{NM} \times FD_{\text{PNM}} + P_{\text{HIT}} \times FD_{\text{PHIT}} \quad (15.14)$$

Because buildings are more susceptible to blast than to fragments, MAE_{BLDG} reduces to MAE_B for vertical weapon impacts.

15.3 INTERPRETATION OF BUILDING DAMAGE

Building damage is a subjective measure, so it is difficult to translate a numerical value such as that produced in Table 15.2 into physical damage.

TABLE 15.2 FRACTIONAL DAMAGE TO BUILDING

MAE Building	
Inputs	
MAE_BLDG	25000
Impact angle	65
Building length	100
Building width	50
Building height	10
REP	50
DEP	25
DPI offset - range	0
DPI offset - deflection	0
Number of iterations	100000
Output	
Fractional damage	0.5583

Compute

Clearly a large value such as 0.95 would correspond to severe damage probably involving collapse, but what does 50% structural damage mean? Does the building collapse? Does whatever function being conducted in the building cease? There are very few data points to link quantitative damage estimates to real buildings, and this is an issue for most methodologies; however, the following examples might provide some guidance.

- The Alfred P. Murrah Federal building in Oklahoma City, shown in Fig. 15.7, was estimated to have sustained 30% structural damage and as a result had to be demolished. Clearly this level of damage ended any functionality in the building, though most of the structure still stood.
- Sometimes the commander's intent is to totally destroy a building and prevent any remaining structure from providing cover for combat troops. This is customarily associated with over 70% structural damage and basically reduces the building to a large area of flat rubble.

Exact building damage is very difficult to predict because some construction methods result in a particularly strong structure whereas small changes in design and codes can produce a weak structure.

Figure 15.8 shows the result of a relatively small gas explosion in a kitchen near the top of the apartment building that demolished a load-bearing wall, resulting in the apartments above it collapsing and then demolishing all those

Fig. 15.7   Building damage in Oklahoma City.

Fig. 15.8   Progressive collapse.

below it. This is called *progressive or disproportionate collapse* and is an example of the difficulty of predicting building collapse.

15.4 JMEM FAST INTEGRATED STRUCTURAL TOOL (FIST)

Although the MAE_{BLDG} method just described served the JMEMs well for many years, the requirement for an engineering-based rather than an empirical model emerged as buildings became a more significant target set for the operational user. This resulted in the Fast Integrated Structural Tool (FIST) program, which addresses buildings, bunkers, and eventually tunnel targets. In this section we will focus just on FIST's capability for determining building damage including external and internal detonations.

FIST is a Monte Carlo–based method that represents buildings in terms of their structural components and their connections. The detonation point is randomized by the delivery error, and the trajectory through the building structure is tracked to the detonation point. Damage to the building structure as well as internal damage to personnel and equipment are also calculated. FIST is a complex program, so a complete description will not be provided here; the *Advanced Weaponeering* textbook contains a more in-depth description.

15.4.1 TARGET MODEL

A target model may be selected from the JMEM database as for the MAE_{BLDG} method or developed using the Smart Target Model Generator (STMG).

For a typical building such as the one shown in Fig. 15.9, the STMG approach requires the user to input as much information about the building as

Fig. 15.9 Sample FIST building description.

is known—for example, the building function, number of stories, floor space dimensions (footprint), exterior wall material, estimated age, number and size of windows, basement, and so on. Using these data together with local building codes, FIST will estimate which building type fits closest and then determine the interior structure of the building including sizing the structural members.

Prior to any structural damage occurring to the building, the loads supported by all the vertical components (walls, beams, and columns) are determined using principles of structural mechanics. Clearly, the STMG must generate a building with sufficient structural integrity prior to sustaining damage by a weapon.

15.4.2 WEAPON IMPACT, PENETRATION, AND DETONATION

FIST actually has two modes available.

1. *Burst mode:* The user *places* the weapon inside the building, which is useful in determining whether the required amount of damage can be achieved if penetration puts the weapon at the chosen location.
2. *Aimpoint mode:* The user specifies an *aimpoint*, usually on the roof of the building, and the actual impact point is defined relative to the aimpoint by the delivery error.

For aimpoint mode, PC Effects is used to predict straight line penetration into the building, as shown in Fig. 15.10. Also for aimpoint mode, Fig. 15.11 shows the variation of impact point about the aimpoint for a specified circular error probable (CEP).

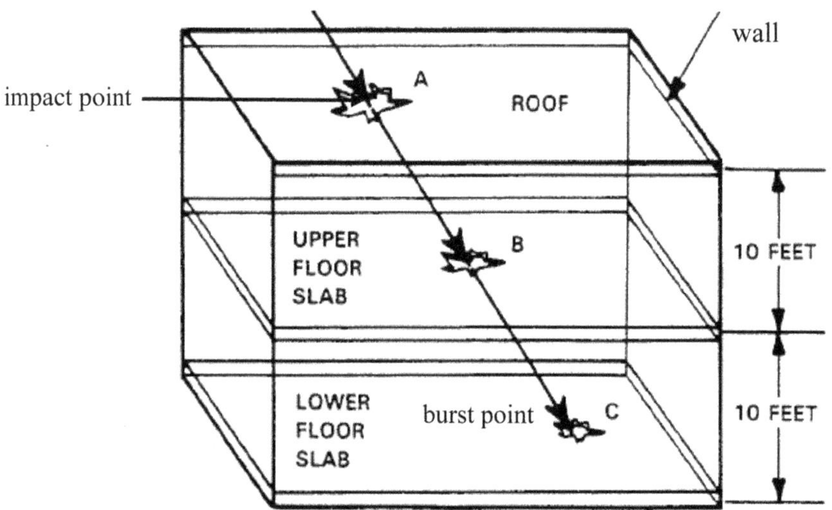

Fig. 15.10   Relationship between impact and burst point.

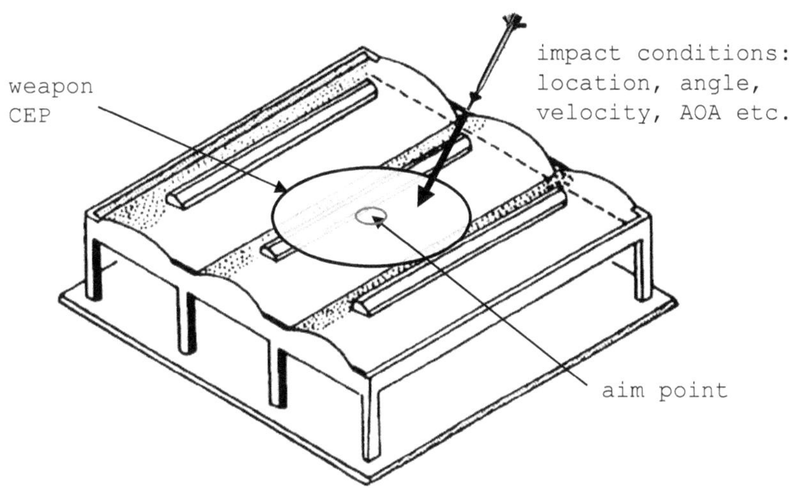

Fig. 15.11   Aimpoint and variability of impact point.

Damage to the building structure can occur due to several mechanisms: flexural failure, breach, spall, or ground shock. Each is discussed in the following sections.

15.4.3 STRUCTURAL DAMAGE DUE TO FLEXURAL FAILURE

Flexural failure is damage caused by vertical supporting elements such as columns or walls failing due to the pressure and impulse applied by an internal or external detonation. Figure 15.12 shows a floor plan of a building with the supporting columns and load bearing walls that support the structure above.

Note that nonstructural wall components are not shown because it is assumed that they do not significantly affect the pressure field inside the floor

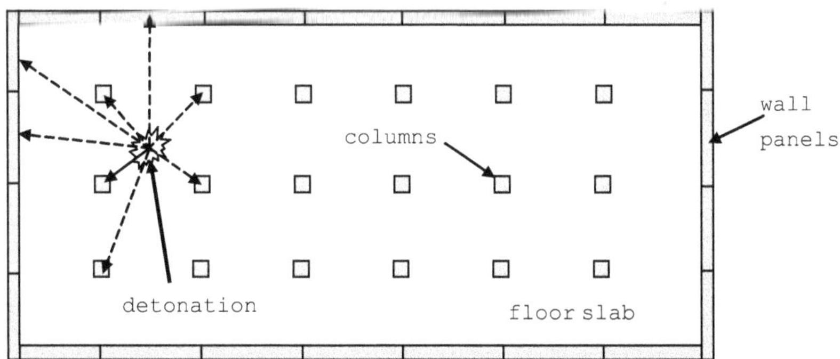

Fig. 15.12   Structural elements supporting a building.

space. Using the Kingery-Bulmash equations, the distance from the detonation point to all the structural elements and the pressure-impulse diagrams from FACEDAP for the columns, we can determine which ones fail.

15.4.4 STRUCTURAL DAMAGE DUE TO BREACH AND SPALL

If a weapon detonates close to a structural wall, the resulting pressure/impulse field imparted on it may cause a section of the wall to be projected away from the detonation point, as indicated in Fig. 15.13. Breach occurs if a hole is made in the wall, whereas spall refers to material on the side of the wall opposite to the detonation being projected into the building without making a hole. If enough material is removed from the wall, it may not be able to support the load it is carrying and may contribute to building collapse.

Instead of using FACEDAP equations for predicting breach or spall and the resulting damage, FIST uses experimental data to predict breach and spall, a sample of which is shown in Fig. 15.14. In this figure, the impacting impulse and wall thickness are used to predict the likelihood of breach and/or spall. Other data are used to estimate the amount of material removed.

Damage is defined as the ratio of volume of material removed divided by the volume of material in the undamaged slab. The following rule is then applied: if this ratio is >75%, then the component is fully damaged and removed from the gravity load path, unless the component is a beam or column, in which case the maximum damage before failure occurs is limited to 30%.

15.4.5 STRUCTURAL DAMAGE DUE TO GROUND SHOCK

Ground shock is the generation of seismic waves in soil due to an underground detonation and may cause damage due to ground motion in the soil supporting structural components. The approach used in FIST is to use a finite element code called FLEX to determine the damage to a particular

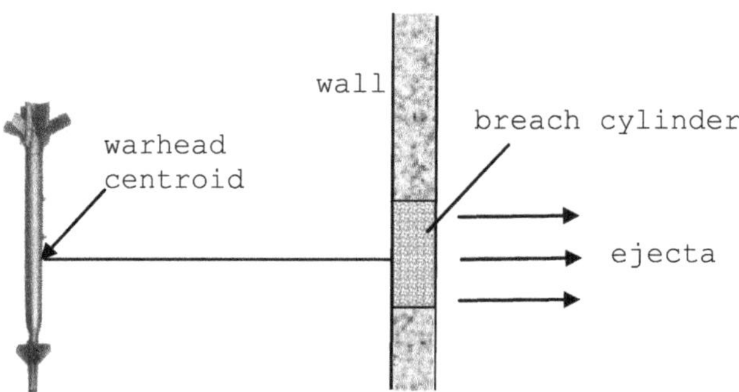

Fig. 15.13   Breach/spall/warhead geometry.

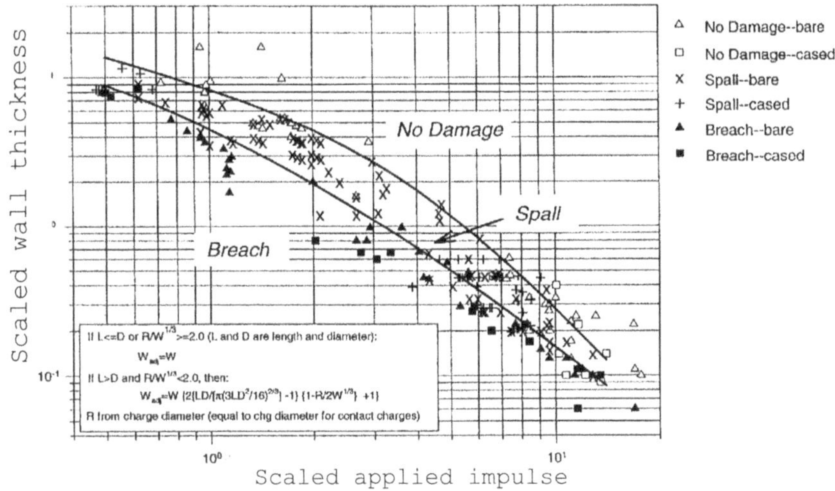

Fig. 15.14   Breach and spall for bare cylindrical charges.

structural member caused by a specific detonation scenario (i.e., warhead size and standoff). This analysis is summarized in the response surface shown in Fig. 15.15.

In this scenario, the weapon may penetrate through the building, pass through an external wall, and detonate in the soil. Alternatively, the weapon may miss the building altogether but be sufficiently large and close enough to still cause damage. For a single Monte Carlo iteration in FIST, the detonation point relative to the building may be determined and the damage calculated from Fig. 15.15.

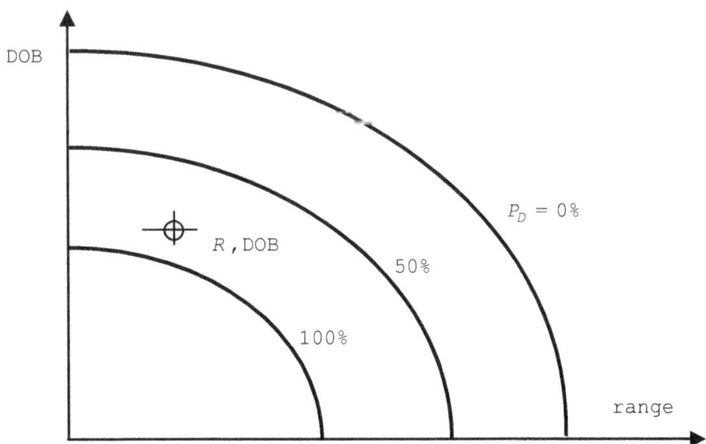

Fig. 15.15   Response surface for component failure.

15.5 BUILDING STRUCTURAL INTEGRITY

Damage to individual components through flexural failure, breach, spall, and ground shock are aggregated to determine which structural components fail. Once this is done, an assessment is made regarding building collapse by removing failed components and redistributing the load path from the top of the building to the ground.

Consider the simple structure shown in Fig. 15.16 consisting of a 6000-lb floor slab supported by four symmetrical columns. Given the symmetry, each column will be under a compressive force of 1500 lbf each.

Now consider what happens when one of the columns fails and is removed from the load path, as shown in Fig. 15.17. One can see that the two remaining columns have their loads doubled to 3000 lb, which may or may not produce failure. The process is therefore an iterative one, in which the initial damage results in a redistribution of loads, increasing the stress in the remaining columns, which may cause them to fail, resulting in a recalculation of loads, and so on.

This is a shortcoming of the FIST methodology. It is important to recognize that the redistribution of loads does not involve equations of force and moment equilibrium on the surviving components as might be expected, because this

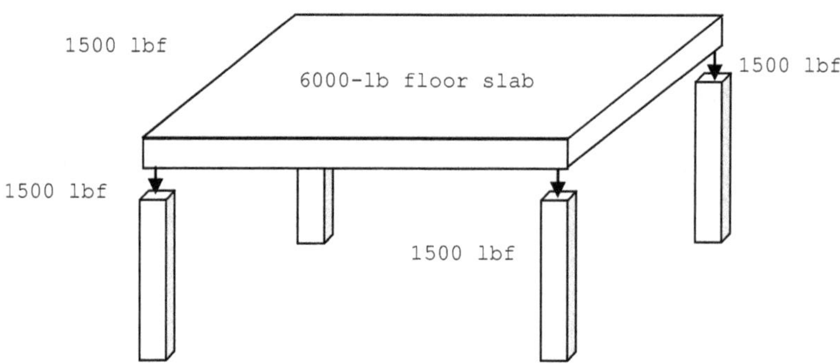

Fig. 15.16   Columns supporting a floor slab.

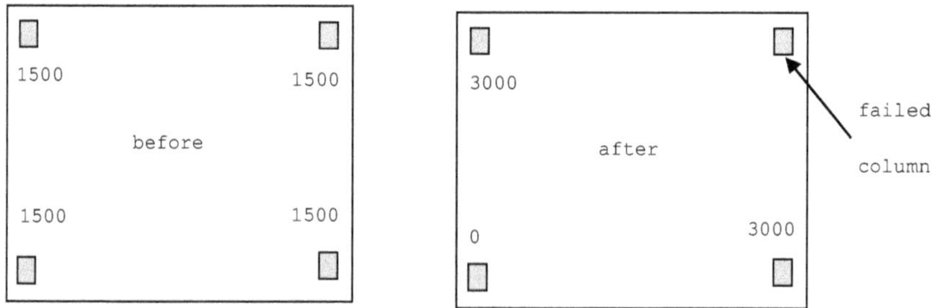

Fig. 15.17   Redistributing loads after support damage.

would take too long to compute. Instead, the redistribution is based on assumptions that the loads are "balanced" or symmetrical. This may be seen in the redistributed load values shown in the three columns in Fig. 15.17 where one might reasonably expect some finite load in the lower left column.

Structural damage, or kill, may be defined using two different methods.

- The percentage of surface area removed (floors, walls)
- The percentage of structural volume removed

Because of the different metrics, significant differences in reported values can be observed for an equivalent amount of damage. For example, consider a two-story building with no interior walls, such as that shown in Fig. 15.18. The target dimensions are 200-ft wide × 150-ft long × 45-ft high. The target has 2-ft-thick concrete roof and floors, and 3-ft-thick concrete exterior walls. The target is attacked with a 2000-lb penetrator, 60-ms fuzing, aimed at the center of the target. This weapon is expected to penetrate the roof and detonate inside the top floor, causing the center floor to collapse. All other members do not fail.

For the area approach, the undamaged surface area of each room is computed as follows:

- Side wall surface area $= 22.5 \times 150 = 3375$ ft^2 ($\times 2$)
- Face wall surface area $= 22.5 \times 200 = 4500$ ft^2 ($\times 2$)
- Floor and ceiling area $= 150 \times 200 = 30,000$ ft^2 ($\times 2$)
- Total room surface area $= 75,750$ ft^2

If the central floor is the only structural member that fails, then the damaged surface area is 30,000 ft^2 for both the top and bottom room. The percentage of structural damage for each room is thus $30,000/75,750 = 40\%$. Because both

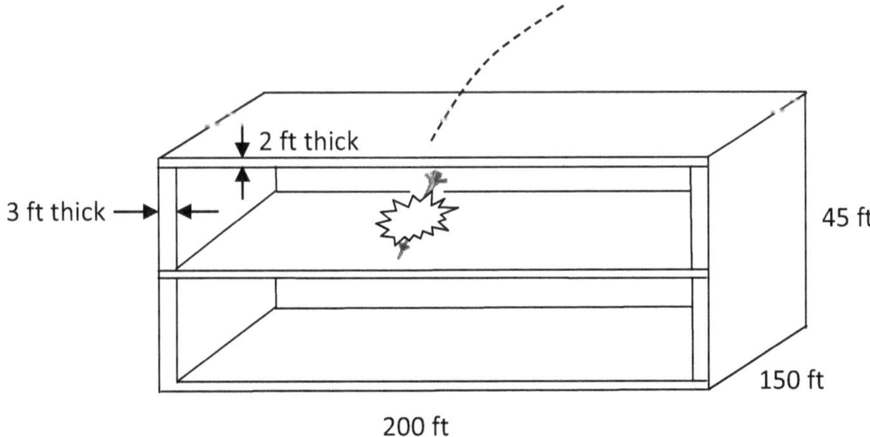

Fig. 15.18   Two-story building model.

the top and bottom floor experience identical damage and have the same surface area, the reported area structural damage would be 40%. Note how the central horizontal slab appears to be counted twice as the floor of the upper room and ceiling of the lower room. This double counting is appropriate because loss of function in both upper and lower floors of the building will occur.

Using the volume approach, however, the structural volume of the target is computed as follows:

- Side wall volume $= 45 \times 150 \times 3 = 20{,}250$ ft^3 ($\times 2$)
- Face wall volume $= 45 \times 200 \times 3 = 27{,}000$ ft^3 ($\times 2$)
- Floor and roof volume $= 150 \times 200 \times 2 = 60{,}000$ ft^3 ($\times 3$)
- Total structural volume $= 274{,}500$ ft^3

If the central floor is the only member destroyed, then the structural volume removed is 60,000 ft^3. The percent volume structural damage for the target is thus $60{,}000/274{,}500 = 22\%$. It is important to recognize that when the concept of volume is discussed, it is the volume of material in the walls, floors, and ceiling that is being referred to, not the internal volume of the structure. Notice the difference in damage measurement: 40% area vs 22% volume.

15.6   *FIST Functional Kill*

So far, structural damage has been the analytical focus, but as mentioned before, in many cases preventing the activity, or function, usually performed in the building may be sufficient. In order to determine this functional kill, information regarding the type, number, and location of critical pieces of equipment needed to perform the function has to be known. At the time of writing, FIST supports the following equipment:

- Primary power and backup power
- Computer equipment
- Communications equipment
- An industrial air compressor
- Transformer equipment
- Machine tools
- Satellite dish placed on roof or outside building

In addition, a functional model of how individual equipment failure affects the overall function must be specified. Such a model is termed the failure analysis logic tree (FALT) or a failure modes, effects, and criticality assessment (FMECA). This is the same process discussed in Chapter 6 regarding materiel target vulnerability (i.e., a physical representation of the components making up the target and a fault tree for their contributing to a failure mode).

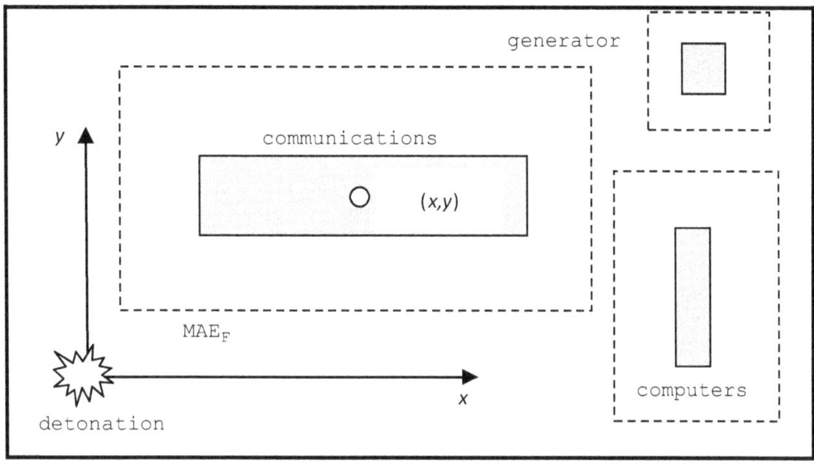

Fig. 15.19   Sample layout of equipment in a building.

For the equipment listed, and for typical weapons used against buildings, the MAE_F is precalculated using General Full Spray Materiel (GFSM) or Joint Mean Area of Effectiveness (JMAE), as described in Chapters 7 and 8, and stored in FIST. The process used to determine equipment failure is as follows:

1. The user places the selected equipment in its correct location, if known, within the building, as shown in Fig. 15.19.
2. For the weapon and target (equipment), select the appropriate MAE_F and represent it by a rectangular cookie cutter.
3. If the weapon detonates inside the cookie cutter, the piece of equipment is damaged; otherwise, it survives.
4. Using the FALT tree, determine if a particular function inside the building (computer system, manufacturing capability, etc.) is stopped.

Clearly, considerable intelligence data are required to be able to construct interior floor plans with this level of detail.

15.7   FIST PERSONNEL INJURY ASSESSMENT

FIST is not intended to assess collateral damage to personnel inside buildings; however, some estimates of personnel injuries are available in the program.

The user begins populating the building with personnel targets by specifying a personnel density appropriate to the building use and time of day. Some assistance may be obtained by using population reference tables from the Oak Ridge National Laboratory (ORNL). This LandScan™ Dataset comprises a worldwide population database and was compiled on a

TABLE 15.3   NOMINAL POPULATION DENSITY DATA

Category	Daytime Population Density (per 1000 ft²)	Nighttime Population Density (per 1000 ft²)
Single-family house	1	3
Restaurant	3	20
Store	4	0
Airport/train/bus station	5	3
School	6	0
Office building	5	0

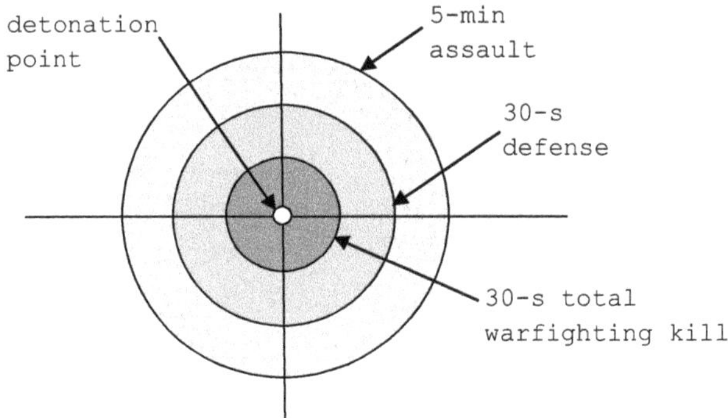

Fig. 15.20   Mean area of injury (MAI).

30-in. × 30-in. latitude/longitude grid. An example of some of these data is illustrated in Table 15.3.

Multiplying the population density and floor area, the total number of personnel in the building is calculated. These personnel may be placed by the user in selected rooms or uniformly distributed over the whole building. Personnel injury is assessed at several levels from blast, fragments, or both by generating MAE_F values, now called a mean area of injury (MAI), from GFSP and representing them as circular cookie cutters, as shown in Fig. 15.20.

The distance from the detonation point to each of the placed personnel targets is calculated, and the number of personnel injured to a specified level is aggregated over the whole of the building.

15.8   SIMPLIFIED BUILDING DAMAGE METHODOLOGY

Although the methods described in the previous sections outline those used in the JWS FIST program, this is of limited distribution and unavailable to many users. However, because the components of the method are available in open literature, it is relatively simple to construct a similar methodology using these components in order to demonstrate FIST's functionality. This is the approach used by Tan [1] and will be explained in this section. The main features of the approach are as follows:

- The user specifies the size of the building by length, width, and height. It is assumed the building is of commercial type.
- Only a few of the more common building types are considered (slab and column, waffle roof), and architectural and U.S. building codes are applied to size the principle components and their placement.
- A grid of 1-ft × 1-ft elements is placed on the floor and a single weapon considered to land in the center of each element in turn.
- The weapon warhead is analyzed to determine the equivalent bare charge explosive weight of TNT.
- For each location of the weapon in the grid, the distance to all columns is determined.
- Using Kingery-Bulmash equations and knowing the distance between each column and the point of detonation, the pressure and impulse loading is determined for all columns on that floor.
- Using FACEDAP pressure-impulse (P-I) curves, the number of columns that fail is determined.

The result is a map of the grid points showing the number of failed columns if the warhead detonated in that cell. Some simple statistics show the probability of one, two, three, or more failed columns for any warhead placement.

Most of these topics have been considered in other chapters, except the building analysis part, so this will be discussed next.

15.9 BUILDING ANALYSIS

The most common construction technique used in the United States for commercial and industrial-use buildings consists of successive floor slabs supported by vertical columns. An example of such a building is shown in Fig. 15.21.

Common variations might include reinforced stairwells or elevator shafts (as illustrated in the figure) or additional support for the floor slab by beams bridging the columns. Whenever a building like this is constructed, sizing the structural components (floor and column) is not performed from first principle for every building; instead, design charts and graphs that are known to work for a particular type of building are used.

Typically, the thickness of the floor slab is defined for whatever use the building is intended (residential, commercial, light manufacturing, etc.), and the resulting live and dead loading will determine the number of columns and their cross-sectional dimensions and spacing.

Fig. 15.21   Typical slab and column construction.

To provide some variation from the simplest slab and rectangular column, Tan [1] included a waffle slab floor and beam-supported flat slab methods designed to support heavier, commercial facilities, such as might be needed for machine tools, computer server facilities, HVAC equipment, auxiliary generators, storage, or high occupancy offices. Examples of these are shown in Fig. 15.22 and 15.23.

These components are usually constructed on site and have pre-tensioned steel reinforcing bars for additional strength. Considering the building shown in Fig. 15.23, it might seem that the columns nearer the ground should be thicker and more numerous because they have to support a greater weight above them; however, for reasons of economy it is usual to make all columns the same. Similarly, columns are uniformly distributed across the floor, rather than placing more in areas known to have a larger load in one region. Components therefore are sized for the largest predictable loading, leading to some overdesign for other cases or locations.

Fig. 15.22   Waffle slab floor/ceiling.

Fig. 15.23   Column and beam construction.

The process of sizing the components tends to be somewhat iterative, because building function will play a part—for example, in selecting the total number of columns. These may have to conform to a minimum spacing in order to have enough space between them to allow unobstructed space of offices, machine shops, computer facilities, parked vehicles, and so on. A typical design chart from [2] for determining the cross section of a column is shown in Fig. 15.24.

In this chart, the tributary area is the total area of the building supported by a single column. For a three-story building with a footprint of 100 ft × 100 ft, the total floor area will be 30,000 ft². If it is supported by 30 columns, the tributary area is 1000 ft² (i.e., total floor area divided by number of columns). The graph shows a range of column sizes depending on the building function, from about 8 in.² for residential to 14 in.² for heavy industrial use. Span spacing is determined from Fig. 15.25.

From this graph, and an assumed column size of 12 in., the maximum span will be between about 27 ft and 45 ft, depending on the method used to reinforce the concrete. If we consider the minimum span of 27 ft and round it down to 25 ft, then along one side (100 ft) of the building, there will be five columns with a column at each corner and 25 ft between interior elements. Because the building is square, there will be a total of 25 columns, and because we allowed 30 columns in the initial design, this is acceptable for the anticipated loads.

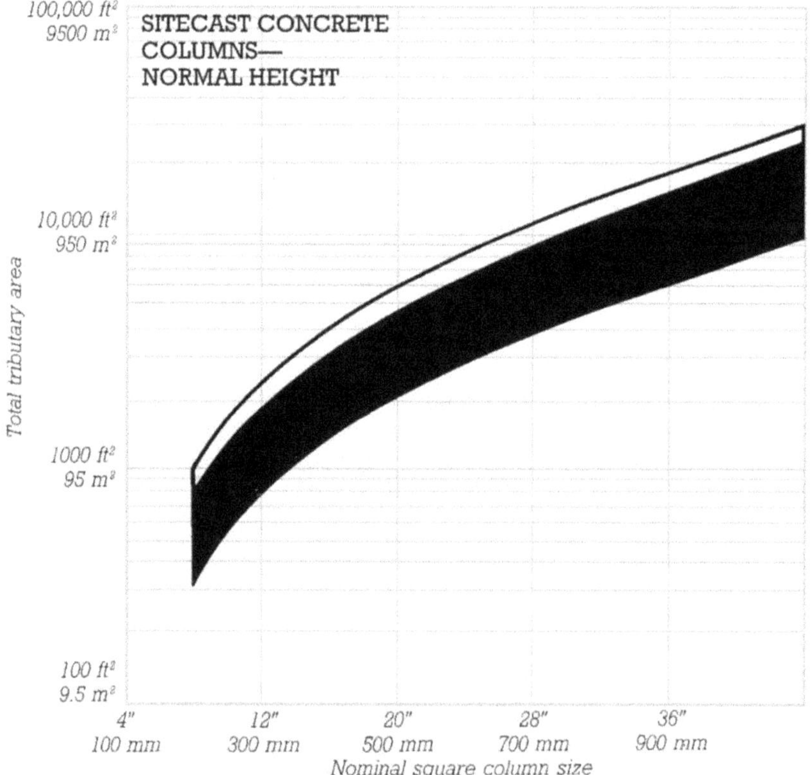

Fig. 15.24   Design chart for column size [2].

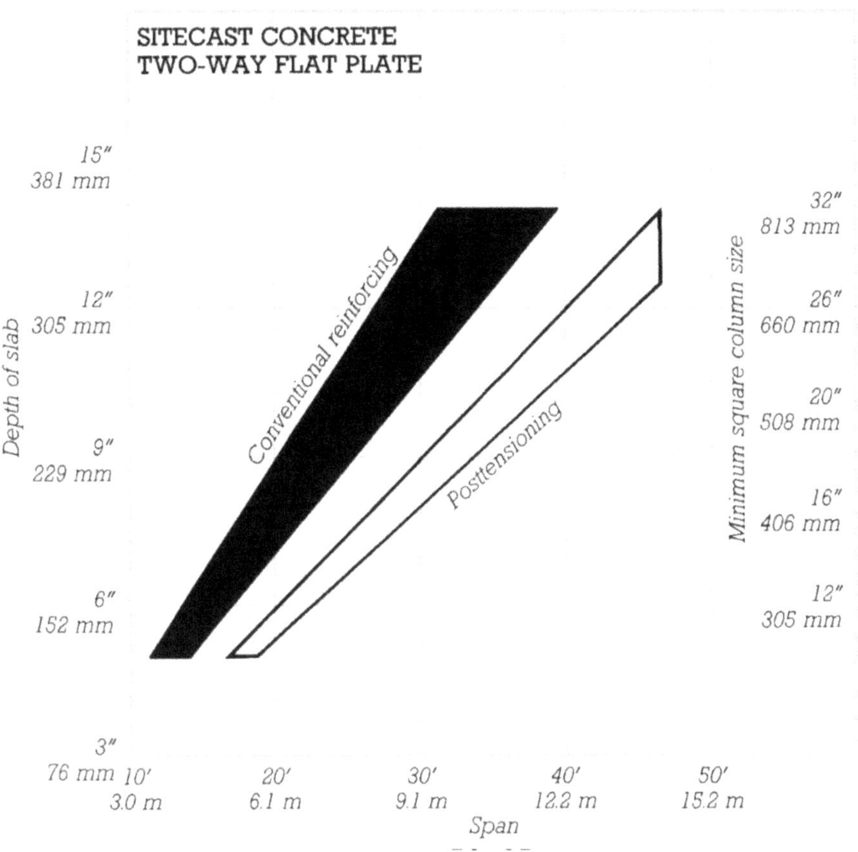

Fig. 15.25   Design chart for column spacing [2].

Therefore, if the function of the building (which determines slab thickness) and the total floor area (footprint and number of floors) is known, then the cross-sectional dimensions of the columns and their spacing (hence number of them) can be determined. Increasing the number of columns to match the floor area will reduce the spacing and therefore provide a factor of safety in the design.

15.10   BLAST FIELD ANALYSIS

In this section we will focus on a common weapon used against buildings: an Mk-84 2000-lb bomb body that usually uses some kind of guidance system (laser or GPS) to impact the building in the center of the roof with as near a vertical impact angle as possible. Fuze setting will determine on which floor detonation occurs.

We begin by using the modified Fano equation to get the equivalent uncased weight of explosive converted to TNT for this weapon using Eq. (14.8) from Chapter 14. Given the bomb contains 945 lb of Tritonal, we get

$$W_U = K \times W \left[0.6 + \frac{0.4}{1 + 2M/c} \right] = 731.7\,\text{lb} \qquad (15.15)$$

We now use the Kingery-Bulmash polynomial fit from Table 14.2, which takes the general form of Eq. (15.16).

$$F = \exp\left[A + BT + CT^2 + DT^3 + ET^4 + FT^5 + GT^6\right] \qquad (15.16)$$

We need the specific impulse and pressure generated by the blast wave at a given distance from the detonation point, and this involves the parameter T obtained from the scaled distance Z.

$$T = \log_n\left(Z\right) = \log\left(\frac{x}{W^{1/3}}\right) \qquad (15.17)$$

For example, using the data in Table 14.2, at a distance of 50 ft from the detonation point, the incident pressure and impulse are $P_{SO} = 160$ psi (10.9 bar) and $i_S = 0.2252$ psi-s, respectively.

15.11 STRUCTURAL FAILURE ANALYSIS

The FACEDAP methodology is used for this phase. Note that one of the structural members for which P-I data are available is for a reinforced concrete (RC) interior column, as shown in Fig. 15.26.

This enables us to determine whether a column located a given distance from the detonation point will fail.

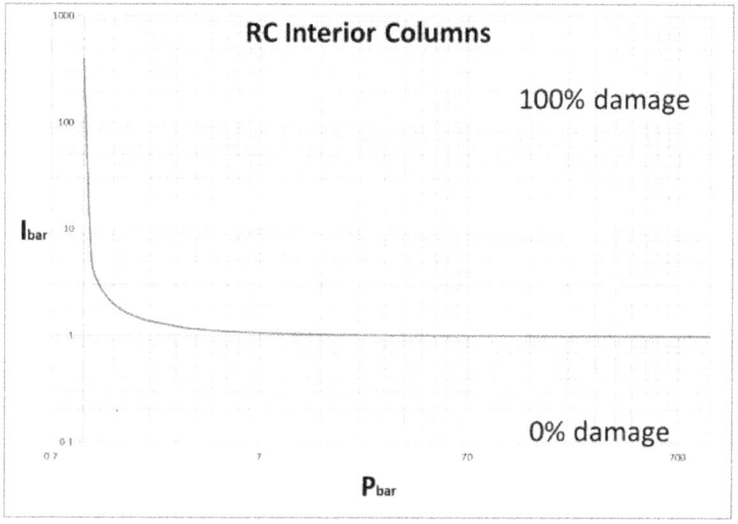

Fig. 15.26   Pressure-impulse failure curve for RC column.

15.12 DAMAGE ASSESSMENT

This process is relatively simple and involves laying out a 1-ft × 1-ft grid on the floorplan of the building to define cells, placing the warhead in each cell in turn, and calculating how many columns will fail. This is shown for a simple structure with four supporting columns in Fig. 15.27.

As the warhead is placed in successive cells, the number of columns that fail for that detonation location is calculated and entered into the cell to produce the lethality map shown in Fig. 15.28.

For this case, detonation anywhere near the center of the floor will cause failure to all four columns and presumably building collapse. For larger footprints, the lethality map becomes repetitive, as shown in Fig. 15.29 for a building with floor slabs supported by 32 columns in an 8 × 4 grid.

Once this lethality map has been derived, an aimpoint may be selected and a detonation point obtained by a random sample about the aimpoint using the delivery error. Averaging over a large number of trials will give the expected number of columns that fail. Once again, the user is left with the difficult task of deciding whether sufficient damage has been done to cause partial or total damage to the building.

Clearly this approach has many limitations and simplifying assumptions, some of which may be seen as follows:

- No shielding of columns by other columns is considered, and the blast is considered to propagate unhindered by interior walls.
- The buildings are of simplistic commercial construction with no attempt at hardening.
- Building collapse is subjectively determined by the user based on the number of supporting columns destroyed.

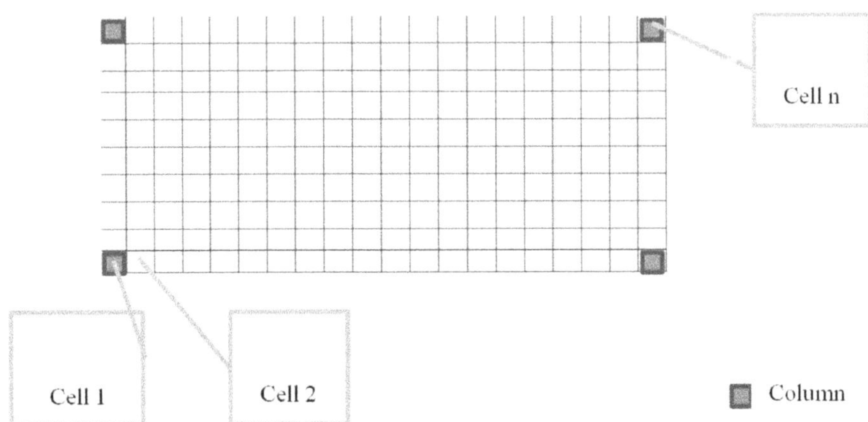

Fig. 15.27   Simple building design with four columns.

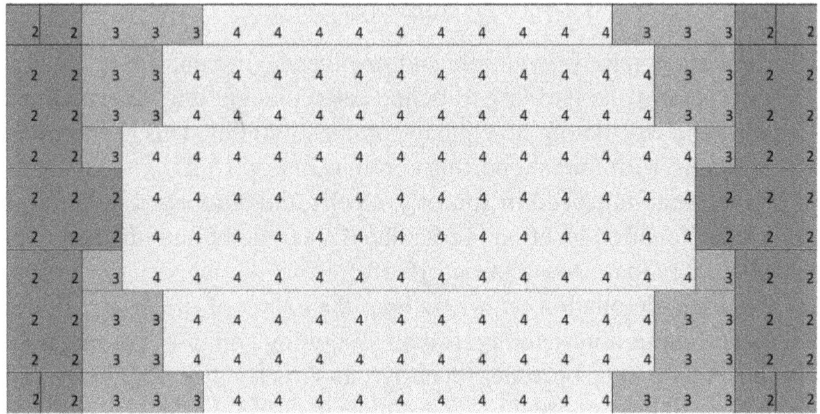

Fig. 15.28   Column failure for discrete detonation points.

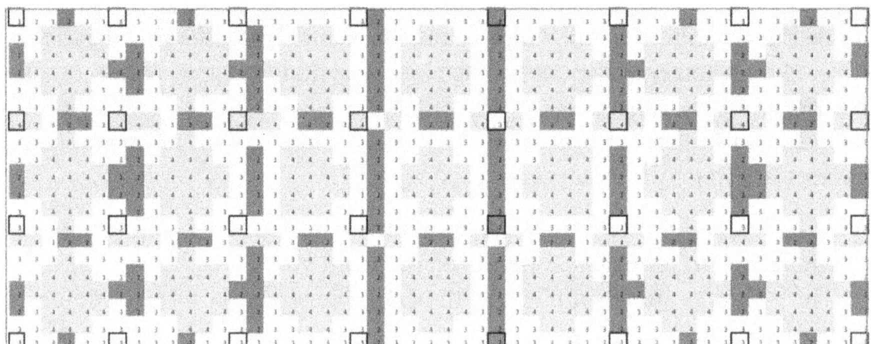

Fig. 15.29   Lethality map for larger building.

With regard to the last point, it should be emphasized that unless precise structural details of a specific building are known, predicting collapse is a difficult task; even then, it may not be obvious what the result of a detonation will be.

15.13   CASE STUDIES OF BUILDING DAMAGE

There are few cases where precise details of damage to buildings due to explosive detonations have been analyzed. Two prominent examples in the United States are the Alfred P. Murrah Federal Building in Oklahoma City and the World Trade Center towers in New York.

15.13.1   ALFRED P. MURRAH FEDERAL BUILDING

A charge of approximately 4000 lb TNT equivalent was detonated approximately 20 ft from the front of the Alfred P. Murrah Federal Building [3] (see Fig. 15.30). In common with many office buildings, the lobby area was higher than one floor; it extended two stories vertically, with the remaining floors supported by transfer beams spanning the lobby. The ground floor layout is shown in Fig. 15.31. Following the detonation of the bomb, the three columns G16, G20, and G24 shown in the layout plan in Fig. 15.31 failed, causing the transfer beam they were supporting to fall and floors three through nine to collapse.

Fig. 15.30   Damage and progressive collapse of Alfred P. Murrah building.

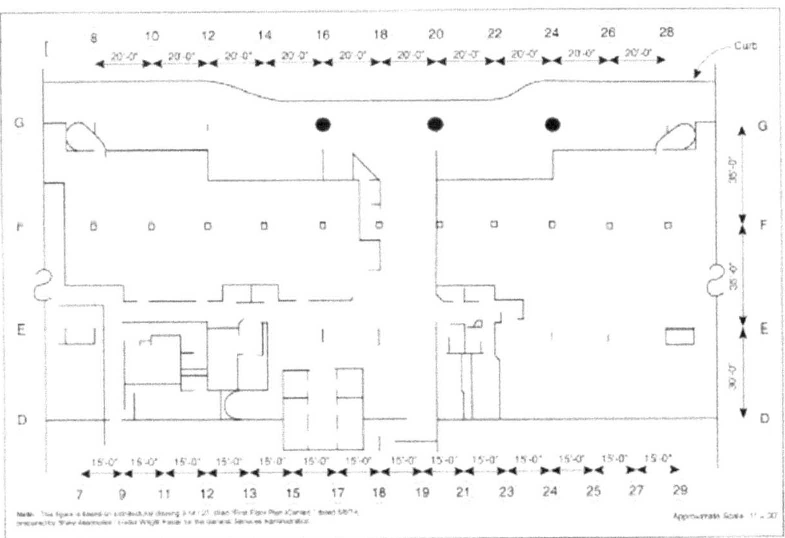

Fig. 15.31   Structural layout of first floor of the Alfred P. Murrah building.

The four substantial corner columns prevented the whole building from collapsing. The structural damage was limited to about a third of the building, which is where the majority of casualties occurred.

15.13.2 WORLD TRADE CENTER BUILDINGS

The design of the two World Trade Center (WTC) buildings consisted of an inner core of 48 large columns, housing services such as stairwells, elevators, and utilities, surrounded by an outer core of 240 columns—basically a box within a box—with trusses joining the two, as shown in Fig. 15.32.

Based on the impact imagery of the sides of the buildings (shown in Fig. 15.33), it was estimated that the aircraft hitting the North Tower destroyed 31–36 outer columns (14%) and an unknown number of inner columns. In the South Tower, 23 outer columns (10%) were destroyed; the aircraft missed the inner columns completely.

The buildings did not collapse immediately after destruction of the supporting columns. Subsequent investigations found the buildings' collapse was due to the failure of the connecting steel trusses due to the effects of fire [4]

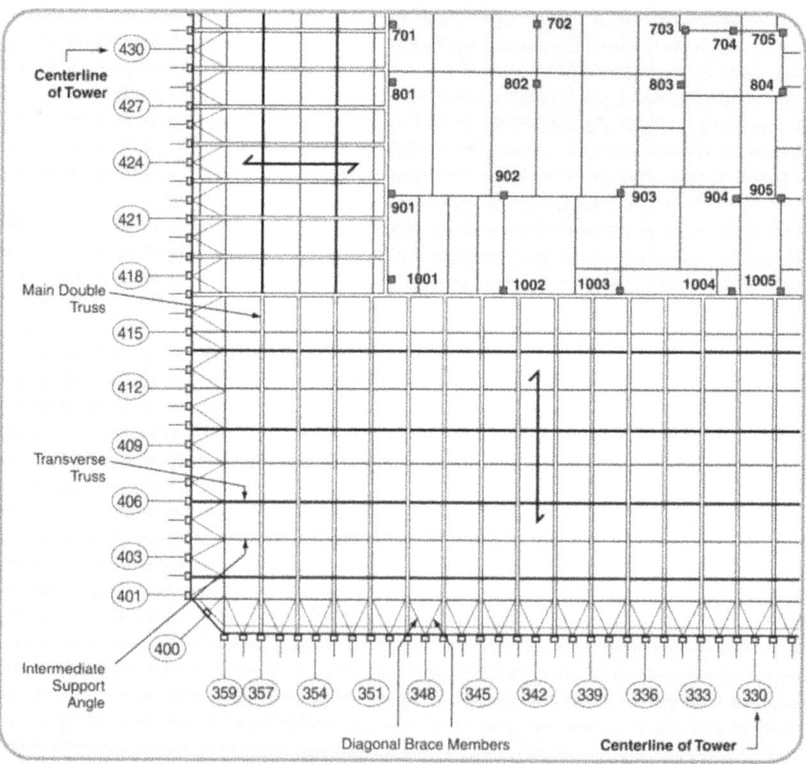

Fig. 15.32   Structure of World Trade Center (corner).

(see Fig. 15.34). This left the core and outer tube essentially unbraced, and any small asymmetric loading would cause collapse.

The lesson from these two building collapse examples is that it is difficult to predict whether collapse will occur unless detailed structural analysis of the actual building is conducted. From a weaponeering perspective, this is usually impractical because the data are not usually available, but even in the cases where they are, the analysis is neither simple nor fast.

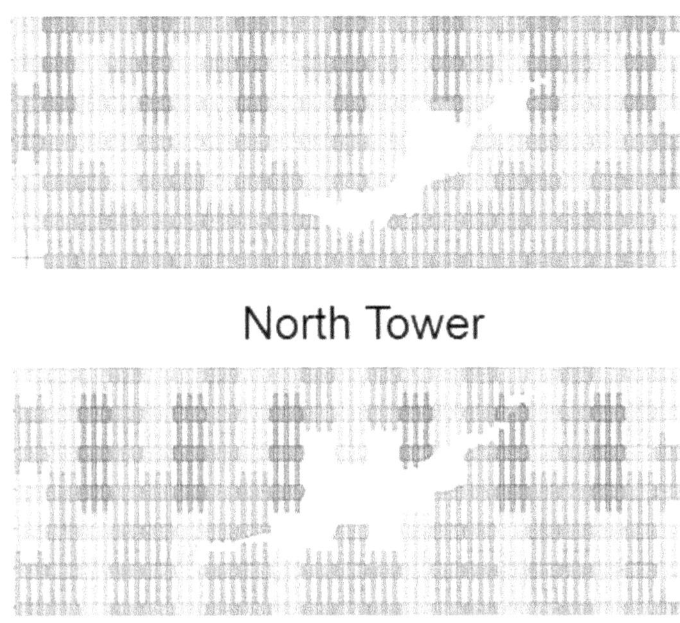

Fig. 15.33   Impact imagery analysis of WTC buildings.

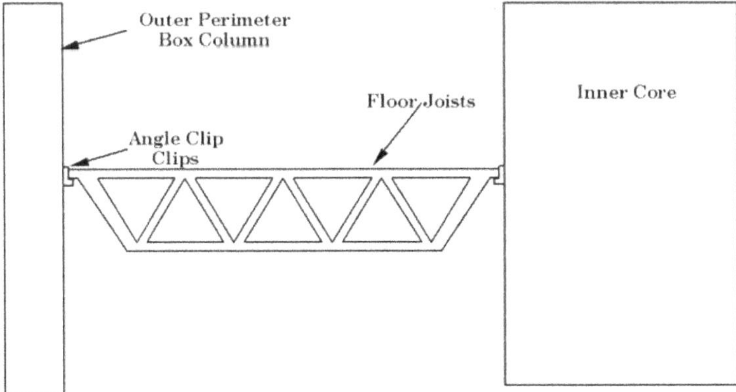

Fig. 15.34   Truss connecting core to outer tube.

15.14 BUNKERS AND HARDENED TARGETS

As mentioned before, hardened bunkers present significant challenges to the weaponeer. It is much easier to build the bunker deeper and cover it with more concrete than it is to design, build, and deliver a weapon to overcome such construction methods.

Calculating the effect of weapons against bunker targets is similar to buildings, allowing users to move easily between the different target types. For this reason, a direct comparison with previous material will be made in the discussion of the methodology behind bunker defeat.

Although they have fewer variants in construction methods, functionally bunkers are treated in a similar manner to buildings, that is,

- A target model is generated.
- Weapon penetration may be analyzed, or a specific burst point is defined.
- Structural and function kill levels are determined.
- A Monte Carlo simulation samples the various components of delivery error, calculates the damage for each sample, and averages over a sufficiently large number of samples.

Most of the differences in weaponeering a bunker compared with a building are in the construction details of the target model. In FIST, the basic bunker model is generated using the IMEA bunker generation algorithms, which may be characterized as follows:

- Predefined bunkers may be imported. These may be of specific target types where the dimensions and construction details are fixed, or generic in nature and are user specified. These include number of rooms and sizes; wall, floor, and roof slab thicknesses; and so forth.
- Targets may be user defined from the Modernized Integrated Database (MIDB)/Physical Vulnerability (PV) CHAR target model database in which seven bunker types are available in FIST.
- Targets may be user defined without reference to the MIDB/PV CHAR database. There are a number of layout footprints, as shown in Fig. 15.35. Within these shapes, the number of rooms, dimensions, and function are fixed.
- Except for some predefined targets, the user may specify the construction materials (masonry, RC concrete, wood, and metal), although hardened targets are usually constructed with reinforced or pretensioned concrete. Column and beam construction methods are also available as for buildings.
- Openings between rooms and the outside are available including several types of protective doors. These openings may be in walls or floors. As shown in Fig. 15.2, berms and burster slabs may be specified.
- Although conventional building codes are not employed for bunker design, FIST does use some design guidelines; for example, specifying the size of

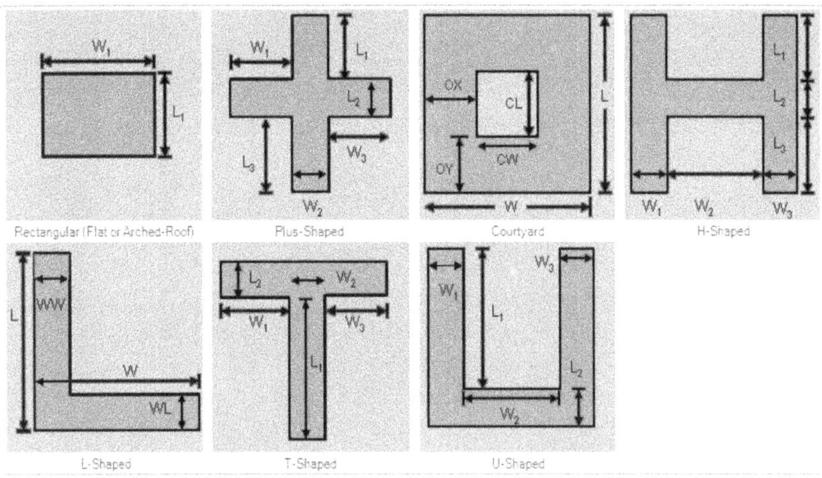

Fig. 15.35   Available bunker shapes in FIST.

a floor slab will determine if it can be supported by the exterior walls or if columns and beams are required. Spacing and size of any rebar is also estimated.

- Because FIST uses the IMEA bunker generation algorithms, any bunker designed in IMEA may be imported into FIST. A sample bunker target is illustrated in Fig. 15.36.

Once the target model has been defined, the damage assessment proceeds in a similar manner as for above-ground buildings: penetration using PC Effects, ground shock and cratering for external bursts, and soil breach of interior and exterior walls and floor.

It should again be recognized that this description relates to the tool at the time of writing. As for any target set of particular interest to weaponeers, the capabilities of the tool will change over time to include higher fidelity and greater capability as computational resources improve.

15.15   CHAPTER SUMMARY

- Various methodologies have been discussed to determine the amount of damage sustained by a building when a weapon detonates inside or outside. These methods include the legacy MAE_{BLDG} approach, the current JMEM FIST methodology, and a simplified structural model exhibiting many of the FIST attributes.
- Underground bunkers may be treated in the same manner as buildings; however, the details of the structure, such as sizing of structural components, has to be specified by the user rather than relying on accepted building codes.

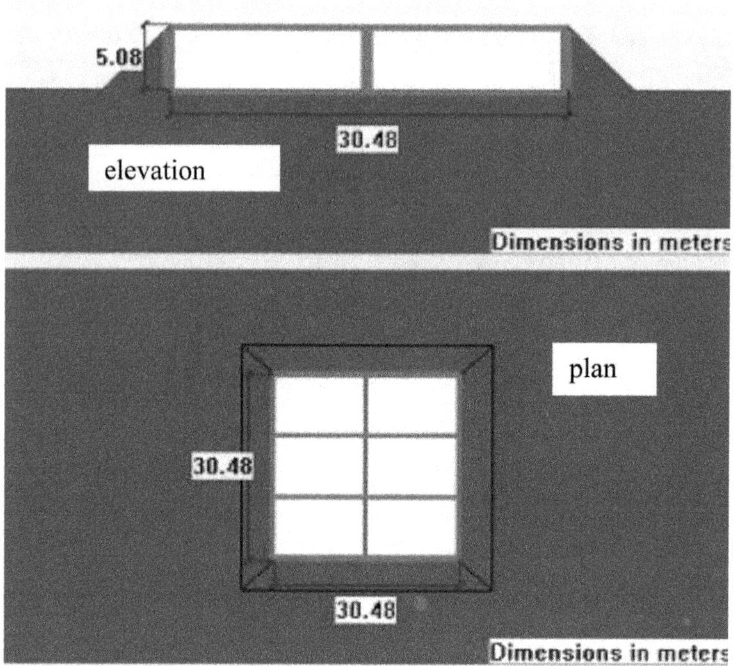

Fig. 15.36   Default bunker target specification.

- A common theme in these analyses is to recognize that accurate prediction of damage, both internal and external, or prediction of collapse is very difficult and is still an active research and developmental topic. A restrictive factor is that for weaponeering applications, a result must be computed quickly, which presents significant challenges for what is a very complex engineering problem.

REFERENCES

[1]   Tan, C. M., "Rapid Estimation of Building Damage by Conventional Weapons," MS thesis, Naval Postgraduate School, Monterey, CA, Sept. 2014.

[2]   Allen, E., and Iano, J., *The Architect's Studio Companion: Rules of Thumb for Preliminary Design*, John Wiley & Sons, New York, 2002.

[3]   Federal Emergency Management Agency, "The Oklahoma City Bombing—Improving Building Performance Through Multi-Hazard Mitigation," FEMA Report #277, Aug. 1996.

[4]   National Institute of Standards and Technology, "Final Report on the Collapse of the World Trade Center Towers," NIST Report NCSTAR 1, Sept. 2005.

Chapter 16

COLLATERAL DAMAGE ESTIMATION

16.1 INTRODUCTION

Collateral damage estimation is the process of determining the unintended damage to nearby buildings, materiel, and personnel when an attack on an intended target is executed. Conflicts are conducted in an environment that increasingly tends to involve civilian personnel and noncombatant infrastructure. The immediacy of news reporting through the media has led to a greater awareness by the public of the results of the unintended consequences of military actions. As a result, increasing attention is being directed toward improving the prediction of collateral damage in order to still damage the intended target while minimizing effects to the civilian population and their surroundings. Estimating collateral damage is a difficult task; too little consideration to noncombatants can lead to high casualty levels whereas too much will be operationally restrictive.

In addition to generating a need for estimating collateral damage, these concerns have affected weapon design directly in the following ways:

- Design of mines/submunitions having fuzes that if not activated within a specific period (about 2 days) render the munition inert
- A trend towards high accuracy, low yield weapons intended for more surgical strikes while minimizing damage to the surrounding area

The methodologies for estimating collateral damage effects have evolved over the years, reflecting the increasing focus on the subject. This chapter will address the following two methods used to estimate collateral damage:

- The collateral damage estimation (CDE) process was defined by the Chairman of the Joint Chiefs of Staff Instruction (CJCSI) 3160, circa 2005. This is a process that uses classified data from the JMEM Weaponeering System (JWS), so its use is restricted to U.S. and selected coalition partners.

* An unclassified approach that uses public domain tools and an unclassified database, allowing students to perform some basic tasks in this subject.

Before these topics are dealt with, however, some basic terminology common to both will be explained.

16.2   COLLATERAL EFFECTS RADIUS AND AREA

Most CDE tasks begin with imagery of the target and its surroundings onto which are placed boundaries, usually circles of calculated radius, centered on the desired point of impact (DPI). This radius, which may be thought of as a cookie cutter area of potential damage, is called the *collateral effects radius (CER)*. An example is shown in Fig. 16.1.

Once this is done, any buildings or other sites of concern close to the target can be identified. These objects are called *collateral objects (COs)*, and again, examples of some COs are shown in the figure. If a collateral object is located inside the CER, then collateral damage might occur.

The obvious issue is how to determine the CER. This is a function of two quantities.

1. The lethal area corresponding to the applicable weapon, target, and damage criterion of interest
2. The accuracy with which the weapon is delivered to the DPI

Notice how the lethal area has been expressed as a circle, as could the delivery accuracy in terms of a circular error probable (CEP). In some instances,

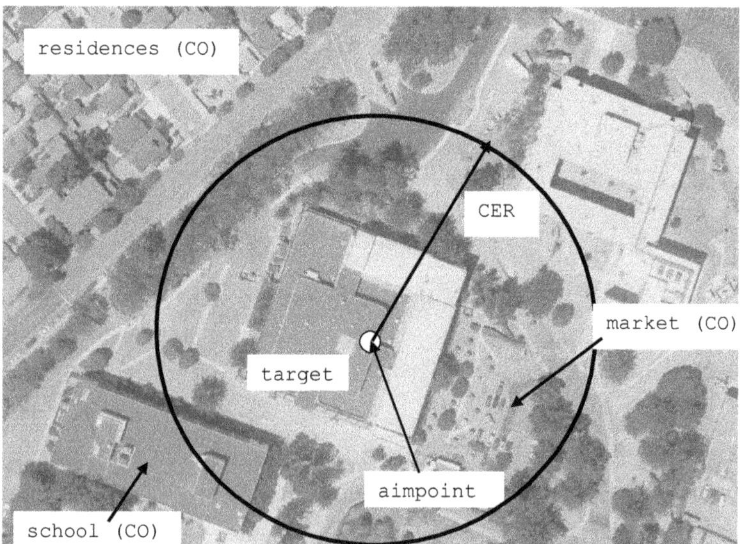

Fig. 16.1   CER centered on the target.

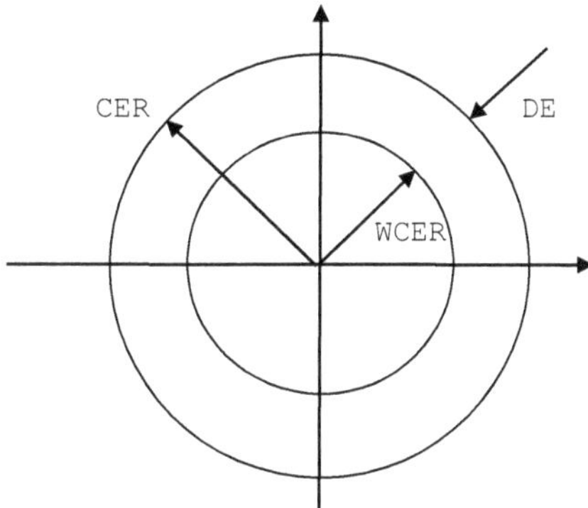

Fig. 16.2 Determining the CER.

this will be a good assumption; for example, blast damage to a building is independent of the weapon impact angle, and many guided weapons are reasonably represented by a CEP derived from a circular normal distribution.

In other attacks, however, the assumption of a circular area of effects is less accurate. For example, the fragment spray pattern does depend on the weapon orientation when it detonates, so a shallow impact angle would produce an area of effects that is wider than it is long. In addition, unguided weapons such as artillery projectiles fired such that the impact angle is shallow tend to have accuracies much greater in range than deflection.

The basic process to determine the CER is to add a measure of the delivery error (DE) to the weapon collateral effects radius (WCER). If both can be represented by circles, this results in the situation shown in Fig. 16.2. The larger area is sometimes referred to as the collateral hazard area (CHA). The obvious question is: How are the WCER and DE calculated?

Suppose we are attacking a target for which we know the MAE_B for the selected weapon. In this case, the WCER is given by the radius of a circle whose area is equal to the lethal area.

$$WCER = \sqrt{\frac{MAE_B}{\pi}} \tag{16.1}$$

If we want the delivery error to contain 90% of the impact points, we can write from Eq. (4.9) in Chapter 4

$$DE = CE_{90} = 1.8227 \times CEP \tag{16.2}$$

The resulting CER is obtained from the following:

$$CER = WCER + DE = \left[\sqrt{\frac{MAE_B}{\pi}} + 1.8277 \times CEP \right] \tag{16.3}$$

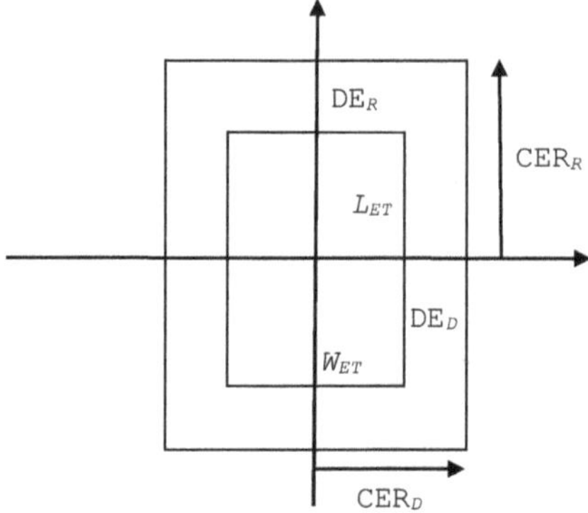

Fig. 16.3   Determining the CER for range and deflection

The CER defines the radius of a circle inside which there is a 90% probability of damage to the target.

Now consider the case in which the damage function is not circular, and neither is the delivery error. For example, if the target is a building of known dimensions, the weapon impact angle is 65 deg, and accuracy in range and deflection is different. Chapter 15, Section 15.2 shows the lethal area is represented by L_{ET} and W_{ET} and may be calculated for given building dimensions and weapon impact angle. Figure 16.3 shows how the CER is calculated for this target.

In this case we have a rectangle for the collateral effects area, although its dimensions are usually still called CERs. If this area is to represent the 90% damage area, we see that in the range direction,

$$\mathrm{WCER}_R = L_{ET}/2 \tag{16.4}$$

From Eq. (4.10),

$$\mathrm{DE}_R = \mathrm{REP}_{90} = 2.483 \times \mathrm{REP} \tag{16.5}$$

Therefore

$$\mathrm{CER}_R = \mathrm{WCER}_R + \mathrm{DE}_R = \left[\frac{L_{ET}}{2} + 2.483 \times \mathrm{REP} \right] \tag{16.6}$$

In the deflection direction we have

$$\mathrm{WCER}_R = W_{ET}/2 \tag{16.7}$$

Again from Eq. (4.10),

$$\mathrm{DE}_D = \mathrm{DEP}_{90} = 2.483 \times \mathrm{DEP} \tag{16.8}$$

$$\mathrm{CER}_D = \mathrm{WCER}_D + \mathrm{DE}_D = \left[\frac{W_{ET}}{2} + 2.483 \times \mathrm{DEP} \right] \qquad (16.9)$$

These areas have to be interpreted with some caution. The values of MAE_B, L_{ET}, and W_{ET} are calculated for the target/weapon combination. If the target is a reinforced concrete frame military building and the weapon is a 500-lb guided bomb, the L_{ET} and W_{ET} relate to that combination. If there is a collateral object inside the CHA comprising a simple wood-framed residential building, the same 500-lb bomb can cause considerably more damage than that sustained by the target, and out to a larger distance also. Basically, the lethal area MAE_B for the target may be substantially different from that of the collateral object. This issue is examined further in Chapter 16.4.

Now that some basic definitions have been covered, we can proceed with describing two processes to evaluate the likelihood collateral objects may be damaged.

16.3 COLLATERAL DAMAGE ESTIMATION (CDE) METHODOLOGY

In 2005, a new methodology was directed by the chairman of the Joint Chiefs of Staff (CJCS) and contained in their instruction CJCSI 3160 [1]. This document codifies the joint standards and methods for estimating collateral damage potential, provides mitigation techniques, and assists commanders with weighing collateral risk against military necessity and assessing proportionality within the framework of the military decision-making process. These joint standards and methods for conducting CDE apply across the range of military operations, and as of the time of writing represent the Department of Defense's (DoD's) collateral damage assessment standard. In addition, any CDE that results from CJCSI 3160 is meant to inform decision makers and commanders and is not a decision itself. CDE helps senior leaders evaluate collateral risk against military necessity during the planning and execution of combat operations.

CJCSI 3160 provides a logical and repeatable five-step process for estimating collateral damage. The technical basis for these CDE levels is a series of munitions collateral effects radii (CER) tables that were developed and accredited by the Joint Technical Coordinating Group for Munitions Effectiveness (JTCG/ME) based on its current data. The CER tables contain collateral damage distances for most air-to-surface and surface-to-surface conventional munitions.

One detailed implementation of the CJCSI 3160 [2] is known as the collateral damage estimation (CDE) method. CDE is a complex process that is classified due to the need to include weapon warhead, delivery platform, and accuracy data from the JWS database. In this section, only an overview is given to indicate the methodology, without reference to specific weapons or

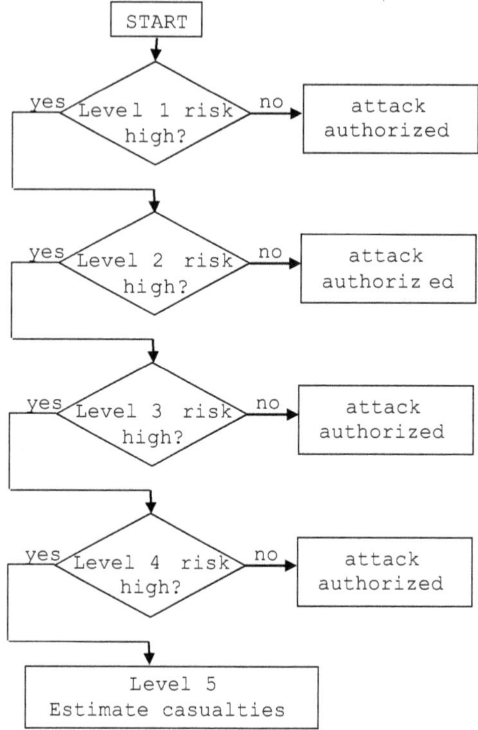

Fig. 16.4   Process flow of CDE methodology.

targets. The methodology consists of five levels, L1 through L5, where each level addresses the assessment with an increasing fidelity and risk. This is shown in Fig. 16.4.

The first question to ask is: Can I positively identify (PID) the target? PID is defined as the reasonable certainty that a functionally and geospatially defined object of attack is a legitimate military target. An answer of yes is required before the analysis begins.

By passing through each level in turn, a determination is made as to whether the CDE risk is high or low. If it is low, the attack may proceed as planned; if the CDE risk is high, the assessment moves to the next level. For example, Level 1 assessment might ask whether there is a collateral object inside any CER values for all conventional weapons. If the largest CER for all inventoried weapons is 1000 ft, for example, the likely answer is yes, resulting in a high risk level, and the process moves on to level 2. Further details of this process may be found in the *Advanced Weaponeering* textbook.

As the procedure progresses and the CER is refined, certain weapon types (e.g., unguided munitions) are eliminated as potential candidates for the

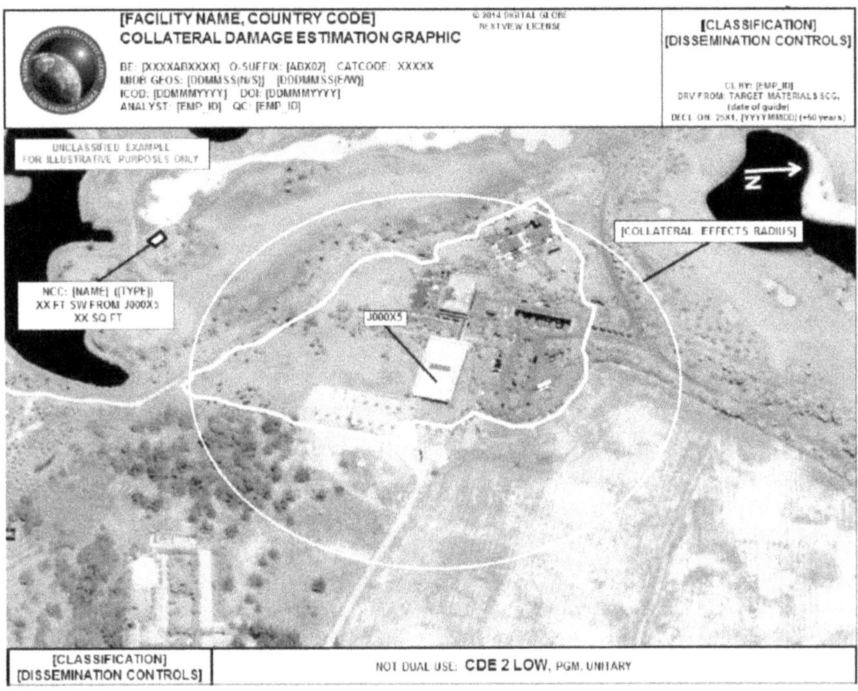

Fig. 16.5 Example of CDE assessment.

attack. This guides the assessor as to which weapons may be used and includes any mitigating effects such as internal building detonations or partially buried warheads. Figure 16.5 shows an example of using the methodology for a specific target where because Level 2 risk assessment was found to be low for the CER indicated, an attack using a unitary warhead, precision guided munition (PGM) was authorized.

When level 5 is reached, the conclusion is that the risk from collateral effects is high, and the number of potential casualties is calculated. This supports any decision made on whether the attack should proceed.

16.4 SIMPLIFIED CDE METHODOLOGY USING OFFSET AIMPOINTS

The process about to be outlined is intended to illustrate how a simplified and unclassified approach may be used to obtain reasonable results for collateral damage estimates, and in some respects has advantages over the process described in the previous section. It begins by revisiting some of the damage calculations we have used in previous chapters, but includes the possibility of aiming a weapon at some point other than the target. The reason for doing this will become apparent shortly.

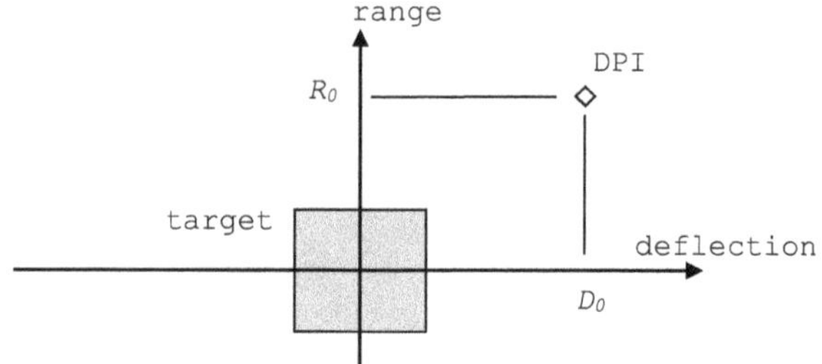

Fig. 16.6   Coordinates of offset aimpoint.

Reconsider Chapter 9, Section 9.4, and in particular the Monte Carlo approach to calculating PD_1 shown in Table 9.1. To determine the impact point of the weapon, we made random draws from a bivariate circular normal distribution defined by the CEP. From Eq. (4.4), we would get

$$\sigma = \sigma_y = \frac{\text{CEP}}{1.1774} \tag{16.10}$$

Because we expect the mean of the impact points in range and deflection to be zero, each impact point will be drawn from a normal distribution of zero mean and sigma, as defined by Eq. (16.10).

If we now aim the weapon at some point other than the center of the lethal area defined by (R_0, D_0), as shown in Fig. 16.6, all we have to do to accommodate this change is set the mean value of the distribution in each of these directions to the offset, that is,

$$\mu_x = R_0 \qquad \mu_y = D_0 \tag{16.11}$$

The spreadsheet shown in Table 9.1 is easily modified to accommodate this, and the revised program is shown in Table 16.1.

Here the DPI is given by the coordinates (10, 10) and shows the value of PD_1 decreases, as expected, from 0.3878 to 0.1311. Similarly, the MAE_{BLDG} method may be modified, specifically the spreadsheet shown in Table 15.2 in Chapter 15, where it will be recalled that these offsets appeared in the spreadsheet but were set to zero.

Now consider the situation shown in Fig. 16.7, showing a building target located close to a collateral object, a school building in this case. If the target is attacked to produce a required level of structural damage, how much damage is sustained by the school?

This attack is against a building, and the figure shows the target (dimensions L_T and W_T) and a school (dimensions L_S and W_S) and the spatial relationship between them, as would be obtained by a mensurated satellite image of the area, as for example shown in Fig. 16.5. Suppose the school is located

TABLE 16.1 SPREADSHEET WITH
OFFSET AIMPOINT

INPUTS	
Target area length	15
Target area width	15
CEP	10
DPI offset - range	10
DPI offset - deflection	10
Number of iterations	100000
OUTPUTS	
PD1	0.1311
Compute	

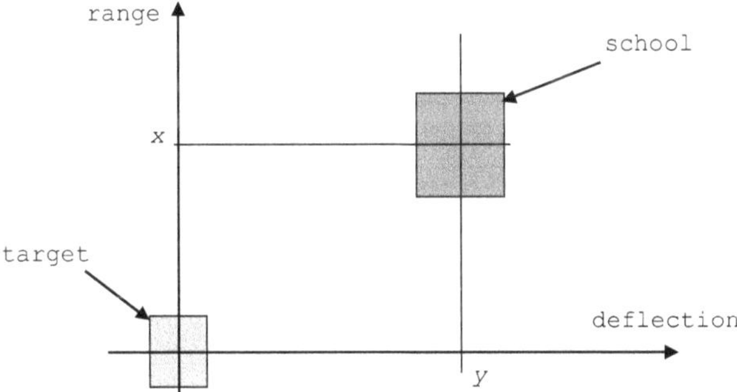

Fig. 16.7   Target in close proximity to a collateral object.

at coordinates (150, 150) from the DPI at the center of the target. The
approach to this problem is as follows:

1. The building is weaponeered in the normal manner using the MAE_{BLDG} for
    the appropriate target building type and the choice of weapon. Offset val-
    ues of zero are used in this case, because the center of the building is the
    DPI. Determine the amount of structural damage due to a single weapon
    and, if insufficient, determine the number of weapons required to achieve
    the fractional damage FD required from Eq. (2.27).
2. Then the school is weaponeered. This time, the weapon type and number
    used would be those obtained from the previous step. For MAE_{BLDG}, we
    would use the building type of the school. This time we set $R_0 = -150$ and

$D_0 = -150$, because this is the point relative to the school at which the weapons will be aimed (i.e., the target). The result of this weaponeering computation is the fractional damage sustained by the school.

Suppose we use a similar spreadsheet to the one shown in Table 15.2, but using data appropriate to this attack. The requirement is for the target to sustain at least 60% structural damage. Table 16.2 shows step 1 with offset DPI options but set to zero, giving the estimated structural damage at 62%; therefore, only one weapon is needed.

For step 2, suppose the school building has dimensions $L_T = 200$ ft, $W_T = 200$ ft, and $H_T = 10$ ft; however, for the same weapon used against the target, the $\text{MAE}_F = 40,000$ ft^2, indicating a weaker structure than the intended target. The same spreadsheet may be used to estimate damage to the school by modifying the data and including the offset DPI of $(-150, -150)$, which is where the aimpoint is located relative to the school. The resulting spreadsheet is shown in Table 16.3.

It is now up to the commander to decide if about 8% structural damage to the school is acceptable. This damage and any resulting casualties may be mitigated by, for example, attacking the target at night.

**TABLE 16.2   MAE$_{\text{BLDG}}$ METHOD
WITH OFFSET AIMPOINT**

MAE Building	
Inputs	
MAE_BLDG	25000
Impact angle	45
Building length	100
Building width	100
Building height	50
REP	50
DEP	25
DPI offset - range	0
DPI offset - deflection	0
Number of iterations	100000
Output	
Fractional damage	0.6165

Compute

**TABLE 16.3   ASSESSING DAMAGE
TO THE COLLATERAL OBJECT**

MAE Building	
Inputs	
MAE_BLDG	40000
Impact angle	45
Building length	200
Building width	200
Building height	10
REP	50
DEP	25
DPI offset - range	-150
DPI offset - deflection	-150
Number of iterations	100000
Output	
Fractional damage	0.0772

Compute

16.5   *IDENTIFYING COLLATERAL OBJECTS FROM IMAGERY*

The methods described in the previous section show how to estimate the damage sustained by collateral objects when a target is attacked; however, this requires detailed knowledge of the target, collateral object(s), and their spatial relationship to each other. How are these data obtained?

- For building targets and collateral objects, knowledge of approximate building dimensions, intended use, and local construction codes is usually sufficient to determine the vulnerability to specific weapons (i.e., the MAE_{BLDG}). In some cases, architectural plans may be available if the structure was built by contractors. Clearly this information may be considered intelligence data.
- Regarding the location of collateral objects close to the target, satellite imagery is a common source, see Fig. 16.5, and each country may have access to its own system to produce these data. To demonstrate how this process works, however, use may be made of commercial products such as Google Earth.

We will use Google Earth to generate imagery of the target area and overlay CER rings that include both warhead effective area and weapon delivery

TABLE 16.4 CER TOOL FOR BUILDING TARGET

COLLATERAL EFFECTS RADIUS TOOL	
Input data	
Warhead Mean Area of Effects (MAE)	5000
Weapon delivery accuracy CEP (CE50)	25
Derived data	
Sigma	21.23
CE90	45.57
CE95	51.97
CE99	64.44
Collateral Effects Radius (CER)	
Weapon only	39.89
Weapon + CE90	85.46
Weapon + CE95	91.87
Weapon + CE99	104.33

accuracy effects. These effects may be calculated in a spreadsheet such as the one shown in Table 16.4.

In this table, the warhead area of effects is input; assuming this can be represented as a circular cookie cutter (blast warhead), the radius of effects due just to the weapon is calculated from Eq. (16.1). Larger rings may be generated by adding on the CE90 [Eq. (16.3)], CE95, and CE99 accuracy metrics.

In detail, here's how the process works:

1. Open Google Earth and navigate to the area of interest. Identify the target and insert a place mark.
2. Use the Ruler tool to select circles, and draw these centered on the target place marker with the radius given in Table 16.4, as shown in Fig. 16.8.
3. Possible collateral objects may now be identified inside the appropriate CER circle.
4. Figure 16.8 shows sample imagery of a commander's area of interest with appropriate CER rings overlaid.
5. In addition to the CER rings, linear measurements may be made by again using the Ruler tool to determine the DPI offset values used in the previous example to determine damage to collateral objects.
6. The Google Earth program also generates latitude/longitude for the target, which might be useful for GPS-guided weapons; however, the accuracy of these coordinates has to be determined (target location error).

Fig. 16.8   Target imagery with CER rings.

16.6   PERSONNEL INJURY

The same techniques used for materiel targets can be used to assess injury to personnel targets just described. Recall Chapter 8, Section 8.7, where the same lethal area matrix, and hence MAE_F, was derived for personnel as was done for materiel targets. This can be input into the spreadsheet shown in Table 16.4. If the weapon detonates inside the target building, the effects of blast and fragments on personnel standing in the open will be less than a detonation outside, but in the latter case we can expect the MAE_F to be substantially larger than for a building, as illustrated in Table 16.5.

The effect on the CER ring size is apparent, reflecting the increased vulnerability of the target elements.

16.7   CHAPTER SUMMARY

- An overview of the joint service U.S. DoD method of calculating collateral damage effects was described. Known as the Chairman of the Joint Chiefs of Staff Instruction (CJCSI) 3160, this multistep procedure uses JMEM weapon lethality and accuracy data and therefore is classified.
- A legacy JMEM method using $\mathrm{MAE}_{\mathrm{BLDG}}$ and offset aimpoints was described.
- An unclassified procedure utilizing user-supplied data and commercial software has demonstrated the basic steps in generating collateral damage estimates for buildings and injury to personnel.

TABLE 16.5 CER TOOL FOR PERSONNEL INJURY ASSESSMENT

COLLATERAL EFFECTS RADIUS TOOL	
Input data	
Warhead Mean Area of Effects (MAE)	50000
Weapon delivery accuracy CEP (CE50)	25
Derived data	
Sigma	21.23
CE90	45.57
CE95	51.97
CE99	64.44
Collateral Effects Radius (CER)	
Weapon only	126.16
Weapon + CE90	171.72
Weapon + CE95	178.13
Weapon + CE99	190.60

REFERENCES

[1] "Appendix G, CJCS Policies on Sensitive Target Approval and Review Process and the Accompanying Collateral Damage Estimation and Casualty Estimation Methodology," *Joint Tactics, Techniques and Procedures for Intelligence Support to Targeting*, Joint Publication 2-01.1, 9 Jan. 2003.

[2] "Collateral Damage Estimation (CDE) Table Development," JTCG/ME report 61 JTCG/ME-04-4, 31 Oct. 2006.

Chapter 17

Computer Implementation of Weaponeering Methods

17.1 Introduction

This chapter begins by reviewing the historical development of various Joint Munitions Effectiveness Manual (JMEM) weaponeering tools. Following that, we develop an integrated computer tool capable of calculating damage for a wide set of targets in conjunction with a database of notional data for the effectiveness indices discussed previously. Because this will be based on methodologies developed in this book, it will allow users to run basic weaponeering examples to demonstrate what they have learned so far.

Keep in mind that the primary focus of these tools has been to support the warfighter; for example, pilots about to fly a strike mission may want to verify the probability of a successful outcome. Any automated process to provide this information should be relatively simple and fast running, and require a minimum of information to derive a result. Operational users often quote the requirement "it is better to get the 70% answer in two minutes than the 99% answer in two hours."

There is, of course, another community that benefits from the work done in providing the warfighter with weaponeering tools, namely the analysts who are mainly responsible for their development in the first place. Analysts tend to operate in a more benign environment compared to the warfighter, typically have more time to run computer programs on up-to-date hardware, and are not usually subjected to life-threatening consequences when they obtain the wrong answer. It is fair to say, however, that the analytical community is well aware of who they are working for and strive to get it right the first time for their customers.

The platforms on which the various tools run have kept pace with computational advances over the last 50 years of development, ranging from manual computations through programmable calculators to the capable personal computers of today. Given the emphasis on warfighter utilization, porting onto alternative, perhaps less capable though more convenient devices such as tablets and cellular telephones has also received attention.

17.2 HISTORICAL DEVELOPMENT OF WARFIGHTER TOOLS

In the initial establishment of the Joint Technical Coordinating Group for Munitions Effectiveness (JTCG/ME) based on consolidating methods to provide effectiveness estimates for air-delivered weapons, the first step was to review the methodologies used by the various services to compute PD_1 and to agree on a single methodology. This resulted in several methods to compute weapon effects for a variety of air-launched weapon systems: unguided and guided bombs against unitary targets and against areas of target elements, sticks, cluster munitions, and projectiles. The basic tools used in the original JMEM methodologies were

- *COVART:* Took target data [target geometric model (TGM) and failure analysis logic tree (FALT)] and generated vulnerable areas
- *General Full Spray Materiel (GFSM):* Combined COVART data with weapon specific Z-data to produce lethal area matrices
- *Matrix Evaluator (ME):* Combined the lethal area matrix with delivery accuracy data to compute a P_K

17.2.1 DEVELOPMENTS 1968–1980

The methods listed were initially implemented on 1960s mainframe computers running the Matrix Evaluator program. They attempted to provide estimates for all inventoried aircraft, weapons, and release conditions.

As might be imagined, this produced considerable amounts of data, which were recorded in orange-covered manuals—the *Joint Munitions Effectiveness Manuals*. These were the original JMEMs and are termed the *precalculated methods*. They focused on air-launched weapons, and much of the original work involved keeping the data current. For example, if a new weapon was inventoried and its warhead characterized, it had to be evaluated against all targets and delivery conditions available, and the resulting data updated in the manuals.

Even though these manuals formed a library of considerable size, there was never the granularity the users desired, and the need arose for a method the individual warfighter or analyst could implement for the specific case of interest. This approach was termed the *open-ended method*, and the emphasis changed to ease of use and speed in getting an answer.

Clearly the legacy programs (COVART, GFSM, and ME) could not be easily used in such an environment, so approximations were made in order to generate a process that could be completed by hand using graphical and tabular data for more complex calculations. It was acknowledged that the accuracy of the results could not equal those of the precomputed data, but the user was willing to sacrifice fidelity for speed of getting a result that was "good enough." The analytical methods described in Chapters 9–12 form the basis of the methods developed.

The fundamental changes to the precalculated methodology were to retain COVART and GFSM to calculate a lethal area matrix, but instead of using ME, the matrix was reduced to an effectiveness index, the EI, that could be represented by either a Carlton damage function or a rectangular cookie cutter. This allowed the analytical methods described in Chapters 10–12 to determine effects for a single bomb or stick to be evaluated against unitary targets or areas of targets, as well as simple projectiles and cluster munitions. The methods were later extended to cover specific target types such as bridges, buildings, and linear targets as well as the effective miss distance, the effectiveness index (EI).

Regarding the implementation platform, the first attempt at a user "tool" was called the *hand methods*. This approach used a multistep, user-completed form for each method (i.e., unguided single weapons against a unitary target, single weapons against an area of targets, sticks, cluster munitions, projectiles, etc.). The form would be completed by hand, step by step, with more difficult calculations being obtained from graphs or nomograms. An example of a nomogram to determine intervelometer settings (Chapter 11, Section 11.2.1) for sticks is shown in Fig. 17.1, and a section of the hand method for clusters is shown in Fig. 17.2.

The reader may conclude that some of the spreadsheet solutions presented in earlier chapters bear a close resemblance to the original hand methods, except, of course, the Excel program is able to perform all the computations that originally had to be done by hand.

The first step to automate the process began in the 1970s when programmable, handheld calculators were used. The Texas Instruments TI-59 calculator shown in Fig. 17.3 used a programmable magnetic strip that could read into the device programs that then used the calculator as an elementary computer.

Having only 960 programming steps and 100 memory registers, clearly simplifying the methodology from Matrix Evaluator to something that would work on this device was a challenge, and this is where the methods described in Chapters 9–12 came into their own. Using deterministic methods, they could be accommodated within the programming limitations and produced results (the 70% solution) in seconds directly on a handheld device.

17.2.2 DEVELOPMENTS 1980–2000

The widespread use in the early 1980s of personal computers with a stable operating system and a common programming environment (FORTRAN, C)

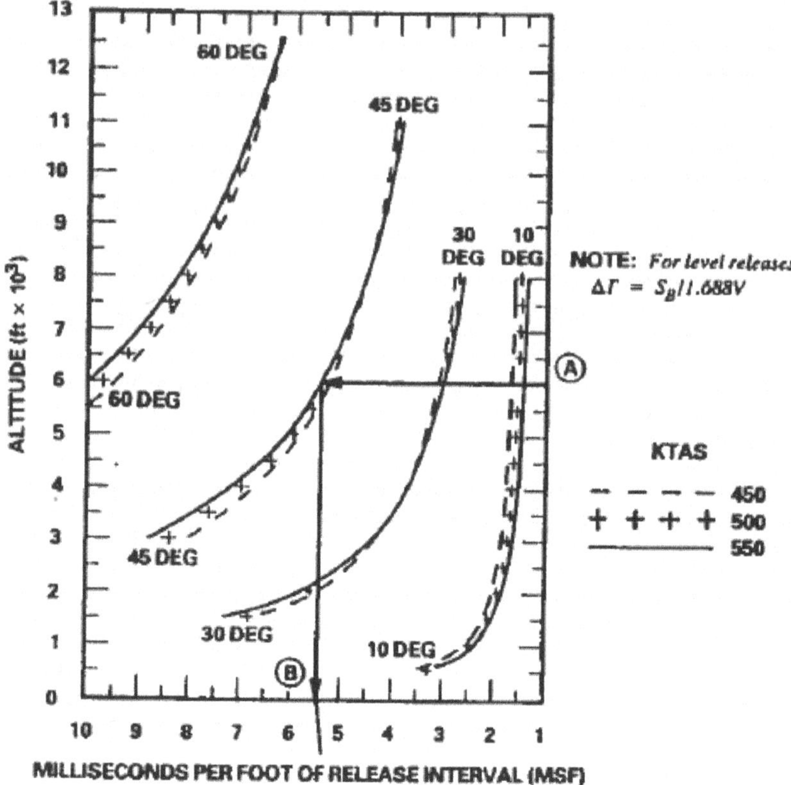

Fig. 17.1　Nomogram for stick spacing.

led to the first PC implementation of the methodologies. The IBM personal computer with Disk Operating System (DOS) was chosen for its widespread acceptability within the Department of Defense (DoD), particularly with the operational community. With the PC's storage capability and execution speed, the methods could be implemented more accurately with fewer simplifying restrictions in order to obtain a better answer.

The original implementation in the early 1980s on a PC platform was called PC OEM and ran under the DOS operating system; however, after Microsoft introduced the Windows operating system in 1985, PC OEM was transitioned to WinJMEM around 1990. WinJMEM was incrementally refined and improved, resulting in the JMEM Air-to-Surface Weaponeering System (JAWS 1.0) being released in 1995. This was the first fully integrated weaponeering tool available to the operational user containing both methodologies and a classified database. For the first time, modified versions of JAWS 1.0 were released to coalition partners such as the Republic of Korea (ROK) and North Atlantic Treaty Organization (NATO).

Both Fuze Types

11.	Functioning Angle, degrees	θ_f	—	—	—	Figure A-III-8 (Sheets 7 thru 12).
12.	Functioning Velocity, ft/sec	V_f	—	—	—	Use CBU-52/B and CBU-58/B system graphs for low and
13.	Ground Range to Functioning, ft	GR_1	—	—	—	high values

III. SUBMUNITION BALLISTICS

14.	Submunition Velocity at Dispenser Functioning, ft/sec	V_D	—	—	—	Actual velocity equals dispenser functioning velocity (Step 12). Low and high values depend on graphs available for the submunition.
15.	Ground Range, ft	GR_2	—	—	—	For CBU-52/58: Interpolate on submunition velocity (Step 14), Figure A-III-8 (Sheets 13 thru 18).
16.	Impact Angle, degrees	I	33	—	—	For Rockeye and APAM: Interpolate on airspeed (Step 4),
17.	Time to Impact, sec	TF_B	—	—	—	Figure A-III-9 (Sheets 4 thru 9).
18.	Outer Pattern Radius, ft	R_p	80	—	—	For APAM: $R_P = (1.15)(\text{Rockeye } R_P)$.
19.	Inner Pattern Radius, ft	R_{pi}	0			For CBU-52: $R_{pi} = R_p - 125$ For CBU-58: $R_{pi} = R_p - 175$ For Rockeye and APAM: $R_{pi} = 0$

NOTE: For 95 percent of the BLU-61 or BLU-63 submunitions to function, TF_B must be greater than the value obtained from Figure A-III-8 (Sheet 19).

IV. TOTAL FLIGHT CONDITIONS

20.	Time of Flight, sec	T_F	6.0	—	—	For CBU-52/58: $T_F = T_D + TF_B$ For Rockeye and APAM: Figure A-III-9 (Sheets 4 thru 9).
21.	Ground Range, ft	GR	2,750	—	—	For CBU-52/58: $GR = GR_1 + GR_2$ For Rockeye and APAM: Figure A-III-9 (Sheets 4 thru 9)
22.	Slant Range, ft	SR	2,903			$SR = \sqrt{(GR)^2 + (Y_1)^2}$

UNCLASSIFIED

Fig. 17.2 Extract from original hand method.

Fig. 17.3 Magnetic card, programmable calculator.

Although the majority of the WinJMEM methods were deterministic, improving computational speed allowed some simple Monte Carlo methods to be incorporated into the suite of programs. An example was the Target Complex (TARCOM) program that modeled a collection of dissimilar (non-homogeneous) unitary targets. (See the *Advanced Weaponeering* textbook for more details.)

Around the same time as JAWS development took place, the surface-to-surface community was included in the JMEM family with the release of the Joint Weapon Effectiveness System (JWES 1.0). This served the indirect fire community with tools such as Superquickie 2 and ARTQUIK and direct fire with FBAR and Passive Vehicle Target Model (PVTM). The Navy Joint Gun Engagement Model (JGEM) was added later.

17.2.3 DEVELOPMENTS 2000–PRESENT

These two parallel developmental streams (JAWS and JWES) continued into the early 2000s with two distinct products:

- The air-to-surface tool JAWS.
- The surface-to-surface program JWES.

Although this approach worked well for the operational users, it did not suit more senior military officers who were more focused on the target rather than the weapon system. For example, if they wanted to attack a target 50 miles away, they could use air-delivered guided bombs or perhaps a salvo of guided multiple launched rockets. To compare the effect of each weapon, both JAWS and JWES had to be run separately, probably on two computers, and the results compared manually. These senior commanders wanted a single combined, target-centric program so that once a target was selected, all available weapons that might have an effect on it were available, irrespective of whether they were air or surface launched.

This requirement resulted in the current product—the JMEM Weaponeering System (JWS). Combining the two service-focused tools led to another round of consolidation of methodologies, which in the prior 20 years of development produced similar but not identical methods. For example, JAWS had a methodology for machine gun effects against ground targets whereas JWES had a different approach. JAWS had a vulnerability model of the T-72 tank and so did JWES. Were they the same? The first step was to incorporate the two programs under a single wrapper, a trivial sounding advance but one that was fraught with problems and resulted in JWS v1.0 being released in 2005. The first fully integrated program, JWS v2.0, was released in 2009 with further refinements continuing to the present day.

Over the last decade, computational speed has increased to the point where the architecture used to generate the precomputed solutions may now be performed in near real time, although the basic models have been improved. This process is shown in Fig. 17.4.

It is perhaps ironic that as JWS has developed into an all-encompassing collection of weaponeering tools giving the user a wide choice of weapon systems and targets to evaluate, its complexity in use may be daunting to the warfighter. As a result, there is now a requirement for table lookup solutions for common weaponeering cases that users encounter, saving the time needed to obtain them directly by running the JWS program. These weaponeering tables are reminiscent of the precomputed solutions of 50 years ago using the same (updated) tools!

The progression of JMEM platforms has matched computer development up to the present time so that the latest versions of the JWS program still run on Windows-based PCs. However, another operational constraint that has governed the level of fidelity of the models is the type of computer available to the operational user. The model developer (analyst) probably has the latest hardware and software available to them, but not so the operational user. These users have computer systems one or two developmental steps behind the analysts; therefore, the models need to run on relatively old systems.

17.2.4 *JWS Releasability*

A significant challenge to current JWS development is to make it releasable to coalition partners. For example, the Desert Storm/Desert Shield coalition included 39 different countries that were expected to fight in a coordinated way against a common enemy.

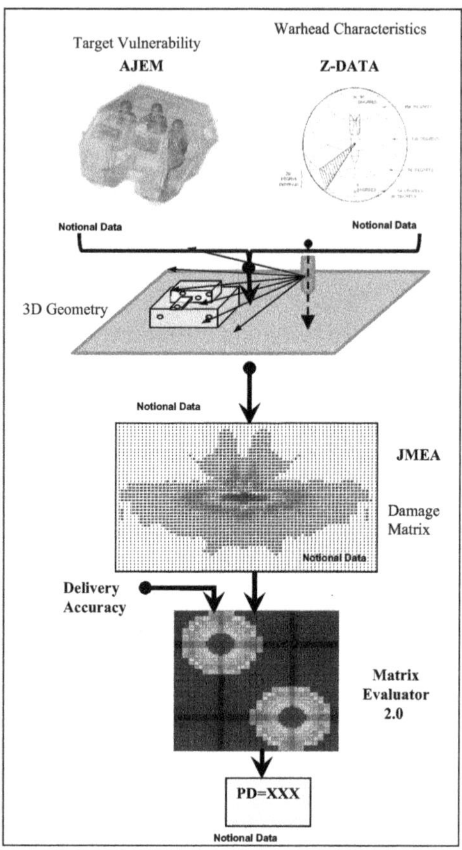

Fig. 17.4   Basic process for computing damage.

The basic argument is: if we fight together, we should train and mission plan together; otherwise, military operations cannot be coordinated effectively. The basic problem is: if our JWS weaponeering tool contains data classified to the secret level, not releasable to foreign individuals or agencies, how can JWS be used to assist coalition partners to mission plan with the United States? In previous conflicts, certain coalition countries had access to use U.S. weaponeering tools in a restricted area only under the supervision of U.S. personnel (i.e., visual use only). This solution is not ideal, because those countries had a certain learning curve to overcome before they could effectively use the tools. A better approach to solving this issue is to write the JWS program in a modular format so that parts may be added or removed for a custom version for a particular country.

The question is: What components should a country-specific version of JWS contain? This depends on the status of the foreign military sales (FMS) portfolio for that country. If a weapon system has been purchased from the United States, that weapon system is retained in the product; otherwise, it is removed. Although this approach is reasonable for weapons, it presents challenges regarding what target vulnerability data may be distributed.

17.2.5 PROGRAMMING LANGUAGE CONSIDERATIONS

The early programs such as Matrix Evaluator and GFSM were written in FORTRAN, and these were promulgated when OEM running under DOS was implemented. Since then, JWS has been written in various versions of the C programming language (C, C++, and C#). JWS is made up of several computational modules, and they are typically developed independently of JWS and incorporated once they have been through a verification, validation, and accreditation (VV&A) process and tri-service approved by the appropriate JTCG/ME working groups.

In some cases, operational users and others in the weaponeering community request the stand-alone version of a module to be used outside of JWS with user-supplied data. The rationale for this is that JWS is a large and complex program requiring strict conditions for its installation and running. If a user is interested in just a small part of JWS, such as bomb penetration through layers of soil, rock, and concrete to breach a bunker, is there a simple stand-alone tool to accomplish this?

The answer in this case is yes (see Fig. 13.4), because the penetration module was developed separately and then integrated into the JWS program by the production contractor team. The question then arises: What programming language is used for initial module development and how easy is it to distribute to customers requiring it?

In the past, module development was usually done in C++ because this provided easy integration into JWS. However, this is not a programming environment many of the operational users have available, and it makes distribution difficult. Similarly, analysts may have used MATLAB for module development because it has extensive function support to perform complex tasks and excellent plotting capabilities to visualize any outputs from the methodology. Unfortunately, MATLAB is not widely available to the warfighter either because anything other than the Student Edition may be prohibitively expensive to purchase and challenging to install on DoD computer systems. In addition, if development used any high-level MATLAB-specific functions, it proves difficult translating these into the native JWS language for later integration.

A suitable compromise has been to use the Visual Basic programming environment to develop modules that users might like to obtain as a stand-alone program. Visual Basic has the following attributes:

- It is a simple programming language to learn, and although it may not have the richness of functions available in MATLAB, it has capable computational and graphical display capacity. Not using intrinsic functions may involve more code but makes translation into JWS native code much easier.

- The main reason it is favored is that it runs under the Microsoft Excel program; therefore, it does not require additional software to be installed to execute. Modules developed with Visual Basic may be distributed to users as a simple spreadsheet that will run under Microsoft Office. Obviously, there are large numbers of DoD computers running Microsoft Office; therefore, distribution of modules is a simple task.

Increasing threats to DoD computer systems from cyberattack will further restrict what can and cannot be loaded onto them, which will make installation of more esoteric programming environments even more challenging in the future.

17.3 WEAPONEERING PROGRAM

Unfortunately, due to the classification issues mentioned earlier, it is not possible to distribute versions of JWS to readers of this book wishing to use the methodologies developed in earlier chapters. Instead, we describe in detail a Weaponeering Program that students may use to evaluate some representative scenarios and, with the use of an unclassified database, calculate damage probabilities. Recall that this program was introduced in Chapter 1, and the air-to-surface module was shown in Figs. 1.9 and 1.34.

Although this program is extremely simple compared to the classified JWS program, it must be appreciated that JWS is classified because of the data it uses, not the methodologies involved in the damage calculation. Therefore, for some cases, the unclassified Weaponeering Program will give the same answers as JWS if the same data are supplied to both. As readers are now aware, this book details methodologies to calculate damage, but the data have to be supplied by the user, a potentially daunting task. Fortunately, from an educational perspective, the JTCG/ME has approved a limited database to use with the Weaponeering Program, allowing the reader to develop case studies in the same manner as would be done when using JWS. We will now investigate and use this Weaponeering Program in more detail.

The program incorporates most of the methodologies discussed in previous chapters dealing with single and multiple weapon attacks and where the delivery errors may be dependent or independent. It is essentially a collection of the methods that have been previously dealt with as individual, stand-alone programs. Students should note, however, that this program also serves the more extended set of methods developed in the *Introduction to Weaponeering* textbook; therefore, not all the options available in the program have been described in this book.

Although many of the methods are applicable to both air- and surface-launched weapons, the program will be divided into these two categories to reflect possible interests of the user base—ground forces and aviation. Also, the data input descriptions use the terminology appropriate to the community for which the program is intended. The program is written in Visual Basic for Applications (VBA) with a Microsoft Excel user interface, allowing widespread dissemination without additional programming language requirements. Any computer running Microsoft Office can run this program.

The Weaponeering Program is a Microsoft Excel spreadsheet that is opened like any other spreadsheet. The user will see four tabs at the bottom corresponding to the following functions:

- The air-to-surface (AS) tool
- The surface-to-surface (SS) tool
- An unclassified database
- A Detailed Results sheet to store plotting data

To begin using the program, the user must select which tool best suits their needs (i.e., air-to-surface or surface-to-surface) and select that tab to bring up the tool interface. Each will be discussed in detail in the following sections.

17.4 AIR-TO-SURFACE WEAPONEERING TOOL

Selecting the Air-to-Surface tab brings up the user interface shown in Fig. 17.5. This screen shows graphically some of the weapon systems available in

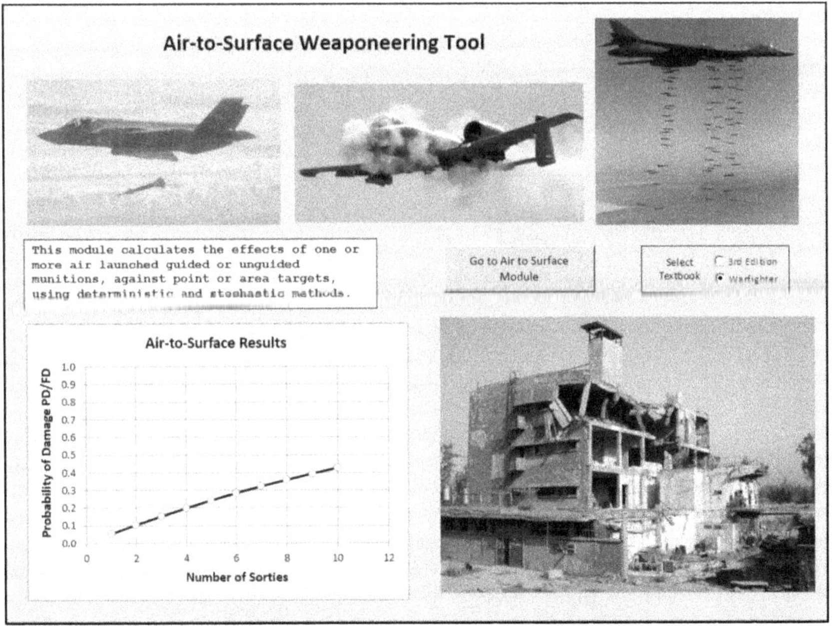

Fig. 17.5   Air-to-Surface tool user interface.

this component together with a graph, the details of which will be described in a moment.

There are two active controls on this screen. First, the user must select which version of the textbook they have. This is because the program will provide references to where in the book the user can find each methodology being used to calculate the result, and these references are different in each book. The user should then click the Go to Air to Surface Module button, which brings up the detailed input screen shown in Fig. 17.6.

This figure shows the default selections and data when the screen is initially displayed. As previously noted, data are grouped into several categories, broadly corresponding to chapters in the book. The main data groupings and

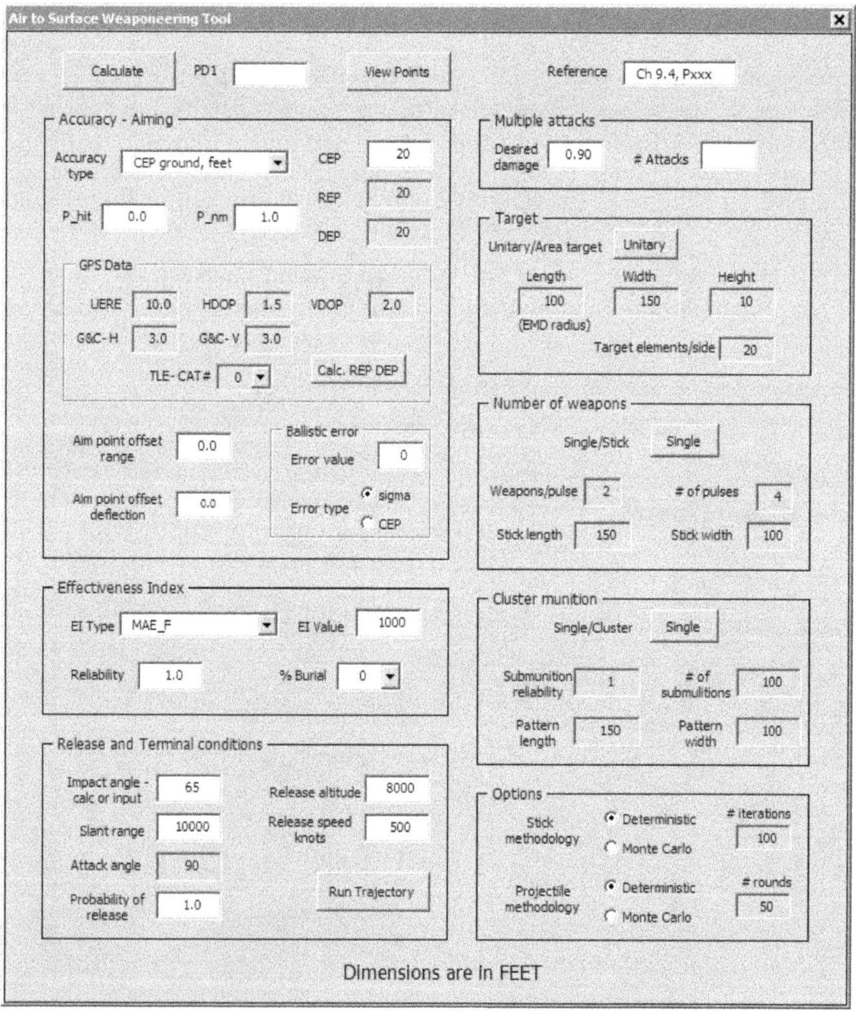

Fig. 17.6   Air-to-Surface module interface.

overall function of the group together with the relevant sections in the book will be discussed in turn.

17.5 AIR-TO-SURFACE PROGRAM INPUTS

Where possible, these will be discussed in the same order as the chapters in the book.

17.5.1 ACCURACY INPUTS

The Accuracy - Aiming group includes the ability to input circular error probable (CEP) or range error probable/deflection error probable (REP/DEP) in either the ground plane (feet) or normal plane (mils), as described in Chapter 4. The selections of accuracy type available in the drop-down menu are shown in Fig. 17.7.

Additional inputs include the P_{HIT} and P_{NM} data allowing for Gaussian and non-Gaussian distributions. All these errors are the aiming or dependent error; ballistic dispersion (independent) errors are input separately. These are assumed circular normal in either the ground or normal plane and may be input as either a sigma value or a CEP. Offset aimpoints are also included, as is a set of inputs for GPS weapons (see Chapter 5).

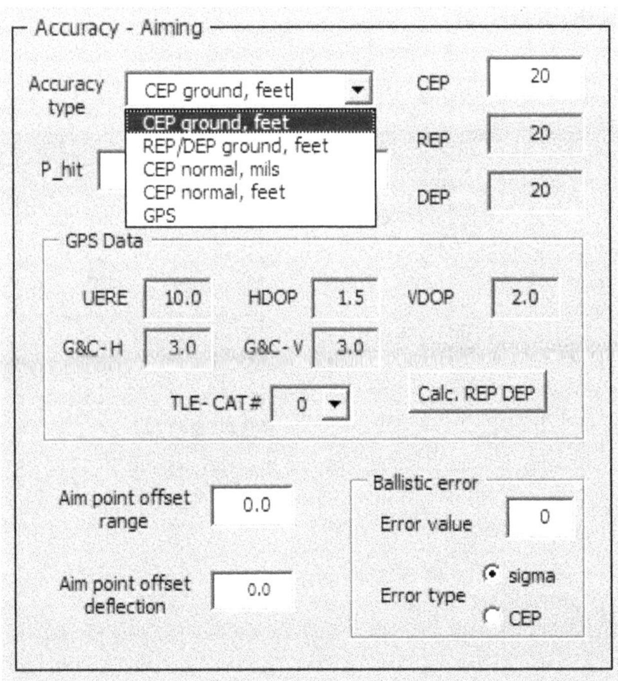

Fig. 17.7 Accuracy options.

If GPS is selected, the GPS data block is activated for user input where the first row corresponds to the NAV error and the second row to the Guidance and Control (G&C) errors, each having a horizontal and vertical component. Note, however, that the target location error (TLE) is specified as a category (CAT) number, which, in conjunction with Table 5.4 in Chapter 5, uses the midpoint of each range for ground plane and vertical values.

In order to convert any normal plane input in mils to equivalent linear measures at the target, the slant range is required, as discussed in Chapter 4, Section 4.6. The slant range is calculated or input in the Release and Terminal conditions block, which will be discussed shortly.

17.5.2 EFFECTIVENESS INDICES

This section follows the types discussed in Chapters 6, 12, and 15, which are listed in the pull-down menu for EI Type shown in Fig. 17.8.

Note that for the EMD EI, a circular or rectangular target may be selected. If circle is selected, the Target Length entry is used for the radius and the width is not used; however, if rectangle is selected, the length and width are both entered. The target height is the same for either selection.

Once the EI type has been specified, its value is input into the EI Value box. Other effectiveness-related terms that may be used in the damage calculation are also specified including weapon (fuze) reliability and the estimated percent burial of the warhead.

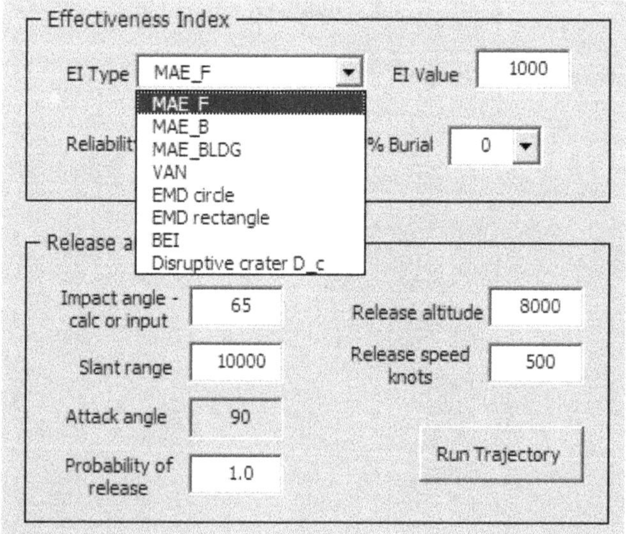

Fig. 17.8   Effectiveness Index options.

17.5.3 RELEASE AND TERMINAL CONDITIONS

This data group contains the parameters needed to determine lethal area aspect ratio and slant range scaling geometry for normal plane accuracies. A zero-drag trajectory model is included where for air-released weapons the user supplies altitude and speed and produces impact angle and slant range, as shown in Fig. 17.9. This model is discussed in Chapter 3, Sections 3.2 and 3.3.

These output variables, which are used to define the terminal conditions for the attack, may be overwritten directly in the data boxes in this grouping. Also in this section is the probability of release or target acquisition, which is usually set to unity; however, if sufficient details of the target and its environment are known, as well as the sensor used to detect it, the methods in the *Advanced Weaponeering* textbook may be used to calculate this value.

Finally, an attack angle option is also included if the linear target EI is selected, because this is a required input (see Chapter 12, Section 12.3).

17.5.4 TARGET PARAMETERS

The Target group allows the user to select a unitary target or an area of target elements, as detailed in Chapter 10, with the appropriate data fields made available or grayed out depending on the selection. If an area target is selected, the dimensions are input. If the MAE_{BLDG}, EMD, BEI, or D_C is selected, appropriate target dimensions are enabled and must be supplied. Even though these EI types correspond to a unitary target, they require dimensional inputs. Typical target area parameters are shown in Fig. 17.10.

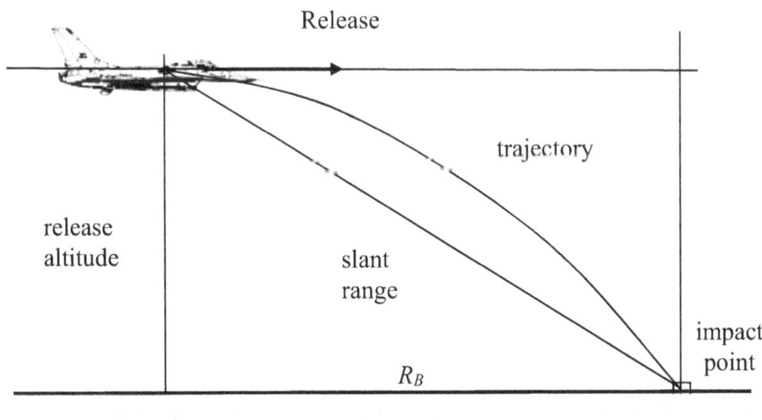

Fig. 17.9   Zero drag, level release trajectory

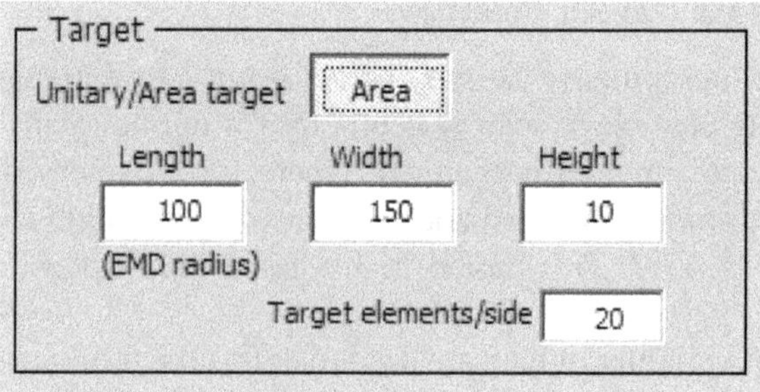

Fig. 17.10   Data entered into the Target group.

The target is defined by its length, width, and height, and all targets are assumed orthogonal to the attack direction with the length aligned with the range axis. In Chapter 10, two methods were described for calculating fractional damage, one in Section 10.3, based on calculating the fractional coverage, and one in Section 10.5, using discrete target elements. If the latter option is used, the number of target elements per side is input in this data block as well.

17.5.5   NUMBER OF WEAPONS

The choices are a single weapon or a stick of unguided bombs. If a stick is selected, it is defined in terms of the stick length, stick width, number of weapons released for each intervalometer pulse, and number of pulses. Stick attacks are described in Chapter 11.

The program considers only a stick of unguided bombs to be used against an area of targets; therefore, if Stick is selected, the Area Target should also be selected in the Target block. Because the methodology used is the Monte Carlo approach described in Chapter 11, Section 11.4, the Target elements/ side variable has to be specified. Figure 17.11 shows data representing a stick of eight bombs released as four sets of two per pulse.

For weaponeering stick attacks against an area target, the View Points option allows the stick geometry and number of weapons to be adjusted to maximize coverage of the target. An example is shown in Fig. 17.12 where a stick of eight weapons (2×4) is dropped against an area of targets. The View Points option allows the user to see the target area overlaid by the stick for a given EI, impact angle, stick length, and stick width inputs. The graphic shows the pattern overlaying the target with no delivery errors and indicates the maximum fractional coverage that may be obtained with this pattern. Although not needed in this case, users may adjust the pattern dimensions to optimize target coverage.

Note that the View Points window has to be closed before further calculations can be performed.

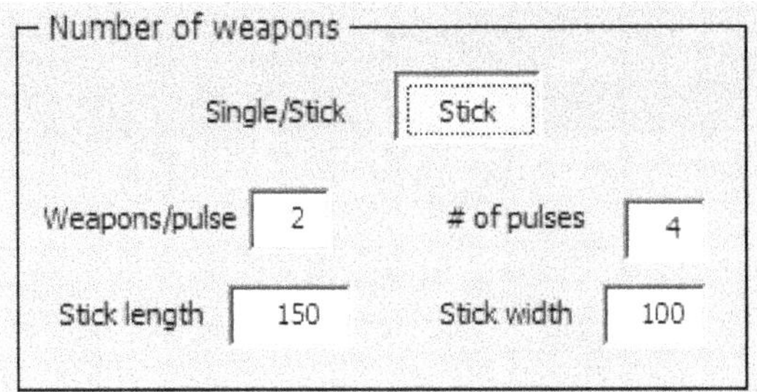

Fig. 17.11   Data corresponding to a stick of eight bombs.

17.5.6 CLUSTER MUNITIONS

Cluster munitions may be specified in a manner described in Chapter 11, Section 11.5 by the pattern dimensions, the number of submunitions, and the submunition reliability, as indicated in Fig. 17.13.

The pattern is assumed to be rectangular. As for stick attacks, only area targets are considered; therefore, that option should be selected in the Target group.

17.5.7 OPTIONS GROUP

For the two weapon systems shown in this group, sticks and projectiles, we have calculated effectiveness using a Monte Carlo-based approach in both

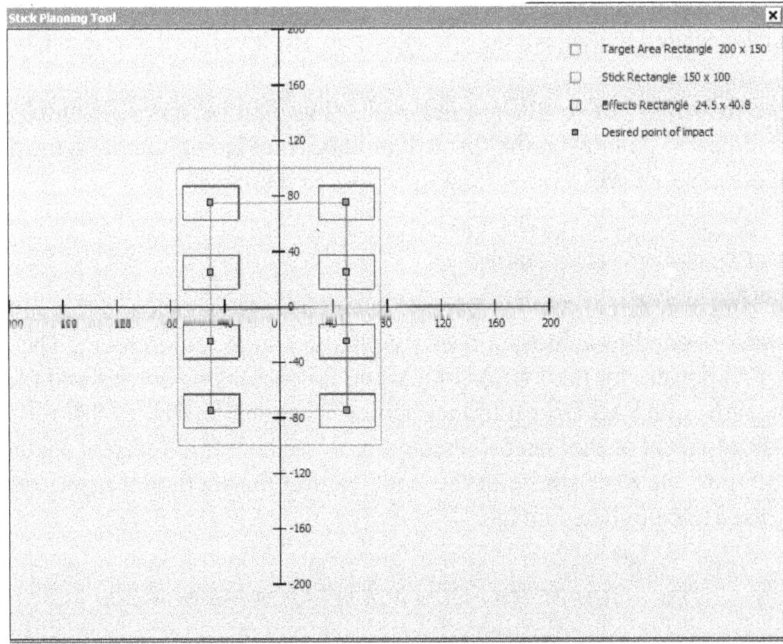

Fig. 17.12   View Points option showing stick and target area.

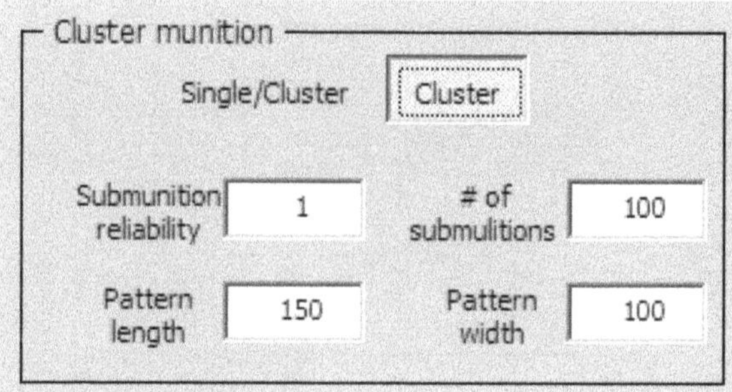

Fig. 17.13   Cluster munition data input.

cases—in Chapter 11, Section 11.4 for sticks and in Chapter 9, Section 9.5 for projectiles. There are deterministic solutions to calculate the effectiveness for both these weapons, which are presented in the *Introduction to Weaponeering* textbook; however, for reasons of simplicity they are not described in this book. Therefore, readers should choose the Monte Carlo option for both sticks and projectiles in this group.

17.5.8 MULTIPLE ATTACKS

The main spreadsheet calculates the effectiveness of a single attack and produces a value of PD_1 or FD_1. The Multiple Attacks section allows the user to input a required damage level, and the program then calculates the number of attacks needed to achieve the specified damage level, as defined in Chapter 2, Section 2.6.

Note that this will result in a decimal value, and because the number of attacks must be an integer, the result should be rounded up to obtain the number of attacks that will give at least the required damage.

17.5.9 PD FOR INCREASING SORTIES

The program calculates the damage due to a single attack and the number of attacks needed to achieve a user-supplied damage requirement. The program also powers up the PD_1 or FD_1 value for increasing sorties and plots a graph on the main AS User Interface screen, as shown in Fig. 17.14.

In the top right of the sheet shown in Fig. 17.6 there is a Reference window indicating to the user where in the book to find the methodology currently being used to calculate damage.

17.6   SURFACE-TO-SURFACE WEAPONEERING TOOL

If the user selects the Surface-to-Surface tab in the main program screen, the initial user interface screen shown in Fig. 17.15 will appear. This is similar

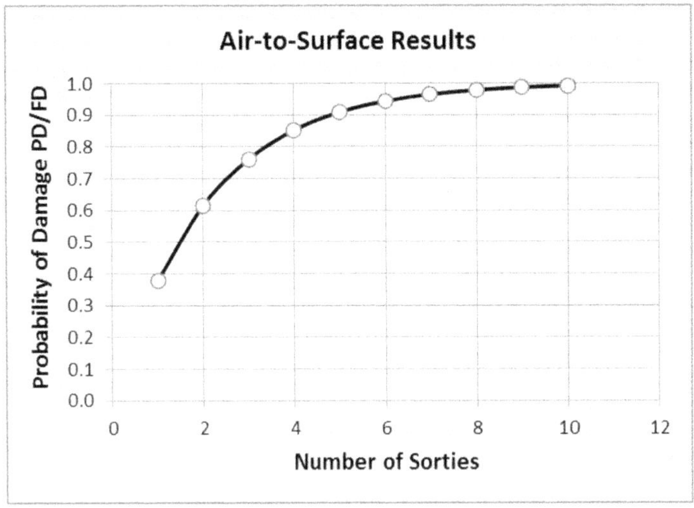

Fig. 17.14 PD for increasing number of sorties.

to the air-to-surface tool interface; however, because both indirect and direct fire systems are supported, two graphs are displayed, one for each modality.

Note that because the Air-to-Surface module comes up first when the program is opened, the choice of textbook is made there, and the same selection applies if the Surface-to-Surface module is opened later. Clicking on the Go

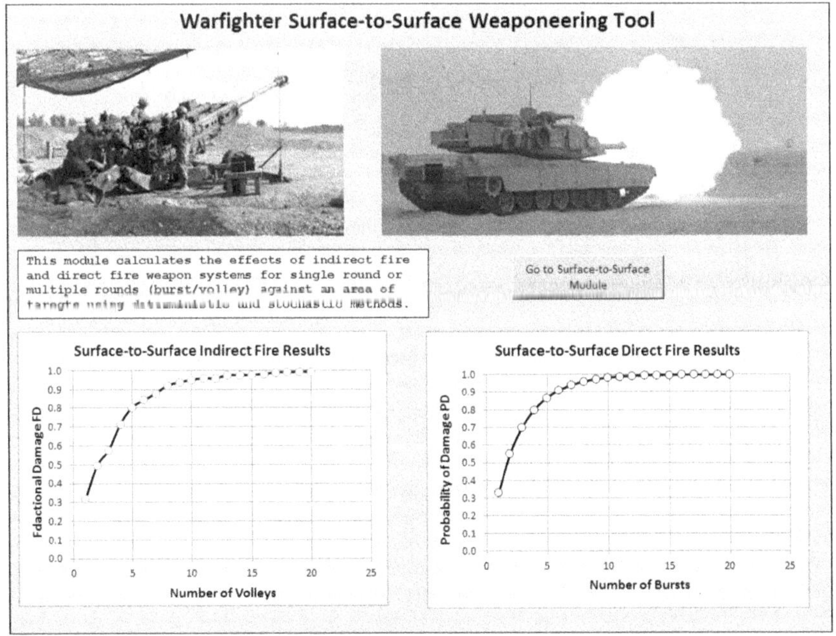

Fig. 17.15 Surface-to-Surface tool user interface.

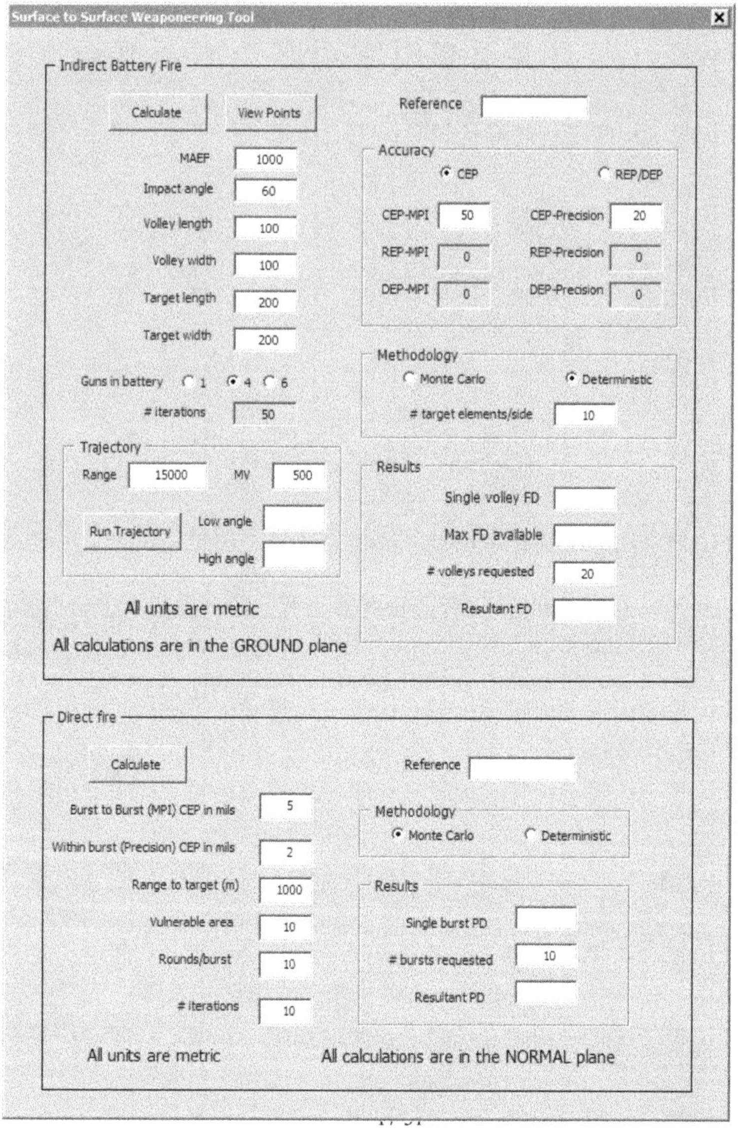

Surface to Surface Weaponeering Tool

Indirect Battery Fire

Calculate   View Points   Reference

MAEF   1000
Impact angle   60
Volley length   100
Volley width   100
Target length   200
Target width   200

Guns in battery   1   4   6
iterations   50

Trajectory
Range   15000   MV   500
Run Trajectory   Low angle
High angle

All units are metric

All calculations are in the GROUND plane

Accuracy
CEP   REP/DEP
CEP-MPI   50   CEP-Precision   20
REP-MPI   0   REP-Precision   0
DEP-MPI   0   DEP-Precision   0

Methodology
Monte Carlo   Deterministic
target elements/side   10

Results
Single volley FD
Max FD available
volleys requested   20
Resultant FD

Direct fire

Calculate   Reference

Burst to Burst (MPI) CEP in mils   5
Within burst (Precision) CEP in mils   2
Range to target (m)   1000
Vulnerable area   10
Rounds/burst   10
iterations   10

All units are metric

Methodology
Monte Carlo   Deterministic

Results
Single burst PD
bursts requested   10
Resultant PD

All calculations are in the NORMAL plane

Fig. 17.16   Surface-to-Surface module interface.

to Surface-to-Surface Module button produces the detailed screen shown in Fig. 17.16.

Each mode, indirect and direct fire, will be described in detail along with the relevant chapters in the book explaining the methodology.

17.6.1 INDIRECT FIRE

This module models artillery battery fire against an area of target elements; there is an option for either a single gun or a four- or six-gun battery. Unitary targets are not usually considered but may be modeled by making the target area dimensions small.

The damage function used is the rectangular cookie cutter where the EI is the MAE_F, and the aspect ratio is determined from the weapon impact angle. Most of the inputs are self-explanatory and are similar to corresponding items in the Air-to-Surface tool, so only a brief description will be provided here:

- *MAE_F:* This is commonly used for unguided artillery shells.
- *Impact angle:* This is either user selected or the high-angle solution from a zero drag ballistics calculation, to be described shortly.
- *Volley length/width:* This may be obtained initially from aimpoint strategy; see Chapter 11, Section 11.2.2 at least as a starting point. Using the View Points option allows the user to adjust for better coverage of the pattern over the target area.
- *Target length/width:* This defines the area of target elements (see Chapter 10).
- *Guns in battery:* This is user selectable—either 1, 4, or 6.
- *Accuracy:* As before, the user can choose CEP or REP/DEP for MPI and precision (see Chapter 4, Section 4.8).
- *Trajectory:* As in the Air-to-Surface module, a zero-drag trajectory model is provided to estimate impact angles for a user-supplied range and muzzle velocity. This produces two solutions, high- and low-angle values, with the program selecting the high-angle solution by default. This calculation is described in Chapter 3, Section 3.4. The user may overwrite the high-angle solution with the low-angle value or input their own impact angle, if known.

The Methodology section contains a choice of Deterministic or Monte Carlo for the effectiveness methodology. As for the Air-to-Surface module, the deterministic method has not been discussed in this book, so readers should select the Monte Carlo option, which implements the methodology described in Chapter 11, Section 11.4.

This requires the user to specify the number of target elements/sides in the enclosing area and the number of Monte Carlo iterations. It is recommended to try small values initially (10/side, 20 iterations); otherwise, long execution times will result.

The outcome of a single volley from all guns in the battery is shown in the Results section as the single volley fractional damage. Recall from Chapter 9, Section 9.6 the difference between dependent and independent sequential attacks. In general, air-delivered sticks of bombs form a sequence of independent attacks, so FD_1 may be powered up using the survivor rule; however, multiple volleys from an artillery battery result in a sequence of dependent attacks, and the survivor rule cannot be used. The double loop Monte Carlo method described in Chapter 9, Fig. 9.12 is used instead. Again, the View Points option may be used to adjust volley dimensions to optimize coverage of the target area, as shown in Fig. 17.17.

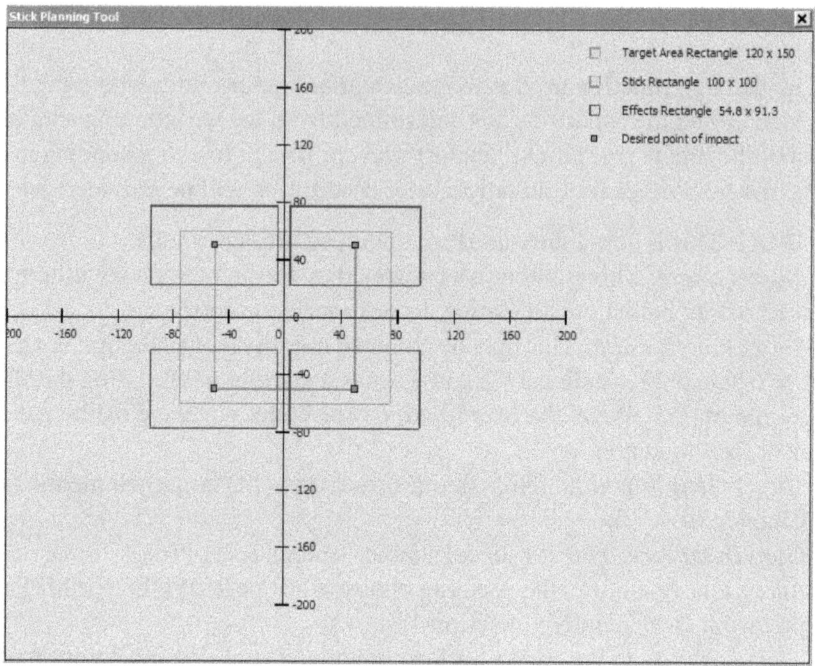

Fig. 17.17   Volley, lethal area, and target geometry.

17.6.2 DIRECT FIRE

This program allows the user to evaluate direct fire weapon effectiveness
against targets represented by a vulnerable area. All calculations are performed in the normal plane and separate burst-to-burst (MPI) and within-
burst (precision) errors to be input using the methods outlined in Chapter 9,
Section 9.6. The single-burst PD is calculated as is the PD corresponding to a
user-supplied number of bursts. Single shots may be simulated by setting
rounds/burst to unity. Again, the PD vs the number of bursts is calculated and
returned to the spreadsheet, as shown in Fig. 17.18.

Note that these programs reflect their status at the time of writing; modifications and additions will occur over time.

17.7 UNCLASSIFIED DATABASE

As previously discussed, the basis of this book is to explain methods to
calculate conventional weapon effects in terms of simple mathematical and
engineering principles, resulting in techniques that are open source with no
classification restrictions. Therefore, the differences between the classified
JWS program and the one just described is that JWS contains classified data
relating to weapon lethality, accuracy, and target vulnerability. In many cases,
entering the same data into both programs will produce the same results.

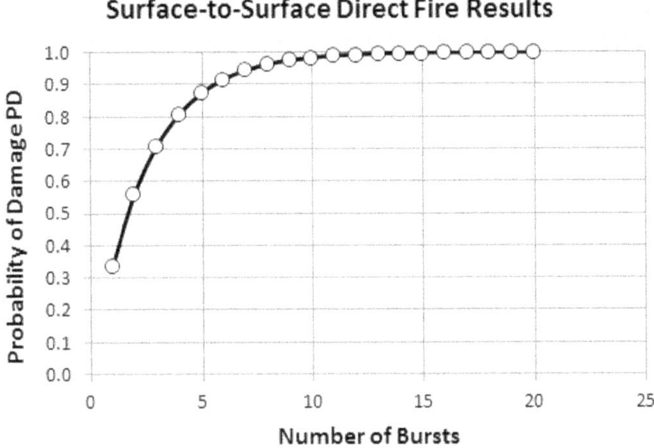

Fig. 17.18   PD for increasing number of bursts.

In order to support promulgation of the weaponeering subject, the JTCG/ME has kindly supplied an unclassified database so that representative results may be obtained from the methods provided in the book. It must be understood, however, that all the data shown are notional; any association with any weapon, target, or their combination should not be inferred, and the data are to be used for educational purposes only.

The database, shown in Table 17.1, contains a variety of surface-mobile targets as well as some infrastructure targets such as buildings, bridges, tunnels, and a Petroleum, Oil and Lubricant (POL) tank. Personnel targets are also addressed. Weapon types are focused on air-delivered weapons and include guided and unguided bombs, rockets, shaped charges, and cluster munitions. Note that not all possible weapon–target pairings have data, because some weapons are not appropriate for all targets (see the Appendix).

Note that this database contains EI values only; in order to compute a PD value, the weapon accuracy also will be required. Many of these data for U.S. weapons are classified, although approximate values may be obtained from public domain sources, as the remaining sections of this chapter will now describe.

17.8   GENERATING DATA FROM OPEN SOURCES: PERSONNEL INJURY

Readers who are connected with the U.S. DoD and have the necessary security clearances may obtain a copy of the classified JWS program and therefore have access to the classified data that it contains. For those not in this position, where can the required data come from? In this section, we

TABLE 17.1 WEAPON–TARGET–KILL LEVEL EI DATABASE

NOTIONAL DATA - FOR EDUCATIONAL PURPOSES ONLY

Target	Kill level	Combined Effects Bomblet	Small High Explosive (HE) Rocket/Missile Warhead <10 lb	Small HE Bomb Warhead	Medium HE Bomb Warhead	Large HE Bomb Warhead	Small HEAT Rocket/Missile Warhead <5" dia.	Large Shaped Charge >5" dia.
		A_V (ft^2), MAE$_F$	MAE$_F$ (ft^2)	MAE$_F$ (ft^2)	MAE$_F$ (ft^2)	MAE$_F$ (ft^2)	A_V (ft^2), MAE$_F$	A_V (ft^2)
MB Tank	K	5 (A_V)	0	750	900	1000	5 (A_V)	30
APC	K	10 (A_V)	50	1500	2000	5000	200	50
SP Howitzer	M	250	200	2500	4000	7000	400	50
MLRS	F	100	300	20000	30000	50000	200	N/A
AA Artillery	F	20	200	5000	10000	15000	100	N/A
Parked Aircraft	PTO-4	600	4000	40000	60000	90000	2000	N/A
Truck	M	100	1500	15000	30000	45000	1000	N/A
Radar	K	0	N/A	2500	4500	10000	N/A	N/A
Bridge - single span grinder	BEI			5	15	40		
Building SS Residential	MAE$_{BLDG}$			4000	7500	15000		
Building MS Military	MAE$_{BLDG}$			500	1000	2000		
Building MS Commercial	MAE$_{BLDG}$			2000	10000	20000		
Personnel standing 5 min assault	MAE$_F$			15000	30000	40000		
Personnel in foxhole 30 sec defense	MAE$_F$			3000	6000	12000		
Tunnel	EMD			10	15	25		
Buried POL Tank	EMD			25	50	75		

investigate a particular case study relating to Table 17.1 as a way to illustrate how open source information may be used to generate approximate vulnerability data for use in the Weaponeering Program.

The case we will investigate is the effect of a general purpose warhead on unprotected personnel targets in the open. From Table 17.1, we see that the lethal area of a large high-explosive (HE) bomb on personnel for the 5-min assault injury criterion is 40,000 ft². If we assume the weapon is the Mk-84 2000-lb. warhead, this corresponds to a circle of radius of about 113 ft. This seems rather small, because one would expect more injuries at a standoff of about 35 m from such a large weapon. Is this number correct? Let us try to generate personnel vulnerability data for this case and compare them to the data provided.

We have seen that injury may be caused by fragments, blast, or both, so we will investigate each damage mechanism in turn. The general approach follows that described earlier in Chapter 8, Section 8.7:

1. *Blast:* Use Kingery-Bulmash data (Chapter 14, Section 14.8) to determine the blast field around the detonation point, and then use the Bowen curves (Chapter 8, Section 8.7.2) to determine the extent of injuries sustained.
2. *Fragments:* Estimate the average number, size, and velocity of fragments generated using the Gurney equations (Chapter 6, Section 6.4), and then use the Sperrazza-Kokinakis approach of Chapter 8, Section 8.7 to determine the probability of injury due to striking fragments.
3. Combine the effects of blast and fragments using the survivor rule.

The process follows the general methods outlined in Chapters 6 and 8 with some approximations and assumptions.

17.8.1 INJURY DUE TO FRAGMENTS

Recall how a particular warhead's fragment characteristics may be defined in terms of the Z-data file. Chapter 8, Section 8.7.1 has the following equation for probability of injury due to each fragment weight group in the Z-data file for the warhead:

$$P_{I-\text{FRAG}(i)} = 1 - \exp\left(-\rho_i A_P P_{K/H(i)}\right) = 1 - \exp A_P\left(-\rho_i P_{I/H(i)}\right) \quad (17.1)$$

Because the Z-data files are restricted, we will represent the fragments in terms of a single average weight, number of fragments, and a single velocity with which they are projected from the detonation point. This simplifies Eq. (17.1) into the following form:

$$P_{I-\text{FRAG}} = 1 - \exp A_P\left(-\rho \times P_{I/H}\right) \quad (17.2)$$

To evaluate this equation, three quantities must be determined:

- The target presented area A_P
- The fragment density ρ
- The probability if incapacitation given a hit $P_{I/H}$

17.8.2 TARGET PRESENTED AREA

We assume a standing man facing the warhead, as shown in Fig. 17.19.
For simplicity, we assume an average height of 5 ft and average width of 1 ft, giving $A_P = 5$ ft^2.

17.8.3 FRAGMENT DENSITY

Just as we consider the blast wave to propagate spherically, we can make the same assumption regarding fragments. Assuming we have a total of N identical fragments and that they are uniformly distributed over the surface of a sphere of increasing radius r, the fragment density in frags/ft^2 is given by the following equation:

$$\rho = \frac{N}{4\pi r^2} \tag{17.3}$$

17.8.4 PROBABILITY OF INCAPACITATION GIVEN A HIT

Based on the work of Sperrazza and Kokinakis, this quantity is given by the following equation:

$$P_{I/H} = 1 - \exp\left[-a\left(mV^{3/2} - b\right)^n\right] \tag{17.4}$$

where a, b, and n depend on the injury criterion of interest (see Table 8.5). Although the presented area is fixed, the fragment density depends on N and

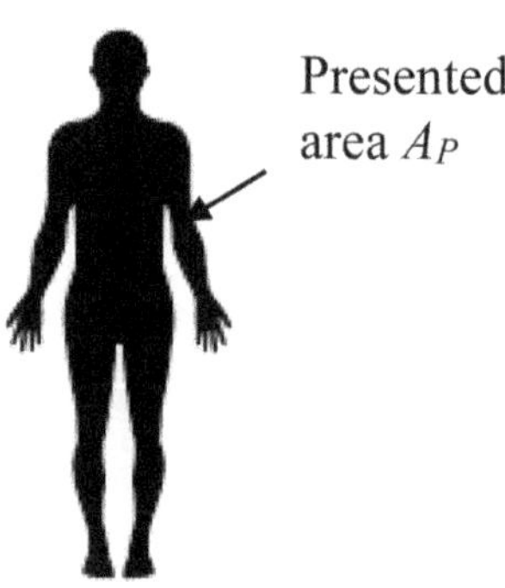

Fig. 17.19 Presented area of personnel target.

TABLE 17.2 PREDICTED FRAGMENT CHARACTERISTICS OF MK-84 BOMB

Gurney Equations for Mk-84 bomb		
Weight of explosive	945	lb
Weight of case	1055	lb
C/M - ratio of expl/case	0.90	
Outside diameter	18.00	inches
Initial velocity	6607.00	ft/sec
Average mass	445.34	grains
Av # frags	16583	
Volume of av fragment	0.220	in^3
Length of equivalent cube	0.604	inch

the distance from the detonation point. Similarly, in Eq. (17.4), the fragment velocity will vary with distance due to drag on the fragment.

It seems that we have to look in more detail at how the fragment velocity changes with distance (i.e., a trajectory calculation), and therefore we also need fragment mass information.

17.8.5 FRAGMENT CHARACTERISTICS

To understand the fragments' characteristics, the Gurney equations described in Chapter 6, Section 6.4.1 are applied to open source data for the weapon under consideration, the general purpose Mk-84 bomb, producing Table 17.2. This shows 16,768 fragments with an average mass of 456 grains and an initial velocity of 6607 ft/s are produced, where each steel fragment may be considered a cube of 0.6-in. per side.

17.8.6 FRAGMENT TRAJECTORY

We now include a trajectory model to calculate the velocity of the fragment as it moves away from the detonation point. The drag force is given by Eq. (3.23), and the resulting equation of motion is solved numerically to find this velocity. A sample trajectory calculation for a representative fragment is shown in Table 17.3.

Notice how the fragment velocity decreases with distance due to the effect of drag. At each specific distance, we can use Eq. (17.3) to determine the fragment density and Eq. (17.4) for the $P_{I/H}$, provided we define a casualty criterion from Table 8.5 in Chapter 8. This table is reproduced as Table 17.4 with the required injury level of 5-min assault shown in bold.

This allows us to extend Table 17.3 to include the density and $P_{I/H}$ as shown in Table 17.5.

TABLE 17.3   TRAJECTORY OF FRAGMENT

Time sec.	Distance ft.	Velocity ft./sec
0.000	0.000	6607.004
0.001	6.547	6547.249
0.002	13.036	6488.570
0.003	19.467	6430.939
0.004	25.841	6374.326
0.005	32.160	6318.706
0.006	38.424	6264.052
0.007	44.634	6210.339
0.008	50.792	6157.544
0.009	56.897	6105.643
0.010	62.952	6054.613

TABLE 17.4   SK PARAMETERS

Casualty Criterion	a	b	n
30-s defense	$8.8771.10^{-4}$	31,400	0.45106
30-s assault	$7.6442.10^{-4}$	31,000	0.49570
5-min assault	$\mathbf{1.0454.10^{-3}}$	**31,000**	**0.48781**
12-h supply	$2.1973.10^{-3}$	29,000	0.44350

TABLE 17.5   INCLUDING FRAGMENT DENSITY AND $P_{I/H}$ CALCULATION

Time sec.	Distance ft.	Velocity ft./sec	Fragment Density	$P_{I/H}$
0.000	0.000	6607.004	N/A	1.000
0.001	6.547	6547.249	30.785	1.000
0.002	13.036	6488.570	7.766	1.000
0.003	19.467	6430.939	3.482	1.000
0.004	25.841	6374.326	1.976	1.000
0.005	32.160	6318.706	1.276	1.000
0.006	38.424	6264.052	0.894	1.000
0.007	44.634	6210.339	0.662	1.000
0.008	50.792	6157.544	0.512	1.000
0.009	56.897	6105.643	0.408	1.000
0.010	62.952	6054.613	0.333	1.000

Finally, we use Eq. (17.2) to determine the probability of incapacitation due to fragments as a function of distance. This is added to the table, a portion of which is shown in Table 17.6.

TABLE 17.6   PROBABILITY OF INCAPACITATION DUE TO FRAGMENTS

Time sec.	Distance ft.	Velocity ft./sec	Fragment density	$P_{I/H}$	P_{inc} frag
0.000	0.000	6607.004	N/A	1.000	1.000
0.001	6.547	6547.249	30.785	1.000	1.000
0.002	13.036	6488.570	7.766	1.000	1.000
0.003	19.467	6430.939	3.482	1.000	1.000
0.004	25.841	6374.326	1.976	1.000	1.000
0.005	32.160	6318.706	1.276	1.000	0.998
0.006	38.424	6264.052	0.894	1.000	0.989
0.007	44.634	6210.339	0.662	1.000	0.964
0.008	50.792	6157.544	0.512	1.000	0.923
0.009	56.897	6105.643	0.408	1.000	0.870
0.010	62.952	6054.613	0.333	1.000	0.811

The results are presented graphically in Fig. 17.20.

It makes sense to terminate the calculations when P_I reaches some small value like 1%, which for this case occurs when $r = 480$ ft. This concludes the section on vulnerability to fragments, so now we turn to blast.

17.8.7   INJURY DUE TO BLAST

In order to calculate the probability of incapacitation due to blast, the Bowen curves shown in Fig. 8.15 will be used. By defining points along the graph, an Excel plot may be reproduced, as shown in Fig. 17.21.

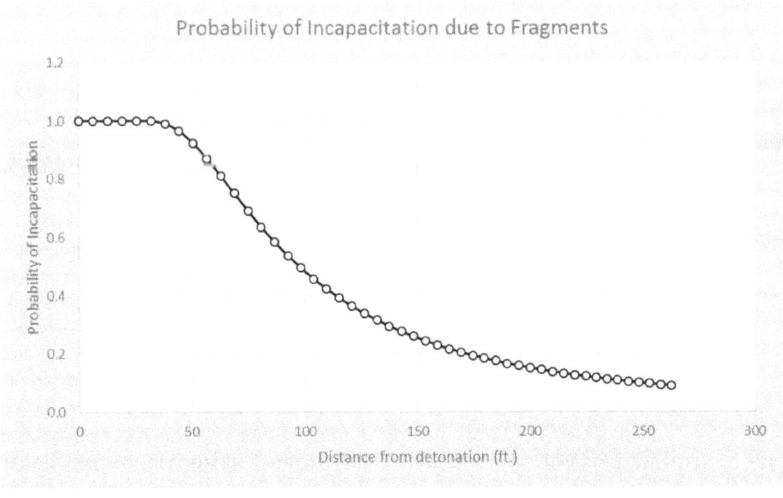

Fig. 17.20 $P_{I/H}$ for fragments as a function of distance.

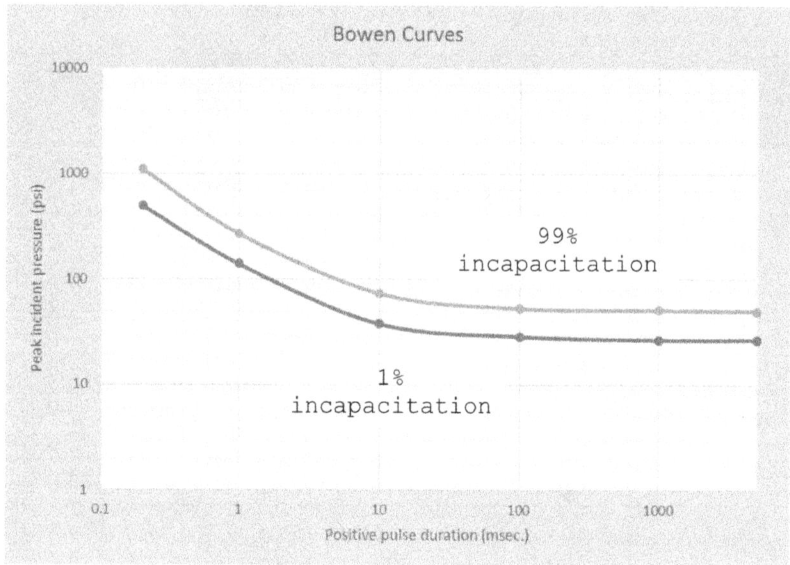

Fig. 17.21   Bowen curves for blast injury.

In this figure, the 99% and 1% incapacitation injury levels will be used rather than the less severe lung and eardrum damage, meaning any area above the lines is total incapacitation, and any area below is no incapacitation. The area between the lines will be interpolated for the injury level.

The peak pressure and positive pulse duration can be obtained from the Kingery-Bulmash curves in Fig. 14.10 and their mathematical form from Tables 14.2. Recall the polynomial forms are defined as follows:

$$Z = \frac{r}{W^{1/3}} \tag{17.5}$$

and we define

$$T = \log_n(Z) \tag{17.6}$$

Then

$$P_{\text{MAX}} = \exp\left[A_1 + B_1 T + C_1 T^2 + D_1 T^3 + E_1 T^4 + F_1 T^5 + G_1 T^6\right] \tag{17.7}$$

$$\Delta t = \exp\left[A_2 + B_2 T + C_2 T^2 + D_2 T^3 + E_2 T^4 + F_2 T^5 + G_2 T^6\right] \tag{17.8}$$

A table similar to Table 17.6 may be developed using the same distances specified to evaluate $P_{I/H}$ for fragments, the first part of which is shown in Table 17.7.

TABLE 17.7   PRESSURE-DURATION DATA

Distance ft.	Z	T	Duration msec	Peak Pressure psi
0.00	0.00	N/A	N/A	N/A
6.55	0.74	−0.30	1.51	1511.12
13.04	1.48	0.39	2.46	544.33
19.47	2.21	0.79	7.46	257.72
25.84	2.93	1.08	14.54	141.73
32.16	3.65	1.29	14.73	86.72
38.42	4.36	1.47	13.97	57.55
44.63	5.07	1.62	13.86	40.71
50.79	5.76	1.75	14.56	30.31
56.90	6.46	1.87	16.03	23.54
62.95	7.15	1.97	18.22	18.94

This pressure-duration data may now be transferred onto the Bowen curves, as shown in Fig. 17.22, where each data point corresponds to a known distance from the detonation. Also shown on the chart are a few significant points.

This allows us to generate another table showing P_I due to blast for the same distances that were used for fragments, the first portion of which is shown in Table 17.8.

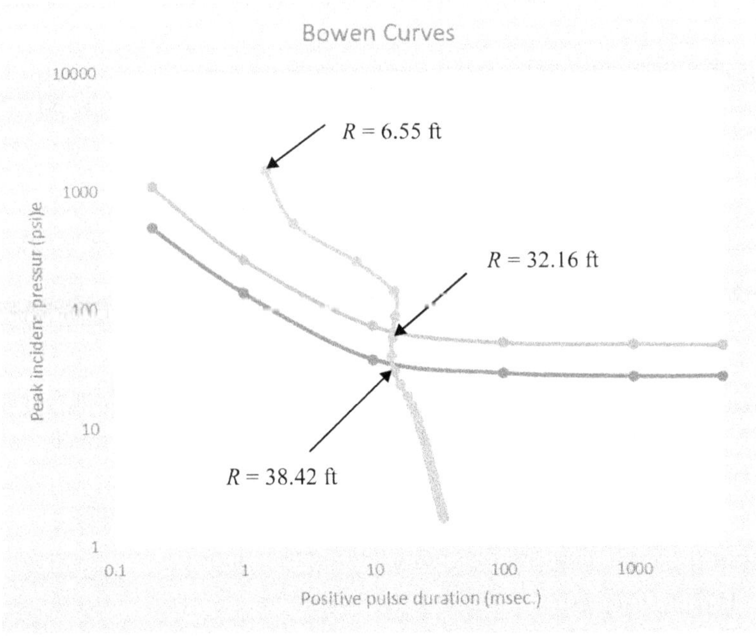

Fig. 17.22   Pressure-duration data superimposed on Bowen curves.

TABLE 17.8 P_I FOR BLAST

Distance ft.	Z	T	Duration msec	Peak pressure psi	P_{inc} blast
0.00	0.00	N/A	N/A	N/A	1.00
6.55	0.74	−0.30	1.51	1511.12	1.00
13.04	1.48	0.39	2.46	544.33	1.00
19.47	2.21	0.79	7.46	257.72	1.00
25.84	2.93	1.08	14.54	141.73	1.00
32.16	3.65	1.29	14.73	86.72	1.00
38.42	4.36	1.47	13.97	57.55	0.67
44.63	5.07	1.62	13.86	40.71	0.11
50.79	5.76	1.75	14.56	30.31	0.00
56.90	6.46	1.87	16.03	23.54	0.00
62.95	7.15	1.97	18.22	18.94	0.00

Observe that where the P-I curve for the blast wave crosses the 99% incapacitation Bowen curve, its position corresponds to RB_1; where it crosses the
1% Bowen curve, the position defines RB_2. Notice also how, for this case,
blast lethality drops quickly and has disappeared at distances greater than
about 50 ft. Now that the effects of fragments and blast have been calculated,
they are combined using the survivor rule.

$$P_I = 1 - \left\{ \left[1 - P_{I-\text{BLAST}}\right]\left[1 - P_{I-\text{FRAG}}\right] \right\} \tag{17.9}$$

This result is generated in tabular form and the results plotted in Fig. 17.23.
Because the effect of blast is small except near the detonation point, Fig.
17.23 is almost identical to Fig. 17.20.

For our assumptions, the fragment propagation is radially spherical about
the detonation point, as is the blast, so the MAE_F may be calculated by direct
integration, as described for the purely blast warhead in Chapter 8, Section 8.6.

$$\text{MAE}_F = \int_{r=0}^{r=\infty} P_K(r) \times \mathrm{d}A = \int_{r=0}^{r=\infty} P_K(r) \times 2\pi r \mathrm{d}r \tag{17.10}$$

Because Excel can represent the curve in Fig. 17.23 in polynomial form, it
is easily integrated to give $\text{MAE}_F = 104{,}153$ ft^2. If we consider this lethal area
to be a cookie cutter, its radius becomes $R_B = 182$ ft.

For this case, here are some observations regarding the results:

- There are three main sources contributing to lethality:

  - The blast wave
  - The lethality of a fragment impact
  - The probability of being hit by a fragment

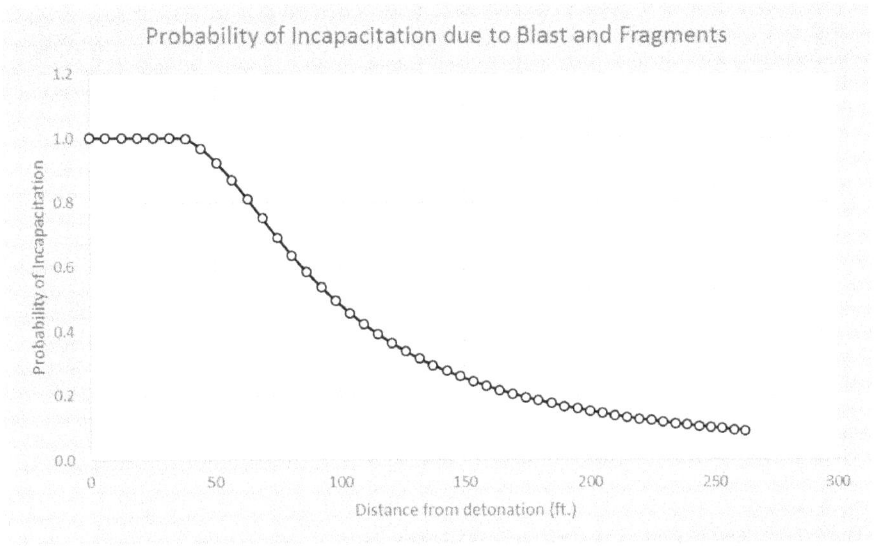

Fig. 17.23 P_I **for both blast and fragments.**

TABLE 17.9   LETHAL AREAS FOR SOME APPROPRIATE WEAPONS

Warhead→	Mk-84	Mk-83	Mk-82	M107	GMLRS	ATACMS
Injury criterion ↓						
30 sec defense	100767	77947	58920	16126	86162	103216
30 sec assault	103978	82476	64266	18362	98759	110072
5 min assault	104153	82923	64904	18696	101255	110750

- The effect of blast is very limited—to within about 50 ft from the detonation point.
- Fragments are lethal out to over 1000 ft; however, the probability of being hit decreases by $1/r^2$; therefore, this is the dominant factor once outside the blast effects distance.
- The value provided in the database of 40,000 ft^2 is an underestimate, so for personnel injury, Table 17.9 should be used.

A similar process can be used for other weapons not provided in the table.

17.9   GENERATING DATA FROM OPEN SOURCES: MATERIEL TARGETS

As we saw in Chapters 6–8, this is a very complex and expensive process that invariably uses classified data to produce lethal areas. For those who do

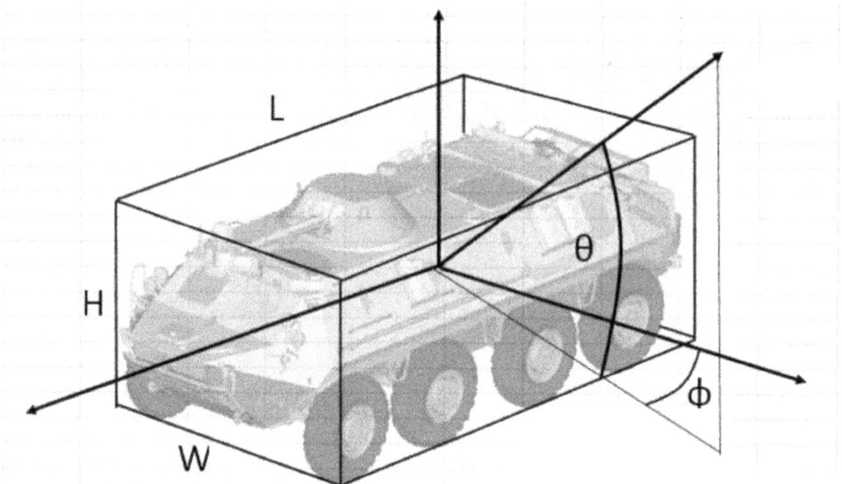

Fig. 17.24   Vehicle target enclosed in a bounding box.

not have access to this type of vulnerability data, can anything be done to give
an approximate answer?

Some methods can and will be proposed, but these are no substitute for the
rigorous techniques explained in earlier chapters. The case to be considered is
surface mobile targets such as vehicles when attacked with penetrating munitions. This covers armored and unprotected military vehicles when attacked with
gun rounds or shaped charge warheads. Consider the target shown in Fig. 17.24.

Here the target is enclosed in the smallest possible parallelepiped, sometimes referred to as a bounding box or shoe box. The next step is to determine
the presented area as seen from the direction from which the penetrator
approaches the target. These directions are consistent with the azimuth and
elevation angles shown in Fig. 7.11; that is, azimuth angles ϕ step in 45-deg
increments, and elevation angles θ step in 30-deg increments. The front, side,
and top areas are given by the following:

$$A_{\text{FRONT}} = H \times W \quad A_{\text{SIDE}} = H \times L \quad A_{\text{TOP}} = L \times W \qquad (17.11)$$

It may be shown that the presented area of the target when viewed from a
direction defined by θ and ϕ is given by

$$A_P = A_{\text{TOP}} \sin\theta + A_{\text{FRONT}} \cos(\theta)\sin(\phi) + A_{\text{SIDE}} \cos(\theta)\cos(\phi) \qquad (17.12)$$

For the target shown, the open source data shown in Table 17.10 may be
obtained. Using Eq. (17.12), presented areas can be calculated for a range of
azimuth and elevation angles, as shown in Table 17.11.

It may be argued that the orientation of the vehicle when attacked may not
be known; therefore, we could average these presented areas over azimuth,

TABLE 17.10   Target Dimensional Data

Dimensions	m
L	6.74
W	2.94
H	2.07
Areas	**m²**
Side	13.93
Front	6.08
Top	19.80

elevation, or both. If we average over both, the result is $A_P = 20.03$ ft². Although the angles are varied over only the first quadrant, the symmetry of this target means no other angles are required.

If we reduce the presented area to a square in the normal plane, then from Chapter 9, Section 9.5, we have the situation represented by Fig. 17.25.

In the figure,

$$L_P = W_P = \sqrt{A_P} \tag{17.13}$$

The probability that one round will fall inside the square is calculated in the usual way.

$$P_{\text{HIT}-X} = \left(\text{ND}\left(L_P/2, 0, \sigma_X, 1\right) - \text{ND}\left(-L_P/2, 0, \sigma_X, 1\right) \right) \tag{17.14}$$

TABLE 17.11   Presented Area of Bounding Box

Azimuth↓	Elevation ↓										
	0	10	20	30	40	50	60	70	80	90	El. Average
0	13.93	17.15	19.86	21.96	23.40	24.12	24.11	23.37	21.92	19.80	20.96
10	14.77	17.99	20.65	22.69	24.04	24.66	24.53	23.66	22.07	19.80	21.49
20	15.17	18.38	21.03	23.04	24.35	24.92	24.73	23.79	22.13	19.80	21.73
30	15.10	18.31	20.96	22.98	24.30	24.88	24.70	23.77	22.12	19.80	21.69
40	14.58	17.79	20.47	22.52	23.89	24.54	24.44	23.59	22.03	19.80	21.37
50	13.61	16.84	19.56	21.69	23.15	23.92	23.95	23.26	21.86	19.80	20.77
60	12.23	15.48	18.26	20.49	22.10	23.03	23.26	22.79	21.62	19.80	19.91
70	10.48	13.76	16.62	18.97	20.75	21.90	22.39	22.19	21.32	19.80	18.82
80	8.41	11.72	14.67	17.18	19.17	20.57	21.35	21.48	20.96	19.80	17.53
90	6.08	9.43	12.49	15.17	17.39	19.08	20.19	20.69	20.56	19.80	16.08
Az. Average	12.43	15.68	18.46	20.67	22.25	23.16	23.37	22.86	21.66	19.80	20.03

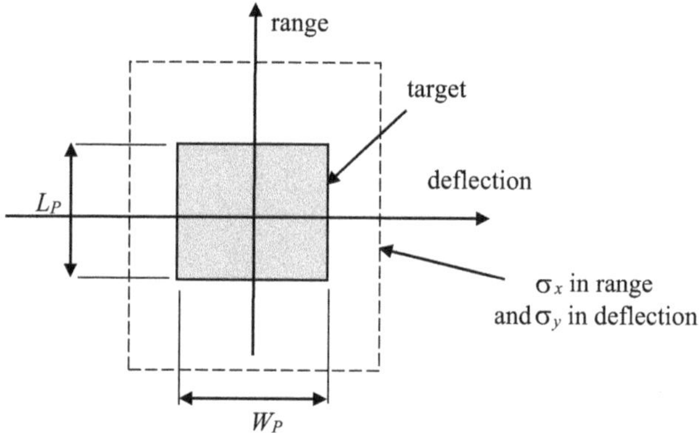

Fig. 17.25   Presented area in the normal plane.

$$P_{\text{HIT}-Y} = \left(\text{ND}\left(W_P/2,0,\sigma_Y,1\right) - \text{ND}\left(-W_P/2,0,\sigma_Y,1\right)\right) \qquad (17.15)$$

$$P_{\text{HIT}} = P_{\text{HIT}-X} \times P_{\text{HIT}-Y} \qquad (17.16)$$

It must be recognized that the volume, and hence presented area, of the target is smaller than that of the bounding box, so P_{HIT} in Eq. (17.16) must be multiplied by a volume ratio factor K_V to adjust for this. With consideration to Fig. 17.24, an estimate of $K_V = 0.9$ seems appropriate. For a target such as the one shown in Fig. 17.26, some judgement is needed in first defining the bounding box and then estimating a value of K_V.

Clearly, P_{HIT} is the probability of hitting the target, not the probability of killing it, so again some judgement is needed regarding the probability of kill given a hit because

$$P_K = P_{K/H} \times P_{\text{HIT}} \qquad (17.17)$$

The target shown in Fig. 17.24 consists basically of an armored box; therefore, a hit almost anywhere might be expected to produce a kill, so a value of $P_{K/H} = 0.8$ might be used. For other targets, however, $P_{K/H}$ might be smaller. Consider the tank shown in Fig. 17.27, where the crew are enclosed in a protected hull inside the outer armor of the tank.

For a target such as this, a value of $P_{K/H} = 0.3$ might be more realistic, depending on the kill definition. The final result, therefore, is given by the following equation:

$$P_K = P_{\text{HIT}} \times K_V \times P_{K/H} \qquad (17.18)$$

Fig. 17.26   Target with a choice of bounding boxes.

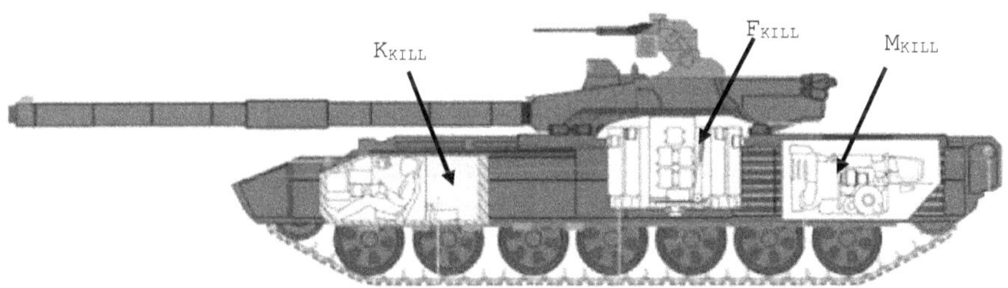

Fig. 17.27   Target with interior protection.

where P_{HIT} can be obtained from an average value of bounding box presented areas, and the other two terms estimated using subject matter expert judgement. For the Armored Personnel Carrier (APC) target shown in Fig. 17.24, we can write for the vulnerable area of a penetrating round

$$A_V = A_P \times K_V \times P_{K/H} = 20.03 \times 0.9 \times 0.8 = 14.4 \mathrm{ft}^2 \qquad (17.19)$$

The number given in the database is 50 ft^2, which seems large given the maximum presented area in Table. 17.11 is around 25 ft^2.

17.10   GENERATING DATA FROM OPEN SOURCES: EI SCALING

Suppose we have an EI value for a particular warhead against a given target and damage level, and we have another warhead for which we do not have an EI value. Can we estimate the unknown EI from the known value? It seems that the EI should scale with the size of the weapon—the larger the weapon, the larger the EI for the same target and damage level. But how does it scale?

Recall from Chapter 14, Section 14.4 how air-blast parameters could be related to different warheads using Hopkinson scaling, where a length or distance R was scaled by dividing by the cube root of the TNT equivalent bare charge explosive weight.

$$Z = \frac{R}{W_e^{1/3}} \tag{17.20}$$

This enabled a particular parameter such as peak pressure to be scaled from known values for a given warhead to an unknown warhead with a different weight of explosive fill. Although this scaling would apply to purely blast warheads, can we scale a more general purpose blast/frag warhead?

Consider any EI that represents an area (ft^2 or m^2) comprising MAE_F, MAE_B, or MAE_{BLDG}. A characteristic length may be defined for these EIs as the square root of the EI value, and this corresponds to the linear distance R in Eq. (17.20). The variable W in this equation is the TNT equivalent bare charge weight, but it could be argued that if the effect of fragments is to be considered in the scaling, W should represent the effect of the case as well as the fill. Therefore, we will scale by the all-up weight of the warhead and not just the weight of the explosive fill.

In order to test this hypothesis, consider the upper part of the target vulnerability database shown in Table 17.1, where MAE_F values are presented for three weapons and a variety of materiel targets. This is reproduced in Table 17.12.

Assuming the weapons correspond to warheads of all-up weights of 500 lb, 1000 lb, and 2000 lb, respectively, the following scaling should apply to all weapons for the same target and kill definition.

$$\frac{EI^{1/2}}{W^{1/3}} = \text{constant} \tag{17.21}$$

Applying this scaling to the values in Table 17.12 should give the same number for all three weapons for a given target and kill definition, the result being in Table 17.13.

Recognizing that in order to keep the original database unclassified, the data are approximated anyway, it appears that this scaling produces reason-

TABLE 17.12   MAE$_F$ FROM THE DATABASE FOR EIGHT TARGETS

Target	Kill Level	Small HE Bomb Warhead	Medium HE Bomb Warhead	Large HE Bomb Warhead
		MAE$_F$ (ft^2)	MAE$_F$ (ft^2)	MAE$_F$ (ft^2)
MB Tank	K	750	900	1000
APC	K	1500	2000	5000
SP Howitzer	M	2500	4000	7000
MLRS	F	20000	30000	50000
AA artillery	F	5000	10000	15000
Parked Aircraft	PTO-4	40000	60000	90000
Truck	M	15000	30000	45000
Radar	K	2500	4500	10000

TABLE 17.13   SCALED EI VALUES

Target	Kill Level	Small HE Bomb Warhead	Medium HE Bomb Warhead	Large HE Bomb Warhead
		MAE$_F$ (ft^2)	MAE$_F$ (ft^2)	MAE$_F$ (ft^2)
MB Tank	K	3.45	3.00	2.51
APC	K	4.88	4.47	5.61
SP Howitzer	M	6.30	6.32	6.64
MLRS	F	17.81	17.31	17.74
AA artillery	F	8.91	10.00	9.72
Parked Aircraft	PTO-4	25.19	24.48	23.80
Truck	M	15.42	17.31	16.83
Radar	K	6.30	6.71	7.93

able values for the given data, and therefore can be used to generate values for other warheads. Using the medium-size bomb as a reference, open source data, and estimated values for warhead weights shown in Table 14.3, the lethal areas for some surface-launched weapons may be calculated; the results are shown in Table 17.14.

For EIs that are not areas but linear measurements (BEI, EMD), these should scale as follows:

$$\frac{EI}{W^{1/3}} = \text{constant} \tag{17.22}$$

In the absence of actual data, EI scaling is a reasonable way to extend known data to other warheads.

TABLE 17.14 ESTIMATED EI VALUES FOR SURFACE-LAUNCHED WEAPONS

Target	Kill Level	Medium HE Bomb Warhead	M107 155 mm Artillery Shell	ATACMS	GMLRS
All up weight lb.		1000	95	1000	400
		Scaled MAE_F (ft²)	MAE_F (ft²)	MAE_F (ft²)	MAE_F (ft²)
MB Tank	K	3.00	187	900	488
APC	K	4.47	416	2000	1085
SP Howitzer	M	6.32	832	4000	2170
MLRS	F	17.31	6241	30000	16273
AA artillery	F	10.00	2080	10000	5424
Parked Aircraft	PTO-4	24.48	12481	60000	32546
Truck	M	17.31	6241	30000	16273
Radar	K	6.71	936	4500	2441

17.11 CHAPTER SUMMARY

- A brief overview of computational methods associated with JMEM development was provided, starting with manual hand methods and progressing to implementation on personal computers.
- An unclassified Weaponeering Program was described embodying most of the methodologies discussed in the book. This program deals with air-to-surface and surface-to-surface attacks separately.
- To support the Weaponeering Program, an unclassified database was presented. The effectiveness indices provided allow many realistic scenarios to be explored for a wide range of weapons, targets, and kill definitions.
- To address the issue of how to generate lethality data if access to restricted sources such as the JWS program is not available, three examples were developed:
  - Personnel injury due to a general purpose warhead
  - Materiel target attacked with a penetrating warhead
  - EI scaling

Chapter 18

APPLICATIONS AND CASE STUDIES

18.1 INTRODUCTION

In this chapter, we will apply the techniques learned in the previous chapters and implemented in the Weaponeering Program discussed in Chapter 17 to several case studies involving a wide variety of weapons, targets, and attack scenarios. The intent isn't to cover all possible cases the reader might encounter, but rather to provide a guide along a common path so that any scenario may be examined by using the program in the most efficient manner.

The basic process in deciding which tool to use and what to enter into the various data fields consists of asking and then answering a series of questions.

1. Can the attack be considered an air-delivered weapon or a surface-delivered weapon? This will determine whether the Air-to-Surface or Surface-to-Surface tool (tab) is selected.
2. How is the accuracy specified? The choices are ground plane, normal plane, in feet/meters or mils, circular error probable (CEP), range error probable (REP) and deflection error probable (DEP), or GPS guided.
3. Does the selected weapon have a unitary blast-fragmentation warhead or a hit-to-kill warhead? This will determine the appropriate effectiveness index (EI) to use.
4. Is there a single target or a collection of targets? The result will be a probability of damage or fractional damage, respectively.
5. Will the attack use one weapon or several weapons? If several weapons are used, are they delivered dependently or independently of each other?
6. How will the terminal conditions be determined? Is a trajectory program available to supply the data, or will the weaponeer estimate impact angle, for example? Is the simple zero drag model in the program accurate enough for this weapon delivery?

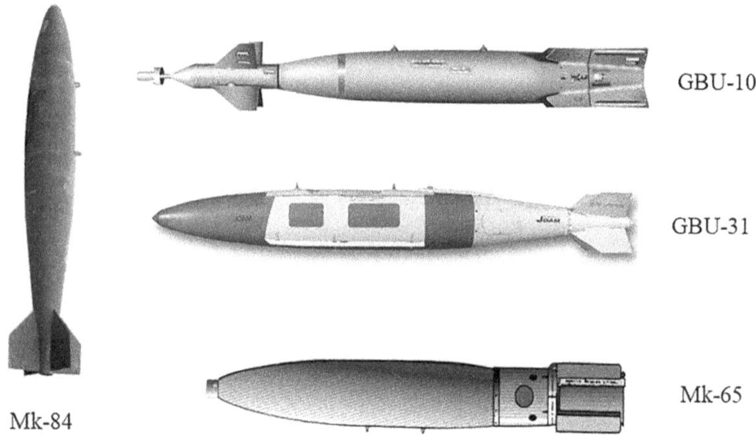

Fig. 18.1   2000-lb bomb variants.

7. Where will the numerical data come from? Are the data available in the unclassified database appropriate for this attack; if not, what process will be used to estimate it?

Clearly, one analytical path will not suit all cases studied. Therefore, for each example some discussion of the train of thought (discussion points) will be provided to illustrate how a particular scenario may be translated into the selection of the appropriate tool and specification of the necessary data in order to compute a damage estimate.

One further point should be made regarding the fundamental weaponeering process itself. In a general sense, weaponeering has been defined as the selection of an appropriate weapon with which to attack a specific target in order to meet desirable requirements—maximize effectiveness, minimize risk to military personnel, minimize collateral damage, and so on. In practice, the selection of an appropriate weapon is somewhat limited, as shown in the table in the Appendix and based on the experience of operational weaponeers who are involved in matching weapons to targets. In many cases, therefore, the choice of weapon may be obvious; however, different characteristics of the variants of this weapon will have considerable impact on the effectiveness.

For example, consider the general-purpose Mk-84, 2000-lb-class bomb shown in Fig. 18.1.

This figure shows the basic, unguided bomb on the left and three variants on the right, consisting of the GBU-10 laser guided, GBU-31 GPS guided JDAM, and Mk-65 Quick Strike sea mine versions. All of these weapons have the same bomb body and therefore we might expect to see the same Z-data file for each; however, the guidance systems are different, resulting in different accuracies. This, in turn, will affect the damage sustained by the target.

EXAMPLE 18.1: GBU-12 LASER-GUIDED BOMB AGAINST A TANK

The example shown in Fig. 18.2 is an A-10 aircraft delivering the GBU-12 against the target.

Discussion points:

- The GBU-12 is a 500-lb-class laser-guided bomb, so the scenario is a single air-launched weapon directed against a unitary target with fuzing set for impact, proximity, or airburst.
- The Air-to-Surface tool will be used. Accuracy is specified in the ground plane. Because we are dealing with a guided weapon, set the ballistic error to zero, and $P_{\text{HIT}} = 0$, $P_{\text{NM}} = 1$ for a Gaussian distribution.
- We could assume a preprogrammed impact angle of 80 deg, however we will use the trajectory calculator for a ballistic release. For example, a release at 10,000 ft at 400 kt gives an impact angle of about 50 deg.
- Assuming clear weather so a line of sight can be established between the laser designator and the target, accuracy will be good, perhaps a CEP of 10 ft.
- From the database, we assume an MAE_F of 750 ft^2, which corresponds to a small bomb/main battle tank/K-kill damage criterion.

We will compose a table of input data for this problem. It is shown in Table 18.1.

The AS module for this example illustrating the data input and calculated result is shown in Fig. 18.3.

The probability of achieving a K-kill for a single attack is 59%. For completeness, the View Points window is displayed in Fig. 18.4 and shows the single weapon lethal area of 19.1 ft × 39.3 ft (= 750 ft^2) associated with the target and the zero error impact point at the origin.

Fig. 18.2   Laser-guided bomb vs tank target.

TABLE 18.1 WEAPONEERING PROGRAM INPUT DATA

Data	Type	Value
Accuracy—ground plane	CEP for guided weapon	10 ft
EI is an MAE_F	HE blast/frag warhead—airburst or PD fuze	750 ft²
Impact angle	Ballistic	50 deg
Target dimensions	Point	Unitary
Number of weapons	Single release	One

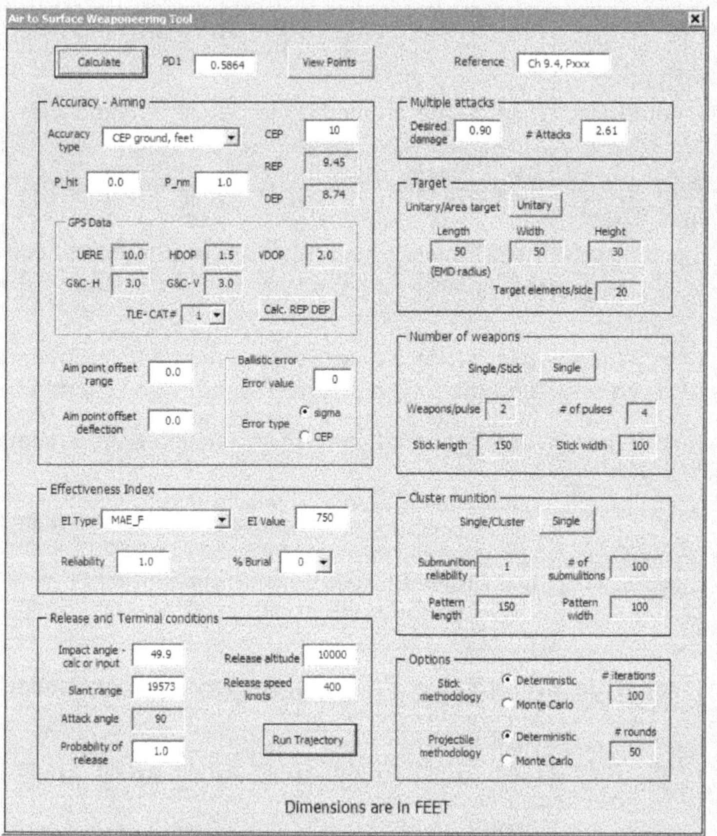

Fig. 18.3 AS module for laser-guided bomb vs tank target.

Although rare for this kind of attack, the effect of multiple independent shots may be seen in Fig. 18.5 as displayed in the main Air-to-Surface Weaponeering tool interface.

The user may vary the accuracy and EI value to confirm the trends expected are those predicted by the model.

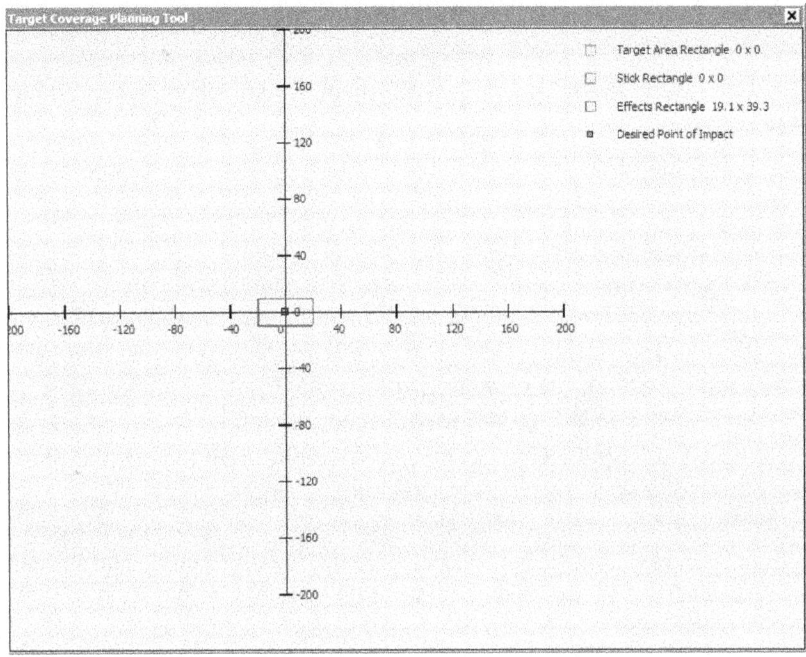

Fig. 18.4   View Points window for single weapon attack.

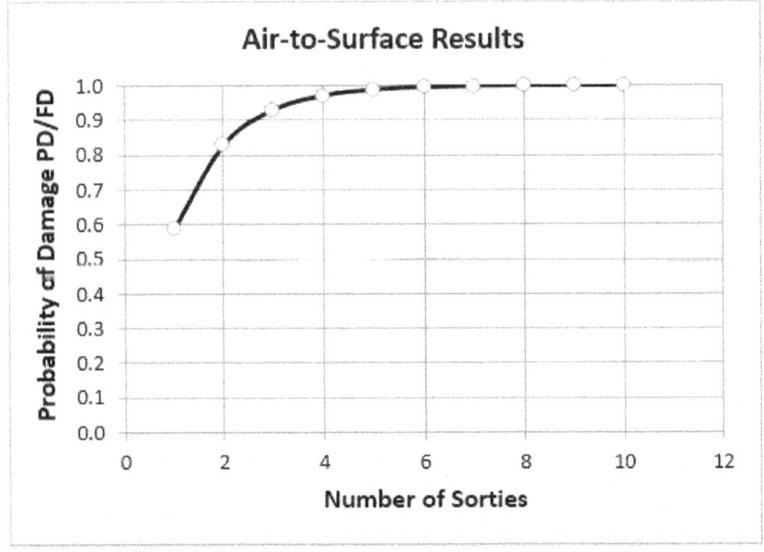

Fig. 18.5   PD for increasing number of attacks.

EXAMPLE 18.2: JAVELIN GUIDED MISSILE AGAINST A TANK

Suppose the same target is attacked with a shoulder-launched, man-portable, Javelin guided missile with a shaped charge warhead, as shown in Fig. 18.6.

Discussion points:

- Again, this is a single weapon used against a unitary target; however, the warhead type dictates a direct hit is required.
- The appropriate EI is the vulnerable area, and the accuracy will be specified in mils—all calculations being done in the normal plane.
- We assume a total accuracy of 1 mil representing the root sum squared (RSS) of mean point of impact (MPI) and precision errors, which is allowable for a single shot. The standoff distance is assumed to be 1000 ft, and a circular normal distribution in the normal plane is a reasonable assumption for a guided weapon.
- From the database, a vulnerable area A_V of 5 ft^2 corresponding to a small shaped charge warhead against a tank is selected. The Javelin is at exactly 5 in. diameter, so either data could be used.
- We use the Surface-to-Surface tool and the direct fire section configured for a single shot.

The data used for this example are shown in Table 18.2. Note that because we are dealing with a single weapon and the burst-to-burst accuracy and within-burst accuracy are combined into a single value of 1 mil, any combination may be used. In this case, the burst-to-burst CEP is set equal to unity, and the within-burst is set to zero.

Figure 18.7 shows the data converted to metric entered into the program, which produce the single shot PD = 0.66. Running the program a few times determines that it executes quickly, and 30,000 iterations are acceptable. The effect of multiple, independent attacks is again

Fig. 18.6   Javelin vs tank.

TABLE 18.2   WEAPONEERING INPUT DATA AND RESULT

Data	Type	Value
Accuracy—normal plane	CEP for guided weapon	1 mil
EI is an A_V	Shaped charge warhead	5 ft^2
Slant range	Linear distance	1000 ft
Target dimensions	Point	Unitary
Number of weapons	Single shot	One

shown in the Surface-to-Surface interface if more than one shot is required.

This solution could also be obtained using the Air-to-Surface module if, for example, a similar weapon were fired from an airborne platform. The appropriate section of the Air-to-Surface module is shown in Fig. 18.8, giving the same result.

The result for a single shot is about 0.66; however, this solution uses a deterministic approach rather than Monte Carlo based, as described by Eqs. (9.15) through (9.18). The results of several shots may be seen in Fig. 18.9.

If a single shot is required but result is too small, either a different weapon could be selected or the firing range decreased, if possible.

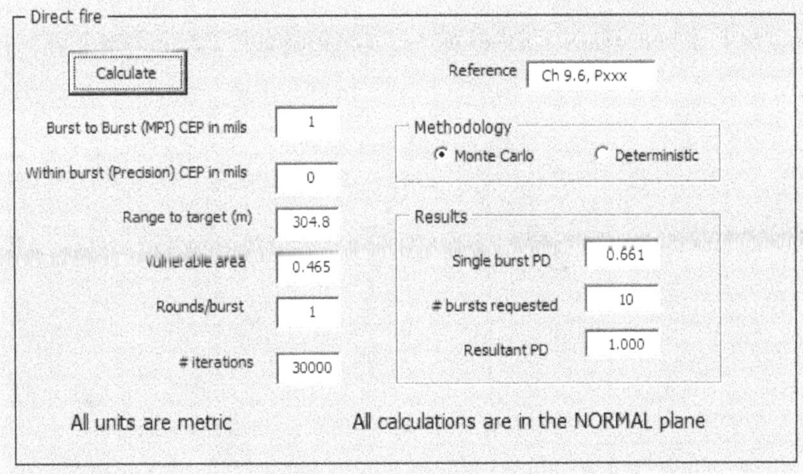

Fig. 18.7   Javelin vs. tank – SS Direct Fire option.

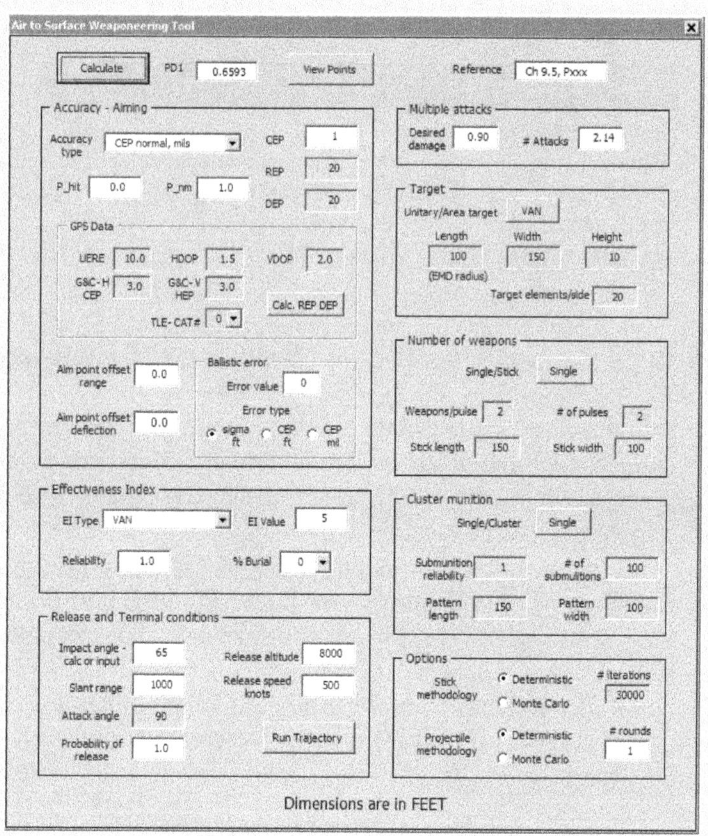

Fig. 18.8   Javelin vs tank—Air-to-Surface module option.

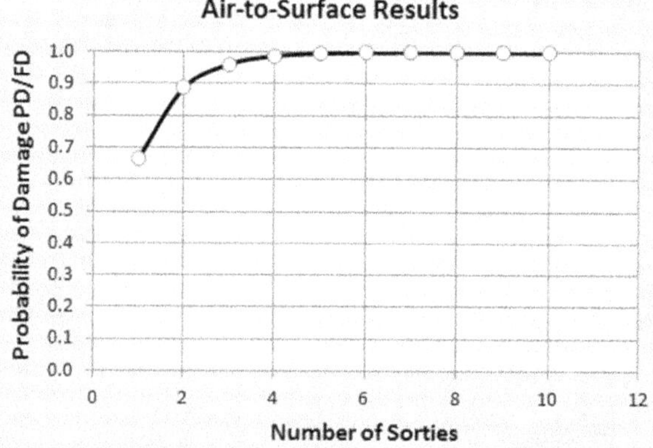

Fig. 18.9   PD for increasing number of shots.

EXAMPLE 18.3: ATTACK AGAINST A MILITARY BUILDING

A satellite image of the target building is shown in Fig. 18.10.
Discussion points:

- Assume the two-story building dimensions (50 ft × 40 ft × 25 ft) are obtained from scaled satellite imagery and photographs, and that the building is constructed to military standard building codes.
- Several weapons could be used to attack this target, but assuming it is beyond the range of surface-launched rockets, this implies an air attack. We will try a GBU-31 Joint Direct Attack Munition (JDAM), which is a GPS/INS-guided bomb (see Fig. 18.11). It may be represented by the large high explosive (HE) bomb warhead listed in the database.

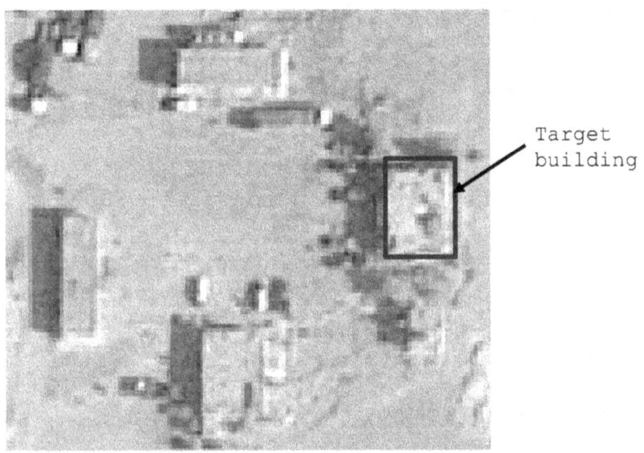

Fig. 18.10   Building target.

Fig. 18.11   GBU-31 JDAM.

TABLE 18.3   AS WEAPONEERING TOOL INPUT DATA

Data	Type	Value
Accuracy—ground plane	GPS with CAT2 level TLE	Calculated
EI is an MAE_{BLDG}	Military	2000 ft^2
Impact angle	Programmed	80 deg.
Target dimensions	Satellite image	50 ft × 40 ft × 25 ft
Required structural damage	Single release	50%

- The appropriate EI is the MAE_{BLDG}, so a value of 2000 ft^2 is selected from the database.
- This is a GPS-guided weapon where the target coordinates are known to within about 25 ft (8 m) corresponding to a CAT2 level of target location error (TLE). A steep impact angle of 80 deg can be achieved.

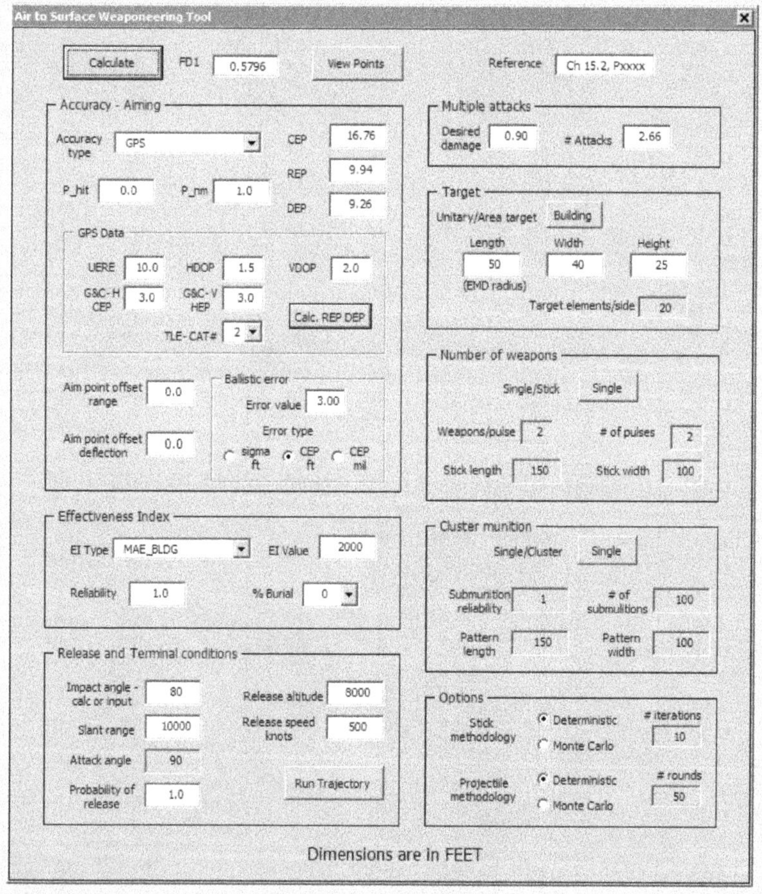

Fig. 18.12   Weaponeering tool with data for building target.

- Structural damage of 50% is the commander's requirement, so the table of input data (Table 18.3) can be constructed.

The building height is estimated to be 25 ft; however, for such a steep impact angle, this value is not critical. Figure 18.12 shows the AS module configured for this example and the result indicates that around 58% structural damage is achievable with a single weapon, which meets requirements.

If the target were in range of other weapons such as the U.S. Army Tactical Missile System (ATACMS) or Guided Multiple Launch Rocket System (GMLRS) rockets, or the U.S. Navy cruise missile system, and we wanted to examine their effectiveness, only the value of MAE_F would be different in the program input shown in Fig. 18.12.

EXAMPLE 18.4: SAME AS EXAMPLE 18.3, BUT WITH COLLATERAL CONCERN

Consider the same scenario as in Example 18.3, but determine the damage to a residential building located close the target. Figure 18.13 shows the relationship between the target building and collateral object.

Discussion points:

- As measured from satellite imagery, the single-story collateral building has a footprint of 80 ft × 50 ft and is built to residential construction codes.
- The center-to-center position has coordinates (−80 ft, −80 ft) relative to the target building.

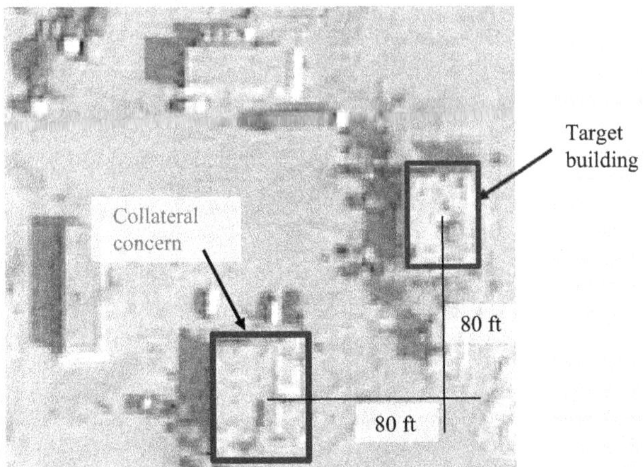

Fig. 18.13   Target and collateral concern.

TABLE 18.4   AS MODULE INPUT DATA

Data	Type	Value
Accuracy—ground plane	GPS with CAT2 level TLE	Calculated
EI is an MAE_{BLDG}	Residential	15,000 ft^2
Impact angle	Programmed	80 deg
Target size	Satellite image	80 ft × 50 ft × 10 ft
Aimpoint offset	Relative to target	−80 ft, −80 ft

- We use the offset aimpoint method described in Chapter 16, Section 16.4, noting that the MAE_{BLDG} will be different for the collateral concern than for the target. The value of 15,000 ft^2 is obtained from the database.
- The inputs for the Weaponeering program for this case are shown in Table 18.4.

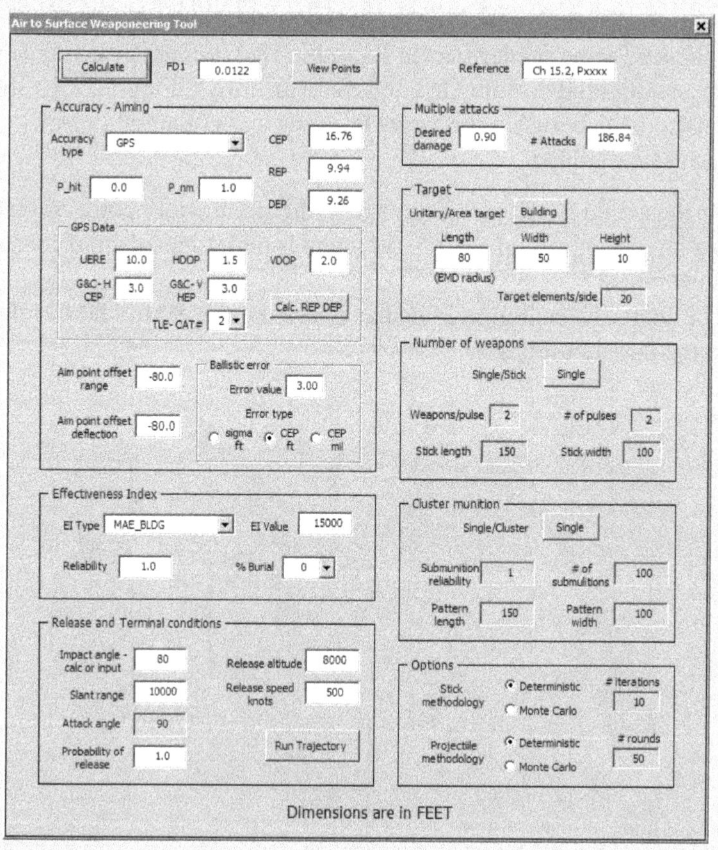

Fig. 18.14   Target and collateral concern.

Note the difference in MAE_{BLDG} for the same weapon against the two buildings, showing the effect of different construction materials, methods and building codes (see Fig. 18.14).

The conclusion is that for a single GBU-31, the target building is demolished, and the residence will suffer 1.2% structural damage. Although not necessary for this case, Figure 18.15 shows the structural damage to the collateral building if more than one attack occurs.

The appropriate level in the command structure must now decide whether this level of collateral damage is acceptable in order to approve the attack on the intended target.

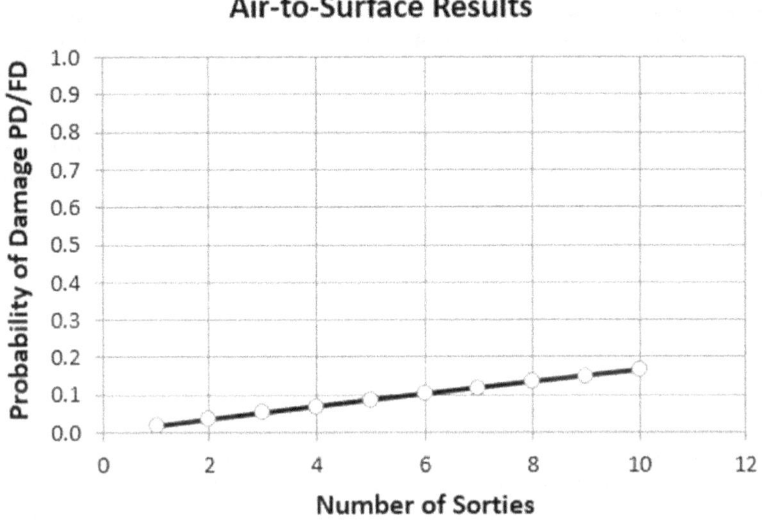

Fig. 18.15   Effect of multiple strikes on collateral damage.

EXAMPLE 18.5: ARTILLERY AGAINST ADVANCING TROOPS

Figure 18.16 shows an open area with advancing troops, which will be countered with a six cannon artillery battery firing 155-mm M-107 unguided HE rounds. How many volleys are required?

Discussion points:

1. Assume the troops shown in Fig. 18.16 are uniformly distributed in an area of 200 m × 200 m.

2. The battery consists of six cannons, each selecting its own aimpoint relative to the rectangle containing the troop targets.
3. The JTCG/ME-approved database does not specifically address the M107 artillery round, so we will use the data derived in Chapter 17, specifically Table 17.9. If the most severe assault injury criterion is selected, we get $MAE_F = 18{,}362$ ft^2 or 1707 m^2.
4. Using the generic aiming strategy of Fig. 11.11, the volley dimensions are calculated to be 100 m $\times$ 100 m from Fig. 18.17.
5. Assume representative values for ground plane MPI and precision errors where REP > DEP for noncircular distributions.

Fig. 18.16 Troops advancing in open terrain.

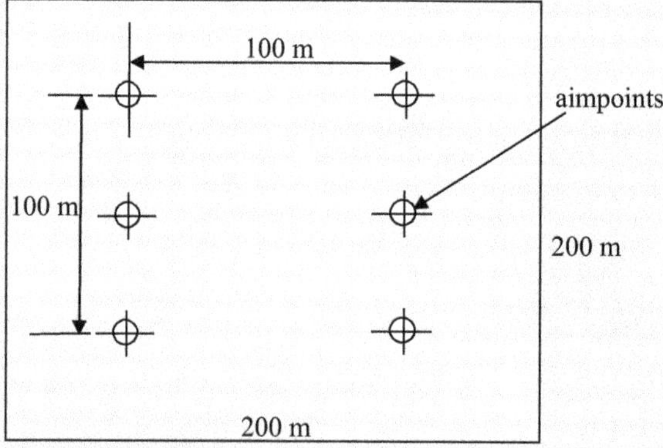

Fig. 18.17 Battery aimpoints.

TABLE 18.5   SS MODULE INPUT DATA

Data	Type	Value
MPI error—ground plane	REP and DEP	100 m, 50 m
Precision error—ground plane	REP and DEP	25 m, 20 m
EI is an MAE_F	HE blast/frag warhead PD or airburst fuze	1707 m^2
Impact angle	Calculated—high angle	78.5 deg
Target dimensions	Area containing troops	200 m × 200 m
Volley dimensions	Standard	100 m × 100 m

6. Assume the battery is located 10 km from the target, run the trajectory model, and assume a high angle value of quadrant elevation (QE).
7. A Monte Carlo methodology is selected with 20 discrete elements and 50 iterations.

The input data for this case are shown in Table 18.5.

These data are transferred to the Surface-to-Surface weaponeering module for indirect fire, as shown in Fig. 18.18, and produce a single volley fractional casualties of about 7%.

The View Points option produces Fig. 18.19, which illustrates the coverage of the target area by the six individual projectile lethal areas, for zero error placement, and represents the maximum fractional coverage achievable for perfect accuracy.

$$F_{C-\text{MAX}} = \frac{6 \times 1707}{200 \times 200} = 0.256 \qquad (18.1)$$

Coverage seems to be optimized without any "wasted" lethality, so not much will be gained by changing the volley geometry. Now investigate the FD or proportion of injuries as a function of the number of volleys, as illustrated in Fig. 18.20.

As a rule of thumb, an infantry assault will be repelled if around 30% casualties are experienced; therefore, about five or six volleys will be needed.

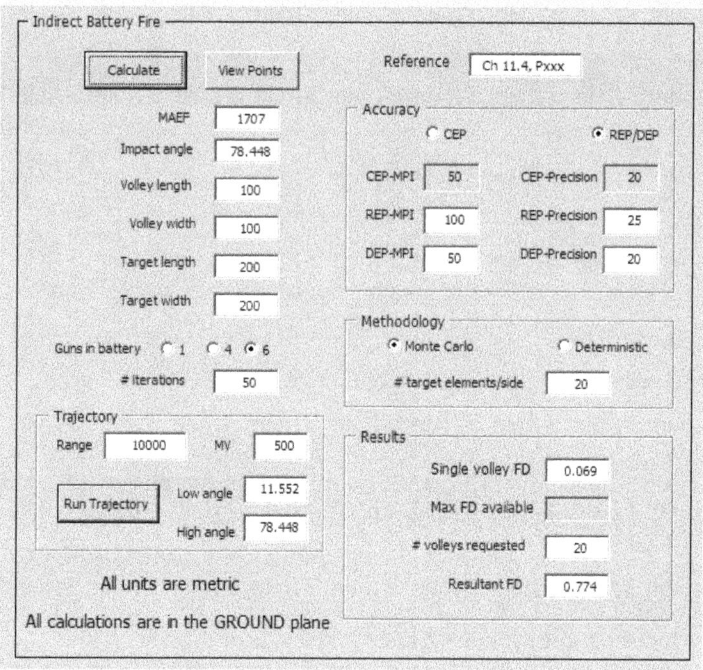

Fig. 18.18   Surface-to-Surface program, indirect fire screen.

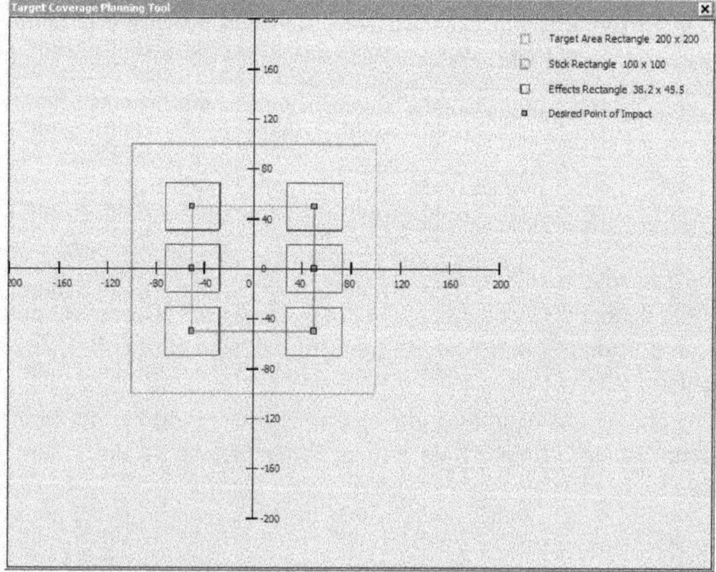

Fig. 18.19   PD for increasing number of bursts.

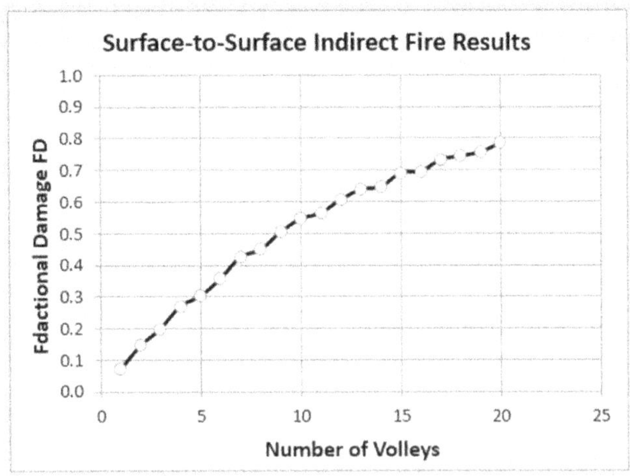

Fig. 18.20   Proportion of injuries as a function of volleys.

EXAMPLE 18.6: BRADLEY INFANTRY FIGHTING VEHICLE (IFV) VS BMP WITH 25-MM GUN

The attacking vehicle and target are shown in Fig. 18.21.
Discussion points:

- For the Bradley, a common doctrine is to fire a succession of 5- to 10-round bursts until the target is defeated.
- The Surface-to-Surface tool is used, in this case the Direct Fire module.
- Assuming the gun fires armor-piercing (AP) rounds with sufficient velocity to penetrate, we investigate how many bursts of these rounds are needed to achieve a specified amount of damage.

Fig. 18.21   Shooter and target.

- The effectiveness calculation will be done in the normal plane, using angular accuracy measures (mil); therefore, the slant range/shooting distance is required. Burst-to-burst and within-burst accuracies will be kept separate.
- The appropriate EI is a vulnerable area. Considering Chapter 17, Section 17.9, and Eq. (17.19) in particular, a value of 14.4 ft^2 = 1.4 m^2 seems appropriate.
- To obtain effectiveness, the double loop Monte Carlo method shown in Fig. 9.12 will be used, starting with a small number of iterations in order to assess execution times.
- The data shown in Table 18.6 are representative of this engagement, assuming a firing range of 500 m.

The Direct Fire Weaponeering section is shown in Fig. 18.22 and indicates that for a single burst of 10 rounds, the PD is about 0.62.

The effect of multiple bursts may be seen in Fig. 18.23. We can see from this figure that two or three bursts would produce about 90% probability of damage.

TABLE 18.6   WEAPONEERING INPUT DATA AND RESULT

Data	Type	Value
Burst-to-burst normal plane error	CEP	2 mil
Within-burst normal plane error	CEP	1 mil
EI is an A_V	API round	1.4 m^2
Range to target		500 m
Rounds/burst		10

Fig. 18.22   Direct Fire Weaponeering program input/output.

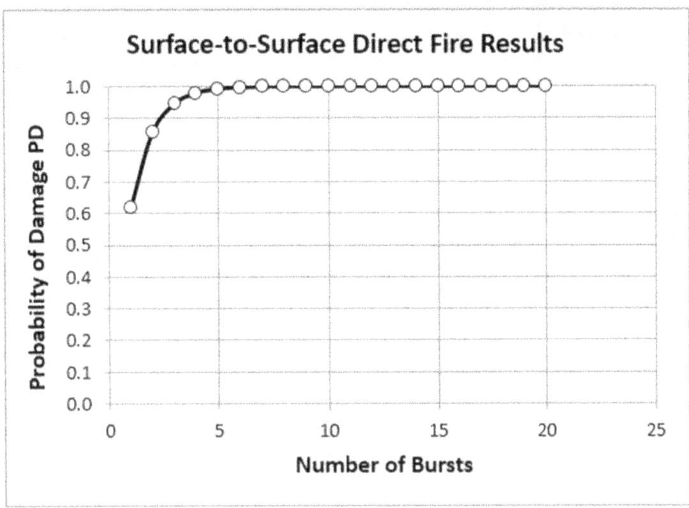

Fig. 18.23   PD for increasing number of bursts.

EXAMPLE 18.7: DEFEAT UNDERGROUND MISSILE SILO

Consider the underground missile silo shown in Fig. 18.24. This target can be defeated by either a direct hit on the silo cover or, if the impact is close enough to the launch tube to collapse the lining, preventing a missile from being fired. Weaponeer an attack that prevents a missile launch, and determine the probability of preventing such an event.

Discussion points:

- Given the image in Fig. 18.24, and assuming the height of the personnel to be 6 ft, the silo diameter is estimated to be 20 ft.
- Tactically, the weapon of choice would be a large 2000-lb guided bomb with penetration sufficient for the substantial cover. A typical choice would be the GBU-24 laser-guided bomb utilizing the BLU-109 penetrator warhead, shown in Fig. 18.25. This weapon will also be capable of penetrating up to 6 ft of the surrounding concrete.
- If bad weather or other impediments to establishing a line of sight (LOS) to the target preclude using laser guidance, there is a variant of the GBU-31 JDAM that also uses the BLU-109 warhead, as shown in Fig. 18.26.
- For both weapons, we assume a CEP of 10 ft; however, LOS issues may affect the LGB, and TLE uncertainty (CAT#) affects the

Fig. 18.24   Underground missile launcher.

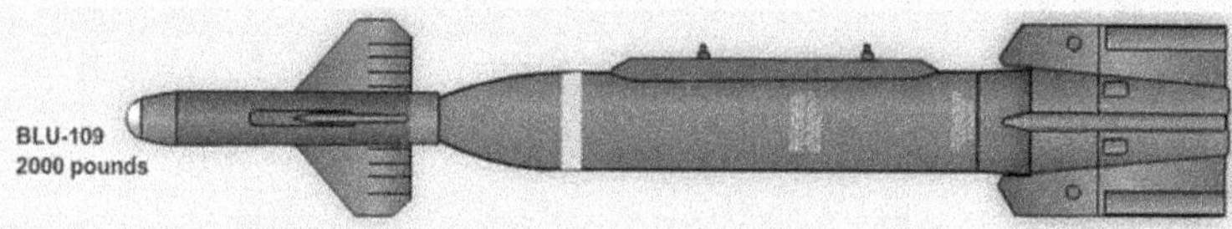

Fig. 18.25   GBU-24 Paveway III LGB penetrator.

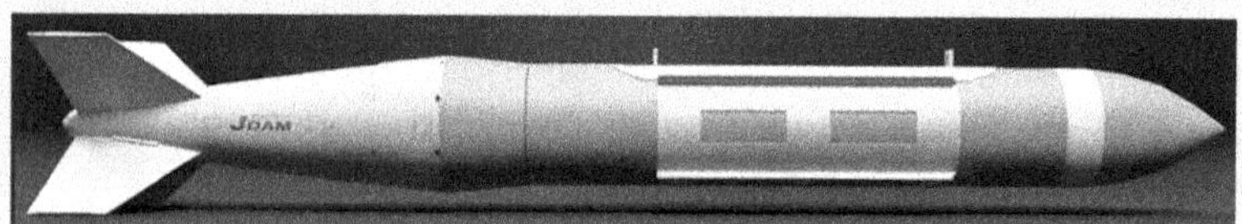

Fig. 18.26   GBU-31 JDAM GPS-guided penetrator.

GPS-guided bomb. In both cases, we assume that the impact angle can be preprogrammed, and 80 deg seems good.

- In the absence of a direct hit on the cover, assume further that for this warhead, a miss of 5 ft from the silo will cause sufficient collapse of the launch tube to prevent firing. This distance may be considered an effective miss distance around a 2-D circular target.
- We will use the Air-to-Surface Weaponeering tool to evaluate the attack, using the 2-D effective miss distance (EMD) effectiveness index, which, together with appropriate accuracies, is shown in Table 18.7.

The Air-to-Surface module inputs for this case are shown in Fig. 18.27. This gives the result for a single attack of $PD_1 = 0.8052$. Note that in this methodology (Chapter 12, Section 12.4), the circular target of radius

TABLE 18.7 WEAPONEERING INPUT DATA FOR MISSILE SILO TARGET

Data	Type	Value
Accuracy—ground plane	CEP	10 ft
Target dimensions	Radius	10 ft
EI is an EMD circle	Radius	5 ft
Impact angle	Preset	80 deg
Fuze	Set for delay	10 ms

10 ft is selected with height set to zero, and these dimensions are entered in the Target data block. Figure 18.28 shows the results for multiple independent attacks. No more than two attacks are needed to be confident the missile cannot launch.

Recall that in the EMD methodology, the circular target is represented by a square of equal area, which means the length of each side of the

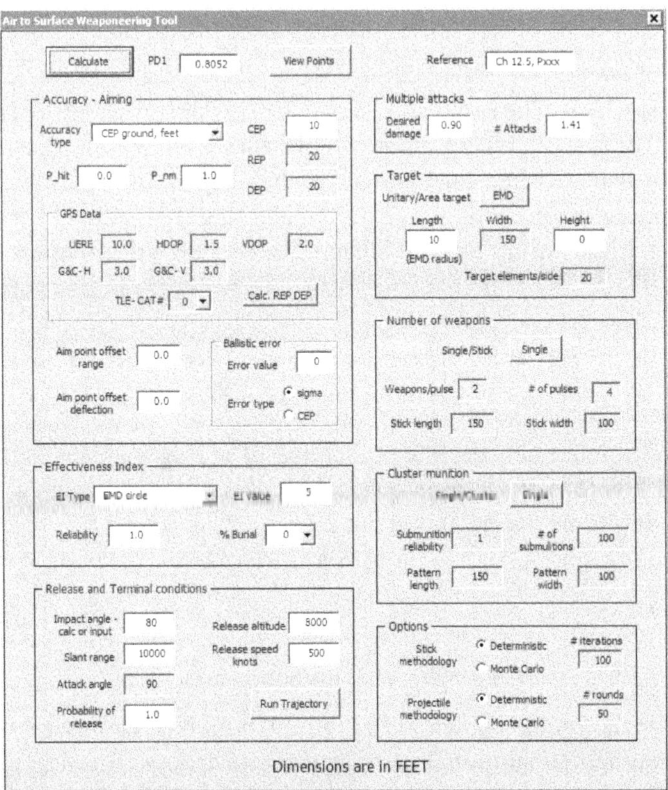

Fig. 18.27 AS module for missile silo attack.

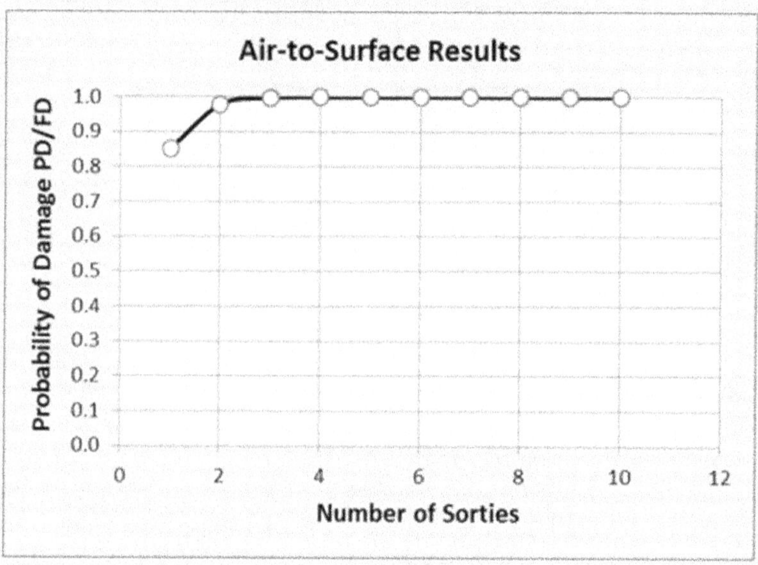

Fig. 18.28   Silo target for increasing number of attacks.

square is 17.7 ft. The result shown of 0.8052 is the probability of hitting a square of side 22.7 ft (i.e., target + 2EMD).

As an aside, we can calculate the probability of hitting a more accurate representation of the target—a circular target of radius 15 ft (target radius + EMD) using Eq. (2.17) from the Rayleigh distribution, where $\sigma = \text{CEP}/1.1774 = 8.5$ ft.

$$P(0 < r < 15) = 1 - \exp\left[\frac{-15^2}{2 \times 8.5^2}\right] = 0.789 \qquad (18.2)$$

The difference between the probability of hitting a square target 22.7 ft $\times$ 22.7 ft (circle -> square + EMD) and circular target of radius $R = 15$ ft (radius + EMD) is about 2%.

EXAMPLE 18.8: BRIDGE ATTACK

Consider the bridge shown in Fig. 18.29, which is a temporary structure designed to support heavy armor and troops crossing a river. What weapon should be used to prevent use of the bridge, and how effective is it?

Fig. 18.29   Temporary military bridge.

Discussion points:

- Based on the dimensions of a typical tank of width 12 ft, the bridge width is estimated to be 25 ft and the length, although probably not critical, is about 600 ft.
- Because the bridge has been recently constructed, it probably cannot be considered an infrastructure target (i.e., its GPS coordinates are unknown).
- The weapon of choice therefore might be a laser-guided bomb with the same aircraft designating the target and releasing the weapon. The question is whether it should be small (500 lb), medium (1000 lb), or large (2000 lb).
- Because the deck of the bridge doesn't offer much resistance to penetration, regular LGB types are appropriate, such as the Paveway II series shown in Fig. 18.30.

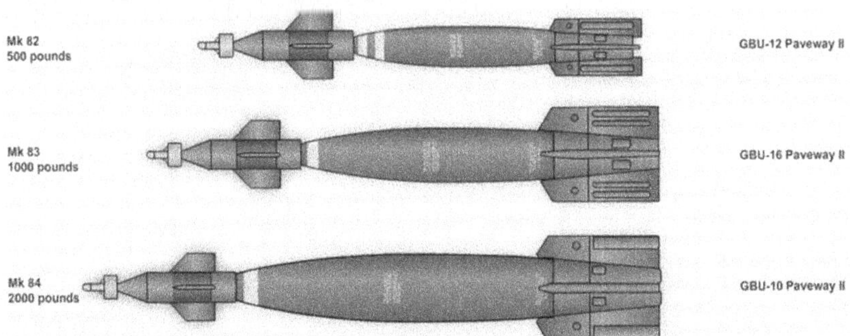

Fig. 18.30   Paveway II series laser-guided bomb.

- For a bridge attack, the fuze is usually set to detonate the weapon beneath the deck to initially lift it up. The subsequent downward motion will then cause the structure to fracture. In this case, however, because the deck is very close to the water surface, a PD fuze is selected because if detonation occurs below the surface, the blast effects will be mitigated.
- It is not known if there are any air defense sensors near the bridge that could, for example, detect the presence of a laser designator and deploy flares to confuse the weapon guidance system. Therefore, a delayed lase mode will be used, as described in Chapter 3, Section 3.8, Fig. 3.15. A ballistic trajectory is therefore assumed from a release at 15,000 ft, 400 kt, level delivery, giving an impact angle of about 55 deg.
- This calculated impact angle is not needed, however, because damage is caused by blast rather than fragments. A review of Chapter 8.6 indicates damage independent of impact angle.
- We assume the accuracy of all three LGBs is the same, say a CEP of 10 ft, and the only difference is the BEI value obtained from the database.
- The Air-to-Surface tool is appropriate for this attack with the data shown in Table 18.8. We will first assume an attack along the bridge length.

The Weaponeering Program with input data and result for the first value of BEI is shown in Fig. 18.31.

Running the program for each weapon in turn produces the results shown in Table 18.9.

If we swap the target length and width data, the same result is obtained. This is expected, because the accuracy is the same (CEP) in any direction. The preferred weapon, therefore, is the GBU-10.

TABLE 18.8   INPUT DATA FOR BRIDGE ATTACK

Data	Type	Value
Accuracy—ground plane	CEP for guided weapon	10 ft
EI is a BEI	HE blast/frag warhead—PD fuze	5 ft, 15 ft, 40 ft
Impact angle	Delayed lase, ballistic	Calculated
Target dimensions	Attack along length	600 ft × 25 ft

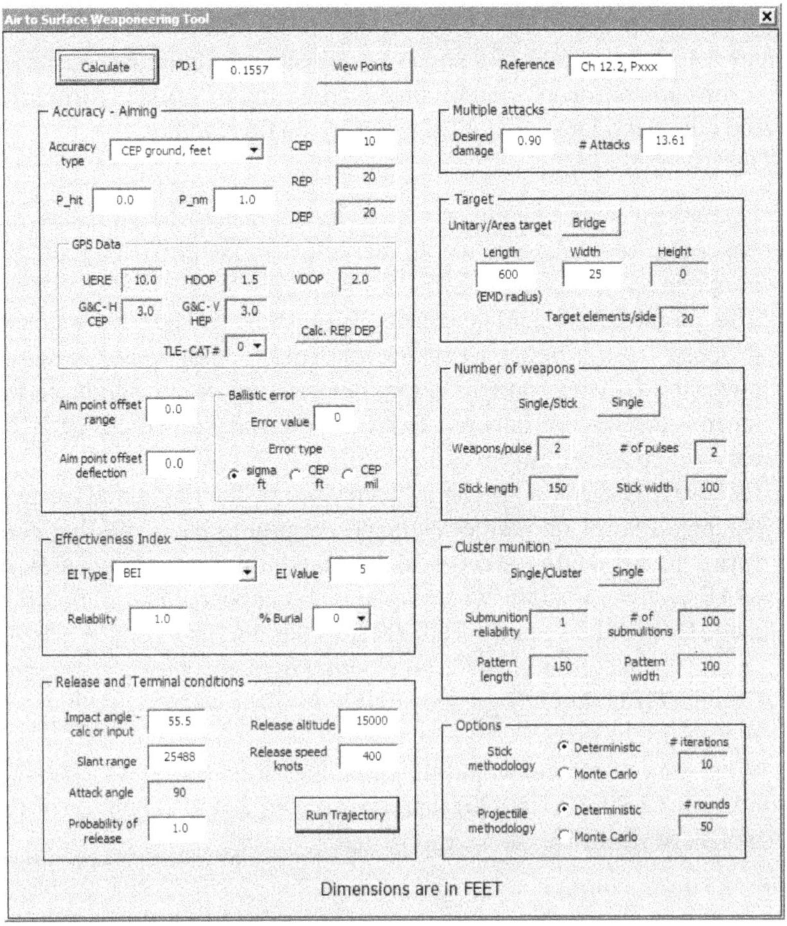

Fig. 18.31   Input data and result for bridge attack.

TABLE 18.9   BRIDGE ATTACK
RESULTS FOR DIFFERENT WEAPONS

Weapon	BEI	PD_1
GBU-12	5 ft	0.156
GBU-16	15 ft	0.388
GBU-10	40 ft	0.686

EXAMPLE 18.9: MILITARY BARRACKS

Figure 18.32 shows a FLIR image of a collection of buildings known to be a military barracks, comprising many similar wooden structures spread out over a large area. Plan an attack on this facility.

Discussion points:

- In attacking such a facility, it may be impractical to use guided weapons because each one would have to be assigned to a unique aimpoint.
- This might be a suitable candidate for a stick of unguided weapons dropped from a bomber because the stick length would certainly cover the target in one direction, but the stick width might be too narrow for bombs released from an internal bay (B-52 or B-1), requiring passes from several aircraft.
- One way to check is to compare the area of the target and a single weapon against one target element. Assuming each building conforms to residential standards, the database suggests an area of $MAE_{BLDG} = 15,000$ ft² whereas the target appears to be enclosed in a 400 ft $\times$ 400 ft square—160,000 ft². Therefore, we need at least 11 accurately placed bombs for maximum target coverage.
- Could a GBU-43 GPS-guided, 20,000-lb-class weapon, as shown in Fig. 18.33, be used?
- What area of effects would this produce? Recall from Chapter 17, Section 17.10, Eq. (17.20) that blast scales as the cube root of the weight of explosive, so we can write

$$\frac{R_1}{W_1^{1/3}} = \frac{R_2}{W_2^{1/3}} \qquad (18.3)$$

Fig. 18.32　Group of buildings target.

Fig. 18.33 MC-130 dropping a single GBU-43.

Suppose the subscript 1 refers to an Mk-84 and subscript 2 refers to the GBU-43. If the uncased explosive weights are $W_1 = 1000$ lb and $W_2 = 18000$ lb, respectively, then

$$R_2 = \frac{W_2^{1/3}}{W_1^{1/3}} R_1 = \left(\frac{18,000}{1000}\right)^{1/3} \times 70 = 187 \text{ ft} \qquad (18.4)$$

This gives an area of effects of about 110,000 ft². Therefore, if the bomb lands accurately at the center of the camp, we might expect about 68% coverage, hence damage. This is probably sufficient to prevent the facility from functioning.

- This weapon is delivered (Fig. 18.33) by dragging it out of the back of an MC-130 cargo aircraft on a sled with a parachute, which is then jettisoned once the weapon is vertical. Based on Fig. 18.34, it seems

Fig. 18.34 GBU-43 MOAB weapon.

that steep impact angles can be achieved for this weapon, so 80 deg will be assumed; however, again recall that blast effects are independent of this parameter.

- This weapon is GPS-guided, so we will assume CAT2-level accuracy for the target coordinates (TLE), which will be at the center of the barracks.

  Table 18.10 shows the data used to evaluate this scenario.

- The corresponding AS tool is shown in Fig. 18.35. Notice that although the individual target elements are a single building, we do not use MAE_{BLDG} because we are considering a uniformly distributed collection of buildings. Instead we use $MAE_B = 110,00$ ft^2 for the individual building damage function and the area containing the buildings is 400 ft by 400 ft. Therefore, we are calculating the fractional damage of the weapon lethal area over a rectangular area of identical unitary target elements (buildings). The result is therefore a fraction damage to the collection of buildings.

Fractional damage of 69% is sufficient to render the barracks inoperative. Figure 18.36 shows the target coverage from the View Points option. Note that the achieved fractional damage is almost exactly the maximum fractional coverage—110,000/160,000—due to the high accuracy of the weapon.

An alternative attack from a bomber (B-1, B-2, or B-52) dropping many individual unguided bombs as a stick, as shown in Fig. 18.37, also should be evaluated.

From the database, a single 2000-lb Mk-84 warhead against a residential building has an MAE_{BLDG} of 15,000 ft^2, and a B-2 carries 16 bombs. However, they all are released from an internal bay, so the stick will be long and narrow. Figure 18.38 shows the Weaponeering Program for this attack.

Figure 18.39 shows the individual lethal areas overlaying the 400 ft × 400 ft target area.

From the View Points window each lethal area is 122.5 ft square so spacing them out along the stick length overlays the target with a strip

TABLE 18.10   DATA FOR BARRACKS ATTACK

Data	Type	Value
Accuracy—ground plane	GPS-guided weapon, CAT2 TLE	Calculated
EI is an MAE_B	HE blast warhead—PD or airburst fuze	110,000 ft^2
Impact angle	Near vertical	80 deg
Target dimensions	Linear distance	400 ft × 400 ft

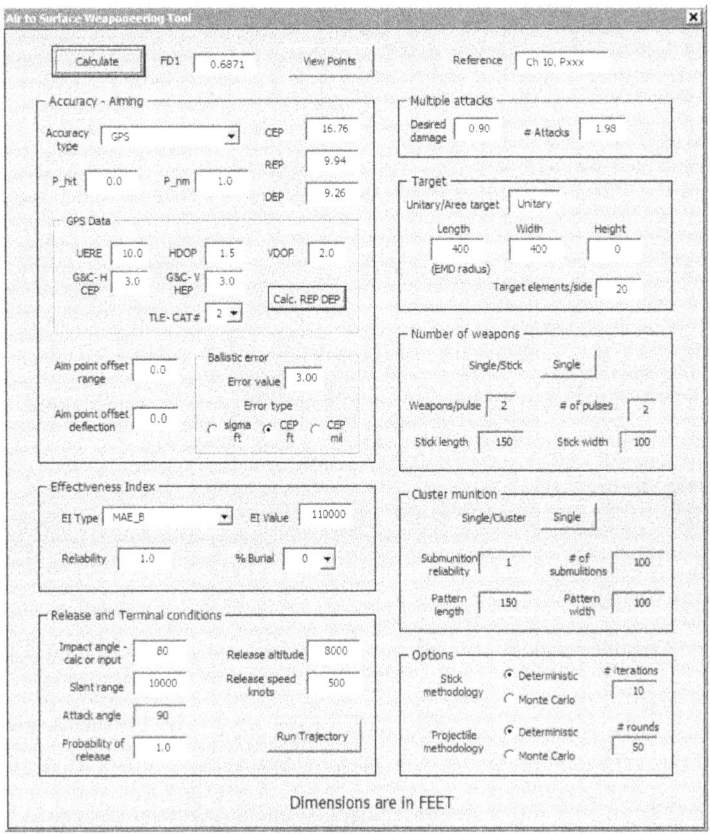

Fig. 18.35   Weaponeering tool for barracks facility attack.

only 122.5 ft wide. Given this fact, the ratio of total weapon lethal area divided by the target area give an estimate of coverage.

$$F_{C-\text{MAX}} = \frac{400 \times 122.5}{400 \times 400} = 0.306 \tag{18.5}$$

This is close to the value obtained but this value assumes perfect delivery accuracy (CEP $= 0$). This confirms that the stick is too narrow to damage most of the target area resulting in fractional damage $<50\%$.

Although the GBU-43 from a lethality perspective seems a suitable weapon for the attack, it can only be dropped by a C-130 cargo aircraft, which is vulnerable to antiaircraft fire on its way to the target. The B-2, on the other hand, is much less vulnerable to the same antiaircraft attack; however, its bomb load and pattern geometry make it less effective.

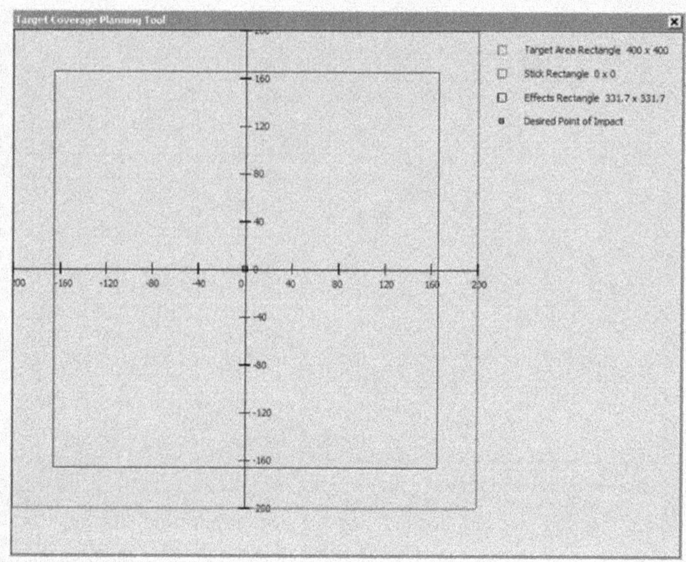

Fig. 18.36 Target coverage.

Re-examining the possibility of using more precise guided weapons, if each can be placed accurately and only 50% of the buildings need to be damaged to remove the threat, then the number of weapons needed is

$$N \times 15{,}000 = 400 \times 400 \times 50\% \tag{18.6}$$

which gives $N = 6$, a workable solution. In this case however, each weapon would have to be individually targeted, so only GPS guided weapons would be practical.

Fig. 18.37 B-2 delivering a stick of unguided bombs.

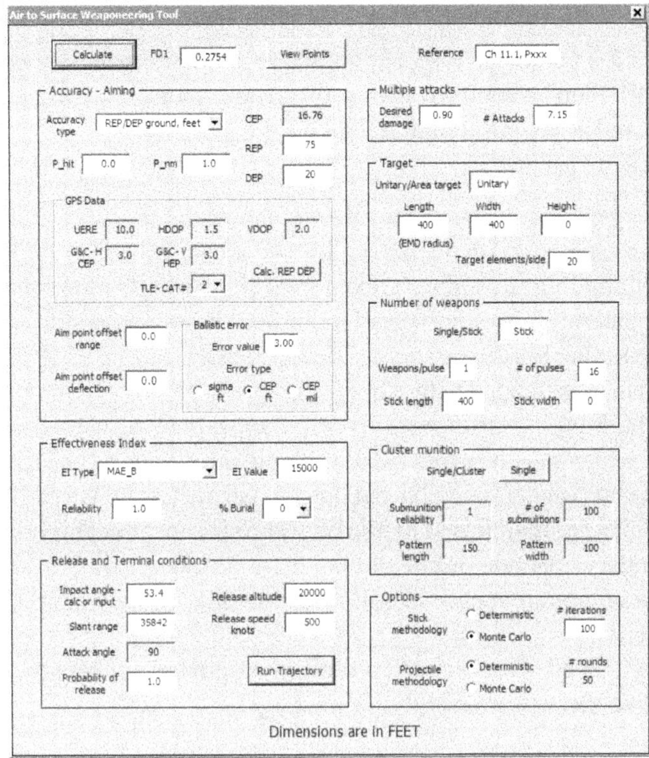

Fig. 18.38   Unguided stick delivery against barracks.

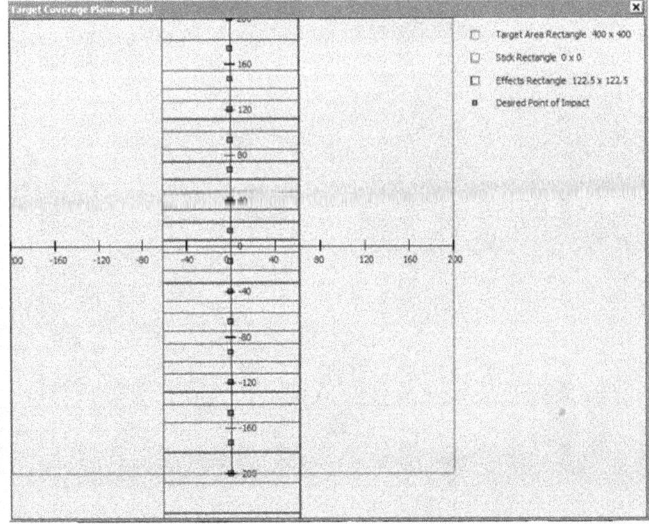

Fig. 18.39   Target coverage by the stick.

EXAMPLE 18.10: HIGH-VALUE PERSONNEL TARGET

A high-value personnel target is known to reside in the house shown in
Fig. 18.40. Plan an airstrike against this target with a high probability of
success, and assess any collateral damage concerns.

Discussion points:

- A Google Earth image of the target is shown in Fig. 18.41. It indicates
  there are collateral concerns (buildings) located about 245 ft from the
  target building. In addition, local intel (Fig. 18.40) shows there may
  be collateral civilian personnel in the vicinity of the target.
- The building is a three-story residence with a footprint of 50 ft × 50 ft
  and has an estimated height of 30 ft.
- The building is surrounded by what appears to be a 15-ft concrete
  wall, which may be expected to mitigate the effects of blast and fragments on collateral objects outside the perimeter; however, the same
  wall may be demolished by the blast, producing large, high-velocity,
  secondary concrete fragments.
- Assume the target and all collateral building objects are residential
  properties.
- Air-launched weapons are preferred to guided rockets to provide
  positive identification (PID) of the target.

Fig. 18.40   Residence of high-value personnel target.

Fig. 18.41   Satellite view of target and surroundings.

- Because local intelligence information seems to be available, we can assume CAT1 TLE if a GPS weapon is used.
- A high probability of success is required for this target, so a 2000-lb warhead will be used.
- So as not to rely on one guidance system modality, two weapons will be used, a GBU-10 laser-guided bomb and a GBU-31 GPS-guided bomb. For a residential building with no expected hardening of the target, there are no special penetration requirements.
- A delay fuze is selected so as to cause detonation inside the building, thereby mitigating the effects of blast and fragments to collateral objects outside the concrete wall.
- Considering the GPS-guided weapon first, the data in Table 18.11 seem reasonable.
- This may be considered a time-sensitive target in that the person of interest can be 100% confirmed to be resident only at a particular time. This might mean bad weather could prevent an LGB attack or an unfavorable satellite constellation could degrade a GPS weapon attack. Figure 18.42 shows the Weaponeering tool inputs and result, confirming total demolition of the building.
- If GPS is jammed, then the attack relies on an LGB weapon. Data for this attack are shown in Table 18.12.

The result of the attack is shown in Fig. 18.43, again suggesting total demolition of the building.

TABLE 18.11   DATA FOR GPS-GUIDED WEAPON

Data	Type	Value
Accuracy—ground plane	CEP for GPS-guided weapon with CAT 1 TLE	Calculate
EI is an MAE_{BLDG}	Residential	15,000 ft^2
Impact angle	Programmed	80 deg
Target dimensions	Scaled satellite image	50 ft × 50 × 30 ft

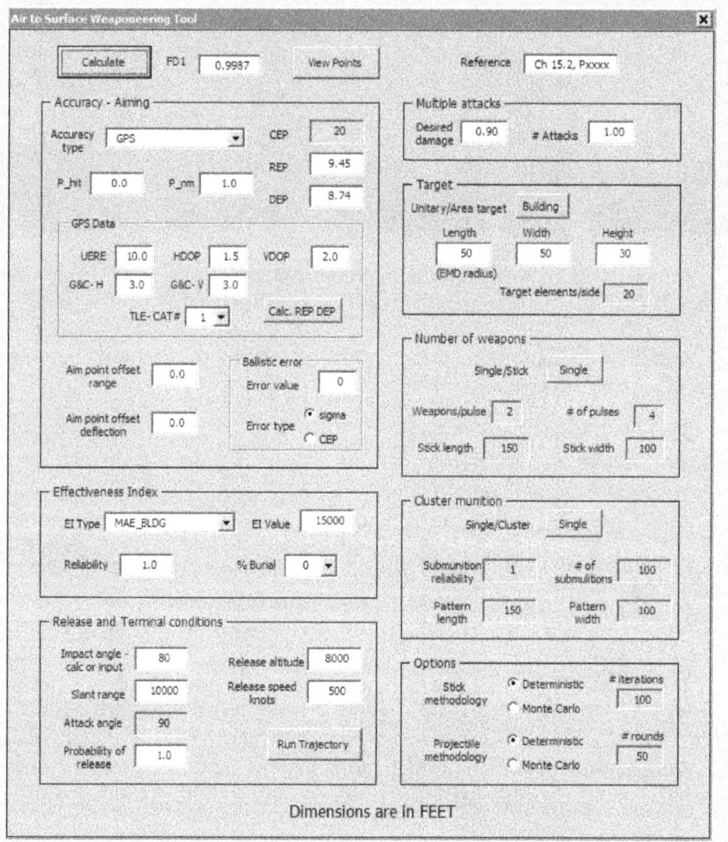

Fig. 18.42   Building attack with GPS weapon.

TABLE 18.12   DATA FOR LGB WEAPON ATTACK

Data	Type	Value
Accuracy—ground plane	CEP for LGB	20 ft
HE blast/frag warhead—PD fuze	EI is a MAE_{BLDG}	15,000 ft^2
Impact angle	Delay fuze	80 deg
Target dimensions	Scaled satellite image	50 ft × 50 × 30 ft

- Clearly, each weapon produces almost 100% structural damage; however, in case one is disabled through weather or jamming, both will be used.
- The damage is confirmed using the View Points option, as shown in Fig. 18.44, in fact, a smaller weapon (1000 lb) may be just as effective.
- Now check for collateral damage effects. Following the offset aimpoint process in Chapter 16, Section 16.4, we calculate the following:

$$\text{WCER} = \sqrt{\frac{\text{MAE}_B}{\pi}} = \sqrt{\frac{15,000}{\pi}} = 69.1\,\text{ft} \qquad (18.7)$$

$$\text{DE} = \text{CE}_{90} = 1.8227 \times \text{CEP} = 1.8227 \times 20 = 36.5\,\text{ft} \qquad (18.8)$$

$$\text{CER} = \text{WCER} + \text{DE} = 69.1\,\text{ft} + 36.5\,\text{ft} = 105.6\,\text{ft} \qquad (18.9)$$

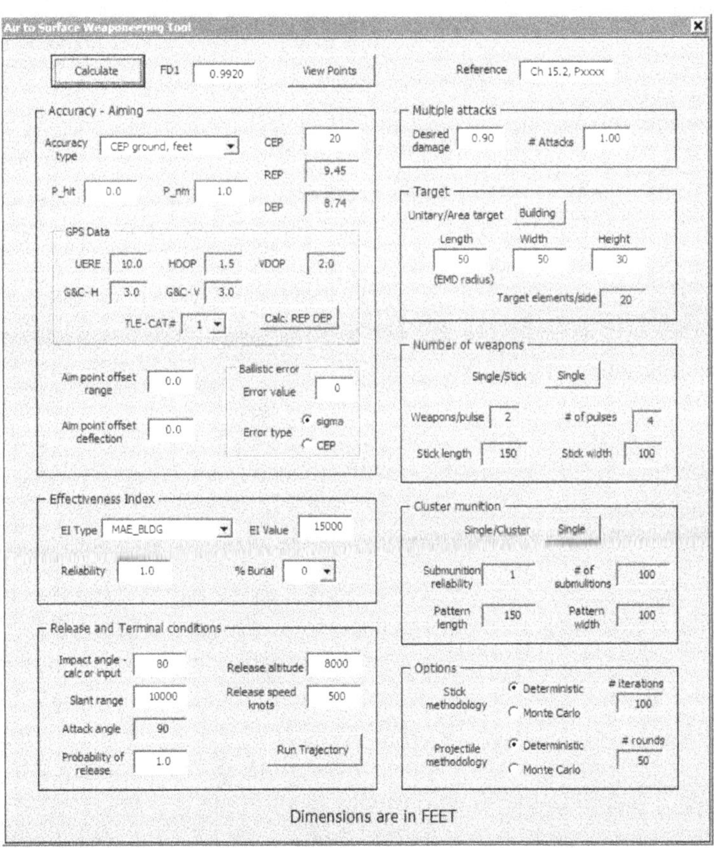

Fig. 18.43   Building attack with LGB.

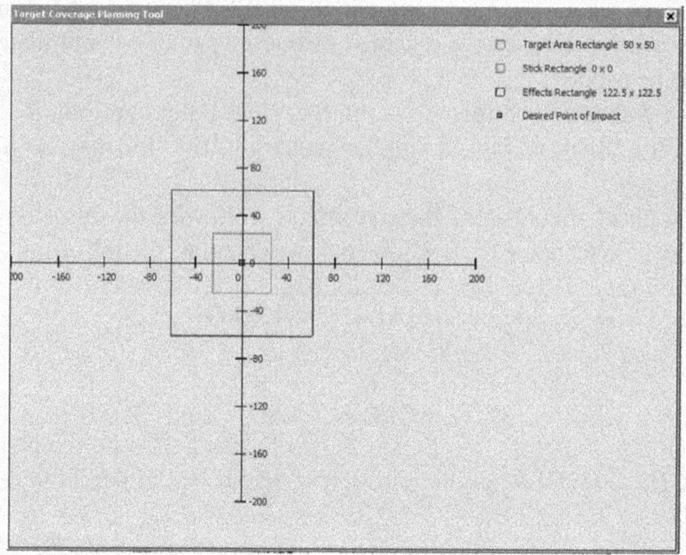

Fig. 18.44   Building overlayed with lethal area.

- Because this is much less than the distance from the aimpoint (245 ft), the collateral buildings should be safe. This can be checked using the offset aimpoint values of 245 ft and a comparable building to the target, as shown in Fig. 18.45.
- Injury to collateral personnel will occur if they are standing close to the target. All that can be done to mitigate this is to time the attack for early morning when there can be fewer expected casual personnel passing near to the target.
- Suppose we consider a worst-case scenario in which bad weather prevents the LGB from being used, and a GPS jammer is in operation near the target. If the attack still proceeds using the GBU-31 only, what is the probability of success, and how does this affect the collateral damage calculation?

It is assumed that the GBU-31 is guided to the target using INS guidance only. The integrated GPS tool shown in Chapter 5, Fig. 5.24 will be used; however, because GPS is jammed, only the lower left portion dealing with this topic is relevant. It is shown in Fig. 18.46.

This assumes a delivery from 10,000 ft at 450 kt, giving a time of flight of the weapon of 25 s. In this time, the CEP has grown to 16.4 m, or 53.8 ft. Running the Weaponeering program (Fig. 18.42) again with this accuracy gives $FD_1 = 0.64$ and a collateral effects radius from Eq. (18.9) of CER = 167 ft. The degradation in effectiveness and risk to collateral concerns is clear.

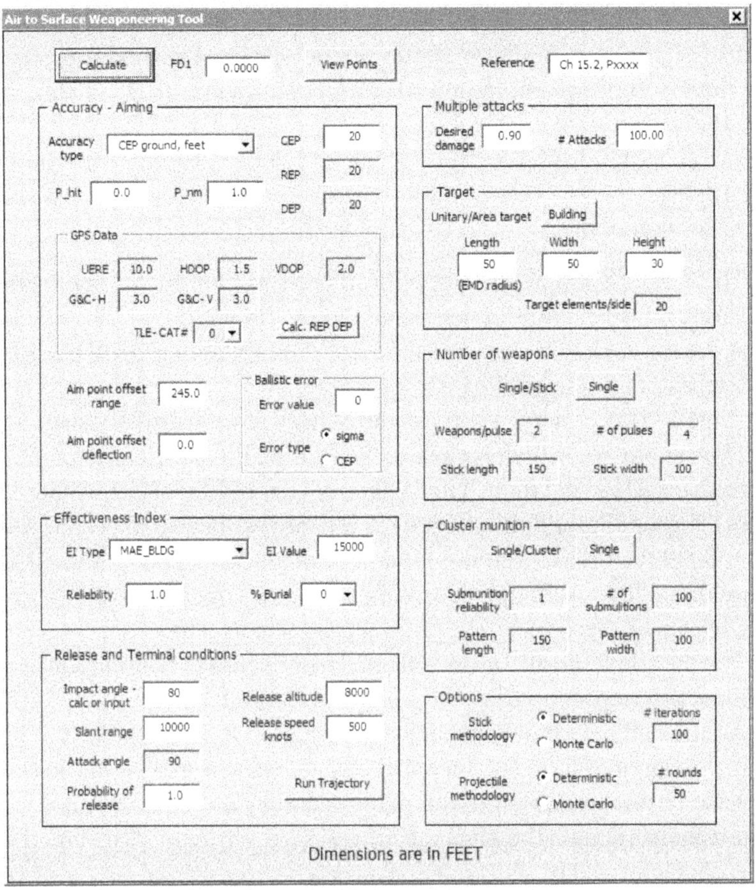

Fig. 18.45 CDE prediction for collateral buildings.

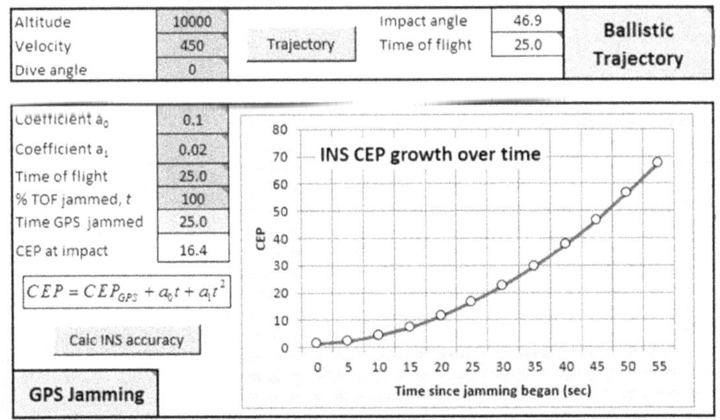

$$CEP = CEP_{GPS} + a_0 t + a_1 t^2$$

Fig. 18.46 GPS jamming tool.

EXAMPLE 18.11: DRONE ATTACK ON PERSONNEL PLANTING IED

In this example, a Predator drone operator has identified a group of individuals digging a hole in a road and planting an IED, as shown in Fig. 18.47.

Plan an attack from the drone using an onboard Hellfire missile (see Fig. 18.48).

Discussion points:

- The Air-to-Surface module will be used to plan the attack using a single missile against a personnel target.
- A common Hellfire variant is the AGM-114R with a 20-lb blast/frag warhead and semiactive laser guidance.
- Table 14.4 in Chapter 14 shows the M107 155-mm artillery shell has a 15-lb explosive, which is close to the Hellfire fill, but clearly fragments from the M107 are more lethal than from the Hellfire. Using Table 17.9 in Chapter 17, we will estimate the Hellfire warhead to have an MAE_F of 8000 ft^2 for 30-s defense injury criterion, corresponding to a lethal radius of 50 ft. Although this might seem small, recall the discussion in Chapter 17, Section 17.8 regarding fragment density.
- The missile will follow the reflected laser beam in continuous lasing mode, so impact angles will probably not be steep—assume 45 deg.
- Assume all personnel are enclosed in a 50 ft $\times$ 50 ft square and treat it as an area of target elements to calculate fractional casualties.
- Assume a reasonable ground plane accuracy of 20 ft CEP; however, testing has shown the linear accuracy is not normally distributed and is characterized by $P_{HIT} = 0.3$, $P_{NM} = 0.65$.

Fig. 18.47   Personnel burying an IED in a road.

Fig. 18.48   Predator drone firing Hellfire missile.

This results in the input data for the Weaponeering program shown in Table 18.13 for this attack.

The program is shown in Fig. 18.49. It predicts fractional casualties of about 82%. This should be sufficient to remove the threat.

With attacks such as this, it is important to estimate the collateral damage effects using Eq. (16.3), reproduced here:

$$\text{CER} = \left[\sqrt{\frac{\text{MAE}_F}{\pi}} + 1.8227 \times \text{CEP}\right] = 50.5\,\text{ft} + 36.5\,\text{ft} = 87.0\,\text{ft}$$

$$(18.10)$$

This is quite large for an urban environment, but the area in Fig. 18.47 looks fairly rural. It could be argued that the collateral effects of the Hellfire warhead are probably less than those of the IED if it were detonated.

TABLE 18.13   DATA FOR PERSONNEL ATTACK

Data	Type	Value
Accuracy—ground plane	CEP, non-Gaussian	$\text{CEP} = 20$ ft, $P_{\text{HIT}} = 0.3$, $P_{\text{NM}} = 0.65$
EI is an MAE_F	30-s defense	8000 ft^2
Impact angle	Continuous lase	45 deg
Target dimensions	Enclosing area	50 ft $\times$ 50 ft $\times$ 0 ft

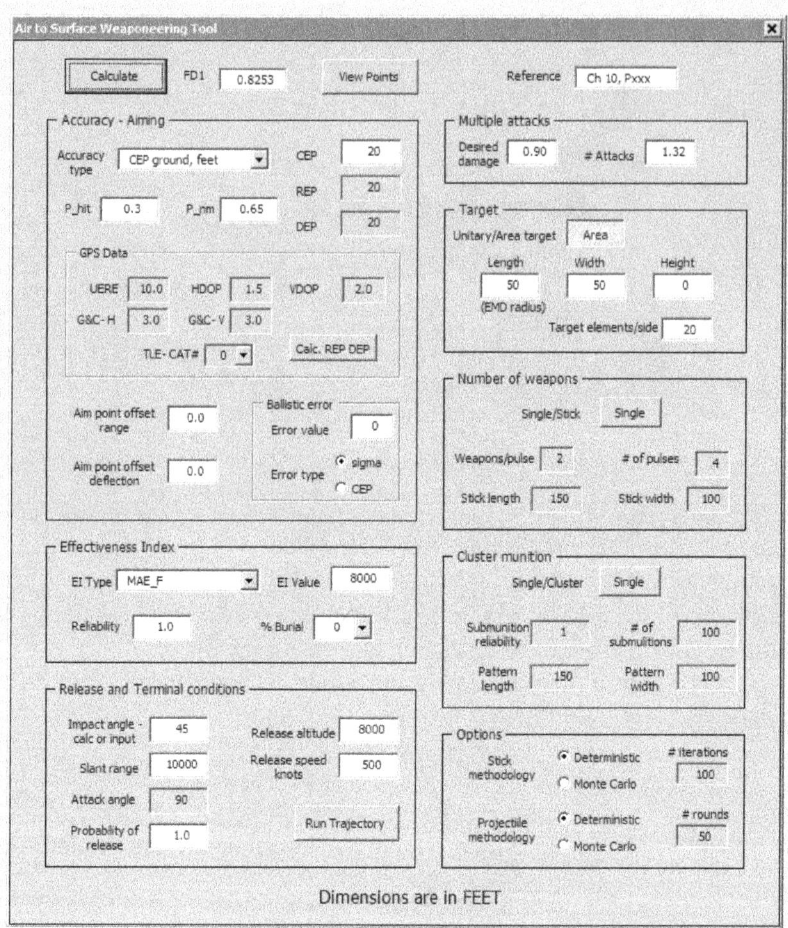

Fig. 18.49   Program input/output for drone attack.

EXAMPLE 18.12 MOBILE ARTILLERY CONCEALED IN A TUNNEL

This is a common scenario in which artillery is concealed in a tunnel to provide cover from air attacks and emerges when required to fire (see Fig. 18.50). Plan an attack on this target.

Discussion points:

- From the photograph, we cannot see the depth of the overburden above the tunnel, so the attack will concentrate on hitting the portal or a small area around it to create sufficient rubble to prevent, at least for a while, exit from the tunnel passage.

Fig. 18.50 Artillery concealed in a tunnel.

- A good weapon for this target would be one that has some glide capability, because a vertical impact may not get a satisfactory result. The GBU-39 small diameter bomb (Fig. 18.51) is a suitable candidate. This weapon has several guidance systems for various phases of flight including GPS, semiactive laser, and, for the terminal phase, millimeter wave radar. It has a 40-lb blast/frag warhead.
- From the size of the target shown, the tunnel portal is estimated to be 15 ft^2. Suppose that if the weapon strikes within 8 ft anywhere around it, enough rubble will be produced to block the tunnel.
- Recall from Chapter 12, Section 12.4 that the EMD effectiveness index was able to model tunnel targets. This is the EI that will be used for this example.

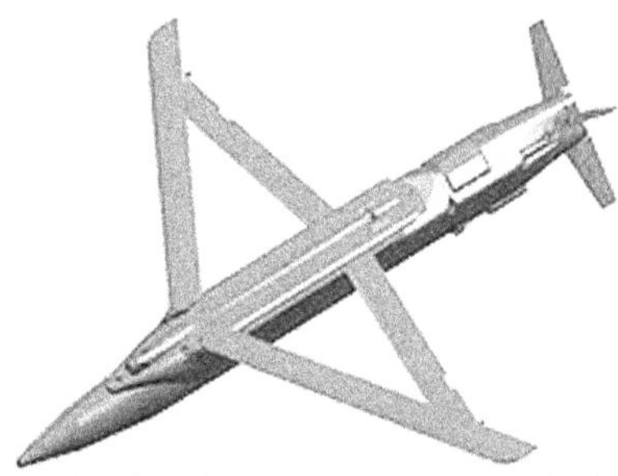

Fig. 18.51 Small diameter bomb: GBU-39.

- Although the target is essentially vertical, height error probable (HEP) and deflection error probable (DEP) are not used to specify the weapon accuracy. The EMD method reduces the target to an equivalent rectangle in the ground plane, so ground plane accuracy must be used. A CEP of 10 ft seems appropriate for this weapon, however (see Table 18.14).

TABLE 18.14   DATA FOR TUNNEL ATTACK

Data	Type	Value
Accuracy—ground plane	CEP	10 ft
EI is an EMD	Rectangle	8 ft
Impact angle	LOS	45 deg
Target dimensions	Tunnel portal	15 ft × 15 ft

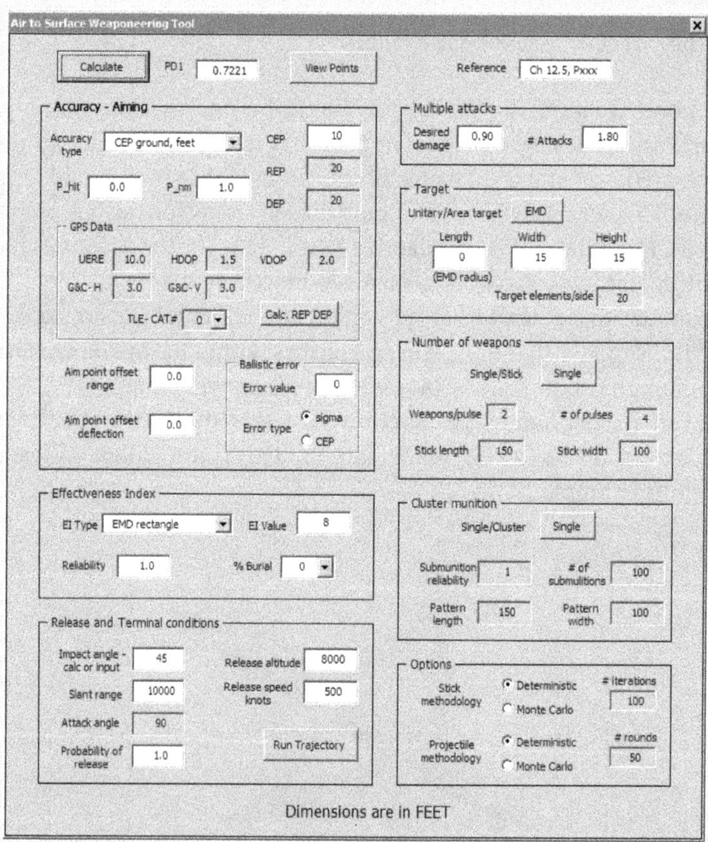

Fig. 18.52   Program input/output for tunnel attack.

The AS module with this input data is used, and the result shown in Fig. 18.52 is $PD_1 = 0.7221$.

The probability that the weapon flies directly into the tunnel opening may be calculated by setting the EMD value to zero, producing the result $PD_1 = 0.39$.

18.2 CHAPTER SUMMARY

- This chapter presented a set of case studies that covered both air- and surface-launched weapons against a variety of ground targets, using guided and unguided munitions, with single and multiple weapon deliveries.
- These cases have been evaluated using the unclassified Weaponeering Program based on the methodologies developed in the textbook. In doing so, a basic and consistent process of asking where to get the data for the program has led to an estimate of target damage for each scenario examined.

Appendix

WEAPON SELECTION BASED ON TARGET AND DAMAGE CRITERIA

The following tables indicate the appropriate weapon selection for specific target types. Also indicated in the tables are definitions of damage criteria that relate to the particular target.

TABLE A.1   TARGET-DAMAGE CRITERIA AND APPROPRIATE WEAPON TYPES

Target	Damage Criteria		Appropriate Weapon Type[1]
1. PERSONNEL	Defense: 30 s Assault: 30 s Assault: 5 min Supply: 12 h	The term *casualty criteria* is used for personnel targets and is comparable to the term *damage criteria* for materiel targets. A casualty results when the capability of the individual soldier to perform military duties is reduced to a sufficient degree, but the criteria depend on the tactical situation. For the analysis of casualties, four tactical roles have been defined, and a time element has been assigned to each: defense, 30 s; assault, 30 s; assault, 5 min; and supply, 12 h. Tactical roles and variations of incapacitation are defined in Joint Munitions Effectiveness Manual (JMEM) *Target Vulnerability*. The time element is the period after which the wound will result in incapacitation to meet the criterion for each role. Chemical weapons and firebombs may have been additional psychological effects on troops, causing an immediate lowering of unit combat effectiveness. Assault $\leq$ 30 s is representative of defense $\leq$ 5 min. Assault $\leq$ 5 min is representative of defense $\leq$ 12 h.	The appropriate weapon type depends on the posture assumed by the target and the cover taken, such as lying prone behind appropriate shielding or under cover in foxholes. *For personnel in the open (standing or prone):* The appropriate weapon types are high-explosive (HE) gun projectiles, fragmentation submunitions, airburst and instantaneous-surface-burst general-purpose (GP) bombs, chemical weapons, and firebombs. *For personnel in foxholes:* The appropriate weapons are fragmentation submunitions in the foxholes, airburst GP bombs, chemical weapons, and fuel–air explosive (FAE) weapons with the fuel–air cloud over the foxholes. *For personnel under light cover:* Cluster weapons dispersing small shaped-charge submunitions (assuming the shaped-charge jet will penetrate the cover) should be effective, as well as rockets or gun projectiles with armor-piercing warheads that strike the covered foxholes. Short-time-delay-fuzed GP bombs that hit the foxhole or strike close enough to crater the foxhole and kill the occupants should be appropriate. Guided weapons will be difficult to use because the target presents limited signature for directing a guided weapon. Chemical weapons should be effective, as well as FAE weapons where the fuel–air cloud is over the lightly covered foxhole.

			For personnel under heavy cover, such as concrete bunkers: Delay-fuzed bombs fuzed for maximum cratering, and guided weapons with large warheads and penetration capability are desirable.
2. ARMORED VEHICLES a. Tanks	Mobility, MC–kill	Damage that causes a vehicle to be incapable of executing controlled movement and is not repairable by the crew on the battlefield. The damage is assumed to occur instantaneously.	Shaped-charge weapons, armor-piercing projectiles, and contact-fuzed GP bombs. Guided bombs, such as electro-optically guided bombs (EOGBs) and laser-guided bombs (LGBs).
	Mobility, M40-kill	Same as M0, but mobility is lost within 40 min.	Same as for M0-kill.
	Firepower, F-kill	Defeat of the main armament, either because the crew has been rendered incapable of operating it or because the armament or its associated equipment has been rendered inoperative and the gun cannot be effectively fired. It is also assumed that the equipment is not repairable by the crew on the battlefield. The damage is assumed to occur instantaneously.	Same as for M0-kill.
	Catastrophic, K-kill	Damage beyond repair or to the extent that repair is not economically feasible. The damage is assumed to occur instantaneously.	Same as for M0-kill.
b. Armored Personnel Carriers (APCs), Reconnaissance and Infantry Combat Vehicles, and Scout Cars	Mobility, M0-kill	Damage that causes a vehicle to be incapable of executing controlled movement is not repairable by the crew on the battlefield. The damage is assumed to occur instantaneously.	Shaped-charge weapons, armor-piercing projectiles, and contact-fuzed GP bombs. Guided bombs, such as EOGBs and LGBs.

(Continued)

TABLE A.1   TARGET-DAMAGE CRITERIA AND APPROPRIATE WEAPON TYPES (CONTINUED)

Target	Damage Criteria		Appropriate Weapon Type[1]
	Mobility, M40-kill	Same as M0, but mobility is lost within 40 min.	Same as for M0-kill
	Firepower, F-kill	Damage to the primary armament, associated equipment, or crewmembers required to operate the armament, rendering the vehicle inoperative. It is assumed that armament and associated equipment are not repairable by the crew on the battlefield. Damage is assumed to occur instantaneously.	Same as for M0-kill.
	Catastrophic, K-kill	Damage beyond repair or to the extent that repair is not economically feasible.	Same as for M0-kill.
	Passenger, P-kill	Incapacitation of transported personnel. The P-kill is a calculated ratio of the number of personnel incapacitated to the total number of personnel being transported in the passenger compartment of the APC. Incapacitation will be based on assault, 5-min wound ballistic criteria for personnel.	Shaped-charge weapons, rockets or gun projectiles with armor-piercing warheads, and contact-fuzed GP bombs. Also guided weapons as specified above.
c. Mobile Assault Guns	Same as tanks.	Same as tanks.	Same as tanks.

3. FIELD ARTILERY			
a. Towed and Self-Propelled	Firepower, F-kill	Damage to the gun that will prevent the accurate delivery of munitions on intended targets and that cannot be repaired in the field. Incapacitation of the crew is not included.	*Against guns in the open*: Contact-fuzed GP bombs or delay-fuzed GP bombs employed for shallow cratering and overturning of guns. If contact fuze is instantaneous and impact angle of GP bomb is carefully controlled, the combined blast-fragmentation can be quite effective against such hard targets. Dispenser weapons using conical-shaped-charge bomblets may be effective (but the vulnerable area of the gun is restricted). Little damage can be expected for holes in gun tube unless they are quite large or located in areas of maximum internal pressure. EOGBs and LGBs, if they can be accurately guided to the target, and guided weapons employing large LSC warheads should be quite effective. *Against guns in revetments*: Contact-fuzed GP bombs or airburst GP bombs detonating at lower altitudes. High-altitude bursts (of the order of 50 ft and above the Mk-82) will be relatively ineffective because of velocity slowdown and low density of fragments. The use of GP bombs assumes that one can hit in the revetment to achieve damage; thus, EOGBs and LGBs will be effective if target characteristics or signatures are such that the weapon can be guided to the target. Also, guided weapons employing the largest LSC warheads should be effective.

(Continued)

TABLE A.1   TARGET-DAMAGE CRITERIA AND APPROPRIATE WEAPON TYPES (CONTINUED)

Target	Damage Criteria		Appropriate Weapon Type[1]
b. Self-Propelled	Mobility, M0-kill Mobility, M40-kill	Same as for tanks. Same as M0, but mobility is lost within 40 min.	Fragmentation bomblets and airburst GP bombs will be useful mainly against the personnel and crew associated with the field-artillery piece. Cluster or dispenser weapons using conical-shape-charge bomblets impacting on the target can be effective against limited areas of the guns and especially effective against the crew of the revetted guns. *Against guns in covered positions*: Delay fuzed bombs, GP bombs fuzed for maximum cratering, and guided weapons with large warheads and some penetration capability would be desirable. Same as for tanks. Same as for M0-kill.
3. ANTIAIRCRAFT ARTILLERY a. Automatic Antiaircraft Guns	Firepower, F-kill	Damage to the weapon that will prevent the accurate delivery of munitions on intended targets and that cannot be repaired in the field. Incapacitation of the crew is not included nor is any off-carriage fire-control equipment, such as a radar or fire-control director. Because various modes of operation from fully automatic to fully manual are prevalent, which can require different components to be damaged, each mode of operation may be evaluated separately. When this occurs, consult TH 61A1-3-1, *Target Vulnerability* (JMEM), for a complete discussion of the damage criteria for each mode of operation.	Same as for field artillery. *For damage to off-carriage fire-control equipment*, such as a radar or fire-control director, airburst GP bombs or instantaneous-contact-burst GP bombs would be quite effective. The same can be said of dispenser weapons with fragmentation or conical-shaped-charge bomblets. For light honeycomb radar vans, firebombs are effective given a direct hit on the target or a near miss in front of the target.

b. Self-Propelled Antiaircraft Guns	M0-kill M40-kill F-kill K-kill	Same as for tanks. (See Armored Vehicles) Same as for tanks. (See Armored Vehicles) Same as for tanks. (See Armored Vehicles) Same as for tanks. (See Armored Vehicles)	Same as for tanks, except that the target has less armor. Some of the self-propelled antiaircraft guns (e.g., ZSU-57/2) are open on the top and will be susceptible to HEI projectiles and dispenser-delivered shaped-charge submunitions.
5. ROCKET LAUNCHERS			
a. Multiround Launchers	Firepower, F-kill	Sufficient damage to all rocket launcher tubes and/or the launcher system that makes the weapon or system inoperable without major repair	GP bombs with contact fuzing for direct hit or near miss or proximity-fuzed GP bombs. Conical-shaped-charge submunitions delivered in dispenser systems should be effective, because small conical-shaped charges striking the rocket warheads may cause high-order detonations. A high-order detonation of one warhead in a multiround launcher should sufficiently damage the remaining launcher tubes for an F-kill. There is also a possibility of other warheads detonating sympathetically. (The probability of this happening depends on the construction of the rocket-launcher tubes and size of the rocket warheads.) Guided weapons with large warheads should be effective.
b. Single-Round Launchers	Mobility, M0-kill	Damage that causes the vehicle to be incapable of executing controlled movement and is not repairable on the battlefield.	Same as for trucks. (See Land Transportation.)
	Mobility, M40-kill	Same as M0, but mobility is lost within 40 min.	Same as for M0-kill.
	Firepower, F-kill	Damage to the rocket or launcher components that prevents successful launching of the rocket.	Same as for tanks. (See Armored Vehicles.)

(Continued)

TABLE A.I   TARGET-DAMAGE CRITERIA AND APPROPRIATE WEAPON TYPES (*CONTINUED*)

Target	Damage Criteria		Appropriate Weapon Type[1]
6. MISSILE SITES a. Surface-to-Air Missile (SAM) Sites	Firepower, F-kill	Damage severe enough to prevent firing missiles effectively against airborne targets for at least 4 h. Accomplished by damage to all missiles/launchers, tracking radars, or fire control vans. Damage to acquisition radars will affect the site's ability to acquire targets but not to fire missiles using external sources of target acquisition.	GP bombs with contact fuzing for direct hit or near miss or proximity-fuzed GP bombs. Conical-shaped-charge submunitions delivered in dispenser systems should be effective, because small conical-shaped charges striking the rocket warhead may cause high-order detonation. Guided weapons should be effective. Proximity or contact-fuzed GP bombs delivered inside the revetments. Antiradiation missiles to temporarily disable the radar, followed by dispenser weapons using shaped-charge or fragmentation submunitions or retarded contact-fuzed bombs. Guided weapons with standoff capability.
b. Surface-to-Surface Missile (SSM) Sites	Mobility, M0/ M40-kill	Same as for tanks. (See Armored Vehicles.)	Same as for tanks. (See Armored Vehicles.)
	Firepower, F-kill	Damage that either prevents the missile from being launched or results in a decision not to launch the missile because of things such as loss of ability of the warhead to produce a nuclear detonation, loss of ability to guide to the target, or loss of proper rocket motor function.	*Against liquid-fueled missiles:* Contact-fuzed GP bombs, proximity-fuzed bombs bursting at reasonably low altitudes, and dispenser or cluster weapons, followed by incendiary munitions. Firebombs delivered after the target is perforated could also be effective. *Against solid-fueled missiles:* Contact-fuzed GP bombs. EOGBs or LGBs and small shaped-charge dispenser weapons.

7. RADAR INSTALLATIONS			
a. Radars: fixed and mobile	Fire-control, F-kill	Sufficient damage to prevent the radar system from performing its intended function for at least 4 h. This applies to radars whose loss of ability to track targets or direct fire prevents missile or gun firings and to radars whose loss of function prevents early warning or ground control intercept.	Airburst or instantaneous-contact-burst GP bombs. Antiradiation missiles are effective if the target is radiating. Dispenser weapons with fragmentation or shaped-charge submunitions. For light honeycomb radar vans, firebombs are effective given a hit or near miss in front of the target. Projectiles (e.g., 20-mm rounds) are effective against radar vans, given a hit. EOGBs or LGBs with fragmentation or linear-shaped-charge (LSC) warheads.
	Mission, MSN-kill	Damage that prevents the jamming unit from completing its mission.	
	Mobility, MO/ M40-kill	Same as for tanks. (See Armored Vehicles.)	Same as for tanks. (See Armored Vehicles.)
	Catastrophic, K-kill	Damage beyond repair or repair is not economically feasible. Damage is assumed to occur instantaneously.	Airburst or instantaneous-contact-burst GP bombs. EOGBs or LGBs with fragmentation or LSC warheads.
b. Communications Jammers	Mission, MSN-kill	Damage that prevents the jamming unit from completing its mission.	Guided and unguided GP bombs. Guided missiles with fragmentation or LSC warheads. Cluster or dispenser munitions. Rocket attacks.
8. AIRFIELDS			
a. Parked Aircraft	Prevent takeoff, PTO-kill	Damage that requires repair other than normal preflight maintenance before level nonmaneuvering flight can be achieved and maintained for at least 5 min. Subcategories are identified by the number of hours of required repairs, as follows: PTO: Repairs requiring at least 5 min PTO_4: Repairs requiring at least 4 h PTO_{24}: Repairs requiring at least 24 h	For both damage criteria—PTO and K-kill: *Against aircraft in the open or in uncovered or lightly covered revetments:* Strafing attacks with guns or rockets; small fragmentation weapons; dispenser weapons with fragmentation or small shaped-charge bomblets; GP bombs fuzed for airburst over the target aircraft; GP bombs with instantaneous contact fuzes that hit within the revetment; EOGBs or LGBs with fragmentation or LSC warheads. Penetrating weapons to open fuel tanks, followed by incendiary munitions, should yield high probability of K-kill. Firebombs would be effective given a direct hit on each aircraft target.

(Continued)

Table A.1   Target-Damage Criteria and Appropriate Weapon Types (Continued)

Target		Damage Criteria	Appropriate Weapon Type[1]
	Catastrophic-on-ground, COG-kill	Nonrepairable damage to the aircraft so as to render it unfit for any purpose except cannibalization and scrap.	*Against aircraft in heavily covered revetments:* Guided bombs or missiles, assuming sufficient target signature for weapons to guide on, and with weapon capable of penetrating through roof and fuzed to detonate after perforation.
b. Runways	Interdiction, I-kill	Surface cratering sufficient to prevent aircraft from taking off or landing. No undamaged part of the runway is long enough or wide enough for use as a takeoff strip.	GP bombs with steel nose plug and short-time-delay tail fuze for runway penetration and cratering. *Note: Do not use nose and tail fuzing because nose fuze may cause explosive to react under impact force and deflagrate on the runway surface rather than penetrate on the basis of the time set on the tail delay fuze. In such case, damage will be minimal.*
c. Hangars	—	Damage is defined as structural damage of a specific level. Hangars that have sustained structural damage of 50% or more should be considered unusable. Aircraft and hangar shops will be severely damaged in the area under the collapsed hangar.	GP bombs fuzed for short delay striking the hangar or impacting close enough to severely damage load-bearing members of the hangar. EOGBs or LGBs with large warheads and short-delay fuzing should be effective.

| d. Hardened Aircraft Shelters | Shelter breach, PTO-kill of aircraft | Damage to the aircraft by causing sufficient concrete spallation due to blast and/or earth shock to achieve the PTO damage level as stated in 8a above. PTO damage may be achieved by blocking aircraft egress from the hangarette because of spall or by jamming or rendering the front blast door inoperable in a closed or partially closed position or by cratering the area in front of the shelter to prevent aircraft egress. PTO damage will result where a large bomb penetrates the overburden to within the breaching radius of portions of the shelter arch wall yet does not penetrate into the interior of the shelter. A detonation within the breach radius of any other point along the shelter will require further analysis to determine if a PTO-kill is produced on the aircraft by spall from the shelter walls or if a shelter destruct kill has been achieved. | Large GP bombs, delay fuzed for maximum penetration. Guided weapons are appropriate against the shelter door provided target signature is adequate for guidance. |
| | Shelter penetration, K-kill of aircraft | Weapon penetrates shelter roof, walls, or doors and inflicts catastrophic damage on the aircraft so as to render it unfit for use and uneconomical to repair. | Any type of penetrating weapon that can place a damage mechanism with sufficient damage potential to kill the aircraft within the target. If the shelter doors are open, this can be achieved by accurate placement of the weapon in the closure. Guided missiles with large shaped charges attacking the doors can be effective in defeating the doors. |

(Continued)

TABLE A.1   TARGET-DAMAGE CRITERIA AND APPROPRIATE WEAPON TYPES (*CONTINUED*)

Target		Damage Criteria	Appropriate Weapon Type[1]
	Shelter destruct	Damage causing shelter collapse that results in an aircraft K-kill (defined in 8a above). Shelter destruct will result where a large bomb penetrates the arch wall and detonates in the interior of the shelter, collapsing the shelter and destroying the aircraft inside.	Shelter destruction can be obtained only with larger delay-fuzed GP bombs utilizing a nose plug and tail fuze.
9. FIELD FORTIFICATIONS	Catastrophic, K-kill	Breaching the walls or overhead cover to cause breakdown of structural integrity and possible filling of the interior by dirt, scabbing, and debris.	GP bombs fuzed for optimum penetration and optimum cratering; guided weapons that can be guided on the field-fortification target and that impact on the target with explosive weights of 190 lb or greater (Mk-82 GP bomb or larger).
10. SUPPLY DEPOTS AND DUMPS a. Stacked Ammunition	Catastrophic, K-kill	Damage resulting in complete destruction of the stack. A high-order detonation of a projectile in a stack is assumed to destroy the entire stack.	Large blast and fragmentation weapons, direct hit by conical-shaped-charge and LSC weapons. (Small cluster fragmenting weapons are not appropriate.) Against wood-crated ammunition, fire and incendiary weapons are appropriate.
b. POL Surface (Aboveground) Storage Tanks	—	Destruction of the tank, ignition of the contents resulting in a sustained fire, or damage to the tank through blast or fragmentation resulting in drainage of the tank within 10 min or less after attack leaving less than 25% of the original contents.	Large blast and fragmentation weapons, direct hit by large shaped-charge or LSC weapons. Small cluster weapons alone are not appropriate.

c. POL Storage Drums	—	Destruction of drums or ignition of contents resulting in sustained fire to 50% of the drums within the unit.	Direct hit by LSC or conical-shaped-charge weapons, small fragmenting submunitions capable of igniting the contents, and all GP weapons resulting in destruction of the drums. Guns using HEI ammunition are effective, resulting in sustained fires (primarily against gasoline). FAE weapons are also effective.
d. POL Underground Storage	—	Destruction of the tanks making them unusable or ignition of the contents resulting in a sustained fire.	GP bombs fuzed for penetration and/or cratering.
11. LAND TRANSPOR-TATION a. Roads	—	Road cratered sufficiently to prevent passage of vehicular traffic for a specified period of time.	Large GP bombs dropped singly or smaller bombs in either salvo or stick release, fuzed for penetration of reinforced concrete or graded earth.
b. Trucks	Mobility, M-kill	Damage that immobilizes a moving vehicle or renders it incapable of controlled movement within a given time. This type of kill is usually conditional based on provisions for repairing the damage.	GP bombs, fragmentation and incendiary submunitions, rockets, shaped-charge weapons, and strafing attacks are effective for inflicting M-, I-, and possible K-kills on light vehicles. Firebombs are effective against vehicles transporting flammable or explosive cargo. Combined attacks with fragmentation weapons that cause fuel to leak, followed by incendiary munitions, are very effective. Dispenser systems containing fragmentation submunitions with high-velocity fragments or shaped-charge submunitions, and guided weapons that can be guided on the signatures emanating from a truck.
	Category A Category B Category C	Vehicle is immobilized within 5 min. Vehicle is immobilized within 20 min. Vehicle is immobilized within 40 min.	

(*Continued*)

TABLE A.1   TARGET-DAMAGE CRITERIA AND APPROPRIATE WEAPON TYPES (*CONTINUED*)

Target		Damage Criteria	Appropriate Weapon Type[1]
	Catastrophic, K-kill	Damage that renders the vehicle unfit for any purpose except salvage.	Same as for M-kill, except that weapons that will enhance fuel fires should be employed. Fragmentation weapons followed by incendiary weapons are effective. Direct hits and near misses with large weapons are needed for K-kill.
c. Railways	Cut	Severe damage to, or undermining of, a section of track so that a train cannot pass until the damaged section of track is repaired or replaced.	Land mines and GP bombs fuzed to produce large craters.
d. Rolling Stock	Mobility, M-kill	Overturn or other damage to the extent that a car is unable to move over the track network on its own truck assemblies without some repair or replacement of components.	Derailment may be caused by short-delay-fuzed GP bombs impacting quite near the car. Damage to trucks and undercarriage will be greater with instantaneous-fuzed GP bombs. Guided weapons with large warheads are effective. Against wooden cars and reactive cargo, use incendiary weapons.
e. Locomotives (1) Moving	S2	Train stops within 5 min and is not derailed; stops within 1 mile.	Instantaneous-fuzed GP bombs striking the locomotive; guided weapons with large fragmentation, LSC, or conical-shaped-charge warheads; cluster of shaped charges striking vital components of locomotive; armor-piercing projectiles of .50 caliber or larger and rockets are effective, given a hit.
	S3	Train stops within 60 min.	Instantaneous-fuzed GP bombs striking the locomotive; guided weapons with large fragmentation, LSC, or conical-shaped-charge warheads; armor-piercing projectiles of 7.62-mm or larger and rockets are effective, given a hit.
	Q1 (quick kill)	Locomotive stops within 1 min.	Same as for S2.

(2) Stationary	R2	Requires 4 to 15 days to repair.	Large GP bombs with instantaneous firing; guided weapons with large HE and LSC warheads.
f. Tunnels	Block	Breaching of ceiling or walls causing complete breakdown of structural integrity and filling by dirt, water, or debris so as to render the tunnel impassable.	HE GP or penetration bombs fuzed for maximum penetration and cratering.
g. Bridges			
(1) Permanent	Drop span	Damage to the extent that it causes a span to drop.	Large GP bombs with instantaneous fuzing for maximum damage on the bridge superstructure, guided weapons with large HE or LSC warheads.
(2) Temporary	Interdiction, I-kill	Interdiction of specified load-capacity categories of traffic as follows:	*Against timber-deck bridges:* HE weapons. *Against cable bridges*: Weapons that produce earth cratering.
	Category 1	Up to 0.5 t: Includes human and animal foot traffic, bicycles, loaded hand carts, etc.	*Against wood or reed pontoon bridges:* Fragments, projectiles, and conical-shaped charges.
	Category 2	Up to 5 t: Includes light passenger vehicles and small cargo trucks.	*Against bamboo bridges:* Weapons delay-fuzed for underwater explosion.
	Category 3	Up to 15 t: Includes cargo trucks and light weapon carriers.	
(3) Tactical Floating	Interdiction	Rupture of adjacent pontoons.	Same as for permanent bridges.
h. Piers	Pier failure	Breaching of piers.	GP bombs with instantaneous fuzing for contact with pier faces above the water and delay fuzing for contact with pier faces under water or in earth.
i. Abutment	Collapse	Damage to the extent that it causes a span to drop.	GP bombs with delay fuzing delivered in the fill behind the abutment from the approach end of the bridge.

(*Continued*)

TABLE A.1 TARGET-DAMAGE CRITERIA AND APPROPRIATE WEAPON TYPES (*CONTINUED*)

Target		Damage Criteria	Appropriate Weapon Type[1]
12. WATER TRANSPOR- TATION			
a. Inland Canals	Block	Reduction of the water cross-sectional area below some critical value, so as to block the canal against vessel passage.	GP, demolition, and penetration bombs, delay-fuzed for maximum cratering in the canal bed.
b. Locks	Pin Shear	Shearing of hinge pins. Deformation or other damage to the gates or caissons so they cannot be moved or cannot retain water in the lock. Pumps and gate-opening machinery damaged, rendering the lock useless. Breaching of levees, dikes, or dams.	GP bombs, delay-fuzed for underwater detonation, air-delivered torpedoes, and floating mines. *Against housed supporting equipment:* GP bombs and rockets with short-delay fuzes. *Against supporting equipment that is not housed:* Instantaneous-fuzed GP bombs and rockets.
c. Port Facilities	—	Blocking port-entrance channels and wharves by sinking vessels. Damage to the port rail- or road-clearance facilities. Damage or destruction of port warehouses and transit sheds or damage to loading cranes on wharves.	Appropriate weapons for attacking rails, bridges, ships, locks, and warehouses are presented in the pertinent sections of this table. GP bombs are appropriate for attacking wharves; when delay-fuzed for cratering, they are suitable against breakwaters and jetties.
d. Commercial Ships	SW-kill, M-kill, Cargo-kill	See Naval Vessels, Surface Combatants for definitions of kills.	Torpedoes, delay-fuzed GP bombs, and guided missiles with penetration capability, delay-fuzed for internal burst in ship.
e. Junks	Sink, K-kill	Sinking kill based on the buoyancy/cargo relationship. Requires structural damage in critical portion of the hull, uncontrolled leaks in all compartments, flooding within 20 min, loss of maneuverability, inability to reach shallow water, and loss by sinking if the junk is motorized or carrying high-density cargo.	Direct hits from large-caliber warheads (5.0 in. or larger). Bombs, delay-fuzed for underwater detonation, bursting near or under the keel of the junk. Hits near the waterline from 20-mm armor-piercing-incendiary (API) or HEI projectiles.

Target	Kill	Damage Criteria	Weapons
f. Sampans		Same as junks.	Same as junks.
g. Barges and Small Craft	Sink, K-kill	Sinking	GP bombs and guided missiles with short-time-delay or instantaneous fuzing striking the target or near the target; guns and rockets (if cargo is reactive).
h. Naval Vessels			
(1) Surface Combatants	Sink	Sinking kill based on the buoyancy/cargo relationship. Requires structural damage in critical portion of the hull, uncontrolled leaks in all compartments, flooding within 20 min, loss of maneuverability. Sinking occurs within 1 h.	Guided weapons with contact fuzes; cluster weapons containing conical-shaped-charge warheads; proximity- and contact-fuzed GP bombs; antiradiation missiles, delay-fuzed GP bombs.
	Mobility, M-kill	Damage such that the ship cannot maintain steerage or all engines are stopped.	Same as sink.
	Firepower, F-kill	Damage to a ship such that its ability to use its offensive armament is lost. Firepower kill is separated into specific types of weapon systems on ships such as long range missiles, short range missiles, surface-to-surface missiles, guns, antisubmarine warfare, or antisurface warfare.	Same as sink.
(2) Submarines	SW kill	Damage that will result in pressure hull rupture and internal flooding of the submarine.	Same as M-kill weapon types.
			Data-link, electro-optical guided weapons with delay fuze; torpedoes; and delay-fuzed GP bombs.
13. BUILDINGS	—	Structural damage of a specified level (percent of roof or floor area). Structural damage is damage to principal load-carrying members (trusses, beams, columns, load-bearing walls) requiring replacement or special support during repair. Buildings that have sustained structural damage of 50% or more are considered unusable.	GP bombs striking the target with short-delay fuzing; guided weapons with LSC and large HE warheads; incendiary weapons (appropriate only for combustible contents).

(*Continued*)

TABLE A.1 TARGET-DAMAGE CRITERIA AND APPROPRIATE WEAPON TYPES (*CONTINUED*)

Target		Damage Criteria	Appropriate Weapon Type[1]
14. POWERPLANTS	—	Damage to the extent that the plant cannot generate or transmit the specified percentage of its output capacity for at least 30 days.	GP bombs with short-delay fuzing; guided weapons with LSC and large HE warheads.
		Damage to the extent that the plant cannot generate or transmit for at least 6 mo.	Same as 30-day criterion.
15. TRANSFORMERS	>1 day	Damage to the transformers so that transmission of electric power is not possible for at least 1 day.	Guided and unguided GP bombs; guided missiles with fragmentation and LSC warheads; cluster or dispenser munitions; rocket attacks.
	>1 mo	Damage to the transformers so that transmission of electric power is not possible for at least 1 mo.	Guided and unguided GP bombs; guided missiles with fragmentation and LSC warheads; cluster or dispenser munitions; 5-in.-rocket attacks.
	>6 mo	Damage to the transformers so that transmission of electric power is not possible for at least 6 mo.	Guided and unguided GP bombs.
16. DAMS			
a. Earthen Dams	—	Breaching of the dam so that erosion significantly reduces the storage capacity of the reservoir.	Largest available GP bombs with delay fuzing for maximum cratering effectiveness.
b. Concrete Dams	—	Breaching or displacing the dam to release impounded water. No reliance on erosion.	Depth bombs with hydrostatic fuzing; large GP bombs with delay fuzing.

17. INDUSTRIAL TARGETS			
a. Missile and Aircraft Factories	—	Damage to components so that the plant is no longer capable of the specified percentage of its output capacity for the times specified.	GP bombs with short-delay fuzing; guided weapons with LSC or large HE warheads.

Percent of Damage	Days to Full Recuperation	Production Loss (Days)
10	96	46
20	134	74
30	147	91
40	149	100
50	150	104
60	150	105

b. Steel Mills	—	Damage to components so that the plant is no longer capable of the specified percentage of its output capacity for the times specified.	Same as for missile and aircraft factories but with larger warheads, if available.

Percent of Damage	Months to Full Recuperation	Production Loss (Mo)
10	9	4
20	18	11
30	23	18
50	24	23
70	25	24

c. Aluminum Plants	—	Undefined	Same as for steel mills.
d. Magnesium Plants	—	Undefined	Same as for steel mills.

(*Continued*)

TABLE A.1   TARGET-DAMAGE CRITERIA AND APPROPRIATE WEAPON TYPES (*CONTINUED*)

Target	Damage Criteria	Appropriate Weapon Type[1]
e. Other Nonferrous —	Undefined	Same as for steel mills.
f. Light Industry —	Undefined	Same as for aircraft and missile factories. Against combustible buildings or combustible building contents: incendiaries.
g. Manufacturing Plants, Machinery, and Electrical Equipment —	*(see sub-table below)*	Same as for missile and aircraft factories.
h. Explosives —	Undefined	*Against raw-material storage, manufacturing, and loading areas:* Small incendiaries supported by GP bombs with instantaneous fuzing. *Against multistory nitration buildings:* GP bombs with short-delay fuzing.

Percent of Damage	Days to Full Recuperation	Production Loss (Days)
Tank and Heavy-Motor-Vehicle Manufacturing Plants		
20	102	54
30	118	64
50	142	88
60	151	100
70	156	111
Heavy Ordnance Manufacturing Plants		
20	68	26
30	76	37
50	90	58
60	94	68
70	97	76

Target		Damage Criteria	Weapon Selection
i. Synthetic Rubber Plants	—	Undefined	*Against massive buildings:* GP bombs with short-delay fuzing. *Against areas with rubber and other chemicals:* Incendiaries.
j. Chemical plants	—		
(1) Oil Refinery		Damage to critical components to the extent that the plant is no longer capable of the specified percentage of its output capacity for the times specified.	GP bombs with instantaneous fuzing; guided weapons with LSC and large HE warheads.
(2) Other	—	Undefined	*Against chlorine and caustic soda, nitrogen, hydrogen, and sulfuric acid plants:* GP bombs with instantaneous fuzing. *Against synthetic ammonia plants:* Rockets and bombs with instantaneous fuzing.
k. Abrasives Factories	—	Undefined	GP bombs with short-delay fuzing to disable cranes and power source.
l. POL Pumping Stations	—	Damage that results in loss of 100% of the pumping capacity and requires at least 1 mo to repair.	GP bombs with instantaneous fuzing.
m. Nuclear-Materials Production	—	Undefined	*Against heavy buildings:* GP bombs. *Against light buildings:* Bombs with delay in nose and no delay in tail. *Against equipment:* Same as for light industry. *Against power source:* Same as for transformers.

Table within "(1) Oil Refinery" row:

Percent of Damage	Days to Full Recuperation	Production Loss (Days)
30	65	45
50	90	75
70	135	130

[1]For additional information, refer to JMEM Target Vulnerability.

INDEX

Note: Page numbers followed by "t" indicate table and "f" indicate figure.

SUPPORTING MATERIALS

A complete listing of titles in the Library of Flight series is available from AIAA's electronic library, Aerospace Research Central (ARC), at arc.aiaa.org. Visit ARC frequently to stay abreast of product changes, corrections, special offers, and new publications.

AIAA is committed to devoting resources to the education of both practicing and future aerospace professionals. In 1996, the AIAA Foundation was founded. Its programs enhance scientific literacy and advance the arts and sciences of aerospace. For more information, please visit www.aiaafoundation.org.

Printed in the USA
CPSIA information can be obtained
at www.ICGtesting.com
JSHW011252271024
72278JS00002B/4